AF572413

Commercial Chicken Meat and Egg Production Fifth Edition

Commercial Chicken Meat and Egg Production

Fifth Edition

Edited by

DONALD D. BELL (emeritus)
Poultry Specialist
University of California
Riverside, California

WILLIAM D. WEAVER, JR.
Professor Emeritus
Department of Poultry Science
Virginia Tech
Blacksburg, Virginia
and
Department Head (retired)
Department of Poultry Science
Pennsylvania State University
University Park, Pennsylvania

SPRINGER SCIENCE+BUSINESS MEDIA, LLC

Library of Congress Cataloging-in-Publication Data

Commercial chicken meat and egg production / edited by Donald D. Bell,
William D. Weaver, Jr.
p. cm.
Rev. ed. of: Commercial chicken production manual / Mack O. North,
Donald D. Bell.
4th ed. c1990.
Includes bibliographical references (p.).
ISBN 978-1-4613-5251-8 ISBN 978-1-4615-0811-3 (eBook)
DOI 10.1007/978-1-4615-0811-3
1. Chickens—Handbooks, manuals, etc. 2. Eggs—Handbooks, manuals, etc. I. Bell, Donald D., 1933– II. Weaver, Jr., William Daniel, 1940– III. North, Mack O. Commercial chicken production manual.

SF487.N77 2001
636.5—dc21 00-045232

Printed on acid-free paper.

In *Commercial Chicken Meat and Egg Production,* the names of many medicinal and other products appear, often with the trade names, which may vary from country to country. But nothing contained herein is to be construed as an endorsement for any named product, either by trade or chemical name, nor is criticism of similar products implied when not mentioned.

Products such as pesticides, rodenticides, disinfectants, drugs, antibiotics, and vaccines are usually licensed for use. These licenses are specific for each country or state, require verification of effectiveness and safety, and must be accompanied with detailed labels explaining the application, dosage, species, safety precautions, and limitations associated with the material. The user is responsible for adhering to these instructions and to use such products only as advised.

This book cannot include all the products licensed for sale, nor can it give the many benefits or limitations associated with their use. Always check with local agricultural authorities regarding the legal use of any product.

Often, materials or processes that are effective in one region may fail in another. Consult with local government agencies, consultants, or university advisors before using any such products.

Contents

Section V. The Breeder and Hatchery Industries

Section VI. The Broiler Industry

Section VII. Poultry Processing

Section VIII. The Table-Egg Industry

Introduction

The poultry industry in the 21st century has evolved from tens of thousands of small independent farms in the post-World War II period to an industry of relatively few large vertically integrated companies, each with multiple farm sites or contract growers, processing, marketing, and feed milling and hatchery capabilities. This change has come about because of the many technologies that have been introduced over the past half-century by the poultry industry with the help of supporting industries and various educational, research, and governmental institutions.

The information and research needs of this dynamic industry grow at an ever increasing rate as individual companies strive to improve performance and efficiencies and to reduce costs. Issues associated with the environment, animal welfare, food safety, business management, and labor have become critical areas for problem-solving efforts.

This edition of *Commercial Chicken Meat and Egg Production* has changed from emphasizing the chicken to emphasizing the business of raising chickens. In so doing, an additional 22 chapters have been added to cover many of the new topics important to the poultry industry. It is also recognized that no single source of information can supply the depth of understanding needed in today's business, and therefore, an extensive list of references and resources has been added in the appendix for further study.

Bibliographies of Contributing Authors

Paul W. Aho
Ph.D., Michigan State University
Areas of Specialty: International agribusiness, poultry economics
Address: 20 Eastwood Road, Storrs, CT 06268

Donald D. Bell
M.S., Colorado State University
Areas of Specialty: Layer management, pullet rearing, economics, flock recycling, house and equipment design, egg quality
Address: Highlander Hall-C, University of California, Riverside, CA 92521

Carol J. Cardona
DVM, Purdue University; Ph.D., Michigan State University
Areas of Specialty: Pathogenesis of viral diseases of poultry, disease prevention
Address: Veterinary Medicine Extension, Surge III. University of California, Davis, CA 95616

Craig N. Coon
Ph.D., Texas A&M University
Area of Specialty: Poultry Nutrition
Address: 0-211 Poultry Science Center, University of Arkansas, Fayetteville, AR 72701

Gregg J. Cutler
DVM, University of California, Davis; MPVM, University of California, Davis
Areas of Specialty: Poultry health, food safety, physiology
Address: P.O. Box 1042, Moorpark, CA 93020

Ralph A. Ernst
Ph.D., Michigan State University
Areas of Specialty: Layer management, environmental physiology
Address: Animal Science Department, University of California, Davis CA 95616

Joan S. Jeffrey
DVM, Ohio State University; MS, Ohio State University
Areas of Specialty: Infectious diseases of poultry, food safety
Address: Veterinary Medicine Teaching and Research Center, University of California, 18830 Rd 112, Tulare, CA 93274

Douglas R. Kuney
M.S., Colorado State University
Areas of Specialty: Layer management, pest management, manure management, environmental protection
Address: 21150 Box Springs Road, Moreno Valley, CA 92557

Michael P. Lacy
Ph.D., Virginia Polytechnic Institute and State University
Areas of Specialty: Broiler management, environmental control, housing design, ventilation, harvesting broilers
Address: Department of Poultry Science, 117 Poultry Building, University of Georgia, Athens, GA 30602

Joseph M. Mauldin
Ph.D., Virginia Polytechnic Institute and State University
Areas of Specialty: Hatchery management, incubation, embryology, sanitation
Address: Department of Poultry Science, 210 Poultry Science Bldg., University of Georgia, Athens, GA 30602

Ronald Meijerhof
Ph.D., University of Wageningen, The Netherlands
Areas of Specialty: Physiology, breeder management, incubation, hatchery management
Address: Hybro B.V., P.O. Box 30, 5830 AA Boxmeer, The Netherlands

---

Thad Morrison III
BA, St. Andrews College (NC)
Areas of Specialty: Hatchery equipment, hatchery automation
Address: 1131 Industrial Blvd. North, Dallas, GA 30132

Gene M. Pesti
Ph.D., University of Wisconsin, Madison
Areas of Specialty: Nutrition, computer applications
Address: Department of Poultry Science, University of Georgia, Athens, GA 30602

Charles J. Wabeck (deceased)
Ph.D., Purdue University
Areas of Specialty: Poultry processing, products, food safety

William D. Weaver, Jr.
Ph.D., Pennsylvania State University
Areas of Specialty: Broiler and breeder management, housing design, environmental control, ventilation
Address: 5168 Weaver Lane, Gloucester, VA 23061

A. Bruce Webster
Ph.D., University of Guelph
Areas of Specialty: Animal behavior, layer management
Address: Department of Poultry Science, 208 Poultry Science Bldg., University of Georgia, Athens, GA 30602

Michael J. Wineland
Ph.D., University of Wisconsin, Madison
Areas of Specialty: Breeders, hatcheries, light management
Address: Department of Poultry Science, North Carolina State University, Raleigh, NC 27695

Gideon Zeidler
Ph.D., Technion–Israel Institute of Technology
Area of Specialty: Food engineering, egg and meat processing
Address: Highlander Hall-C, University of California, Riverside, CA 92521

Acknowledgments

A book of this size and scope would not have been possible without the whole-hearted support of all its authors. Many deadlines came and passed, multiple editing was tiring and many re-writes were oftentimes necessary, but the book finally got done. Everyone pulled together and we hope the effort will finally be well received by the present and future poultry industries, both in the United States and around the world.

The editors would like to express their appreciation to Mrs. Joann Braga, secretary to Don Bell, for coordinating and working with the many (62) manuscripts during all stages of development. We would also like to thank Doug Kuney, regional poultry advisor for Southern California, for making his time available during the assembly stages of the book and for his assistance with the reference and appendix sections.

Preface

Since the publication of the fourth edition of the *Commercial Chicken Production Manual* in 1990 (the first edition was published in 1972), the senior author, Dr. Mack O. North has passed away, and Professor Donald D. Bell has assumed the role of senior editor and contributing author. In addition to serving as overall coordinator for the text, Professor Bell has focused on nutrition, poultry health, and all aspects of the production and processing of table eggs.

For the fifth edition Dr. William D. Weaver, Jr., Professor Emeritus from Virginia Tech and retired head of the Department of Poultry Science at Pennsylvania State University has joined the publication as co-editor and author. He has contributed primarily in the areas of housing, ventilation, hatchery, breeder and broiler management, and meat processing. The name of the text has also been changed to *Commercial Chicken Meat and Egg Production* to better reflect the contents included in the new edition.

Possibly the most significant change in the new text, however, involves the addition of sixteen new chapter authors. Each author was selected based upon his or her knowledge and experience in a particular phase of poultry production, and then was asked to provide the latest information available in that area. In doing so, the text has not only been extensively revised, but in many instances has been expanded to include totally new material. Topics such as US and global economics, business management, processing chicken meat and eggs, computer applications, chicken behavior, ventilation, water quality, waste and by-product management, hatchery planning, design, and construction, inspection, grading and quality standards for poultry meat and eggs, hazard analysis critical control point (HACCP) programs, and models of integrated and company-owned broiler and egg complexes are included for the first time in the new edition.

While the fifth edition has incorporated many changes, it has not lost its focus on commercial chicken production. Rather, the industry approach to production has been greatly enhanced with discussions on economic integration, contract production, processing and marketing, and the many global aspects of the industry.

Although the authors have focused their attention primarily on US production practices, globalization within the industry has made much, if not most, of the material presented relevant to the production of broilers and table eggs around the world. For instance, with similar genetic stocks, competitively priced feedstuffs, and poultry production and processing equipment and technology readily available in most locations in the world, poultry management practices globally are much more homogeneous today than ever before.

The new edition includes more than 400 tables, figures, and pictures in an attempt to further enhance the reader's understanding of commercial chicken production. Primary users of the text include various segments of the commercial chicken industry, such as producers, service personnel, and others responsible for the production and processing of chicken meat and eggs, as well as students in colleges and universities studying poultry science.

For the convenience of its readers the authors have again used both English and metric standards for weights and measures. When costs of production and other economic parameters are discussed, US dollars are normally used for making comparisons.

In an attempt to better serve users of the text the authors are providing means whereby additional information and technology can be obtained. A complete listing of authors can be found at the beginning of the book. In addition, the authors have developed a web site where contact procedures are listed and where updates and new information will be made available to subscribers. To take advantage of this service, purchasers of the text are asked to register their purchase of the book at:

Commercialchicken.com

For information about obtaining copies of the book or other questions directed to the publisher, contact:

Kluwer Academic Publishers

List of Tables by Chapter

CHAPTER 10

CHAPTER 11

CHAPTER 12

CHAPTER 13

CHAPTER 14

CHAPTER 15

CHAPTER 16

CHAPTER 17

CHAPTER 18

CHAPTER 19

CHAPTER 20

CHAPTER 21

CHAPTER 22

CHAPTER 33

CHAPTER 34

CHAPTER 35

CHAPTER 36

CHAPTER 37

CHAPTER 38

CHAPTER 39

CHAPTER 41

CHAPTER 42

CHAPTER 43

CHAPTER 46

CHAPTER 47

CHAPTER 48

CHAPTER 49

CHAPTER 50

CHAPTER 51

CHAPTER 52

CHAPTER 57

CHAPTER 58

CHAPTER 59

CHAPTER 60

CHAPTER 61

List of Figures by Chapter

CHAPTER 10

CHAPTER 11

CHAPTER 12

CHAPTER 13

CHAPTER 15

CHAPTER 16

CHAPTER 17

CHAPTER 18

CHAPTER 19

CHAPTER 21

CHAPTER 22

CHAPTER 25

CHAPTER 26

CHAPTER 28

CHAPTER 29

CHAPTER 30

CHAPTER 31

CHAPTER 32

CHAPTER 33

CHAPTER 34

CHAPTER 36

CHAPTER 37

CHAPTER 38

CHAPTER 39

CHAPTER 40

CHAPTER 41

CHAPTER 42

CHAPTER 43

CHAPTER 45

CHAPTER 46

CHAPTER 47

CHAPTER 48

CHAPTER 49

CHAPTER 50

CHAPTER 51

CHAPTER 52

CHAPTER 53

CHAPTER 54

CHAPTER 55

CHAPTER 56

CHAPTER 57

CHAPTER 58

CHAPTER 59

CHAPTER 60

CHAPTER 62

List of Abbreviations

alternating current	ac
ampere	A
ante meridiem	AM
average	avg
Board foot	bd ft
body weight	BW
British thermal unit	Btu
bushel	bu
calorie, large	C
calorie, small	c
candella	cd
cent	¢
centimeter	cm
cubic feet per minute	cfm
cubic meters per minute	cmm
cycle per minute	c/min
day	d
decibel	dB
degree Celsius	°C
degree Fahrenheit	°F
deoxyribonucleic acid	DNA
direct current	dc
dollar	$
dozen	doz
each	ea
east	E
foot	ft
foot-candle	fc
foot per minute	ft/min
gallon	gal
gallon per minute	gal/min
gram	g
gram each	g/ea
horse power	hp
hour	h
hundred weight	cwt
hydrogen ion concentrate	pH
inch	in
infrared	IR
joule	J
kilo	k
kilogram	kg
kilometer	km
kilowatt	kW
kilowatthour	kWh
liter	L
lumen	lm
lux	lx
metabolizable energy	ME
meter	m
mile	mi
mile per hour	mi/h
milligram	mg
milliliter	ml
millimeter	mm
millimeter mercury	mmHg
minute	min
month	mo
north	N
ounce (avoirdupois)	oz
ounce each	oz/ea
ounce per dozen	oz/doz
parts per million	ppm
percent	%
pint	pt
plaque-forming units	pfu
post meridiem	PM
pounds per square inch	PSI
probable error	pe
protein	pro
quart	qt
relative humidity	rh
revolutions per minute	rpm

revolutions per second	rps
ribonucleic acid	RNA
second	s
south	S
tablespoonful	tbsp
teaspoonful	tsp
therm	thm
total sulfur amino acids	TSAA
United States	US
United States of America	USA
US gallon	US gal
US Pharmacopeia	USP
volt	V
watt	W
watthour	Wh
week	wk
weight	wt
west	W.
yard	yd
year	yr

Section I. General

1

The World's Commercial Chicken Meat and Egg Industries

by Paul W. Aho

The world chicken industry is a growing part of global agribusiness and also one of the most dynamic parts of world agribusiness trade. As trade in agricultural products between nations becomes less restricted in the future, global competitiveness will determine the success or failure of individual poultry companies. This chapter provides an introduction to the world commercial chicken meat and egg industries and also provides guidelines for determining competitiveness in a world where globalization is rapidly becoming an important issue.

1-A. GROWTH IN WORLD CHICKEN MEAT AND EGG PRODUCTION

During the 1990's, world chicken meat production has grown from 29 million metric tons (MMT) to an estimated 50 MMT, an increase of 72% or 2 MMT per year. Coincidentally, the world chicken egg industry will also end the 1990's with total production of an estimated 50 MMT, an increase of 56% from 32 MMT at the beginning of the 1990's. Total world chicken meat and egg production increased at the rate of 4 MMT per year and will end the decade with approximately 100 MMT of total production. Since the capital investment necessary for an increase in production is roughly $1 per kilo of production for both eggs and broilers, the investment for new facilities in the poultry industry has been $4 billion annually worldwide. During the 1990's a total of $40 billion will have been invested in the world chicken industry.

In the first decade of the 21st century the world increase in chicken meat and egg production will continue but not at the same rapid pace. Production increases are likely to decrease to a 3 MMT per year pace since the extraordinary increases in both egg and broiler production in Asia have recently slowed. World broiler production will outpace world egg production in the next century as total chicken meat and egg production rises to 130 MMT by the year 2010. Because of the increasing technology used in chicken production, the 30 MMT increase of the first decade of the next century will require an additional $40 billion investment. Therefore, between 1990 and 2010 a total of $80 billion is likely to be invested in the world's chicken meat and egg industries.

The $80 billion investment in the world poultry industry from 1990 to 2010 will be invested unevenly around the world. There is a more rapid growth in developing countries than in the developed world because of increases in per capita disposable income, particularly in Asia. During the 1990's Asia accounted for nearly half the world production increase in broiler meat and more than half of the world's production increase in eggs. In the next decade, production increases will be spread more evenly around the developing world while production continues to rise slowly in the developed world.

Trade in poultry meat is relatively small but growing rapidly. A strong local production base exists in almost all countries, largely as a result of the local market structure, which tends to be based on fresh chicken. Since internationally traded chicken is mostly frozen, there is a limit to the acceptance of imported chicken. In addition, most countries have tariffs and/or other restrictions on the import of poultry products. Despite these barriers, world trade has increased in recent years thanks to new trade agreements. Certain countries like Japan and Russia have begun to place an increasing reliance on imported poultry products. In 1990 only 7% of all chicken meat was traded internationally. For example, by the year 2010 it is likely that 17% of all chicken meat will be traded between countries. The percentage of US chicken meat exported was already 15% of production in 1998 (Figure 1-1).

Trade in eggs is still very small, averaging less than 1% of world production with a slight upward trend. Trade is projected to reach 2% of world egg production by the year 2010.

Relative increases in the trading of poultry products will make competitiveness a key issue in the future. In "broad brush" and oversimplified terms, competitiveness will depend primarily on the political will to embrace free trade in grain and poultry products in each country. The poultry industries in countries with free trade in both grain and poultry products will be competitive although their market share may be small. They will be competitive because their grain costs will be no greater than the world price of grain and the local industry will be competing with the poultry industry of the rest of the world. Countries that chose to impose high grain and/or poultry product tariffs will be uncompetitive by choice to "protect" the local industry from the need to be competitive. The following matrix describes the policy choices. In the matrix the word "barrier" is used because there are a variety of policy choices besides tariffs including

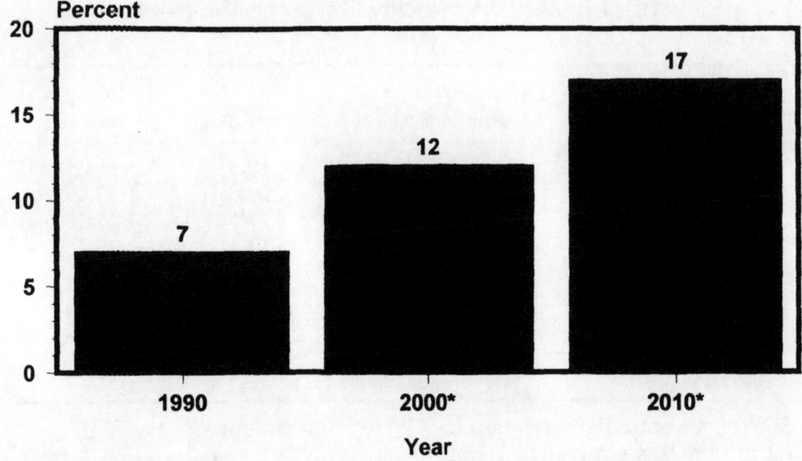

Source: USDA Livestock and Poultry World Markets and Trade.
* Projected

Figure 1-1. Percentage of Chicken Meat Traded Internationally—1990 to 2010

non-tariff barriers, taxes, and other means which all constitute barriers to trade.

Policy Matrix—Trade Barriers, Competitiveness and the Cost of Chicken Production

	Grain Trade No Barriers	*Grain Trade Barriers*
Poultry Trade No Barriers	Competitive Lowest Cost	Less Competitive Higher Cost
Poultry Trade Barriers	Less Competitive Higher Cost	Not Competitive Highest Cost

The interesting point made by the policy matrix is that grain supply is not as important as once believed. It has been common knowledge that local grain production was the most critical issue in poultry competitiveness. However, in the real world of complicated and numerous trade barriers, the lack of local grain production is sometimes less important than the level of barriers to trade. The elimination of trade barriers in countries without local grain production can create a competitive chicken industry. At the same time a country with a generous supply of grain can create a non-competitive chicken industry through the use of trade barriers.

The technology and structure of the competitive world chicken industry is trending toward convergence of best practices. Technology is readily accessible on a worldwide basis and growth in individual countries is not

Table 1-1. Total Livestock Meat and Poultry Consumption per Person. (US, 1960–2000)*

Year	Beef	Pork	Lamb & Mutton	Veal	Fish & Shellfish	Young Chicken	Turkey
1960	59.8	48.9	3.1	4.2	10.3	16.2	4.9
1965	69.5	43.8	2.4	3.7	10.8	20.4	5.9
1970	79.8	48.6	2.1	2.0	11.7	25.2	6.4
1975	83.2	38.5	1.3	2.8	12.1	25.2	6.7
1980	72.2	52.6	1.0	1.3	12.5	31.5	8.3
1985	74.7	48.1	1.1	1.5	15.1	35.2	9.2
1990	64.1	46.8	1.0	0.9	15.0	41.1	13.9
1995	64.0	49.1	0.9	0.8	14.9	48.5	14.1
2000**	62.4	48.3	0.8	0.5	14.5	57.7	14.1

Source: USDA, Economic Research Service (1999)
* Expressed in boneless equivalent weight
** Estimated

likely to be constrained by lack of access to technology. However, government intervention in many countries prevents the adaptation of best industry practices. An example of this is the conspicuous absence of vertical integration in certain countries, normally the least cost structure for broiler and egg production. Lack of a vertically integrated structure in a chicken industry may be a sign of government intervention.

1-B. THE WORLD CHICKEN MEAT INDUSTRY

Although chicken meat is still only the third most popular meat in the world after pork and beef, chicken meat accounts for an increasing share of world meat consumption. While that share was just 10% in 1970, by the end of the century it is projected to reach 20%. The US chicken meat or broiler industry is one of the most dynamic of all the animal industries (Table 1-1).

Who Eats Broiler Chicken Meat?

Most people on the planet have tasted young chicken. However, it is important to remember that broiler chicken meat remains a rare luxury for about half the world's population. In China for example, about 40% of the population have an income of over $3,000 per year (using the World Bank purchasing power parity, PPP, numbers). It is approximately that top 40% of the population by income that has sufficient purchasing power to participate in the cash market for broiler chicken meat. Therefore, China has a total population of broiler chicken meat buyers of approximately 475 million people. Using the same method, that is, choosing the world population with a PPP of more than $3,000, there are a total of about 2.5

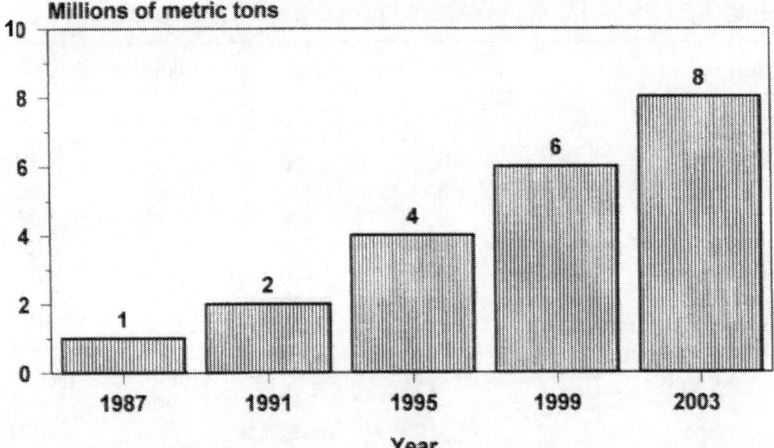

Source: USDA World Markets and Trade and projection by Paul Aho.

Figure 1-2. China Broiler Consumption—1987 to 2003

billion potential market economy chicken meat buyers in the world. Of course there is a big difference between being *able to buy* and actually buying. In India, for example, 50% of the population are vegetarians and in Germany most of the population prefers pork to chicken.

Since the total population of the earth is approximately 6 billion, about 45% of the world's population are potential consumers of chicken meat. However, that percentage has been rising. There is a high income elasticity of demand for chicken meat in developing countries, meaning that a small increase in income results in a large increase in the consumption of chicken meat. Given the overall increase in the wealth of the world's population over the last few decades, it is not surprising that there have been increasing expenditures for chicken meat. The experience of China shows what can happen to chicken consumption as wealth increases. Total chicken consumption in China doubled from 1 to 2 million metric tons between 1987 and 1991, a period of rapidly rising income in China (Figure 1-2). Although tens of millions of people became new market economy chicken consumers each year during that period, there was a vast difference between wealthy (mostly urban) and poor (mostly rural) household consumption rates. Urban household consumption was estimated to be five kilos per person while rural household consumption was only one kilo.

In the four years between 1991 and 1995, consumption of chicken doubled again from 2 million to 4 million metric tons as rural China began to participate more in total national chicken consumption. Production will not double every four years however, as the next doubling is projected to take eight years. Production, though, can be expected to reach 8 million metric tons by 2003 as China continues to prosper and an ever-greater percentage of the population participates in purchasing chicken meat, even as China's rate of economic growth slows.

Table 1-2. World Chicken Meat Production—1985 to 2000

Country	1985	1990	1995	2000*
	Thousands of metric tons			
China	900	1,400	4,700	7,300
Russia	1,000	984	340	480
India	200	312	500	665
Eastern Europe	1,510	1,200	1,250	1,750
Indonesia	295	410	850	1,000
Brazil	1,490	2,356	3,800	5,000
Other Latin American	925	1,315	1,610	2,100
Other Asia	1,261	1,760	2,455	2,890
United States	6,242	8,360	11,435	13,700
Canada	472	572	695	850
Argentina	310	305	700	900
Africa	950	875	1,050	1,240
Thailand	393	575	780	950
Mexico	735	990	1,435	1,700
Middle East	1,485	1,885	2,350	2,875
Western Europe	3,780	4,493	5,284	5,600
Japan	1,270	1,332	1,171	1,000
Total	23,218	29,124	40,405	50,000

Source: USDA Foreign Agricultural Service (1985 to 1995) and Paul Aho (2000)

Income elasticity works in both directions. When the world economy shrinks and there is an overall decrease in the wealth of the world's population, chicken consumption drops as well. The same is true for both regions and countries. An example of decreasing consumption was observed in parts of eastern Asia in 1998 where economic conditions caused a temporary decrease in the consumption of chicken meat in an area that had seen rapidly rising chicken meat consumption for decades. Broiler slaughter rates in Indonesia dropped from 18 million per week in April of 1997 to 4 million per week in April of 1998.

World Broiler Production

Table 1-2 shows projections of world chicken meat production thru the year 2000. The countries that are growing the fastest are those that are in transition from a system of central government control to that of a market economy. In that group can be placed China and the countries of Central and Eastern Europe. Russia is also in this group, but represents a special case, as broiler chicken production dropped 70% between 1985 and 1995 and then began to recover at the end of the decade. Other countries with rapid growth will be found in Asia, such as India and Indonesia, and in Latin America, particularly Brazil. Growth in East Asia slowed temporarily in the late 1990's due to financial and currency problems. Production

Table 1-3. Egg Production in Billions (selected countries)—1990 and 1998

Country	1990	1998	1998
	Billions of eggs*		Table eggs
China	158.9	363.0	n/a
United States	63.2	79.9	67.4
Japan	40.3	42.2	n/a
Russia	47.5	35.0	n/a
France	14.6	16.3	n/a
Brazil	13.5	13.6	n/a

Source: USDA, Foreign Agricultural Service
* Includes hatching eggs

in the European Union is expected to grow slowly due to reduced subsidy payments for exports and environmental pressures. Production in Mexico will be held back somewhat by competition from the US while high costs in Japan will continue to reduce the size of that industry.

1-C. WORLD CHICKEN EGG INDUSTRY

The chicken egg industry has reached maturity in many parts of the world and is therefore not growing at quite the same rate as the world chicken meat industry. Part of the reason for this is that the elasticity of demand is lower in many cases for chicken eggs compared to chicken meat. Nevertheless, world egg production has been growing. During the decade of the 1990's the world's egg production climbed from 32 to approximately 50 million metric tons, an increase of nearly 56%. Most of the increase has taken place in one country, China. Out of the total world increase of 18 million metric tons during the 1990's, 12 million or 66% of that total represents the astounding increase in *documented* production in China.

Relatively few eggs are traded internationally. Only 1% of all eggs were traded between countries in the year 2000, about the same percentage of eggs traded internationally in 1990. It is interesting to note that world egg production and world broiler production were expected to be about the same in the year 2000 at 50 million metric tons each.

While the US dominates world production and consumption of poultry meat, China dominates world egg production and consumption (Table 1-3). In 1998 the US produced 79.9 billion eggs while China produced 363 billion eggs, up from an officially reported 158.9 billion eggs in 1990. Total egg production in the rest of the world is increasing at a much slower rate than in China. It must be noted that world egg statistics usually include hatching eggs. In the US, for example, table eggs for consumption represent 84% of total egg production. On a world basis a somewhat higher percentage would apply.

1-D. COMPETITIVENESS IN THE WORLD CHICKEN INDUSTRY

Worldwide, 85 to 90% of all commercial poultry meat is produced and consumed in the same country. However, thanks to trade agreements such as the Uruguay round of the General Agreement on Trade and Tariff (GATT) that led to the World Trade Organization (WTO) and regional agreements such as the European Union, North American Free Trade Agreement (NAFTA), and Mercosur, there is likely to be an increasing proportion of poultry meat and eggs involved in world trade.

Given this growth in the international trade of broiler meat and the potential future growth in trade of eggs, it is important to answer the competitiveness question. How does a country (and company within a country) establish and maintain competitiveness in the world broiler and egg market?

How To Establish and Maintain Competitiveness

Regardless of location on the planet, establishing and maintaining competitiveness requires either a competitive advantage or some type of subsidies from taxpayers. The poultry industries of many countries rely on government intervention, subsidies, and marketing orders to make the issue of competitiveness unnecessary or irrelevant. The alternative is to develop a poultry industry that is competitive without government subsidy.

What about the US? Although there are agricultural commodities in the US that receive significant and even outrageous subsidies, poultry commodities do not fall into that category. In fact, the subsidies given grain and soybean farmers is a form of "negative subsidy" to the US animal feeding industries, as such subsidies drive up the price of local feedstuffs, thereby increasing the costs for US producers of meat and eggs. The subsidies given to the poultry industry have consisted primarily of a small amount of export subsidy. In that regard, the US poultry industry can be thankful because the world is moving toward freer (not free) trade in agricultural goods. Uncompetitive subsidized commodities in the US and elsewhere, sooner or later, are likely to lose the battle to continue receiving subsidies.

The US poultry industry, as well as the rest of US agriculture, has received a large, extremely valuable form of government assistance over the years. That government assistance has come in the form of research and development by state agricultural experiment stations at land grant universities and the US Department of Agriculture. This research and development has been part of a cooperative public and private agricultural ef-

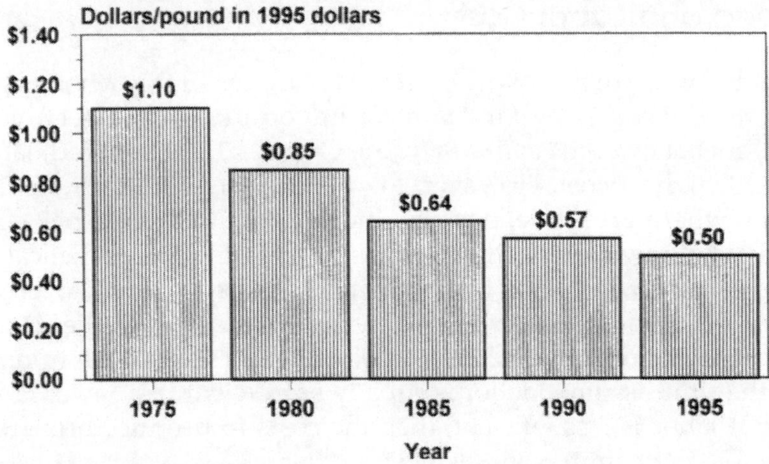

Source: USDA Data deflated using US Commerce Dept. Consumer Price Index.

Figure 1-3. US Broiler Cost of Production—1975 to 1995

fort that goes back to 1862 when President Abraham Lincoln signed into law the Morrill Act which established the Land Grant University system.

The US has made a large research investment in agriculture and the investment has paid off. As a result of this investment, the poultry industry has enjoyed a steadily decreasing cost of production. From 1975 to 1995, for example, the cost of production as measured by the USDA dropped by 54% (Figure 1-3). This production cost decline helped fuel the explosion of poultry demand in this country, as well as increasing levels of exports. It should be noted that not all poultry research is government funded. A large and increasing percentage of research related to the poultry industry is paid for and conducted by private firms. Nevertheless, the government portion of agricultural research has been notably large during the 20th century and the cooperation between public and private research, particularly fruitful.

Since taxpayer supports to agriculture are in a long-term downward trend almost everywhere in the world, government intervention will become less important and competitiveness will, of necessity, become increasingly important. There are four prerequisites for being a real competitor:

Competitive Poultry Producing Countries

a. Low feed and labor cost
b. Good business climate
c. Vertical integration and economies of scale
d. Access to technology

Low Feed and Labor Cost

Feed and labor are two of the most important cost items when producing poultry meat. Feed is by far the most important. The subject of feed is covered in *Feed and the Poultry Industry*, Chapter 13. After feed, labor cost is the next highest cost. However, low labor costs will not help in those situations where grain is expensive. It costs about 45 cents per pound to produce whole eviscerated broilers in the US. In a hypothetical higher grain cost and lower labor cost country, it costs 49 cents to produce a pound of eviscerated whole broilers where the feed ingredients are 50% more expensive and labor is 50% less expensive (Table 1-4). It is important to note that it is assumed labor is equally as efficient in a lower labor cost country. If labor is less efficient, then the costs to produce broilers in the lower labor cost country will be even higher.

The ideal competitor would have both low feed and low labor costs. However, as can be seen in the example, since feed represents such a large portion of the cost of production, lower feed costs can more than compensate for higher labor costs. Therefore, relatively high labor cost countries with low feed costs (exporters of grain) can be competitive. In many cases, added competitiveness can be gained by substituting capital and technology for labor in high labor cost countries. Also, those countries with low labor costs that import grain at no more than world prices will be competitive.

Good Business Climate

Although difficult to quantify, a good business climate is essential for a competitive poultry industry. Some of the factors that lead to a good business climate are:

1. clear title to land
2. a well designed and functioning infrastructure of communications and transportation

Table 1-4. Whole Eviscerated Broiler Meat Costs—US vs Other Countries

	United States	Higher grain / lower labor country
	Cents per pound*	
Feed	18	27
Labor	10	5
All other	17	17
Total	45	49

Source: Estimates by Paul Aho
* Ready-to-cook broiler meat per pound

3. a healthy banking system
4. a light tax and regulatory burden.

A clear title to land is essential for the financing of agriculture in general and poultry in particular. Some of the problems faced in Russia in the 1990's were related to the slow development of clear titles to land. Title to land provides the ability to use land as collateral to finance agriculture.

The benefits of a well-designed and functioning infrastructure are particularly important in poultry production because of the large geographic area over which poultry complexes must operate. A healthy banking system provides the needed financial stability. Last, but not least, a light tax and regulatory burden round out the most important factors essential to a good business climate.

Vertical Integration and Economies of Scale

Vertical integration and economies of scale are important to the operation of a competitive poultry industry. Vertical integration is the ownership and/or coordination of the various stages of poultry production, processing and marketing in a single business enterprise. A vertically integrated poultry firm typically includes the feed mill, hatchery, and processing plant involved in a chicken business under the ownership of a single person or corporation.

Vertical integration is important because it coordinates production in each stage of the company, establishes a single profit center, and controls quality from beginning to end. Coordination of capacity utilization means that each stage of production such as a feed mill or a hatchery is producing its products at full capacity and at an optimal level for and in coordination with the next stage of production. This principle extends from the rearing of breeder replacements through to the final marketing of products. The vertically integrated company operates like a single factory spread out geographically over a wide area.

The alternative to a vertically integrated structure is the system under which independent feed mills, hatcheries, and processing plants each sell their products independent of each other, with each designing its own level of production and with each adding a measure of profit to the sale price of its various products. Production costs of non-integrated systems usually surpass those of vertically integrated systems and tend to survive only in areas of government intervention and/or subsidy where competitiveness is irrelevant. See *A Model Integrated Broiler Firm*, Chapter 42 and *A Model One Million Hen In-line Egg Production Complex*, Chapter 50 for descriptions of vertically integrated companies.

Economies of scale are extremely important in the broiler industry. This means that a broiler company must process enough broilers at its pro-

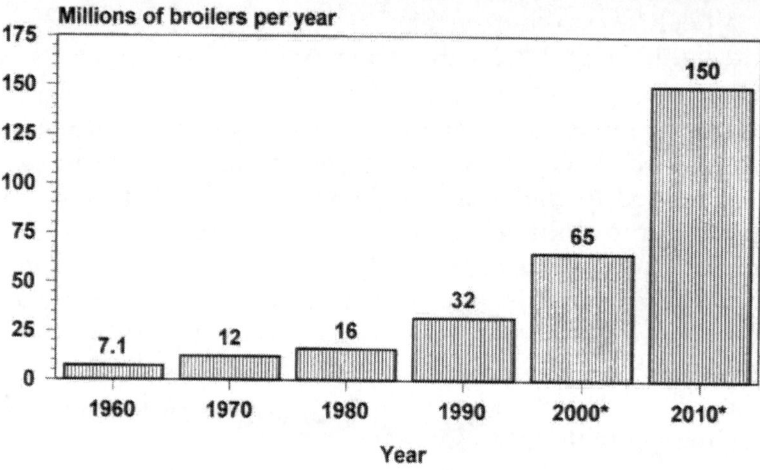

Source: University of New Hampshire (1959).
*** Estimated by Paul Aho**

Figure 1-4. Appropriate Scale of Operation—US

cessing plant to get the lowest possible cost per pound of production. At any given moment in time there is a point at which there is no advantage to making a processing plant any bigger and the economies of scale will, in economic terms, have been "captured." For example, in 1959, a company processing 7.1 million broilers per year was the point at which most of the economies of scale were captured, according to a University of New Hampshire study at the time. It could be said that 7.1 million birds per year was the scale of operation required for a technically and economically efficient broiler production complex in 1959. Twenty years later that appropriate scale of operation had more than doubled to reach 16 million birds per year according to a Michigan State University study in 1982. At the end of the 20th century, the appropriately sized new plant must process approximately 65 million birds per year to capture most of the economies of scale. By the year 2010, an appropriately sized new plant may be 150 million birds per year as new technologies are developed and implemented (Figure 1-4).

In the egg industry, a similar transformation has taken place. In about 1935, a 10,000 hen operation was considered a large but attainable number for a single farm. About a generation later in 1965, 100,000 hens became a normal number of hens for a single farm. After another generation and another 30 years, 1,000,000 hens on a single farm became common in 1995. It would not be surprising if 10,000,000 hens on a single farm became standard by the time the next generation of farmers takes over by the year 2025. The increase in bird numbers in the egg industry responds to the same search for the appropriate economies of scale that drive ever larger broiler operations. As new technologies are developed and implemented, "the bar is raised higher and higher."

Access to Technology

Access to technology is not normally a problem in the world's chicken industry, except in countries that are extremely underdeveloped technologically or in those nations that isolate themselves from the rest of the world. The normal sources of technological information for the world include the primary breeders, equipment suppliers, feed ingredient suppliers, suppliers of animal health products, and research institutions including Universities.

Breeding companies are always interesting in having their customers fully express the genetic potential of their stock. In their own interest, breeding companies have provided a generous amount of technical support around the world. These companies not only pass along needed technical information but also generate technical knowledge through their own research. Breeding companies are known to be excellent sources of free information, particularly in areas of live production.

Suppliers of equipment, feed ingredients and animal health products are another important source of technological information. Equipment companies maintain a close relationship with the industry while developing new products. These products are demonstrated at trade shows held regularly throughout the world. Suppliers of feed ingredients such as premixes and amino acids are normally an excellent source of nutritional advice. Companies that provide these inputs tend to be large international organizations that provide both products and information on a worldwide basis. Finally, the providers of animal health products are also large multinational corporations that are a source of research in the area of animal health from both their own research facilities as well as from the general scientific community.

Last, but not least, are the worldwide network of Universities and research laboratories that are funded by industry, government, and private sources. The information generated at these sites is generally easily available through the scientific literature and educational meetings presented by a number of scientific societies such as the Poultry Science Association in the US, the World's Poultry Science Association, and other societies devoted to poultry and associated areas of research.

1-E. GLOBALIZATION

As important as economies of scale and vertical integration have been in the last 50 years, they will pale in comparison with the next challenge for the chicken industry, that of globalization. The next era, the era of globalization, will be characterized by the development of global strategic alliances with suppliers, major customers, and even sometimes with competitors. Globalization is driven by increasingly liberalized world trade in conjunction with rising world income.

Globalization brings with it both increased opportunity and increased risk. Economic opportunities are being created by profound social change as peasants join the middle class and agricultural societies industrialize. However, risk abounds internationally. Companies will have to choose foreign markets that promise political and economic stability, and be prepared to ride out inevitable periods of instability. The silver lining to increased risk is the potential of rewards commensurate with that risk.

When using the broiler chicken industry as a model, some examples of how companies will operate in this new era are already visible:

- There are innovative cost-plus arrangements between fast food companies and broiler integrators that involve the production of entire processing plants. In such an arrangement, the idea is that an integrator and a customer work closely together in a way that makes both firms more profitable.
- The export of chicken products will be assisted in the future by cooperative ventures between US companies and organizations in other countries, perhaps even competitors that wish to offer a branded product that is only available from a US company.
- US companies are already involved in chicken production in other counties. This trend will continue as large US companies become global producers and advertise branded products around the world with direct satellite communication to consumers. Some will go beyond chicken to become global food suppliers.
- Strategic alliances with suppliers will be established as well. For example, contract linkages to obtain identity-preserved grain will become established. In this way an integrator can not only better control the quality of the grain, but request specific types of grain with specific nutritional characteristics. (See *Feed and the Poultry Industry*, Chapter 13).
- Stronger alliances with contract broiler growers are being forged so that integrators and growers can work more closely together and develop a true partnership to make both entities more profitable.

Conclusion

The world chicken meat and egg industries are an important part of world agribusiness. In the next century this large and growing industry will transform itself through the process of globalization. The surviving companies will consist, for the most part, of companies that use the best practices available including vertical integration and economies of scale to remain competitive. An uneasy coexistence with competitive chicken enterprises will be non-competitive enterprises supported (to a greater or lessor degree) by protective local governments.

2

Components of the Poultry and Allied Industries

by Donald D. Bell

To many, the poultry industry consists of production and processing. In reality, there are many separate but related businesses that comprise today's modern poultry industry. Without the dedicated support of the entire allied industry, the poultry industry would not have been able to make the advancements it has made during the last half century.

The variety of elements in the poultry industry are an indication of the opportunities that exist to express one's skills and imagination. The tens of thousands of individuals who have found a lifelong source of employment in this industry represent hundreds of separate vocations. The industry needs new members who will challenge the old ways of doing things and who can think of more efficient systems for the future.

The following discussion is an attempt to briefly describe the various segments of the industry and to illustrate their relationships to one another. Specific details usually relate to US industries except where noted.

2-A. POULTRY BREEDING

The so-called primary breeding industry for poultry is comprised of fewer than two dozen major companies worldwide for meat and egg type chickens. These companies maintain the foundation and great grand parent stock required for the production of commercial lines of poultry.

Poultry breeders work on the improvement of their stocks for several years before they are marketed. Improvement consists of the evaluation of important performance characteristics of present lines, careful analysis of industry needs, selection of lines which carry the required characteristics for improvement, and the accumulation of enough grand parent stock to support the market need for parents.

The major primary breeding firms usually have staffs which include geneticists, veterinarians, nutritionists, and general management specialists. These professionals act as a team to develop new strains and the management techniques for optimizing their performance. Breeders commonly publish management guidelines with standards of expected performance for purposes of comparison. In general, these standards are quite accurate and usually represent attainable production results from better than average producers.

2-B. HATCHERIES (see *Operating the Hatchery*, Chapter 40)

Commercial hatcheries usually hatch either meat or egg type chicks, rarely both. Hatcheries are normally one of three types:

1. Owned by a large integrated poultry company, usually a chicken meat producer, with in-company use only.
2. Independent with parent stock purchases or a franchise with a major primary breeder.
3. Owned by a primary breeder with direct sales on the open market.

Today, it is estimated there are 358 commercial chicken hatcheries in the US, having a total one-time capacity of 825 million eggs. These hatcheries produce some 8 billion broiler chicks per year and 220 million female pullets for the table egg industry (1998). The worldwide need for day-old chicks is estimated to be 25 billion meat-type and 3 billion egg-type birds.

A hatchery operation commonly includes parent stock rearing and hatching egg production farms. These may be company owned or contract. Day-old breeders, male and female, are reared to sexual maturity on isolated rearing farms and transferred to hatching egg farms for their laying cycle. (See *Managing the Breeding Flock,* Chapter 34). It is estimated that the US has a year-around population of approximately 52 million broiler breeder females and 2.6 million table egg breeder females. These are the birds that produce the hatching eggs, which produce the commercial stocks used in the industry for meat and egg production. National statistics of hatching egg production indicate that the each broiler and egg-type female produced 155 and 161 un-sexed chicks, respectively, in 1997.

The 451 million egg-type chicks hatched in 1999 produce 225 million day-old pullets. These, minus mortality, would result in 215 million pullets to provide the replacements to maintain an average national layer count of between 265 and 270 million layers—an 80% replacement rate, (it is estimated that between 75 and 80% of all layers in the US are molted and kept for a second cycle of egg production). Countries that do not molt their flocks would require proportionally more chicks depending upon the age at sale.

With egg-type chicks, the local independent or breeder-owned hatchery usually makes the contact with the buyer, assists with the delivery, and provides continuing service and assistance where needed. Some hatcheries and / or breeders even provide detailed management advice, often in computer print-out form. Emphasis is on achieving flock standards for the strain.

2-C. FEED SERVICES (see *Feed and the Poultry Industry*, Chapter 13)

The world's chicken industry consumes some 200 million metric tons of poultry feed annually, assuming a ratio of 2:1 for feed:product (meat and eggs) conversion. Feed represents the greatest single cost of production and an efficient and competitive feed industry is of critical importance to the success of the poultry industry.

The relationship of source of feed production, transportation costs for feed vs product, and the location of the consuming population are all key determinants when locating the poultry production industry. Poultry feed sources for many countries may be a continent or more away. In the US, feed ingredients may be transported 1,500 miles (2,400 kilometers) to production sites and product may be transported the same distance to markets. Such distances can create inefficiencies in production.

The US feed industry is comprised of several different types of feed companies:

1. Large international companies with facilities in several countries
2. National firms with multiple mills in various states
3. Companies with more than one mill but limited to one state or a smaller region
4. Single independent mills
5. Cooperatively owned mills
6. Producer owned mills, quite often on a poultry production site.

Feed mills vary in the capacity of their equipment and their weekly tonnage, degree of computerization and associated labor efficiencies, type and number of products produced, handling methods for incoming feedstuffs, and storage capacity, and as a result of these differences, costs also vary. Mills for the chicken meat industry must have the ability to pellet feeds, while those associated with the table egg industry can use all-mash feeds, thereby reducing mill investment costs.

Critical to any mill's success, though, is the competitiveness of its ingredient purchasing system and its success in buying ingredients delivered to the mill at the least cost. Fluctuations in the market demand active involvement in futures trading for grains and protein meals. This, in turn, requires highly trained professionals.

In addition to ingredient purchasing skill, the feed company must also have flexible feed formulation policies which can quickly respond to market or ingredient changes, access to excellent nutritional advice, and a mill capable of producing a consistent quality product.

A related sector of the poultry business is the feed ingredient industry. Ingredients can normally be traced back to the original supplier (elevators and growers), which usually include commodity suppliers and brokers, manufacturers of micro-ingredients (vitamins, minerals, medications), pre-mix manufacturers, and the transporters of these ingredients.

2-D. BREEDER AND REPLACEMENT PULLET REARING (see *Managing the Breeding Flock*, Chapter 34, and *Cage Management for Raising Replacement Pullets*, Chapter 51)

A chicken meat production complex for 1.3 million broilers per week (see *A Model Integrated Broiler Firm*, Chapter 42) with breeder replacements grown by contract growers requires eight 2-house farms for rearing replacements. In general, a table egg layer complex requires 1 rearing house for every 3 to 5 houses of similar capacity—depending upon its replacement policy.

The production of replacement birds represents a major investment and a very important one in regards to the effect it may have on subsequent company profits. Errors in management during this stage can seriously affect the efficiencies of adult birds and their progeny.

The rearing programs associated with the chicken meat industry are usually under the complete control of the contracting company who supplies the chicks, feed, service (vaccination and catching) crews, and general management supervision. Flocks are visited by company personnel and records of growth, mortality, and feed consumption are maintained and reviewed on a routine basis. Growers are retained over time on the basis of their ability to produce high quality pullets compared to other growers.

In years past, table egg producers relied more heavily on outside growers for their replacement pullet needs. A large "started pullet" industry grew birds either on order or on speculation. In recent years, this practice has decreased as the owners of large egg production companies sought to gain more control over the way their pullets were reared. Today, most major companies have their own rearing farms.

2-E. BROILER GROW-OUT (see *Broiler Management*, Chapter 43)

The majority of broilers are raised by growers on contract with a company that controls chick supplies, feed milling, bird processing and marketing. The grower supplies the facilities and labor; the integrator (contractor) provides chicks, feed, service, and technical expertise. In some regions of the US and areas of the world, integrators also own the grow-out farms and employ their own workers. In the US, this system is definitely less commonly used.

The company with a production of 1.3 million broilers per week (see *A Model Integrated Broiler Firm*, Chapter 42) will require 400 grow-out houses, each with a capacity of 27,500 birds. This represents a need for 100 to 200 farms depending upon the number of houses per farm (4 or 2). It would require 118 such complexes to accommodate the 8 billion broilers produced annually in the US.

2-F. EGG PRODUCTION (see *Introduction to the US Table Egg Industry*, Chapter 49)

Table egg farms are commonly company owned with a smaller fraction on contract. In 1997, the *American Egg Board* estimated the US had 329 companies with more than 75,000 layers. The number of individual farms this represents is unknown.

Egg farms commonly receive new pullets at ages from 16 to 20 weeks. Layer flocks are kept through one cycle of production and sold at 75 to 80 weeks of age, or are kept for two cycles of production and sold at 105 to 110 weeks of age. Some producers may keep their flocks through 3 or more cycles with sale at 125 to 150 weeks of age (see *Flock Replacement Programs and Flock Recycling*, Chapter 54).

Farms have either single aged birds, common with contract farms, or multiple aged birds. The trend in the US is to the large (500,000 to 1 million +) in-line complexes with on-site egg packaging or breaking facilities and feed mill. In 1991, it was estimated that the US would have sixty 1-million hen complexes by the year 2000.

2-G. POULTRY PROCESSING (see *Processing Chicken Meat*, Chapter 46)

The poultry processing industry is made up of:

a. Plants owned by integrated poultry companies, e.g., broiler industry.

b. Independent companies that process poultry from other suppliers., e.g., fowl processors.

In 1998, federally inspected poultry processing plants in the US slaughtered 7.9 billion broilers, 103 million table egg fowl, and 66 million broiler breeder fowl. The total reported slaughter for 1998 was listed as 8.07 billion with a total live weight of 17.9 million metric tons.

2-H. SHELL EGG PACKAGING (see *Processing and Packaging Shell Eggs*, Chapter 58)

The shell egg packaging industry is structured in one of three ways:

a. Egg packaging on the site of production—commonly used with an in-line system where eggs are transported directly to the plant from the production houses on conveyor systems
b. Off-site packaging of eggs within the same company that owns or contracts for the production
c. Independent companies that purchase their nest-run supplies directly from producers.

Plant capacity is usually geared to either the one- or two-shift capacity of the egg handling equipment. An example of this is given in *A Model One Million Hen In-Line Egg Production Complex*, Chapter 50.

The processing plant is responsible for cleaning, grading, sizing, and packaging eggs. The in-line plant requires daily operation, while the other two systems can operate on a 5 or 6 day per week basis.

In 1998, some 67 billion table eggs were produced in the US and it is estimated that approximately 54% of these were cartoned for household use, 15% were loose packed for the institutional trade, 1% were exported, and the remaining 30% were diverted to the processed egg industry.

2-I. BREAKER EGG PROCESSING (see *Further Processing Eggs and Egg Products*, Chapter 59)

The USDA lists about 70 egg breaking plants in the US that broke 58.5 million cases (30 dozen eggs/case) in 1999. Products produced in these plants include fresh liquid, frozen, and dried eggs. Most of these products are provided to the food processing industry in these forms for user convenience or specific needs for yolks or albumen. Other processed eggs are sold to further processors who produce specialized consumer products which utilize either yolk, albumen, or whole eggs. By law, these products are pasteurized.

2-J. PRODUCT MARKETING, PROMOTION, AND ADVERTISING

Marketing is a major component in both the poultry meat and egg industries. Most large companies have marketing departments and, in addition, may rely on outside companies and consultants to assist them with their marketing needs. Such departments may include marketing specialists as well as home economists who help to promote a company's products.

Private broker firms facilitate the movement of products into the retail industry and help to move supplies from producers with excess products to ones in need of products. The table egg industry has its Egg Clearinghouse Inc. which facilitates movement of surpluses to companies needing eggs. A side benefit of this type of operation is that trading is public and the information generated can be used to substantiate the strength or weakness of the market relative to price discovery.

Product promotion and advertising is done privately by companies with "branded" products and industry-wide (regional or national) with generic advertising and promotion of poultry and egg products. Generic advertising is commonly done by trade associations or with the use of state and national marketing orders normally requiring special legislation which allows for industry-wide assessments of funds.

2-K. PACKAGING

The packaging industry serving the poultry industry in the US is huge, as many products are packaged twice—individual consumer packs placed in outer protective containers. Most types of poultry and egg products require some type of outer packaging for shipment to retailers, further processing sites, or storage. Other merchandising systems may only require single packaging. For example, whole body chickens may be placed directly into a display case for retail sales. Eggs may be cartoned but delivered and displayed on racks or may be sold loose on filler flats for the restaurant and hotel trade.

Packaging comes in a variety of forms and appearances. Poultry meat is commonly packed in trays with a clear overwrap plus appropriate labeling. Eggs are packed in either pulp or foam cartons. Outer containers for both poultry and eggs are commonly designed to protect the inner containers, facilitate storage and handling, and are constructed of material suitable for transportation.

2-L. HOUSING AND EQUIPMENT MANUFACTURERS

Today's modern, technology-driven poultry industry is highly dependent upon housing and equipment designers and manufacturers and a wide variety of suppliers of specialized equipment. New concepts in housing, processing, and management are constantly being developed and put into use in every segment of the industry. Every specialized need is being addressed by numerous competing companies on a world-wide basis, resulting in almost yearly advancement in the way things are done.

Housing is usually constructed by companies within a geographic region using local crews and materials. Plans are often provided by equipment manufacturers and University Extension Specialists. Different needs for weather protection, systems of waste handling and labor availability commonly dictate the type of housing utilized within a local industry.

Equipment tends to be more similar within a country but may vary between countries because of national regulations which require specific management systems. Differing space requirements, especially as it relates to cages in the egg industry, place totally different financial constraints on the industry in the European Community compared to other countries with fewer regulations of this sort. Equipment variation appears to be greater in the layer industry than in the chicken meat industry.

The development of much of the equipment used in poultry meat and egg processing plants is so rapid that it often makes 5-year old equipment obsolete. But, most importantly, cost savings with new designs are of such magnitude, that if change isn't made, a production unit or processing plant can become uncompetitive quite rapidly.

2-M. VACCINE, DRUG, CHEMICAL AND FEED ADDITIVE MANUFACTURERS

This group manufactures a wide assortment of products used to meet the health and nutritional needs of poultry flocks. Many worldwide companies, often based in Europe or the US, operate internationally and provide their products for poultry meat and egg producers all over the world.

Poultry health products include vaccines and pharmaceuticals for the prevention and treatment of the dozens of bacterial, viral, fungal, and parasitic problems faced by the poultry industry. Feed additives include synthetic vitamins, amino acids, fermentation products, and other products necessary for poultry flocks.

Companies that manufacture these products also have large research staffs who determine the needs of the industry, develop the products, tests them for safety and effectiveness, and develop the manufacturing processes that will guarantee they are dependable and affordable.

2-N. PRIVATE LABORATORIES AND CONSULTANTS

Large production firms cannot always afford to hire their own staffs with expertise in all the needed areas. Numerous consulting firms and individuals are available to provide this very necessary function. Outside advisors can apply their experience, concentrated effort, and specialized knowledge to a specific problem—oftentimes at great reward to poultry firms. The consultant's approach may be different from that of the company staff and can frequently shed new light on a given problem.

Consultants include veterinarians, nutritionists, economists, engineers, computer applications specialists, financial experts, pest control advisors, processing specialists, labor relations advisors, government interaction consultants, marketing advisors, and general management specialists. Some consulting firms may handle several of these topics with on-board staff or with outside individuals who they retain to address specific problem areas. Private health and nutrition laboratories are considered an important part of this group.

2-O. FINANCIAL INSTITUTIONS

Many times the sources of investment and operating capital are not included as part of the poultry industry, but they are an integral part and without them the wheels of progress would cease to turn. Few poultry firms can operate or should operate without the close partnership of their financial institution. Annual below cost of production periods, which require carry-over funding, and cyclic periods of "boom and bust" require imaginative financing arrangements. Rapidly changing technology and its need for new capital all require close coordination with this segment of the business community.

Sources of capital include private banks, cooperative financing institutions, major corporations, and other segments of the poultry industry. Institutions that routinely finance industry needs require documentation of the borrower's expertise and ability to repay a loan, and are generally extremely knowledgeable about the intricate operations of the industry in question. Realistic cash flows using realistic price and cost projections are required of the borrower (see *Computer Applications*, Chapter 33).

2-P. GOVERNMENT

Local, state, and national governments play an enormous role in today's society in practically every country throughout the world. Their involvement with the poultry industry includes regulatory, research, information, protection, quality, financial, marketing, and safety services—to name a few. Within the poultry industry, the government's role is most noticeable in the processing plant where inspectors are responsible for ensuring the health and safety of consumers of poultry and poultry products.

A major, while less noticeable role, is in research, extension and economics. Without these activities, many of the important problems facing the poultry industry would not be addressed or technology would not be transferred from its source to the users. Throughout the world, government research stations are responsible for important discoveries which directly help the poultry industries of the world.

2-Q. UNIVERSITIES AND OTHER RESEARCH AND TEACHING INSTITUTIONS

Universities and other educational institutions are mandated to teach, and in most cases to conduct, research. In the US, a third responsibility is added, to extend itself to society in general through Cooperative Extension. The Land-Grant system was established in the 1860's in the US and has served agriculture ever since. Today, about a dozen Universities have significant poultry programs with undergraduate and graduate education, poultry research centers and a full program of research. An additional 20 or so have poultry staff within an animal science department. Worldwide, poultry research is commonly done in poultry research centers with government sponsorship.

2-R. PUBLISHERS

The authors and publishers of textbooks, trade newsletters and newspapers, scientific journals, and trade magazines are an important part of the communications and technology transfer network which is so vital to the industry. The services they provide not only educate our students, but provide an on-going source of ideas and information which help the industry to progress. Many reporters are especially competent in sorting out important ideas from those that have little application or may be of questionable value.

2-S. TRADE ASSOCIATIONS

The poultry industry has numerous volunteer-served organizations which function to represent the industry in its interaction with government and the public. These associations observe the political climate relative to their industry and seek to include sensible provisions in new legislative proposals and to exclude those which have no basis in fact, and therefore may harm the industry and ultimately the consuming public. A core of dedicated individuals within each industry invests countless hours on committee assignments in debate of various industry-related issues.

These organizations are also heavily involved in conducting educational programs for producers and consumers, workshops for their members, and funding student scholarships.

2-T. TRANSPORTATION

The poultry industry has an enormous need for various forms of transportation that are required to transport feedstuffs from regions of production to the feed mills and farms, and for transporting finished products to the marketplace. Countries deficient in the production of grain and protein meals rely heavily upon ocean transport of feedstuffs from major exporting countries such as the US and Brazil. Regions within a country rely on rail, barge, and truck transportation from the areas where it is grown or port of entry to production sites.

Shipment of poultry meat and eggs is usually done by company owned or leased trucking and often involves shipments extending over 1,000 miles (1,600 kilometers). Poultry meat is commonly shipped from regions in the southern US to California—over 2,000 miles away (3,200 kilometers). Eggs are regularly transported from Iowa to California—a distance of 1,800 miles (2,900 kilometers) at an estimated cost of $.10 per dozen.

Summary

The $22 billion poultry industry (1999) in the US represents people with all sorts of training and background. This chapter is dedicated to all of you who have helped make this industry such a challenging environment for the rest of us.

3

Modern Breeds of Chickens

by Donald D. Bell

During the past two centuries more than 300 pure breeds and varieties of chickens have been developed. However, few have survived commercialization and therefore, are used by modern chicken breeders. Many of the earlier breeds are kept for exhibition purposes only, some have been lost forever, and others are maintained by private or government breeding stations so they will be available to breeders if necessary. These gene pools are important because they maintain certain genetic characteristics found in these rare breeds.

3-A. VARIETIES USED FOR MODERN BREEDING

In the early days of the commercial poultry industry, most of the chicks sold represented pure breeds or varieties. Breeding practices at that time were confined to improving the economic potential of these pure lines. Gradually, however, two or more breeds were crossed to improve productivity. Eventually, particularly in the case of birds bred for the production of meat, new synthetic lines were developed incorporating important characteristics from two or more breeds. Although many pure breeds were used in their development, these new lines do not represent any specific former breed or variety. They were new and different, and are continually being produced to meet expanded market demands.

Most of the breeds and varieties of chickens used in today's breeding programs, or used to develop new commercial lines, are included in the following discussions.

Single Comb White Leghorn

The Single Comb White Leghorn is one of several varieties of Leghorns, but the only one used for commercial egg production. All Leghorns have yellow skin and lay eggs with white shells. Although only one variety is used, there are many strains in existence.

Single Comb Rhode Island Red

The Rhode Island Red has a long block-like body, a single comb, and lays a brown egg. It has yellow skin, and the feathers are red with some black in the tail, hackle, neck, and wings. Several years ago many strains of this variety were in existence, most of which were excellent egg producers. Today, a good many of the commercial brown-egg layers are the result of crossing special strains of Rhode Island Reds and Barred Plymouth Rocks. The offspring are excellent producers of large brown eggs.

New Hampshire

The New Hampshire has a light-red color, yellow skin, a single comb, and produces a light-brown egg. At first the New Hampshire was known for its high egg production, but later it became recognized as a bird with good meat qualities. For several years it was the leading breed for the production of broiler chicks. Later, New Hampshire females were crossed with males of other meat-type varieties to produce crossbred broiler chicks.

The New Hampshire has been used in developing many of the new lines of meat-type chickens and is still used for this purpose. Its ability to produce a large number of eggs that hatch well has made it a valuable asset to many breeding combinations.

White Plymouth Rock

The White Plymouth Rock has yellow skin and a single comb. Although a pure variety was used by early broiler parent breeders, it now makes up the background for many synthetic lines. The white feathers are beneficial to broiler production and commercial processing plants, which do a better job of picking chickens with white as opposed to colored feathers.

Cornish

Cornish chickens have pea combs, lay a brown egg, and have yellow skin. They have a body type very different from most other breeds. The legs are short, the body is broad, and the breast is very wide and muscular.

The Cornish features are desirable from a meat standpoint, but the birds lay only a few small eggs with poor hatchability. In order to utilize the strain's meat qualities, Cornish males are crossed with females from such breeds as Barred Plymouth Rock, New Hampshire, and White Plymouth Rock, forming new lines to produce meat-type birds.

Barred Plymouth Rock

The Barred Plymouth Rock has feathers with bars of white and black running crosswise, giving the bird a gray appearance. It has a single comb, yellow skin, and lays a brown egg.

Today, the breed is mainly used to produce the female that is mated with a Rhode Island Red male to produce chicks for the production of commercial brown eggs for the table egg market.

Light Sussex

The Sussex is predominantly a British meat-type breed with several varieties, of which the Light Sussex is the most popular. It has white skin, lays a brown egg, and is a good meat producer. In England and some European countries, broiler chickens with white skin are preferred to those with yellow skin.

3-B. PRESENT-DAY EGG PRODUCTION LINES

Egg production lines are those used to produce pullets for the production of commercial table eggs with either a white or brown shell. The birds are relatively small in size, lay a large number of eggs with sound shells, live well, and produce eggs economically (Figure 3-1).

White-Egg Lines

Today, practically all commercial white-egg lines of chickens are Single Comb White Leghorns. In the early days the lines were pure; that is, they were not cross bred with other lines. Today, however, most breeders cross birds of two or more lines to produce the commercial pullet.

Single line. The breeder normally uses a closed flock, continually selecting the better birds in each generation and breeding from them. Only a small percentage of the better birds are used in the matings. Normally, the pullets are kept in egg production for a year in order to measure factors responsible for economic production of quality eggs. Selection of the best birds is made at the end of the first year of egg

Figure 3-1. Modern White Leghorn Egg-type Hen (courtesy of Hy-Line International)

production. Many traits will be considered simultaneously in the selection such as:

- body weight
- growth rate
- livability
- pullet quality
- age at sexual maturity
- egg weight
- egg production
- eggshell quality
- interior egg quality
- ability to convert feed to eggs

Hybrid vigor. Individual birds within certain breeds and varieties of chickens breed truer for some of the above traits than for others. When mated, the variability of the offspring is increased, new dominant genes are brought into the gene pool, and the offspring are superior to the parent lines. This so-called hybrid vigor, or heterosis, implies a physical well-being as the cause of the improvement in the offspring, but actually it is due to the increased genetic complexity of the stock. Recessive genes—those that generally produce poorer results—are masked by the more desirable dominant genes.

Male line and female line. It is obvious that in making any cross between two egg lines, a male from one line must be mated with a female from another line. The male offspring in the female line and the female in the male line are destroyed at 1 day of age because they are not needed. However, with the production of the commercial pullet, the cockerel chicks are also destroyed.

Strain cross. Rather than select for superiority of all good traits within a single strain, many breeders resort to a technique of selecting for only a few in a line, then crossing two or more lines to produce the commercial pullet.

Two-line cross. Crossing two or more lines increases heterosis in the offspring, defined as a marked improvement in vigor or capacity for increased productivity. In order to assure as much improvement as possible from the cross, one parent line is bred to excel in only certain qualities; the other parent excels in others. A simplified example of a two-line cross would be as follows:

Male line (Bred for superior)	*Female line (Bred for superior)*
livability	egg production
body weight	shell quality
egg weight	interior egg quality

Although there may be several other factors involved with each line, the above listing represents the major ones. When the two lines are crossed, the resulting pullets would be used for the production of commercial eggs. These pullets would have positive production traits derived from both parents, however at a lower level than is present in the individual parent lines:

good livability	good egg production
efficient body size	good shell quality
good egg size	good interior egg quality

Three-line cross. Three lines are developed, each with different qualities. Two lines are crossed, and the offspring from these two are crossed with the third line. Although additional lines generally add to the cost of producing the commercial pullets, the advantages may outweigh the additional expense.

Four-line cross. Four lines are developed. Two of the four lines are crossed; then the remaining two are crossed. The male offspring from one of the above crosses is mated with the female offspring of the other cross to produce the commercial pullet.

Strains used for crossing must nick. Two lines of chickens, which when mated together complement each other, are said to nick. The poultry breeder will develop many lines of egg-laying strains, and will mate many of them together. Some of the crosses will give improved results in the offspring, others will not. A few of those that do nick will be used for the production of commercial pullets. In this way it is also possible to develop commercial birds that excel in only one, or at least a few, particular traits. For example, the breeder may develop a commercial line that lays exceptionally large eggs. Another line, resulting from another cross, may live unusually well. In each of these cases, the nickability is especially involved with one particular trait.

Inbred crosses. Some breeders resort to heavy inbreeding within certain lines by mating brothers and sisters or other closely related individuals, for several generations, then two of the inbred lines are crossed to produce a commercial pullet. This increases purity (homozygosity) within the inbred lines that in turn improves uniformity. Although inbreeding decreases performance, it is more than restored when inbred lines are crossed.

Brown-Egg Lines

While it is known that shell color has no effect on the nutritive value of eggs, shell color is a consumer preference in certain localities. In the US (except for the New England region) and Germany white shells are preferred, while brown shells are preferred in France, the United Kingdom, and the Far East.

Several breeders have developed special lines and crosses for the production of commercial pullets that lay eggs with brown shells. In some instances, two breeds or varieties are used to make the cross. Not only do the offspring lay brown eggs but the chicks may be *sexed* at 1 day of age by differentiation in the color of their down.

Body size. Today, birds producing brown-shelled eggs are 15 to 25% larger than those producing eggs with white shells. This larger size increases the feed cost to produce eggs because a larger bird consumes proportionally more feed than a smaller bird. This is attributable to higher maintenance nutrient requirements for the larger bird.

Egg production. Most lines of birds producing commercial eggs with brown shells lay as well as those producing eggs with white shells (see *Egg Production and Egg Weight Standards for Table-Egg Layers,*

Figure 3-2. Modern Broiler Chicken (courtesy of Ross Breeders Ltd.)

Chapter 55, Table 55-1). In most instances the brown egg lines lay eggs that are significantly larger than those produced by white egg lines, with only minor differences in shell and internal egg quality. The only exception to this is that brown egg layers usually lay considerably higher percentages of eggs with blood and meat spots.

3-C. PRESENT-DAY MEAT PRODUCTION LINES

Certain varieties and lines of chickens have been bred with emphasis on the production of meat rather than eggs. They are capable of producing economical gains in weight when raised as broilers or roasters. Generally, it is impossible to breed a single line of chickens that will produce both eggs and meat in abundance as there is a significant negative genetic correlation between egg production and growth. Therefore, when strains are selected for high meat production, their ability to lay a large number of eggs decreases.

Female Meat Lines

In the past, breeders of meat-type birds specialized in developing the necessary line for either the male or female parent of the mating to produce commercial broiler chicks. Today, however, most, but not all, meat-line breeders develop both the male and female sides of the mating.

While the primary emphasis of geneticists is on growth, a secondary emphasis of selection in female lines is on hatching egg production. Consequently, White Plymouth Rock, a superior egg-producing breed, and white Cornish, a superior meat breed, are commonly used to produce the female and male parents for broiler chicks, respectively.

Male Meat Lines

Male meat lines grow very rapidly, are well fleshed, and have good feed conversion. To acquire these traits within a meat strain, both egg production and hatchability have been sacrificed.

Today, such male lines predominantly incorporate genes necessary for meat production, conformation, and ease of processing with little emphasis on egg production and hatchability.

Cornish used for meat lines. Probably all meat lines on the male side include genes derived from the Cornish (English) breed. Varieties of this breed give the modern broiler a broad breast, short legs, and a plump carcass.

White-feathered male meat lines. Not only do the birds from these male meat lines have white feathers but when males are mated with colored females the offspring have white or nearly white feathers. This is a decided advantage at processing time because it is easier to pick white rather than dark feathered chickens. Genetically, the male lines are dominant white for plumage color.

Yellow and white skin color. Consumers in most countries have a preference for broilers and roasters with yellow skin. Practically all current male and female broiler breeding strains have yellow skin. The exception is England and some European countries where white skin is preferred. The practical way to produce white-skin broilers is to mate a white-skin male with a yellow-skin female. The Light Sussex with white skin is predominantly used for the male line, and is mated with yellow-skin females. The offspring from such matings have white skin as white skin is dominant to yellow skin.

Special Lines for Meat Production

Sex-linked meat lines. Certain feather colors and patterns and speed of feather growth can be linked with the sex of the bird. When gold (buff or red) males are mated with certain silver (white) females, the offspring female chicks are gold or buff and the male chicks are silver (white). Similarly, if fast-feathering males are mated with slow feathering females, the characteristics are reversed in the offspring and the fast feathering characteristics can be observed in the wings of the newly hatched female chicks. Either of these matings makes it possible to determine the sex of the chicks at 1 day of age. The procedure is used to produce what are known as sex-linked chicks and makes it possible to easily segregate males and females at hatching time. There are many such matings used today.

Lines for roaster production. Roasters are larger than broilers and require special lines of birds that will grow efficiently to the heavier weights. Primary breeders have developed special strains or crosses that produce chicks with this desired trait. Roasters are commonly grown to 6 to 8 pounds (2.7 to 3.6 kg) or more.

Squab broilers. Squab broilers are usually sold at 2.0 to 2.5 lb (0.9 to 1.1 kg), live weight. They can be raised either straight-run (sexes not separated) or as sexed females and males grown to different ages to meet market demands. Although somewhat of a misnomer (they are typically a Rock Cornish cross and they may be females, however, they are not game chickens), the popular name of Rock Cornish Game Hen is often used to merchandise the processed squab broiler. They are sold as a whole bird and therefore are never cut up.

3-D. THE PACKAGE DEAL

Today, most of the meat-line primary breeders produce both male and female parent lines. In such cases, the breeder has the ability to sell both cockerel and pullet day-old chicks as a package in which the customer would receive 12 to 15 cockerel chicks with every 100 female chicks delivered. In Europe, South America, and a number of other international locations, parent-line breeders are normally marketed as a package with both males and females originating from the same breeder. However, the US market is different in this respect, as in may instances, broiler companies will purchase males from one primary breeder and females from another.

Primary breeders of table-egg lines, producing eggs with either white or brown shells, produce both the male and female parent lines needed for the production of the commercial egg-type pullet. This is necessary because of the intricacies involved in making the matings to produce the parent males and females used in the breeding programs. As mentioned earlier, only certain combinations will properly nick. Therefore, male and female egg-type breeder parents almost always come from the same breeder, and are shipped to the customer as a package.

3-E. NATIONAL POULTRY IMPROVEMENT PLAN

The National Poultry Improvement Plan (NPIP), which is specific for the US, began in 1935 as a voluntary program administered by agreement among the states, the US Department of Agriculture, and poultry producers.

There are two objectives of the plan:

1. To improve the production and market quality of poultry.
2. To reduce losses from certain diseases commonly disseminated by hatcheries and breeder flocks, in particular:
   a. pullorum
   b. fowl typhoid
   c. *Mycoplasma gallisepticum*
   d. *Mycoplasma synoviae*
   e. *Mycoplasma meleagridis* (in turkeys)
   f. *Salmonella enteritidis.*

A program of blood testing breeders to determine if they are carriers of any of these diseases has been established and is generally administered by individual state diagnostic laboratories in conjunction with the US Department of Agriculture.

Those interested in the many details of the plan should secure a copy of "National Poultry Improvement Plan" from the National Poultry Improvement Plan, 1500 Klondike Road, Suite A-102, Conyers, GA 30207-5115.

4

Anatomy of the Chicken

by Donald D. Bell

The chicken is a warm-blooded vertebrate, evolving from reptiles. Although there are many similarities between the two, there are also vast differences.

Reptiles are poikilotherms. That is, they are cold blooded, meaning their body temperature is not regulated to a specific temperature and therefore is usually that of the environment. Chickens are homeotherms. They are warm blooded, meaning their deep body temperature is relatively high and usually almost constant. They are also endotherms. They have the ability to generate deep body heat to increase body temperature.

Both reptiles and chickens lay eggs that are incubated outside the body. However, the female reptile buries her eggs in the sand or soil, and the surrounding temperature is adequate for the growth of the developing embryo. During natural embryonic development, eggs of the chicken are covered (set) by the hen and they are maintained at a temperature close to her body temperature for the entire incubating period. Also, most birds can fly while reptiles cannot.

4-A. SURFACE OF THE CHICKEN

The chicken's body is covered with a combination of skin, feathers, and localized scales, with the latter being a derivative of reptiles (Figures 4-1 and 4-2).

Feathers

Birds are almost completely covered with feathers, making them different from other vertebrates. During the evolutionary process of the chicken,

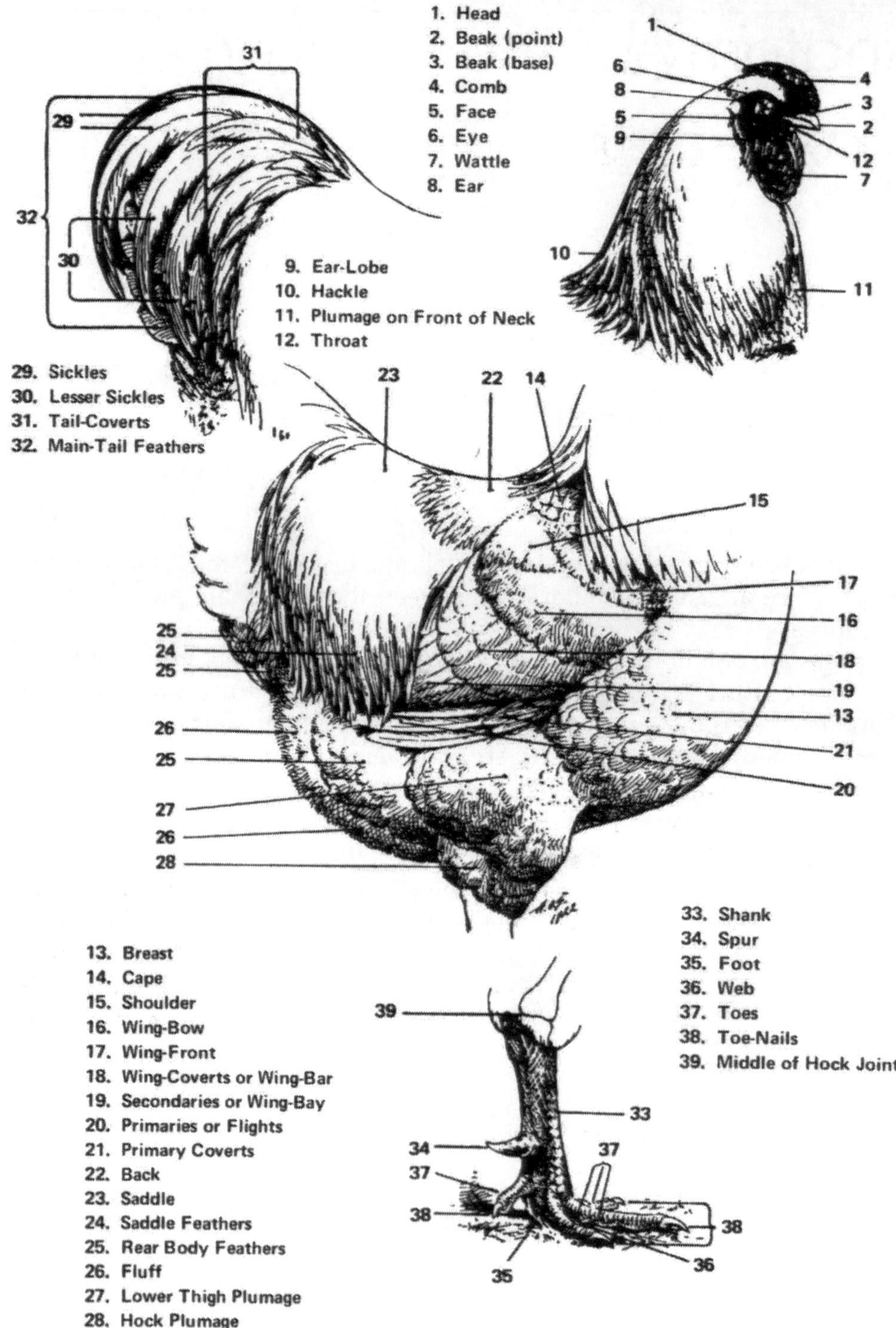

Figure 4-1. Nomenclature of the Male Chicken (courtesy of the American Poultry Association)

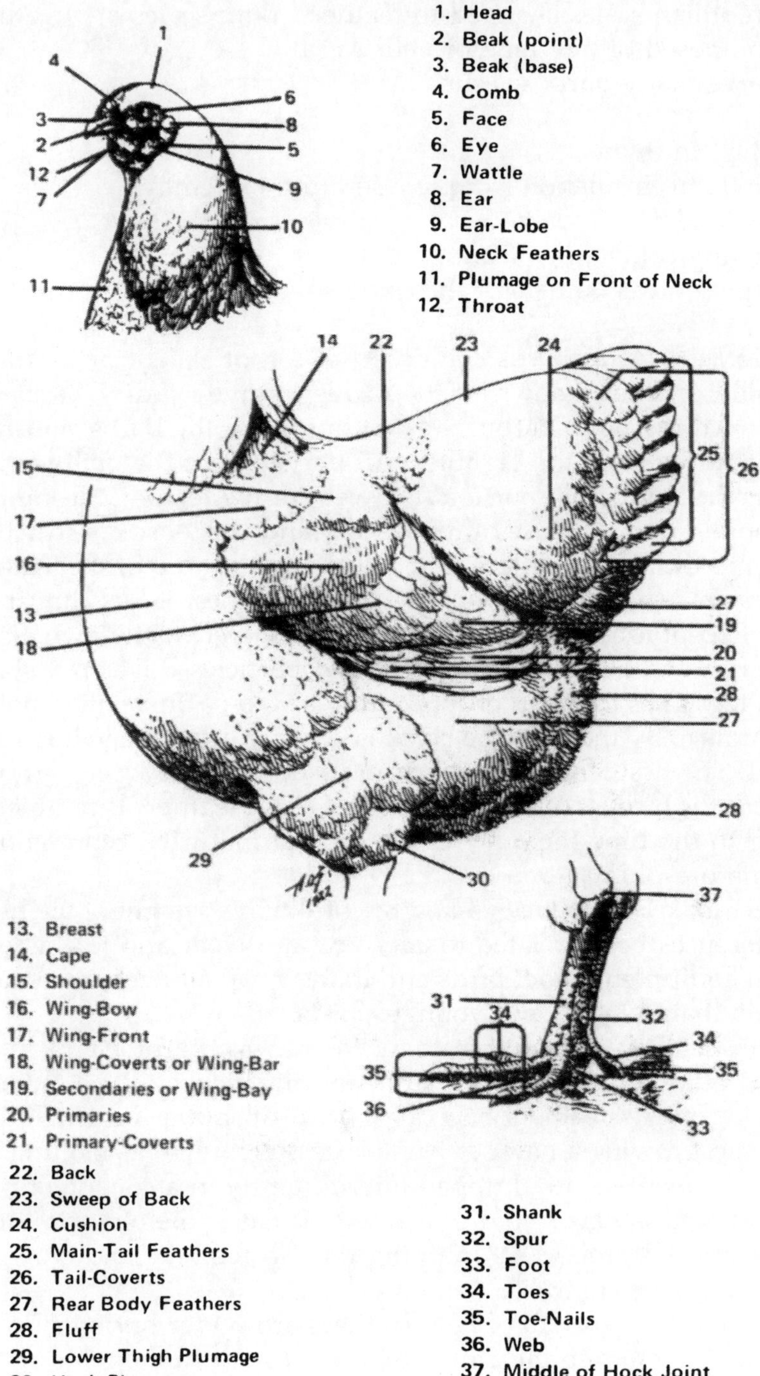

Figure 4-2. Nomenclature of the Female Chicken (courtesy of the American Poultry Association)

most of the reptilian scales changed to feathers. Both scales and feathers are chiefly composed of the same protein, keratin.

Feathers serve many purposes such as:

1. aiding in flight
2. providing insulation from temperature extremes
3. repelling rain and snow
4. creating camouflage from predators
5. helping attract others of the same species

Parts of a feather. A feather is composed of a root called the calamus; a long quill or shaft, known as the rachis to give rigidity; barbs extending from the quill; barbules extending from the barbs; and barbicels extending from the barbules. All parts except the quill tend to mesh together in the flat portion (the web) of the feather. Meshing is not pronounced at the base of the feather and the loose construction gives rise to fluff, often different in color than the web of the feather.

How feathers are replenished. When the chick hatches, it has almost no feathers. Except for the wings and tail, it is covered with down. Soon the down grows longer, and most of the particles develop a shaft. Within a few days the shaft erupts, and the web of the feather makes its appearance. By the time the chick is 4 to 5 weeks of age it is fully feathered. The first feathers are then molted, and a new set is grown by the time the bird is 8 weeks old. A third set of feathers is completed just prior to the time the bird reaches sexual maturity, representing its first mature plumage.

Feathers make up between 4 and 8% of the live weight of the bird, with differences being related to age, sex, and wear and tear associated with equipment; older birds and males have a lower percentage of feathers than females and younger birds.

The annual molt. Because adult feathers wear away, become broken, or are pulled out, nature has provided the adult chicken with a method of renewing all its feathers once a year by dropping its remaining feathers and growing a new set. The process is known as molting. In the wild, the feathers are dropped intermittently in a consistent pattern so the bird is never void of feathers; it has some old and some new. The normal process of dropping the old feathers and growing new ones requires from 3 to 4 months.

Molting and the growth of new feathers are under hormonal control. To molt, a chicken must initiate new growth in the buds at the base of the feathers that in turn forces the old feathers out.

The hormone levels that induce egg production and cause broodiness inhibit feather-bud growth. Consequently, hens that are molting are seldom producing eggs. If egg production is curtailed by artificial means, such as reducing feed intake, the molt may be precipitated in a more rapid and complete manner (see *Flock Replacement Programs and Flock Recycling*, Chapter 54).

Shape of the feather. Not only do feathers vary greatly in their size over the surface of the body, the shapes of certain feathers are associated with the sex of the bird. Gonadal hormones play an important part in this sex variation. They lengthen and narrow the hackle, saddle, sickle, and lesser sickle feathers of the male.

Feather tracts. Feathers do not uniformly cover the body, but rather grow in rows producing feather tracts in specific areas over the body. The ten major feather tracts are:

shoulder	abdomen
thigh	leg
rump	back
breast	wing
neck	head

The order and time of the appearance of the various feather tracts are as follows:

Shoulder and thigh	2 to 3 wk
Rump and breast	3 to 4 wk
Neck, abdomen, and leg	4 to 5 wk
Back	5 to 6 wk
Wing coverts and head	6 to 7 wk

Color of feathers. Feathers can have many colors and color patterns. In some instances, differences in color vary according to the location of the feathers on the body. Color patterns can be different on the male and female. Feather colors and feather patterns are the result of genetic differences (feather color is sex-linked) and the presence of gonadotropic hormones.

Waxy coating on feathers. The uropygial, or preen, gland is located on the dorsal area of the tail, and is the only secretory gland located on the surface of the chicken. It secretes an oily wax that the bird spreads over its feathers with its beak. The material makes the feathers water resistant; they do not absorb water, and water quickly runs from coated feathers.

Head

The head of the chicken includes the following parts:

Comb. There are several types of combs, but only the first three of the following list are common in commercial chickens. The various comb types include:

single	strawberry
rose	walnut
pea	V
cushion	buttercup

Comb type is the result of gene interaction, but comb size is associated with gonadal development and the intensity of light, either natural or artificial.

Eyes. Chickens have the ability to discern various colors and have superior ability to focus and to detect movement. Sturkie (1986) credits the avian eye as "the finest ocular organ in the animal kingdom." (See *Fundamentals of Managing Light for Poultry,* Chapter 10).

Eyelids

Eye rings. Inner margin of eyelids.

Eyelashes. Bristle feathers composed of a straight shaft.

Ears. Avian species are know for their keen sense of hearing. Their voice production and ability to imitate sounds infers an exceptional degree of pitch discrimination.

Earlobes. Either red or white.

Wattles

Beak. The beak is a multi-functional appendage of considerable importance. It is involved with procuring food, defense and aggression in social behaviors, courtship, nest-making, grooming, and communications. Its normal functions are oftentimes adversely affected by improper beak trimming. (See *Cage Management for Raising Replacement Pullets,* Chapter 51).

Feet and Shanks

The shanks and most of the feet are covered with scales of various colors. Yellow is due to dietary carotenoid pigments in the epidermis when melanic pigment is absent. Varying shades of black are the result of melanic pigment in the dermis and the epidermis. When there is black in the dermis and yellow in the epidermis, the shanks have a greenish appearance. In the complete absence of both of these pigment, the shanks are white. Important parts of the shank and foot are:

Hock

Shank

Toes. Most chickens have four toes on each foot, but there are a few breeds with five toes.

Skin

Most of the chicken's body is covered with a thin skin. With the exception of the uropygial gland (preen gland) the skin is void of glands. The absence of sweat glands makes it impossible for the bird to cool itself by evaporation from the surface of the body.

The skin has a coarser texture in the areas of the comb, wattles, earlobes, beak, scales, spurs, and claws. Except for certain specialized areas, the color of the skin is either white or yellow. The density of the yellow color is directly correlated with the amount of xanthophylls in the diet and inversely correlated with the intensity of egg production.

4-B. SKELETON

The skeleton is the frame that supports the body and to which the muscles are attached. The rib cage protects some of the vital organs. Close scrutiny shows that the bones found in the skeleton of mammals are also found in the skeleton of chickens. However, some of the bones in the chicken are fused or elongated. Others are hollow to aid in flight. Figure 4-3 illustrates the major bones found in the skeleton of the chicken.

The vertebrae of the neck move freely, but unlike mammals, the remaining portion of the vertebral column is rigid, containing many fused bones. Several of the thoracic vertebrae are united to form a firm base for the attachment of the wings and their muscles. There is also a heavy keel. The wings correspond to the arms and hands of humans with the legs containing the same bones as found in the legs of man. The bones of the metatarsus, common to the human foot, have been fused and elongated to form the shank.

Bones found in the skull, humerus, keel, clavicle, and some vertebrae are hollow and connected to the respiratory system, with air continually moving in and out of these specialized bones. Most bones are light in weight, yet very strong. There is also a soft, spongy bone material known as medullary bone present in varying amounts in the long leg bones (femur and tibia), and certain other bones of the skeleton of females in egg production. This medullary bone is used to store calcium for later use in eggshell formation. The amount of calcium stored in these specialized bones is highly variable, depending on the length and rate of egg production. Most of the calcium needed for the production of eggshells is not stored but comes directly from the feed eaten each day.

4-C. MUSCLES

Muscles are categorized by their function as *voluntary* or *involuntary*. Voluntary muscles are used for movement and flight while involuntary muscles (smooth muscles) are used in the functioning of organs such as the heart, intestines, blood vessels, and others.

The muscles that move the bird are especially important, yet those that control the action of the heart, blood vessels, intestines, and other vital

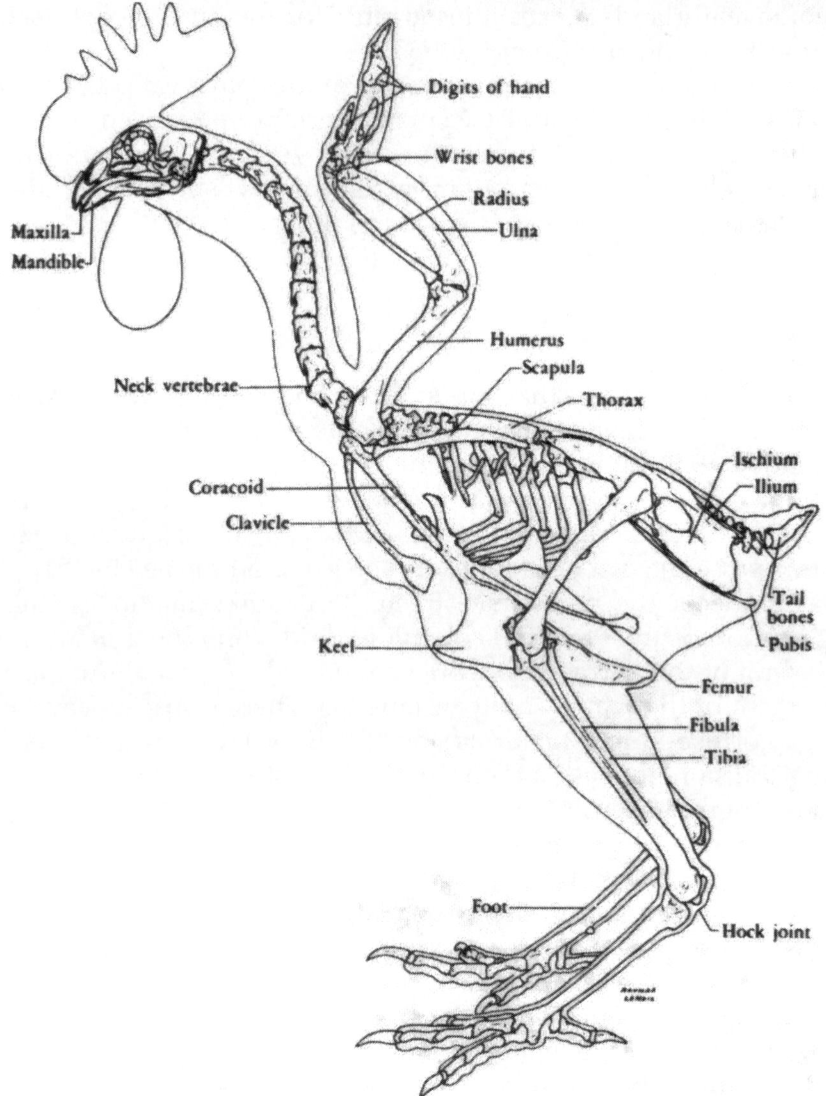

Figure 4-3. Skeleton of the Chicken

organs cannot be overlooked. Muscles that move the wings are attached to the keel (breastbone). These muscles also support the vital organs of the abdominal cavity. These muscles are well developed in most birds, but especially in meat-type broiler strains as genetic selection has produced birds with larger breasts.

Chickens have both white muscle and red muscle giving rise to light and dark meat. More fat and myoglobin, an iron and oxygen carrying compound, are found in red meat than in white. Usually the activity of the muscle determines its color. In the chicken, those of the leg are darker than those of the breast because there is constant stress on the leg muscles to keep the body upright when the bird is standing. In wild flying birds, the breast muscle is darker because greater stress is placed on it during flight. Broiler-type chickens have muscle fibers that are larger in diameter and lighter in color than those of layer types.

4-D. RESPIRATORY SYSTEM

The respiratory system of chickens consists of:

nasal cavities	bronchi
larynx	lungs
trachea (windpipe)	air sacs (9)
syrinx (voice box)	air-containing bones

Lungs of the chicken are small compared with those of mammals. They expand or contract only slightly, and there is no true diaphragm. The lungs are supported by nine air sacs and a group of hollow, air-containing bones. There are two pairs of thoracic and two pairs of abdominal air sacs, and a single interclavicular air sac.

While air freely moves in and out of the air sacs, only the lungs are responsible for the exchanging of oxygen and carbon dioxide occurring during respiration. Both the lungs and air sacs function as cooling mechanisms as moisture evaporates from their surfaces and is exhaled as water vapor.

The respiratory rate is governed by the carbon dioxide content of the blood; increased levels increase the rate, which ranges between 15 and 25 cycles / min in the resting bird.

4-E. DIGESTIVE SYSTEM (see *Digestion and Metabolism*, Chapter 14)

Figure 4-4 shows the digestive system of the chicken. The various parts are discussed below.

Mouth

The chicken has no lips, soft palate, cheeks, or teeth, but rather has a horny upper and lower mandible (beak) which it uses to pick up food. The upper mandible is firmly attached to the skull while the lower mandible is hinged. The hard palate is divided by a long narrow slit in the center that is open to the nasal passages. This opening and the absence of a soft palate make it impossible for the bird to create a vacuum to draw water into its

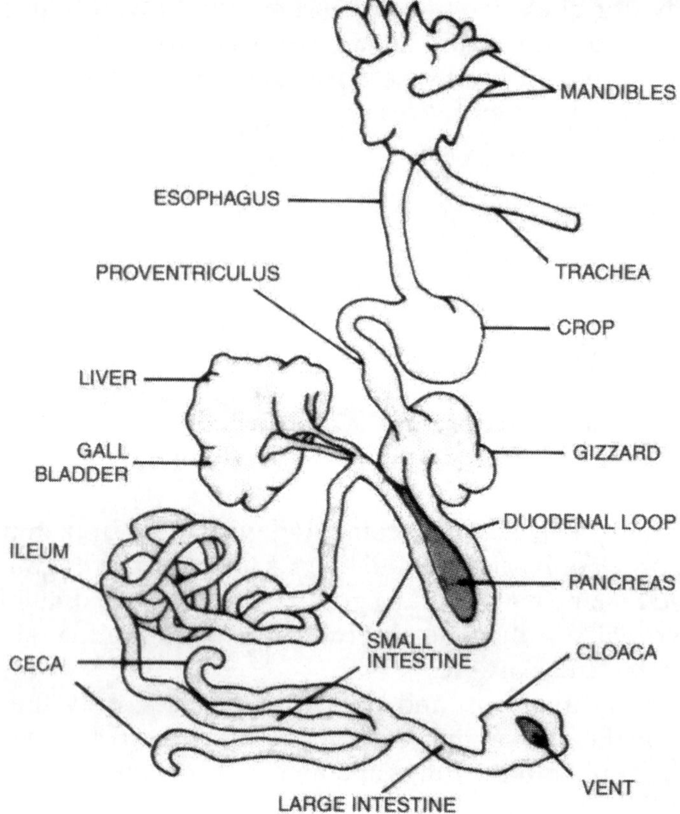

Figure 4-4. Digestive System of the Chicken

mouth. Therefore, in order to drink, the bird must elevate its head to allow the water to run down the esophagus by gravity.

The chicken has a dagger-shaped tongue that has a very rough surface near the back which helps to force food into the esophagus. Saliva, with its enzyme amylase which is used to convert starches to sugars during digestion, is secreted by the glands in the mouth. Another function of saliva is as a lubricant to help with the transport of food particles from the mouth down the esophagus and into the crop.

Chickens have fewer taste buds than mammals; the human has about 9,000 compared to only 250 to 350 for the chicken. The chicken's taste buds are located in several areas of the mouth and beneath the tongue. Chickens are considered to have relatively poor taste acuity, but do respond to specific tastes such as salt and sugar. Birds appear to have a wide tolerance for acidity and alkalinity of their drinking water. Birds can discriminate between drinking water temperatures of as little as 5°F (3°C) and will refuse to drink water at temperatures above 110°F (38°C). (See *Consumption and Quality of Water*, Chapter 22).

Esophagus

The esophagus or gullet is the tube through which the food passes on its way from the back of the mouth (pharynx) to the proventriculus. It is composed of two regions; the upper part (between the mouth and the crop) is approximately 8 inches (20 cm) long in the adult chicken, while the lower part (between the crop and the proventriculus) is about 6 inches (16 cm) in length.

Crop

Just before the esophagus enters the body cavity it extends on one side into a pouch known as the crop, which acts as a storage place for food. Little or no digestion takes place here except for that involved with the salivary secretion of the mouth, which continues its activity in the crop.

Proventriculus

An enlargement of the esophagus just prior to its connection with the gizzard is known as the proventriculus, sometimes called the glandular or true stomach. It is here that gastric juices are produced and secreted. Pepsin, an enzyme needed for the digestion of protein, and hydrochloric acid are secreted by the glandular cells. Because the food passes quickly through the proventriculus there is little digestion of food material here, but the secretions pass into the gizzard where the enzymatic action occurs.

Gizzard

The gizzard, sometimes called the muscular stomach, lies between the proventriculus and the upper portion of the small intestine. It has two pairs of very powerful muscles capable of exerting great force and a very thick mucosa, the surface of which constantly erodes and sloughs off. The gizzard is inactive when empty, but once food enters, the muscular contractions of its thick walls begin. The larger the particles of food, the more rapid the contractions. When fine material enters the gizzard it leaves in a few minutes, but when the food is coarse it can remain in the gizzard for several hours. If the gizzard contains an abrasive material, such as grit, rock, gravel, etc., food particles can be ground more rapidly prior to entering the intestinal tract. Finer feed grinding practices and the use of larger particle size calcium sources for layers have practically eliminated the need for grit in today's commercial diets.

Small Intestine

The small intestine is comprised of two major sections, the duodenal loop and the ileum. Within the duodenal loop lies the pancreas that secretes pancreatic juices containing the enzymes amylase, lipase, and trypsin. Other enzymes are produced by the walls of the small intestine, further aiding with the digestion of protein and sugars. In the adult chicken, the small intestine is approximately 55 inches (140 cm) long. The small intestine is the primary site of nutrient absorption.

Ceca

Between the small and large intestines lie two blind pouches known as ceca. Each cecum is about 6 inches (15 cm) long in the adult bird. The exact function of the ceca is not well defined, but it has been concluded they have little to do with digestion and only minor functions associated with water absorption. A small amount of carbohydrate and protein digestion and the microbial fermentation of dietary fiber also takes place in the ceca.

Large Intestine

The large intestine is a relatively short extension of the small intestine in the chicken, being only 4 inches (10 cm) long in the adult bird. It is about twice the diameter of the small intestine. It extends from the end of the small intestine to the cloaca. The large intestine is involved in water resorption, and in doing so assists with maintaining the water balance in the bird.

Cloaca

The bulbous area at the end of the alimentary tract (from the mouth to the vent) is known as the cloaca. Cloaca means "common sewer," and in the case of the chicken, the digestive, urinary, and reproductive tracts all empty into the cloaca.

Vent

The vent (anus) is the external opening of the cloaca. Its size varies greatly in the female, depending on whether or not she is producing eggs.

Supplementary Digestive Organs

Certain organs are closely associated with digestion because their secretions empty into the intestinal tract and aid with the breaking down and absorption of food material.

Pancreas. The pancreas lies within the duodenal loop of the small intestine. It is a gland that secretes enzymes into the duodenum by way of the pancreatic ducts. These enzymes aid in the digestion of starches, fats, and protein. The enzymes, also know as pancreatic juices, neutralizes the acid condition created in the proventriculus.

Liver. The liver is composed of two large lobes. Among its functions is the secretion of *bile,* a slightly sticky yellow-green fluid containing bile acids. These acids enter the small intestine at the lower end of the duodenum and aid with the digestion of fats. The bile secretions contain no digestive enzymes. Its chief function is to neutralize the acid condition and to assist with the digestion of fats by forming emulsions. In addition, the liver is involved in the metabolism of fats, proteins, and carbohydrates.

Gallbladder. While the chicken has a gallbladder, some birds do not. As discussed under *Liver* above, two bile ducts are used to transfer bile from the liver to the intestines. The right duct, through which most of the bile passes and is temporarily stored, is enlarged forming the gall bladder. The left duct is smaller, therefore only a small amount of bile passes through it directly into the intestines.

4-F. URINARY SYSTEM

The urinary system consist of two kidneys that are located just behind the lungs. A single ureter connects each kidney with the cloaca. The urine of chickens is mainly uric acid, the end product of protein metabolism, which is mixed with the feces in the cloaca and evacuated in the droppings as a white pasty material.

4-G. CIRCULATORY SYSTEM

The purpose of the circulatory system is to carry oxygen (O_2) from the lungs and nutrients that have passed through the intestinal walls to the cells (*arterial blood*). The *venous system* carries carbon dioxide (CO_2) back to the lungs and waste products from metabolism back to the kidneys for excretion from the body. The heart of the chicken, as in mammals, has four chambers: two atria and two ventricles. It beats at a comparatively rapid rate of about 300 pulsations per minute. The smaller the bird, the more rapid the contractions. Chicks show an increased rate as they age. Birds in bright light have a faster heart rate. The rate of individual chickens is highly variable and may often double as the result of excitement alone.

Composition of blood. Blood is composed of plasma, salts, and other chemicals, plus erythrocytes (red cells) and leukocytes (white cells). In the chicken, the erythrocytes contain a nucleus in contrast to those of mammals. The blood of a chicken contains about 3 million erythrocytes per cubic millimeter.

The spleen serves as a storage site for the erythrocytes, and expels its contents into the circulatory system as needed. Blood constitutes about 12% of the weight of a newly hatched chick, and about 6 to 8% of the mature chicken.

Function of blood. Blood has numerous functions, including the following:

1. It moves O_2 to body cells and removes CO_2 from them.
2. It absorbs nutrients from the alimentary tract and transports them to tissues and cells.
3. It removes the waste products of cellular metabolism.
4. It transports hormones produced by the endocrine glands to various sections of the body.
5. It helps regulate the water content of body tissues.

Blood pressure. Blood pressure of chickens of all ages is normally measured as mmHg. Even the pressure of the developing embryo can be recorded. As with humans, there are two separate measurements:

1. systolic pressure (arterial)
2. diastolic pressure (as the blood returns to the heart).

Following are the recognized blood pressures of adult chickens:

	Systolic Pressure (mmHg)	*Diastolic Pressure (mmHg)*
Adult female chicken	145–180	133–160
Adult male chicken	186–203	154

Source: Sturkie (1986)

4-H. NERVOUS SYSTEM

The nervous system controls all body functions and consists of many parts. The brain represents highly concentrated nerve cells and serves as the base for all nerve stimuli. Hearing and sight are well developed, with the chicken being able to distinguish between various sounds and colors. The chicken's ability to distinguish various smells, however, is not well developed.

Chickens have an ability to learn; they can be trained to follow certain physical procedures. Furthermore, they learn to recognize large numbers of pen mates at a young age, and their ability increases with age.

4-I. ENDOCRINE GLANDS

Within the body are certain endocrine glands and tissues that produce chemical products known as hormones. Hormones pass directly into the bloodstream and have an effect on cells and organs in many parts of the body. Hormones are primarily derived from proteins and perform a variety of functions. Some increase the activity of certain organs, others depress organ activity, still others have an effect on metabolic processes.

The glands producing hormones include:

thyroid	pineal
parathyroids	adrenals
testes	ultimobranchial body
ovary	islets of Langerhans
pituitary	pancreas
hypothalamus	

In addition to glands, hormones are also produced by the gastric and intestinal mucosa and a number of local sites throughout the body.

The functions and interaction of hormones are varied and great in number. Thyroxine, produced by the thyroid, helps regulate the metabolic rate. Parathyroid hormone from the parathyroids influences calcium and phosphorus metabolism. The hormones of the pituitary, a small gland at the base of the brain, are many. Some aid in growth; others affect the thyroid and parathyroids; while others have a pronounced effect on ovulation, the oviduct, broodiness, and egg laying in the female, and semen production in the male.

Hormones of the ovary influence fat deposition, increase the release of calcium from the medullary bone, and cause ovulation. Chemicals produced by the adrenals aid in retention of glycogen by the liver and affect mineral metabolism. The islets of Langerhans and some cells of the pancreas produce insulin and glycogen, which regulate the utilization of glucose and its level in the bloodstream. Hormones of the gastrointestinal tract increase the production of gastric juice, pancreatic juice, and bile.

4-J. REPRODUCTIVE SYSTEMS

Male

The male reproductive system consists of two testicles in the dorsal area of the body cavity, just in front of the kidneys. The many ducts of the testes lead to the vas deferentia and vas deferens, which carry the semen from the testicles to the papillae in the dorsal area of the cloaca and then to the copulatory organ located in one of the folds of the cloaca. Normally, semen is stored in the vas deferens where it is diluted with lymph fluid; both are ejaculated as a mixture during copulation.

The penis of the male chicken is quite small. Lymph enters the penis to form a mild erection, but it does not penetrate the cloaca. Rather, during mating, the cloaca of the female opens to expose the end of the oviduct where semen is deposited. Once it has entered the oviduct, it travels up the duct to pouches, known as semen storage sites, where it is held prior to fertilization.

Spermatozoa from the male chicken have a long pointed headpiece that is attached to a long tail. The pH of semen is between 7.0 and 7.4. The volume of semen ejaculated during one mating may be as high as 1.0 ml at the beginning of the day, but decreases to as little as 0.1 ml after many matings (see *Managing the Breeding Flock*, Chapter 34).

Female

The female reproductive system consists of one functional ovary and oviduct. These are described in detail in *Formation of the Egg*, Chapter 5.

4-K. HOW A CHICKEN GROWS

The body of the chicken consists of a large number of cells that are about the same size in all breeds, regardless of ultimate mature body weight. Most early embryonic increases in growth occur as the result of cell multiplication: 1 cell divides into 2, 2 into 4, 4 into 8, 8 into 16, and so on. But this rhythmic increase does not continue indefinitely. Soon there is cell specialization which is necessary to form different body components. Growth rate and rate of division among the various specialized cells varies depending on function and age. The older the bird, the lower the daily increments of increased body weight.

After hatching, when the number of muscle fibers (single cells) no longer increases, growth of muscle and nerve cells is the result of cell enlargement rather than cell division. Muscle fibers have a maximum dimension, controlled mainly by the genetic makeup of the bird, but can decrease or increase in size with varying amounts of activity. Both protein synthesis and protein degradation are involved. Both synthesis and degradation operate simultaneously, with the net result determining whether muscles increase or decrease in size. The muscles of the breast are exceptionally well developed in birds because these muscles are used to move the wings during flight.

The degree of fatness of a chicken rests entirely on the number of fat-containing cells. Some breeds and lines of chickens have a greater number of fat cells than others, an indirect consequence of breeding birds for larger sizes and plumper carcasses. Fat cells reach their maximum number in the early growing period. The ability of a broiler to gain weight rapidly is principally the result of fat deposits in the fat cells rather than increases in the growth of the skeleton or muscle fibers.

4-L. BODY CHANGES DURING EGG PRODUCTION

During the time female chickens are laying and during the time they are molting certain changes occur in their appearance.

Molting

Some layers may produce a few eggs after the molt begins, but generally cessation of egg production preempts the molt. The length of the molting period varies. Good egg producers molt late in the season, and rapidly; poor egg producers molt early, and slowly.

Order of the molt: During the molt, feathers are dropped from the various parts of the body in a definite order, that is:

1. head
2. neck
3. breast
4. back
5. fluff
6. abdomen
7. wings
8. tail

Many times a flock will experience a temporary stress resulting from a disease or change in environment, causing a partial molt of the feathers from the head and neck and a few feathers from the wings. If the cause can be corrected, it should not interfere with the primary annual molt.

Yellow Pigmentation

The yellow color in the skin and shanks of yellow-skinned chickens is due to several xanthophylls (hydroxycarotenoid pigments) that are deposited in the fat layer under the skin. The bird's only source of these xanthophylls is from the diet it eats. The more xanthophylls in the feed, the denser the yellow skin color. Xanthophylls in this fatty layer continually undergo chemical breakdown, but are replenished from the feed.

Bleaching. Xanthophylls are also responsible for the yellow color of egg yolks. However, when a pullet starts laying at a fast rate most of the xanthophylls in the feed go to the egg yolks. Not enough is left to replenish those being chemically lost in the skin, and it begins to bleach. The longer a bird lays, the greater the bleaching. When the bird has laid about 180 eggs the skin will have a blue-white color.

Other Changes Resulting from Egg Production

There are other changes in the bird during the course of egg production, namely:

1. Vent becomes large and moist.
2. The pubic bones become thinner.
3. Space between the pubic bones increases.
4. Distance between the pubic bones and end of keel bone increases.

These changes are used to determine the egg-laying status of individual birds.

5

Formation of the Egg

by Donald D. Bell

The avian egg consists of a minute reproductive cell quite comparable to that found in mammals. But in the case of the chicken, this cell is located on the surface of the yolk and surrounded by albumen, shell membranes, shell, and cuticle. The ovary is responsible for the formation of the yolk; the remaining portions of the egg originate in the oviduct.

5-A. OVARY

At the time of early embryonic development, two ovaries and two oviducts exist, but the right set atrophies, leaving only the left ovary and oviduct at hatching. Prior to egg production, the ovary is a quiet mass of small follicles containing ova. Some ova are large enough to be visually seen; others require magnification. Several thousand are present in each female chicken, many times the number that will eventually mature into full-size yolks necessary for egg production during the life of the bird.

Formation of the Yolk

The yolk is not the true reproductive cell, but a source of food material from which the minute cell (blastoderm) and its resultant embryo partially sustain their growth.

When the pullet reaches sexual maturity, the ovary and the oviduct undergo many changes. About 11 days before she is to lay her first egg, a sequence of hormonal changes occur. The follicle-stimulating hormone (FSH) produced by the anterior pituitary gland causes the ovarian follicles to increase in size. In turn, the active ovary begins to generate hormones: estrogen, progesterone, and testosterone (sex steroids). Higher blood plasma levels of estrogen initiate development of the medullary bone, stimulate yolk protein and lipid formation by the liver, and increase the size of the oviduct, enabling it to produce albumen proteins, shell membranes, calcium carbonate for shell formation, and cuticle.

The first yolk (ovum) to begin maturing does so as major amounts of the yolk material produced in the liver are transported by the circulatory system directly to the developing ovary. A day or two later, the second yolk begins to develop, and so on, until at the time the first egg is laid, five to ten yolks are in the growth process. About 10 days are required for an individual yolk to mature. Deposits of yolk material are very slow at first and light in color. Eventually the ovum reaches a diameter of 6 mm at which time it grows at a greatly increased rate, with the diameter increasing about 4 mm per day. A greater number of yolks are under development at one time in the broiler breeder hen than in the egg-type hen, but the broiler breeder hen does not have the ability to produce as many complete eggs.

The color present in the yolk is xanthophyll, a carotenoid pigment derived from the diet. The pigment is transferred first to the bloodstream, then quickly to the yolk, as well as other parts of the body. Consequently, more is deposited in the yolk during the hours when the hen is eating than during dark hours when she is not. This gives rise to deposits of dark and light layers of yolk material, depending on the dietary pigment available. From seven to eleven concentric rings are found in each yolk. Yolk formation is rather uniform and the total thickness of both dark and light deposits during 24 hours is about 1.5 to 2.0 mm.

Egg yolk is composed mainly of fats (lipids) and proteins, which combine to form lipoproteins, of which 60% of the dry yolk weight is of low density lipoproteins (LDL), and are known to be synthesized by the liver through the action of estrogen. In the laying hen, LDL is removed from the blood plasma as intact particles for direct deposition in the developing ova.

What influences growth rate of the yolk? Yolks vary greatly in size between individual chickens in the flock at the same age, and are usually associated with body weight differences. Yolk size is not associated with rate of lay, but probably more with the length of time required for the ova to reach maturity. The yolks from an individual hen increase in size over the production cycle. Furthermore, the first egg laid in a clutch will usually contain a larger yolk than the remaining ones. Eggs laid later in the day are 0.5 grams lighter for each additional hour in the day; this is also associated with smaller yolks. The inclusion of added fat and protein in the diet has also been shown to increase the size of the developing yolk.

Location of the germinal disc. The yolk material is laid down adjacent to the germinal disc that continues to remain on the surface of the globular yolk mass. Once the egg is laid, the yolk rotates so the germinal disc remains in the large end of the egg.

Ovulation

At maturity the ova are released from the ovary to enter the oviduct by a process known as ovulation. Each ovum hangs on the ovary by a narrow stalk containing the arteries that supply the blood to the developing yolk. The arteries undergo much branching in the surface membranes of the yolk and the follicle appears highly vascular except for the stigma, a narrow band surrounding the yolk that is almost void of blood vessels.

When an ovum is mature, the hormone progesterone, produced by the ovary, stimulates the hypothalamus to cause the release of the luteinizing hormone (LH) from the anterior pituitary, which, in turn, causes the mature follicle to rupture at the location of the stigma releasing the ovum from the ovary. The yolk is then surrounded only by the vitelline membrane (yolk membrane).

Delaying first ovulation. Sexual maturity, as indicated by the first ovulation, may be accelerated or retarded. Restricting feed or decreasing day lengths during the pullet's growing period are the two main procedures used (see *Cage Management for Raising Replacement Pullets*, Chapter 51, and *Managing the Breeding Flock*, Chapter 34).

What initiates ovulation? It is not known what sets the hour for the bird's first ovulation, but both the nervous system and hormonal secretions are of primary importance. The second ovulation is regulated by oviposition (laying) of the first egg and occurs about 15 to 40 minutes after the first egg passes through the vent. Future ovulations occur at about the same frequency after subsequent eggs are laid.

Eggs laid in clutches. Chickens lay eggs on successive days known as clutches, after which none are laid for one or more days. The length of the clutch may vary from 2 days to more than 200 before a day is missed, but most commercial egg-type chickens can produce more than 50 eggs in succession without a pause during the early stages of production in the first lay cycle. The length of clutches is quite consistent with individuals; poor producers have shorter clutches, good producers have longer clutches. Once the clutch length is established, the hen will not ovulate for one or more days and then will produce another clutch. Poor egg producers have a longer rest period between clutches than do good producers.

Time necessary to produce an egg. The time necessary for an egg to transverse the oviduct varies with individuals. Most hens lay successive eggs with time intervals of 23 to 26 hours. If the time is greater than

24 hours, each successive egg will be laid later in the day, and the ovulation of the yolk for the next egg will also occur later in the day. Eggs laid in the afternoon have spent several more hours in the oviduct than those laid in the morning. Eventually eggs are laid so late that the rhythm is broken and an ovulation is skipped.

Time of ovulation. Hens that produce long clutches lay their first egg of a clutch early in the day, an hour or two after the sun rises or the artificial lights are turned on. Ovulation of the next yolk comes quickly after an egg is laid, with only a slight time lag. Those hens with shorter clutch lengths lay their first egg of the clutch later in the day, ovulation of the next yolk is slower, and the time lag for laying is greater. Most ovulations occur during the morning hours, as it is not natural for ovulations to occur in the mid to late afternoon.

Egg production at start of lay. During the first week of lay, ovulation is quite irregular; as the hen's hormonal mechanism is not in balance. Often, only two to four eggs are produced in the first clutch. But by the second or third week, ovulation is progressing at its peak rate, only to drop slowly each week throughout the remainder of the laying cycle.

Light and ovulation. Light, either natural or artificial, has an effect on the pituitary gland, stimulating it to secrete an increased quantity of the follicle stimulating hormone (FSH), which in turn, activates the ovary. Both duration and intensity of light are important. The procedure for correctly lighting a flock of laying hens is complicated and is discussed in *Cage Management for Layers*, Chapter 52, and *Fundamentals of Managing Light for Poultry*, Chapter 10.

Nesting as an indication of ovulation. On most occasions the hen seeks a nest about 24 hours after ovulation, leading scientists to theorize that nesting can be used as an indicator of ovulation. Evidently, the presence of a fully formed egg in the cloaca has nothing to do with the hen's desire to seek a nest. For example, some hens will ovulate, but because of a malfunction, or for some other reason, the ovum does not reach the oviduct, these hens will still seek a nest a day later.

Double ovulation. Normally, only one yolk is ovulated per day, but occasionally two may be released and on rare occasions there may be three. If two are ovulated at the same time normally only one enters the oviduct, but if both are picked up simultaneously by the oviduct, a double yolk egg will result. About two-thirds of the double-yolk eggs are the result of ovulations within 3 hours of each other. If there is a great difference in ovulation time, two eggs may be produced on the same day, but usually the second is soft-shelled.

Double-yolk eggs are more common during the first part of the egg production period because of an overactive ovary, and are more often associated with meat-type strains than with egg-type ones. The incidence is an inherited trait since some birds produce higher percentages of double-yolk eggs than others. Spring- and summer-housed pullets also produce a greater number of double-yolk eggs than fall- and winter-housed pullets.

Defective Eggshells

When the normal interval of about 23 to 26 hours between ovulations is broken, more eggs are produced with defective shells, including those with sandpaper texture, white bands, calcium splashing, and chalky white deposits. The occurrence is greater in meat-type than in egg-type breeds. From 5 to 7% of the eggs produced have some form of defective shells. These defects are mostly associated with the age of the flock, with some strains more prone to the problem than others. Various egg shell defects are described in *Egg Handling and Egg Breakage,* Chapter 56.

Yolk Size Affects Egg Size

The size of the completed egg is more closely associated with yolk size than with any other factor, although variations in albumen secretions in the oviduct have some influence. The yolk-albumen relationship changes throughout the laying cycle. Eggs produced at the beginning of the laying period have yolks that comprise about 25% of the total weight of the egg, while yolks make up about 30% of egg weight when hens are near the end of their laying period. In other words, as egg size increases, yolk weight increases more rapidly than the weight of albumen. In younger flocks when egg size is small, increasing the level of protein in the diet may increase the total weight up to 1.5 oz/doz (3.5 g/ea).

Blood Spots and Meat Spots

Often, when the yolk sac ruptures along the stigma, small blood vessels near the area of the rupture are broken, leaving a clot of blood attached to the yolk. The frequency of hemorrhages can be related to a number of factors: genetics, feed, age of the hen, and others. Blood spots are two to three times more common in brown-shelled than in white-shelled laying hens.

Any tissue sloughed from the follicular sac or the oviduct can be included in the developing egg as it passes through the oviduct. These bits of tissue darken with age and are known as meat spots. Many blood spots darken too, and are often incorrectly classified as meat spots. This problem is especially prevalent in brown-shelled eggs where 15% or more of the eggs can be affected, compared to less than 1% in white shell eggs (Carey, 1988).

5-B. PARTS OF THE OVIDUCT

The oviduct is a long tube through which the yolk passes and where the remaining portions of the egg are secreted. Normally, the oviduct is relatively small in diameter, but with the approach of the first ovulation its size and wall thickness expand greatly. The segments of the oviduct and their purpose are summarized below and are illustrated in Figure 5-1.

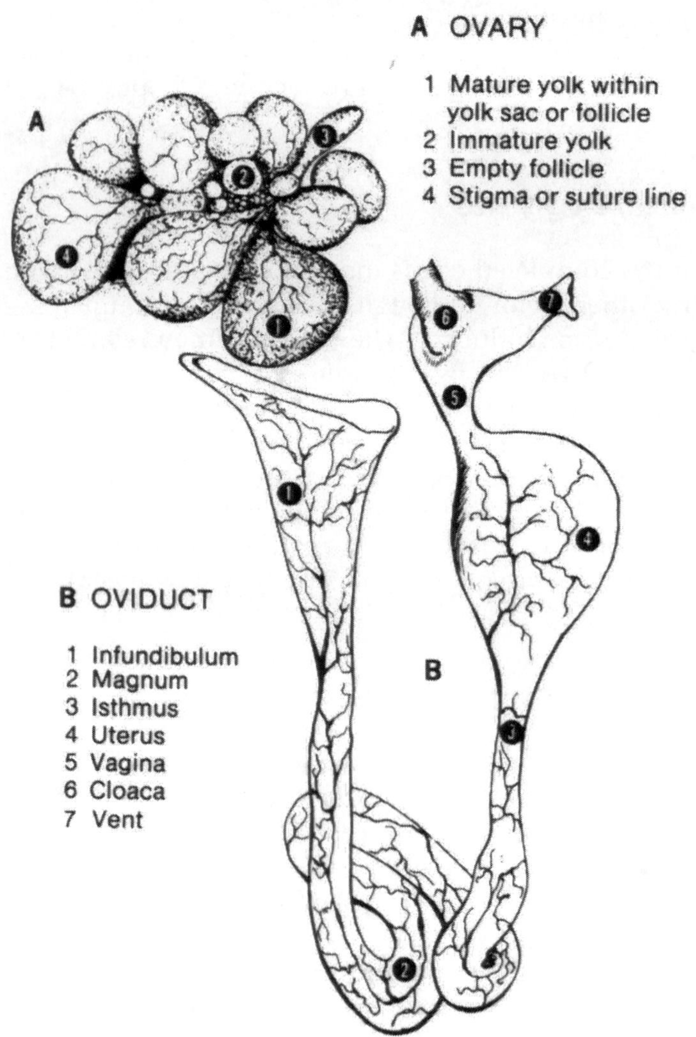

Figure 5-1. Ovary and Oviduct

1. *Infundibulum*

The funnel-shaped upper portion of the oviduct is the infundibulum. When functional, its length is approximately 3.5 inches (9 cm). Normally inactive except immediately after ovulation, its purpose is to search out and engulf the yolk causing it to enter the oviduct. After ovulation, the yolk drops into the ovarian pocket or the body cavity, from which it is picked up by the infundibulum. The yolk remains in this section for only a short period of about 15 minutes, then is forced along the oviduct by multiple contractions.

Malfunction of the infundibulum. To be completely functional, the infundibulum should pick up all the yolks dropped into the body cavity. However, it has been found that an average of 4% are not drawn into the infundibulum, but remain in the body cavity where they are reabsorbed within a day. The percentage varies with strains of chickens, some of which retain up to 10% of their yolks in the body cavity. Meat-type birds are more often affected than egg-type strains.

Internal layers. Sometimes the infundibulum loses its ability to pick up a high proportion of the yolks, and they accumulate in the body cavity faster than they can be reabsorbed. Such hens are known as "internal layers," although the term does not define the condition well. The abdomen in such layers becomes distended, and the hen stands in an upright position.

2. *Magnum*

The magnum is the albumen-secreting portion of the oviduct, and is about 13 inches (33 cm) long in the average laying hen. It takes approximately 2 to 3 hours for the developing egg to pass through the magnum.

Albumen. The albumen in an egg is composed of four layers (see *Shell Eggs and Their Nutritional Value,* Chapter 57). The names and percentages are:

Chalazae	2.7%	Dense white	57.3%
Liquid inner white	16.8%	Outer thin white	23.2%

While all four are produced in the magnum, the outer thin white is not completed until water is added in the uterus.

Chalazae. Upon breaking an egg, one notices two twisted cords, known as chalazae, extending from opposite poles of the yolk through the albumen. The chalaziferous albumen is produced when the yolk first enters the magnum, but the twisting to form the two chalazae seems to occur much later as the egg rotates in the lower end of the oviduct. Twisted in opposite directions, the chalazae tend to keep the yolk centered in the egg after it is laid.

Liquid inner white. As the developing egg passes through the magnum, only one type of albumen is produced, but the addition of water plus the rotation of the developing egg gives rise to the various layers, one of which is the liquid inner white.

Dense white. The dense white makes up the largest portion of the egg albumen. It contains mucin that tends to hold it together. The amount of thick white generated in the magnum is large, but the breakdown of mucin and the addition of water as the egg moves through the oviduct tend to reduce the amount of thick white while increasing the amount of thin white. At the time the egg is laid, it has about one-third of its original content of thick white, but what remains still comprises over half the albumen in the egg.

Egg quality deterioration. After laying, there is a constant change in the internal contents of the egg. The thick white gradually loses its viscous composition and its volume decreases, while the thin white becomes more watery, and the amount increases. These conditions are affected by holding temperature, relative humidity, time and certain diseases. The increasing amount of thin white is one of the best indicators of the age (freshness) of the egg.

3. *Isthmus*

Next, the developing egg is forced into the isthmus, a relatively short section approximately 4 inches (10 cm) in length, where it remains for about 75 minutes. Here the inner and outer shell membranes are formed in such a manner as to represent the final shape of the egg. The contents at this time do not completely fill the shell membranes, and the egg resembles a sack only partially filled.

The shell membranes are a papery material composed of protein fibers. The inner membrane is laid down first, followed by the outer membrane, which is about three times as thick as the inner membrane. The two membranes are held closely together until the egg is laid; then at the large end of the egg, the two membranes separate to form the air cell. In a small percentage of the eggs, the air cell will form in the small end or on the side.

Air cell is important. When the egg is first laid there is no air cell. However, it soon appears and increases in diameter to about 0.7 inches (1.8 cm). As the egg ages, moisture within the egg evaporates through the shell pores and the air cell increases in diameter and depth. The size of the air cell can be affected by various storage conditions. High surrounding temperature and/or low humidity increase the size of the air cell. The size of the air cell, as determined by candling, is used in grading programs to judge the age of the egg. Larger air cells are indicators of poorer interior quality.

Shell membranes act as a barrier. The shell membranes act as a barrier to the penetration of organisms such as bacteria. Eggs laid by young hens have thicker shell membranes than eggs laid by older hens.

4. *Uterus (Shell Gland)*

The uterus is from 4.0 to 4.7 inches (10 to 12 cm) long in the laying hen. The developing egg normally remains in the uterus from 18 to 20 hours, much longer than in any other section of the oviduct.

Outer thin white deposited after shell membranes. When the egg first enters the uterus, water and salts are added through the shell membranes by the process of osmosis to plump out the loosely adhering shell membranes and to liquefy some of the thin albumen to form the fourth layer, the outer thin white.

The shell. Eggshell calcification begins just before the egg enters the uterus. Small clusters of calcium appear on the outer shell membrane just before the egg leaves the isthmus. These are the initiation sites for calcium deposition in the uterus. Their number is probably inherited and plays a part in the amount of calcium deposition later. They disappear a short time after the egg enters the shell gland.

The first shell is deposited over the initiation sites to form the inner shell, a layer composed of calcite crystals, a sponge-like material. This layer is followed by the addition of the outer shell which is made up of a layer of hard calcite crystals that are chalky and about twice as thick as the inner shell surface. The longer the calcite columns, the stronger the shell. The completed eggshell is composed almost entirely of calcium carbonate ($CaCO_3$), with small amounts of sodium, potassium, and magnesium.

Source of calcium for eggshell. There are only two sources of calcium for eggshell production: the feed and certain bones which act as storage sites in the body. Normally, most of the calcium for egg formation comes directly from the feed, with some being derived from the medullary bone which serves as the calcium reservoir. The reservoir is particularly important at night when the bird is not eating and eggshell is being deposited.

Formation of calcium carbonate. Calcium carbonate is formed when calcium ions from the blood and carbonate ions from both the blood and the shell gland combine in the shell gland. Anything that reduces the supply of either of these ions interferes with $CaCO_3$ formation and eggshell development, many times resulting in poor shell quality. It is thought that high environmental temperatures may also contribute to this problem inasmuch as eggshells are thinner during hot weather.

Poor shell quality. Many factors may cause a deterioration in eggshell quality, and their influence may or may not be due to an inadequate supply of calcium or carbonate ions. Shell quality is generally defined as the shell's ability to withstand shock, its overall appearance, and smoothness. Shell strength can be measured by several different techniques, including resistance to breaking, specific gravity, shell deformation, and shell thickness (see *Shell Egg Quality and Preservation*, Chapter 60). Several factors lower eggshell quality; for example:

1. Quality is reduced as the bird ages and continues to lay, as the hen cannot produce as efficiently an adequate quantity of calcium carbonate to cover the larger eggs produced during the latter part of the laying cycle.
2. Increased environmental temperatures.
3. Eggs laid in the morning have poorer shell quality than those laid in the afternoon.
4. Stress experienced by birds in the flock.
5. Practically all misshapen eggs and eggs with body checks are laid between 6:00 and 8:00 a.m.
6. Certain poultry diseases (Infectious Bronchitis, Newcastle disease).
7. Certain drugs.

Calcium requirements are high during production. The demand of the laying hen for calcium is extremely high. A 4-lb (1.8-kg) hen producing 250 2-oz (56.7 g) eggs per year requires about 1.25 lb (0.56 kg) of calcium. Since this is about 25 times the amount of calcium in the bird's skeleton, it is evident that the dietary need for calcium is great. Most laying rations contain from 3 to 4% calcium to meet the requirements and to allow for the inefficiencies of absorption.

Pores in the eggshell. Both the inner and outer shell layers contain small openings called pores. There may be as many as 8,000 per egg. Through these pores, air passes into the egg to supply oxygen to the developing embryo. Also, carbon dioxide and moisture is removed from the egg by passing through these same pores. In the freshly laid egg, the pores are almost completely closed, but as the egg ages or is washed, the number of open pores is greatly increased.

Color of eggshell. Eggshells are predominantly white or various shades of brown. However, a South American breed, the Araucana, produces eggs with green or blue shells. Pigments produced in the uterus at the time the shell is produced are responsible for the color. The shade of coloring is quite consistent for each bird, with the color intensity being a derivative of the genetic makeup of the individual. Some strains of birds lay eggs with very dark brown shells, while others may vary all the way to pure white. The brown pigment in eggshells is porphyrin, uniformly distributed throughout the entire shell.

The cuticle. The cuticle is laid down on the outside of the shell in the uterus and represents the last of the concentric layers of egg formation. The cuticle is composed primarily of organic material. Containing a high percentage of water, it acts as a lubricant during the laying process. But once the egg is laid the cuticle material soon dries, sealing many of the pores of the eggshell to help prevent too rapid an exchange of air and moisture and to aid in preventing bacteria from entering the egg. Various shell cleaning processes (washing and sanding) will reduce the effectiveness of the cuticle. To counteract this, egg processors commonly apply a coating of mineral oil to the shell's surface during processing. The mineral oil helps to slow down the loss of moisture and maintain interior egg quality (see *Processing and Packaging Shell Eggs*, Chapter 58).

5. *Vagina*

The final section of the oviduct is the vagina, which is about 4.7 inches (12 cm) in length in a bird during egg production. Normally, the egg is held in the vagina for only a few minutes, but in some instances may be held there for several hours. The vagina has no role in egg formation and only serves to expel the egg once it leaves the shell gland.

Eggs are laid large end first. Although the egg transverses the oviduct small end first, if the hen is not molested or frightened, the egg will rotate horizontally just prior to oviposition and will be expelled large end first. The rotation requires less than 2 minutes, and makes it possible for the uterine muscles to exert greater pressure on more surface area during oviposition. However, if something disturbs the bird prior to rotation, the egg will be laid quickly and forced through the vent small end first.

See *Shell Eggs and Their Nutritional Value,* Chapter 57, for more information on the composition and characteristics of the chicken egg.

6

Behavior of Chickens

by A. Bruce Webster

Behavior is an important subject in the management of commercial flocks. The behavior of chickens characterizes the species as much as does any anatomical attribute and is the means by which chickens cope with the environments in which they live. A chicken's flexibility in dealing with different situations is limited by its inherent behavioral characteristics. Commercial production systems must accommodate chicken behavior or fail to achieve performance objectives. In fact, production systems rely to a great extent on the behavior of chickens, e.g., feeding behavior, sexual behavior, egg laying behavior, etc. On the other hand, many of the problems that occur in intensive production systems arise from behavior which, unfortunately, is harmful to the flock or is inappropriate to performance objectives.

6-A. BEHAVIOR OF FREE-RANGING FOWL

Although the chicken was probably domesticated over 8,000 years ago, only in recent times has poultry husbandry changed from the practice of maintaining free-ranging flocks in farm yards to that of housing birds in intensive confinement systems. Therefore, despite the long history of domestication, the behavior of the chicken has probably changed relatively little over time. Genetic selection in the twentieth century, while intensive and resulting in changes in production and morphology, has done little to alter the fundamental behavioral characteristics of the chicken. The most notable behavioral changes have been the reduction of escape tendencies, and in some breeds decreased broodiness, increased aggressiveness, or changes in appetite. In free-ranging situations, domestic chickens adopt social structures similar to those of wild jungle fowl, the species from which the domestic chicken was originally derived.

It is instructive to briefly review the characteristics and social organization of wild populations of jungle fowl and chickens to shed light on the behavior of domestic chickens in commercial production systems. The following discussion summarizes observations by Collias, et al. (1966), Collias and Collias (1967), McBride, et al. (1969), Duncan, et al. (1978), Savory, et al. (1978), and Wood-Gush, et al. (1978).

The Red Jungle Fowl, generally thought to be the primary ancestor of the domestic chicken, is found in the foothills of the Himalaya Mountains in the north of India to tropical Southeast Asia. Frosts are not uncommon during winter in the northern part of the Red Jungle Fowl's range. The species' adaptability to a variety of climates helps explain the ability of today's domestic chicken to survive in northern temperate regions when given some shelter and protection. Jungle fowl live in a variety of forested habitats which have good sites for roosting, an adequate food supply, and sufficient cover in the forest to offer protection for their young.

Social Coordination

Wild fowl and feral chickens live in small groups, generally comprised of a dominant male and one or more hens. Subordinate males may maintain a loose association with these groups. In situations of high feed availability and limited space, larger groups may form, although generally less than 20 birds, with separate dominance hierarchies existing among males and among females. Neither wild fowl nor feral chickens are active at night and, once old enough to fly, spend the night roosting in trees. Large groups may break up during the day into subgroups consisting of a male and some females. Subordinate males also may form small groups.

Individuals in groups tend to stay together and synchronize their activities, i.e., they forage, rest, and preen at the same times. Fowl use a location call, or "Ku" call (Konishi, 1963), consisting of a gentle drawn-out low frequency vocalization often interspersed with several shorter calls of similar intensity, to maintain contact with other group members in the underbrush. This call corresponds to the "singing" of hens commonly heard in commercial layer houses.

Flying

Most traveling is done on the ground rather than by flying. Adult jungle fowl generally fly only to escape an immediate threat, surmount a physical obstacle, or reach a roost or perch. They seldom fly farther than a few hundred yards, and prefer to withdraw from danger on foot if possible.

Modern breeds of chickens vary in ability to fly, generally in accordance with their body weight. Light bodied varieties of egg laying stocks can fly quite well. Heavy bodied meat stocks have lesser flight capability and generally use their wings only to support jumping activity, e.g., to add thrust, aid balance, and soften landing when jumping between raised positions and the floor.

Predator Avoidance

Wild fowl are wary and difficult to approach. Many predators live in areas inhabited by these birds. When a threat is perceived, fowl typically take cover and become still, or retreat on foot. They will fly to escape close danger. Jungle fowl can habituate, however, to humans or animals that prove not to be a threat. Commercial breeds of domestic fowl vary in escape tendencies. White Leghorn varieties generally exhibit the greatest flightiness. Medium weight egg-laying varieties and meat-type chickens have more moderate escape reactions to humans. Some varieties of chickens can be quite curious of humans and have very little aversion to human presence.

When a bird becomes aware of a potential predator that is not an immediate threat it gives warning calls consisting of a series of short, wide frequency vocalizations often followed by a vocalization of longer duration (Konishi, 1963). This vocalization is frequently referred to as the ground predator call because it typically is produced in response to non-flying predators. The call usually is repeated while the predator is in the vicinity and will be taken up by other fowl nearby. The contagious influence of ground predator calling is often evident in commercial cage layer houses when a person (the perceived predator) stops to work at a location in the midst of a flock. A few hens near the person will begin ground predator calling, followed by hens in an expanding radius until dozens or even hundreds of hens in the vicinity are calling, producing a loud racket.

A drawn out, low frequency squawk is produced in response to flying aerial predators (Collias and Joos, 1953). Chicks respond to aerial predator calls by becoming immobile or running for cover, and adult birds may become silent and alert. The aerial predator call is easily elicited from domestic chickens by objects tossed through the air. The call can be heard in commercial houses, although the reason for the call may not be evident because the cause is not the human observer.

Foraging and Feeding

Feral chickens spend about half their time foraging and feeding, and make an estimated 14,000–15,000 pecks at food items and other objects in the course of a day. Free-living fowl eat a wide variety of foods, including grass, leaves from broad leaved plants, seeds, small fruits, invertebrates (e.g., slugs, insects, spiders), and carrion.

Home Range

A home range is the geographic area within which an individual or group generally stays. Groups of fowl may have home ranges varying in size from 12.5 acres (5 hectares) in relatively open, dry forests to less than an acre (0.4 hectare) in areas with plentiful food and high population density. During the non-breeding season, the ranges of different groups may overlap, but groups seldom intermingle except at sites with high concentrations of food.

Behavior during the Breeding Season

1. *Territoriality*. The breeding season begins during the spring when the photoperiod lengthens, at which time dominant males establish territories which they defend against other territorial males. Subordinate males, usually young individuals, may be tolerated within territories but are kept away from hens by the dominant male. Crowing is done from the roost or other elevated positions and also when dominant males meet along territorial boundaries. Fights may occur when a dominant male intrudes into the territory of another dominant male. When population densities are high, making group sizes too large for individual males to maintain discrete harems of females, adult males establish dominance hierarchies and more than one male may be able to mate with hens in a group.

   Non-broody hens generally travel with the dominant rooster, thus remaining within his territory. The male plays an active role in keeping the group together. Tidbitting behavior by the rooster attracts hens to the male. This behavior involves the rooster producing food calls, consisting of a rapid series of short, wide frequency vocalizations (Konishi, 1963, Marler, et al., 1986a,b), while standing in place and pecking at or picking up and dropping bits of food. The dominant rooster often stands alert while the hens in his group feed after having been led to a food source by his behavior.
2. *Courtship*. Waltzing is performed by males during court-

ship. It involves a male walking in a circular direction around another bird with the outside wing lowered. Tidbitting also is performed by the male when courting a female, although in this case, small non-food items may be picked up and dropped. Copulation occurs when a hen crouches. Crouching may occur without the rooster having obviously courted the hen. Waltzing and tidbitting also may be performed during aggressive encounters between males.

3. *Nesting, Laying, and Incubation.* Hens select secluded sites to build nests in which to lay their eggs. It is important that the nest be well hidden because both hen and eggs are vulnerable to predators during the period of incubation. While the nests themselves are simple in construction, hens have strong predispositions for certain prelaying and laying actions, e.g., in relation to activity associated with nest approach and nesting actions at the nest (these predispositions still exist in commercial stocks although stocks have retained specific tendencies to different degrees). Approach to the nest, particularly for a hen incubating eggs, is indirect and cautious. Such behavior in response to a potential predator in the vicinity of the nest functions to minimize the chance that the nest will be discovered. If the hen is on the nest, she will sit still. If not on the nest when the predator is discovered, the hen will give ground predator calls and move away from the nest.

   The hen remains a member of the adult group while accumulating her clutch of eggs. It is not unusual for a male to lead the hen to investigate prospective nesting sites in response to a laying call (Konishi, 1963) given by the hen when the time comes to lay an egg. The action by the male may be related to the "cornering" behavior that has been observed in domestic roosters. Hens that do not have a male as an escort may be chased by subordinate males, which make forceful attempts to copulate. Hens typically give cackles after egg laying which resemble ground predator alarm calls (Konishi, 1963). According to McBride, et al. (1969), the cackle serves to attract a male to escort the hen back to the flock and other hens are able to distinguish the egg laying cackle from the ground predator call.

   A hen becomes solitary when she begins to incubate her clutch and does not return to the adult group until she leaves her offspring (discussed below). During incubation, the hen remains on the nest for long periods, leaving only to defecate and feed.

4. *Brooding and Rearing.* Newly hatched chicks have little ability to maintain their body temperature by metabolic processes and must be brooded frequently by the hen to keep warm. A brooding hen will crouch to allow chicks to huddle against her in the angle between her body and the ground, or underneath her wings. The time spent brooding is greatest during the first week after hatching and during inclement weather. Brooding frequency declines as the metabolic thermoregulatory capacity of chicks develops and a first generation of feathers is grown to replace the down. A hen will also brood chicks after having encountered moderate disturbance or danger.

In the first day after hatching, chicks undergo a form of learning called "imprinting." This involves a strong predisposition to approach a prominent moving object, memorize its characteristics and follow it wherever it goes. In natural circumstances this object would be the hen, but chicks also imprint to some extent on other chicks. Imprinting has decided survival value. By following the hen, chicks have ready access to warmth and shelter, and are led to sources of food and water. The hen helps chicks to identify food by giving food calls (Collias and Joos, 1953, Sherry, 1977) while scratching the ground and pecking items of food (tidbitting behavior). Chicks are attracted to these signals and peck vigorously in the area of the hen's focus. Newly hatched chicks are strongly attracted to broadcasts of recorded broody hen food calls, and can be stimulated to initiate earlier and more synchronized feed consumption if food calls are broadcast from speakers located next to the feed supply.

The behavior of following familiar individuals and pecking at things in response to the pecking actions of other individuals makes it possible to raise chicks successfully in commercial brooding environments. When one chick finds food or water, many others do as well.

The hen leads the brood of chicks around for several weeks, generally keeping them separate from other broods and adult groups. The hen actively keeps the brood together by tidbitting behavior or by performing a display to attract the chicks which consists of a short run with or without wing-flapping. The hen also clucks as she travels, helping chicks to track her progress through vegetation. Clucks are short, double-pulse, low frequency vocalizations given at a rate of 1 to 3 per second (Collias and Joos, 1953), and are attractive to chicks. Chicks also are attracted to repetitive tapping sounds having a spectrographic resemblance to the sound of clucks or food calls.

Young chicks which get separated from the hen, become cold, or otherwise become distressed, give peep calls (Collias and Joos, 1953). Peeping elicits food calling by the hen (Hughes, et al., 1982) and also attracts her so that contact is restored between hen and the chick. Juveniles gradually range farther from the hen and develop more independence as they grow older.

At first the hen roosts with her chicks on the ground. After the second generation of feathers has grown, chicks gain some ability to fly at 6 to 8 weeks of age and gradually both dam and brood begin to roost in the branches of bushes and small trees.

In the commercial chicken breeding industry where adult breeder hens are expected to lay eggs in nest boxes on raised slatted areas, it may be important for pullet chicks to learn to use perches or roosts during the growing period, otherwise when adults, they may lay many floor eggs in the breeder house (Appleby, et al., 1988). Even low nest boxes may be avoided by untrained hens. Sections of slatted flooring material placed across low sawhorses, 12 to 18 inches (30–45 cm) in height, or alternatively, at the height of the slatted areas in the breeder house for which a given flock is destined, are sufficient to allow pullets to learn to perch. The greatest learning effect appears to be achieved by providing the perches by the time pullets are 4 weeks of age. Stocks of domestic chickens may differ in their tendencies to jump up to raised areas (Faure and Jones, 1982).

The time when hens leave their broods permanently is variable, generally occurring when the offspring are 6 to 12 weeks of age. The hen then returns to reproductive condition and rejoins the adult group. The brood continues to forage and roost together as a unit for several weeks. It is not unnatural, therefore, for juvenile chickens to live in independent groups from a young age until sexual maturity. Around 18 to 20 weeks of age, cohesion of the brood dissipates as individuals become sexually mature and begin to integrate into adult social groups. This coincides with the change to adult appearance due to feather molt and enlargement of comb and wattles, and the adoption of adult behavioral characteristics.

6-B. NEUROMUSCULAR CONTROL OF BEHAVIOR

Behavior has been defined as the "action of a living organism, either instigated by the organism or imposed by external circumstances" (Hurnik, et al., 1995). In vertebrates, extraneous forces are seldom of interest as a cause of physical motion; rather, our interest is in physical actions resulting from the contraction of muscles. Muscular contractions occur in response to signals from control centers in the central nervous system (brain and spinal cord) passed through nerves which transfer signals from the central nervous system to the periphery of the body. These control centers, in turn, process information received as signals from the periphery of the body through nerves connected to a variety of sensory receptors.

Some reflexive actions may be controlled within the spinal cord without directly involving the brain. Each type of sensory receptor is specialized to produce nerve signals in response to specific qualities of the environment, e.g., light, sound, heat, touch, etc.

Large numbers of sensory receptors are distributed around the body surface of a chicken, with concentrations occurring in specialized organs to support specific sense modes, e.g., eyes, ears, and nares for sight, hearing and smell, respectively, or in structures essential for the bird's exploitation of its environment, e.g., the beak and feet. Any quality of the environment which causes a sensory receptor to produce a nerve signal is a stimulus. The array of stimuli received by an animal determines what the animal is able to know of its environment. These stimuli, however, represent only a portion of the total qualities of the environment which impact the animal because the information that the animal receives is limited by the nature of its sensory receptors. The animal's actual comprehension of its circumstances is further limited by its psychological ability to process the information that it does receive.

Some sensory receptors are located deep within a bird's body, e.g., in the gastrointestinal tract, to provide information regarding its internal state. Therefore, in addition to stimulation from the external environment, behavior is influenced by internal factors which are important to the animal's condition, such as blood sugar levels and hormone concentrations.

6-C. SENSES

All animals depend on information gained from the environment to make appropriate behavioral choices. This information is obtained through stimulation of various sensory modes, e.g., sight, hearing, smell, taste, touch. Sensory mode sensitivity differs among species. To understand how the domestic chicken perceives the world around it, one must know something of the nature of its senses. The following material draws from discussions in Appleby (1992), Fischer (1975), and Wood-Gush (1971).

Vision

The chicken's retina contains a higher density of cone receptors than does the human retina, indicating the importance of color vision in its eyesight. The range of color perceived by chickens is similar to that of humans. Differences in spectral sensitivity between the chicken retina and the human retina suggest that chickens see better in the red and orange

range of the light spectrum, somewhat better in the green and ultraviolet ranges, but a little less well in the blue range (Nuboer, et al., 1992).

Newly hatched chicks demonstrate unlearned color preferences which are maximal at long and short wavelengths (orange and violet ranges) and minimal at intermediate wavelengths (green). This capability might aid discovery of food items during early foraging behavior in natural circumstances.

The chicken retina also contains rod receptors, and dark adaptation of vision has been demonstrated for this species.

The chicken's eyeball is flatter than the mammalian eyeball and, unlike mammals, little movement of the eyeball is possible. To compensate, the chicken retina, also unlike mammals, is nearly equidistant from the lens at all points so visual acuity is uniform throughout most of the field of vision. The chicken's field of view is about 300 degrees, with binocular vision possible in a narrow span of about 26 degrees directly forward of the head. The great mobility of the chicken's neck and the bird's habit of performing frequent head movements in many directions expand the chicken's effective field of view.

Chicks show unlearned perception of depth, which aids movement through spatially complex environments. Chickens respond to visual illusions in similar fashion to humans. Images presented in such a way as to appear larger to the human eye than other identically sized images are also judged to be larger by chickens. Chicks can discriminate size, shape, and pattern complexity at an early age, as evidenced by their preferences to peck at small round objects ~0.1 inch (3 mm) in diameter and to approach fairly large, angular objects 4 to 8 inches (10–20 cm) in diameter. This ability is important for initiation of feeding and establishment of social bonds.

Hearing

Chickens appear to hear quite well, although the sense of hearing may not be as important as in many mammalian species. A variety of calls are used for communication in different contexts, and chickens respond differently to different vocal sound patterns.

Smell

The chicken has an olfactory epithelium and its olfactory nerve can develop electrophysiological responses to odorants. The sense of smell is believed, however, to be poor and relatively unimportant to behavior. Chickens do not appear to notice many odors that are prominent to humans.

Taste

Chickens are sensitive to salt and bitter substances, and will reject solutions containing even relatively low concentrations of these compounds. The aversion to salt solutions makes it important in commercial production systems to ensure that the water provided to a flock contains little dissolved salt. Water refusal leads to reduced feed consumption, causing sub-optimal production performance. Chickens can sense acidity and alkalinity, but tolerate a wider range of pH than mammals generally do. Chickens evidently can taste sugar dissolved in water but generally show relatively little preferential response to common sugars. Sensitivities to flavor tend to be greater for liquids than for solid foods, possibly related to the fact that solid food is swallowed whole.

Integumentary (Outer Surface) Sensitivity

The integument of the chicken (skin and accessory structures, e.g., the beak) contain many sensory receptors of several types allowing perception of touch (both moving stimuli and pressure stimuli), cold, heat, and noxious (painful or unpleasant) stimulation. The beak has concentrations of touch receptors forming specialized beak tip organs which give the bird sensitivity for manipulation and assessment of objects.

Beak trimming, done to reduce the damaging effects of feather pecking and cannibalism by adult birds in commercial production systems, removes or damages the beak tip organs. Beak trimming affects the sensory experience of a chicken in more than one way (Hughes and Gentle, 1995). It deprives the bird of normal sensory evaluation of objects when using the beak. It has been found, on the other hand, that in birds beak trimmed as young as 5 weeks of age, neuromas may develop in the beak where nerves have been severed, and these neuromas have been shown to produce abnormal nervous activity analogous to that which causes phantom limb pain in human amputees. Behavioral evidence indicates that chronic pain may persist for at least six weeks after beak trimming in these chickens. It is possible that neuromas do not develop in chickens beak trimmed at less than 3 weeks of age. See *Cage Management for Raising Replacement Pullets, Chapter* 51.

6-D. BEHAVIORAL CHARACTERISTICS OF DOMESTIC CHICKENS

Behavioral Repertoires and Time Budgets

Chickens use an array of behavioral actions to meet their needs in the environments in which they live. These actions together form their behavioral repertoires. Behavioral actions generally occur in sequences called behavioral patterns which fall into broad behavioral classes serving specific functions, e.g., foraging and feeding behavior, aggression, dominance / subordinance-related behavior, reproductive behavior, sleep, etc. Many of the needs of a chicken recur on a periodic basis, so a bird must divide its activity during the course of a day among different classes of behavior in accordance with the immediate importance of specific needs and demands of the circumstances. Both the environment and the genetic makeup of the chicken influence the behavior it manifests and the amount of time it devotes to different behavioral classes.

Egg-type stocks of hens, both light hybrid varieties (White Leghorn) and the heavier medium hybrids based on Rhode Island Red crosses, are recognized as having relatively high levels of overall activity. This is evident in Table 6-1, which summarizes various studies of the behavior of chickens in cages and floor pens. Adult laying hens in confined environments spend a lot of time on their feet, and spend much of the lighted portion of the day apparently surveying their surroundings (head movements), in feed-related activity, or in preening. This tendency for high level of activity differs little from that of the chicken's wild relative, the Red Jungle Fowl. The discrepancies in resting between different studies is due to differing definitions of resting behavior (sitting or crouching vs. somnolence) and different times of day selected for observation.

Relatively little scientific information is available regarding behavioral time budgets of broilers and broiler breeders. The study of broilers depicted in Table 6-1 suggests that full-fed broilers are much less active than laying hens, as is generally understood by people who are familiar with commercial stocks of chickens. Broilers spend less time standing and apparently less time in feed-related activity than layers. Broiler breeders with unrestricted access to feed are relatively inactive, similar to broilers; they show greatly altered behavioral time budgets when feed is restricted. Specifically, feed-restricted broiler breeders tend to exhibit increased levels of non-nutritive pecking and considerably less resting (Table 6-1).

The example for broiler breeders in Table 6-1 also illustrates that chickens will alter their behavior so as to perform important activities when circumstances dictate and engage in other behavior at other times. A limited amount of feed was provided to these birds every morning and the drinkers were cut off in the afternoon to prevent overconsumption of water. Intense feeding activity caused the feed to be consumed within 30 min of its presentation. Thereafter, a lot of drinking but only a moderate level of preening occurred during the morning. Preening increased in the afternoon when feeding and drinking were not possible. Clearly, the behavioral time budgeting of commercial broiler breeders would also be influenced by the nature of feed and water restriction, e.g., skip-a-day vs. limited daily ration.

Table 6-1. Time Budgets of Chickens. Percentages of Time Spent in Different Actions

	Stock	Housing	Density	Head Movement	Eat	Drink	Preen	Non-nutritive Peck	Walk	Still	Rest	Stand	Ground Scratch/ Dust Bathe
1. Bareham, 1972	WL	1/cage	880 cm^2/bd	—	18	6	4	—	—	—	—	—	—
		30/pen	2000 cm^2/b	—	7	4	6	—	—	—	—	—	—
	RIR	1/cage	880 cm^2/bd	—	17	4	9	—	—	—	—	—	—
		30/pen	2000 cm^2/b	—	7	4	8	—	—	—	—	—	—
2. Eskeland, 1977	WL	5/cage	864 cm^2/bd	—	20	4	8	9	13	—	19	81	—
		10/cage	432 cm^2/bd	—	17	3	5	20	10	—	15	85	—
		36/pen	1666	—	17	2	11	—	12	—	15	85	9
		72/pen	833 cm^2/bd	—	18	1	6	—	12	—	12	88	3
3. Mench, et al., 1986	WL-N-line	2/cage	697 cm^2/bd	—	26	—	18	1	1	—	20	—	—
		25/pen	1394	—	33	—	15	4	8	—	14	—	—
4. Murphy & Preston	Broiler	Commercial	694 cm^2/bd	—	11	5	—	—	—	—	—	36	—
5. Savory, et al., 1992	RBB-f	15/pen	3840	—	10	2	6	0	—	—	45	—	—
	RBB-r (a.m.)	26/pen	2215	—	0	25	5	42	—	—	0	—	—
	RBB-r (p.m.)	26/pen	2215	—	0	0	13	49	—	—	9	—	—
6. Webster & Hurnik,	WL-Stock	2/cage	731 cm^2/bd	31	20	3	10	5	2	3	6	78	—
	WL-Stock	2/cage	731 cm^2/bd	28	19	3	10	5	2	3	7	75	—
7. Webster, 1995	WL-W77	2/cage	516 cm^2/bd	23	22	2	10	—	0	9	0	67	—
	MH-HB	2/cage	516 cm^2/bd	40	25	3	8	—	0	6	0	89	—
8. Webster, (unpubl.)	WL-W36	3/cage	342 cm^2/bd	26–32	30–40	5–6	7–9	1	0–1	9–12	5–6	87–93	—

1. Direct visual observations. 2. Direct visual observations, 0600 h–2000 h. 3. Direct visual observations. 4. Direct visual observations, 0950 h–1730 h. 5. Direct visual observations; full-fed birds (f)—behavior data averaged for morning and afternoon observations; feed-restricted birds (r), (a.m.)—morning observations after feed consumed, (p.m.)—afternoon observations. 6. Video records, 0800 h–1700 h. 7. Video records, 1700 h–2000 h. 8. Video records, 1200 h–1400 h.
WL—White Leghorn, MH—Medium hybrid, RIR—Rhode Island Red, W36—Hy-Line W-36 variety, W77—Hy-Line W-77 variety, HB—Hy-Line Brown, RBB—Ross Broiler Breeder.

Daily Cycles (Rhythms) of Behavior

It has been known for thousands of years that chickens are active during the day and inactive at night, i.e., their activity recurs predictably according to a diurnal (24-hour) cycle. Specific actions, such as feeding, preening, and nesting, may occur predominantly at certain times and not at others. For chickens, the daily cycle of darkness and light appears to be the strongest environmental stimulus influencing the timing of behavior. The timing of activity also may be controlled by factors internal to the chicken. Circadian rhythms of behavior are those which recur on a 24-hour cycle without requiring an external stimulus. Since the timing of most chicken behavior is influenced by external factors such as light, true circadian rhythms of behavior have seldom been demonstrated. It has been shown, however, that newly hatched chicks which had been incubated in darkness and then housed in continuous light exhibited 24-hour cycles of motor activity which evidently were under the control of an internal circadian timing mechanism (Miller, 1980).

1. *Factors Which Set the Timing of Behavioral Rhythms*

The transition between light and darkness appears to be the most powerful time-setting stimulus for diurnal rhythms of physiological change, e.g., body temperature, heart rate, and respiration rate; and of behavior, e.g., egg laying, locomotor activity, and feeding (Savory, 1980).

Diurnal rhythms of behavior are also subject to the relative lengths of the light and dark portion of the day in a way that suggests that the effects of the usual time-setting stimuli may be modified by internal factors. Wood-Gush (1959) studied two flocks under natural lighting, one in winter with 8 h light and 16 h dark and the other in summer with 18 h light and 6 h dark. The winter flock did very little sleeping during the short lighted portion of the day and, contrary to the common perception that chickens are inactive at night, began to come off their roosts to feed at least 2 hours before dawn. In summer, the flock did not leave the roosts until after daylight and were back on the roosts and asleep in the evening while it was still light. The summer flock also had a small peak in sleeping behavior around midday. The total amount of time spent sleeping did not differ greatly between summer and winter flocks. For behavioral states such as sleep, therefore, the chicken may have a minimum daily requirement, with the photoperiod length influencing how the bird distributes the behavior throughout the day. When the period of darkness is long and daylight is short, the chicken does all its sleeping at night, and may start to perform other activities before it is light. Conversely, when the night is short and the daylight long, the chicken will spend some time asleep during the day.

2. *Timing of Feeding Behavior*

Much of the research on diurnal rhythms of behavior has focused on eating patterns (Savory, 1980). It is unusual for chickens to eat in darkness if they receive 8 to 16 hours of light per day, but it does occur on occasion, perhaps when the nature of the environment, e.g., cold (Wood-Gush, 1959), or the nature of the bird, e.g., growing broiler (May and Lott, 1992), dictate.

Time Setters of Feeding Behavior in Continuous Light (24-Hour Photoperiod)

Flocks of chickens kept in constant light tend to eat at a constant rate, unless other cues such as temperature variation, changes in noise levels, or periodic presence of a human worker stimulate birds to establish a diurnal rhythm. These cues cause broilers in commercial flocks given continuous artificial light, or just a brief period of darkness at night, to tend to have higher levels of eating and drinking during the light portion of the natural day.

Patterns of Feeding Behavior During the Lighted Portion of the Day

Table 6-2 summarizes the results of 30 papers reviewed by Savory (1980). He concluded that chickens tend to eat more at the beginning or the end of the light period, or both, but not in the middle of the day. Laying birds tend to have feeding peaks in morning and evening or in the evening only, but not in the morning only. Non-layers show the opposite pattern, tending not to feed mostly in the evening, but having feeding peaks in the morning only, or both morning and evening. Egg laying and inability to predict the coming of darkness were identified as major influences on the timing of feeding activity.

Table 6-2. Diurnal Patterns of Feeding Behavior of Domestic Chickens. Number of Studies with the Indicated Pattern. (Adapted from Savory, 1980)

	No Trend	Highest at Start of Day	Highest in Middle of Day	High Morning and Evening, Low Midday	Highest at End of Day
Laying birds	3	3	1	9	10
Non-layers	3	10	1	9	4

Egg-Laying Effects on Feeding Activity

Feed consumption declines 2 to 3 hours before an egg is laid. During this prelaying period, hens often become restless and preoccupied with nesting-related behavior. Most eggs are laid within a few hours after lights come on, so feeding activity of hens with eggs to lay would tend to be depressed during the morning, setting the stage for higher feed consumption in the afternoon and evening. Since commercial laying hens and broiler breeders usually are kept on long photoperiods, which encourages a period of rest or sleep at mid-day, eating would tend to be further concentrated later in the day.

Laying birds increase feed intake for several hours after an egg enters the shell gland, which usually happens in the afternoon. This increased feed intake is probably due to increased demand for dietary calcium because hens selectively eat oyster shell around the time shell calcification starts, if given the chance to do so. With the usual high calcium diet fed to laying birds, therefore, feed intake would tend to increase toward the end of the photoperiod on most days due to the effects of egg laying and egg formation, especially on days when eggs were both laid and formed.

Anticipation of Darkness

Many chickens have difficulty predicting a sudden onset of darkness (as opposed to darkness after a period of dusk). Since non-laying birds do not undergo the physiological and behavioral effects of egg laying and egg formation on feeding motivation, they have less stimulation than layers to eat late in the day independently of an anticipation of darkness. For birds such as breeder and layer pullets, which are given substantial dark periods, failure to fill the crop during the evening because of inaccurate prediction of darkness would elevate hunger in the morning and cause feed intake to be high at that time. Dusk, which cues the coming of darkness, can stimulate chickens to increase feeding during the evening.

At least some stocks of chickens can learn to anticipate the coming of darkness. May and Lott (1992) demonstrated that 8-day-old broilers learned within a 2-day period to increase their feed consumption prior to darkness when switched from continuous lighting to a 12 h light:12 h dark cycle. It took up to 5 days for these broilers to suppress this learned anticipation and to begin feeding at a constant hourly rate when returned to continuous light at 29 days of age. Continuously lit broilers, on the other hand, did not learn to anticipate a regular period of feed withdrawal. If a lighting program or other environmental cue causes a flock of broilers to develop cyclic feeding patterns, carcass contamination in the processing plant could be a problem if end-of-flock feed withdrawal programs are not adjusted to accommodate the feeding pattern of the flock.

Other Factors Affecting Feeding Behavior

Other factors associated with social interaction, housing and management can influence the timing of activities, such as the following:

1. *Behavioral synchrony.* Chickens tend to synchronize activity with each other, particularly feeding behavior (Hughes, 1971, Webster and Hurnik, 1994). The social facilitation of feeding is so strong that the feeding activity of a hungry hen can induce a less hungry hen to resume feeding, increasing its overall feed consumption. This effect, coupled with the opportunistic feeding habits of chickens (stemming from instinctive feeding tendencies wherein food is eaten upon encounter during foraging activity), provides the rationale for the use of multiple feed delivery cycles during the day to stimulate feed consumption of commercial laying hens when necessary to improve flock performance. Examples of this include intermittent or meal feeding broilers and starting feeders to stimulate layer feed consumption.
2. *Cage shape/feeder space.* Group diurnal feeding patterns tend to be less pronounced for hens in deep cages than for those in shallow cages. The former cages, having less feed trough space per bird, evidently prevent hens from feeding in synchrony and thus cause flock-feeding activity to be spread out. Limitation of feeder space could be expected to have a similar effect on floor-housed flocks.
3. *Photoperiod length.* Long photoperiods result in greater variability of feeding activity from hour to hour, whereas short photoperiods induce birds to spend proportionately more time at feed troughs and to eat more per hour.
4. *Feed particle size.* Feeding patterns are more distinct when chickens are given feed in pellet form as opposed to crumbs or mash. Feeding pellets therefore, allow birds to grasp and swallow food quickly so that meals can be ingested in relatively short periods of time.

7

Behavioral Genetics

by A. Bruce Webster

The behavior of a chicken is controlled by its genetic makeup in interaction with its environment. An individual bird's genetic makeup limits the degree to which it can adjust its behavior to cope with environmental challenges. Chickens from the same genetic stock, however, may behave differently in the same situation, sometimes with associated differences in performance. Behavioral differences indicate that birds in the same stock may have differing genetic control of behavior. The existence of genetic variation for behavior, in fact, is of immense value to the commercial chicken industries. This variation has made possible the development of stocks which not only have the morphological and physiological characteristics necessary for commercial performance objectives, but also have behavioral attributes which allow them to perform well in intensive housing environments. This same genetic variation should make it possible for continued adaptation of chickens to the housing systems and management used by commercial producers. Such adaptation is important not only to sustain productivity but also to address modern-day animal welfare concerns.

7-A. DOMESTICATION AND GENETIC CHANGE IN BEHAVIOR

Commercial production environments differ greatly from the natural environments where the ancestors of modern-day domestic chickens lived. As a result, the broiler and egg industries have been criticized for housing chickens in situations to which they are not adapted, and thus causing them to suffer. In *Behavior of Chickens*, Chapter 6, it was pointed out that domestic chickens still share many behavioral attributes with jungle fowl.

Taking an alternate view, it is commonly accepted that the ancestors of the chicken had certain behavioral attributes which favored domestication of the species. These ancestral stocks were in many ways pre-adapted to animal agriculture. The behavioral attributes are:

1. hierarchical social group structure
2. promiscuity
3. precocious young
4. unspecialized dietary habits
5. limited agility
6. ground living habits
7. ability to adjust to a variety of environments.

This is not to claim that chickens are fully adapted to commercial production environments, but that they have the basic adaptation necessary to survive, grow, and reproduce under such conditions. Some behaviors which would have been essential to a wild bird are no longer necessary in commercial stocks, e.g., incubation and brooding behavior, and have been reduced through selection. Others, such as foraging behavior and nesting behavior, are no longer needed to the extent required of a free-living population in a natural environment, but may not have been altered to the point that the expression of foraging motivations (litter eating, feather pecking) and nesting motivations (floor laying, prelaying pacing in caged layers) are completely adjusted to commercial settings.

Siegel (1993), in a discussion of the genetic distance of domestic chickens from jungle fowl, pointed out that molecular genetic research involving DNA fingerprinting has indicated that both commercial broilers and layers are quite removed from their jungle fowl ancestors, at least at the gene loci examined. Even so, since domestic fowl and jungle fowl interbreed easily, the genetic difference between the two is small on an evolutionary scale.

During domestication, behavioral changes probably occurred as correlated responses to selection of birds for non-behavioral characteristics. Siegel (1989) reviewed how his own selection of lines of chickens for high and low body weights led to correlated changes in mating activity, aggression, docility, and appetite. Behavioral changes arising in this manner could be said to be adaptive if they help a stock to achieve the performance objective targeted by the selection program or benefit well-being by better adjustment of the bird to its circumstances, but correlated behavioral responses to selection cannot be presumed to promote adaptation. For example, the increased appetite of meat stocks helps these types of chickens achieve their genetic potential for growth, but also forces producers to restrict the amount of food provided to breeder flocks to prevent obesity. Neither a disposition for obesity, and the high rates of mortality associated with it, nor a chronic state of hunger could be said to indicate a state of adaptation.

Other behavioral differences between stocks of chickens, particularly in the case of those selected for similar production objectives, e.g., egg laying or meat production, may have occurred as a result of genetic drift, which involves random changes in gene frequencies in small breeding populations over successive generations. For the most part, these behavioral differences are not directly linked to performance, but under some circumstances may affect performance by influencing the adaptation or well-being of the bird or its flock mates.

7-B. GENETIC STOCK DIFFERENCES IN BEHAVIOR

People in the broiler and egg industries are well acquainted with the fact that commercial varieties of chickens behave differently. The most obvious case is the difference between meat stocks and egg stocks. The former have large appetites coupled with efficient conversion of feed into the building of body mass, and when mature will overeat to the point of becoming obese. When full-fed, these stocks are relatively inactive and docile. Sexual behavior of meat-type roosters tends to be less well elaborated than that of egg-type roosters, and when feed restricted to control body weight, meat-type males may become excessively aggressive toward hens and even toward humans. Egg-type stocks, on the other hand, have much less tendency to overeat, particularly the light hybrids (White Leghorn varieties), and generally maintain higher levels of activity. Among the egg stocks, the light hybrids (generally white egg layers) tend to have more pronounced escape reactions, "flightiness," than the medium hybrids (brown egg layers).

A wide range of behavioral differences has been found between stocks of chickens, involving for instance, early innate behavioral tendencies (imprinting), feeding behavior, social behavior (aggression), emotionality (fearfulness, hysteria), behavioral problems such as feather pecking and cannibalism, stereotypic behavior (prelaying restlessness), vocalization, and the amount of time devoted to different actions in the general array of behavior manifested by individual birds (Table 7-1). Many of these behavioral differences between stocks were found to be associated with differences in performance traits.

Actions such as feeding behavior and activity directly impact physiological processes underlying growth and feed efficiency. One might expect selection for improved feed efficiency to result in lower levels of activity, as seen for General Behavior in Table 7-1. The association of productivity with activity, however, depends on the type of production and the genetic stock of the birds. For example, some egg-type stocks housed in cages become restless when egg laying is imminent due to the activation of nest search and nesting motivations. Better producing birds in these stocks show higher levels of activity (Table 7-1, General Behavior). Stocks of hens

Table 7-1. Genetic Stock Differences in the Behavior of Chickens

Behavioral Trait	Selected References	Synopsis of Results
Aggression	Choudary, et al., 1972; Al-Rawi, et al., 1976	Found differences in aggression among five White Leghorn strains housed in cages (not the same five strains in each study). The earlier study reported that the strain with the lowest aggression had the highest hen-housed productivity and best survival. The later study did not find a relationship between aggression and productivity.
General Behavior	Braastad and Katle, 1989	Individually housed White Leghorn hens from a line selected for poor feed efficiency exhibited increased food pecking, walking, pacing, escape effort, and aggression, whereas hens from a line selected for good feed efficiency spent more time resting and sleeping, and showed no prelaying pacing.
	Webster and Hurnik, 1990a	Caged White Leghorn hens derived from matings of a commercial male parent line to an unselected stock of females performed more head movement, displacement of cage mates, aggression and cage wall climbing, and had higher egg production than hens derived from matings of a different commercial male parent line to the same female stock.
Fearfulness	Phillips and Siegel, 1966	Two closely-related lines of White Plymouth Rock chicks differed in responsiveness to a novel, frightening sound.
	Jones, 1977; Faure, 1979; Jones and Faure, 1981; Jones and Mills, 1983	Compared several genetic stocks of chickens at different ages in a variety of standardized tests. Found numerous differences among stocks for various measures of fear-related behavior.
	Ouart and Adams, 1982	Of two commercial varieties of White Leghorn hens compared, the one which demonstrated lower fear-related behavior in cages had better feathering, higher egg production, and fewer checked eggs.
	Cunningham and Ostrander, 1982	Two commercial varieties of caged White Leghorn hens differed in egg production, egg size, and feed conversion, but did not differ for fear-related behavior.
	Craig, et al., 1983; Craig, et al., 1984; Okpokho, et al., 1987; Craig and Milliken, 1989	Of two White Leghorn stocks, each selected for increased part-year egg mass, the one which exhibited lower fearfulness and had better feathering tended to have poorer egg production performance. White Leghorn hens from strains selected for increased part-year egg mass resumed movement more quickly in tonic immobility tests than did unselected control hens.

Feather Pecking/ Cannibalism	Hughes and Duncan, 1972	Found differences among three commercial layer strains in the development of feather pecking and in the incidence of cannibalism to 20 weeks of age.
	Craig and Lee, 1990	Found differences in feather loss and mortality due to cannibalism among caged, non-beak-trimmed hens of three commercial White Leghorn varieties. The variety with the best feathering also had the lowest cannibalistic mortality, but the variety having the worst feathering did not have the highest cannibalistic mortality.
	Craig and Muir, 1996	Noted genetic stock differences in the preferred body site for cannibalistic pecking.
Feeding/Appetite	Nir, et al., 1978	After the first few days post hatch, light breed chicks voluntarily limited their food intake to less than the gut capacity. The relative weight of the digestive tract was greater in light breed chicks than in heavy breed chicks. Heavy breed chicks showed much less tendency to voluntarily limit feed intake, consistently consuming feed in amounts that approached gut capacity. Force feeding stimulated skeletal growth of light breed chicks, as measured by shank length, but actually reduced skeletal growth of heavy breed chicks.
	Dunnington, et al., 1987	Chicks from White Plymouth Rock lines selected for high body weight manifested greater compensatory feed intake on the feeding day of an alternate day feeding program than those from lines selected for low body weight. White Leghorn chicks demonstrated an intermediate ability in this regard.
Hysteria	Hansen, 1976	Two stocks of White Leghorn pullets reared in intermingled flocks had different tendencies to develop hysteria as adults when housed in single stock groups in large community cages.
Imprinting	Graves and Siegel, 1969	Four lines of chickens selected for traits unrelated to imprinting behavior (mating activity, body weight) were found to differ in their response as day-old chicks to a standardized imprinting stimulus.
Pre-laying Behavior	Wood-Gush, 1972; Mills and Wood-Gush, 1985; Mills, et al., 1985b	A light hybrid (White Leghorn) strain characteristically performed restless pacing before laying, suggestive of frustration, whereas a medium hybrid brown egg laying strain tended to sit during the same period. Heart rates of the light hybrid hens rose steadily during the prelaying period, but those of the medium hybrid strain were low and relatively constant until the point of lay.
	Heil, 1984	Reported differences in prelaying restlessness among five strains of White Leghorn hens.
Vocalization	Stone, et al., 1984	Two commercial egg-type stocks of hens housed in high density, battery cage systems differed in the sound frequency range of the vocalizations each emitted. The stock with the broader range of frequencies gave out more calls indicative of disturbance. This same stock also exhibited more disruption within cages involving stepping on and physical displacement of hens.

which do not experience pre-laying restlessness (Table 7-1, Pre-laying Behavior) would show a different relationship between productivity and activity.

Behaviors such as cannibalism or hysteria also have a direct impact on flock productivity, not by being an integral part of psycho-physiological mechanisms determining the expression of some aspect of productivity, but by creating harmful or stressful circumstances which prevent birds from being productive.

The nature of the relationship of behavior to measures of production is not always clear. Sometimes a relationship between behavior and production is seen in one situation but not in another, as noted in Table 7-1 for aggression and fearfulness. It can happen that different studies find different relationships between behavior and production, e.g., between fearfulness and egg production performance (Table 7-1, Fearfulness).

Table 7-1 also illustrates variation among stocks in behavioral traits that may affect or indicate the well-being of chickens.

- Genetic stock differences in flightiness due to fearfulness may have welfare implications for egg-type stocks of hens prone to osteoporosis. Exaggerated escape reactions could cause increased rates of bone breakage during flock removal.
- Feather pecking and cannibalism impact well-being on levels ranging from discomfort to mortality. (Beak trimming, commonly carried out to reduce the negative impact of feather pecking and cannibalism, has its own negative influences on bird well-being, emphasizing the desirability of stocks which do not engage in harmful pecking behavior in modern commercial housing systems.)
- Hysteria has a decidedly negative welfare impact on flocks, particularly in floor-housed pullet or breeder flocks, where it can cause dramatic declines in growth and egg production and increased mortality.
- Various types of prelaying behavior, e.g., pacing, sitting, vacuum nest building actions, are performed in repetitive stereotypic fashion by egg-type stocks in cage systems, leading to the suggestion that some stocks experience frustration of nesting motivation in cage environments.
- The vocalizations of stocks of chickens in production environments may indicate their behavioral status or well-being. If the vocal patterns produced by a flock indicate excessive disturbance, a producer might try to reduce light levels in the house to minimize disruptive activity.

Our knowledge of behavioral genetics and of the relationships of behavior to measures of performance or well-being is too limited to develop many expectations regarding the effects of various behavioral characteristics on chickens in commercial production systems. Some of the behavior-performance/well-being relationships noted in Table 7-1 may have been specific to the individual stocks involved. When this is true of a stock, it may be necessary to adjust flock management protocols to accommodate the specific characteristics of the variety of birds being housed. For instance, careful control of light levels may be needed for a layer stock with strong feather pecking or cannibalistic tendencies to minimize stress and injury of birds. A stock with lesser tendencies in this regard may tolerate a more variable light environment. When studies find differing associations between behavior and performance or welfare-related variables, some of the apparent associations may not reflect meaningful biological linkages.

Glossary of Terms

Bi-directional selection	Selection of two lines of birds from a common foundation stock for opposite expressions of a trait, e.g., high activity vs. low activity.
General behavior	The overall set of behaviors normally performed in the course of day, e.g., eating, drinking, preening, head movements, rest, etc.
Hansen's score	A measure of fearfulness wherein caged birds are assessed for level of nervousness when an observer stands in front of the cage.
Heritability (h^2)	An estimate of the proportion of the total variation of a given trait in a population that is due to additive genetic effects. The heritability value ranges from 0 to 1, with high values indicating that the trait should respond well to selection.
Imprinting	A learning phenomenon in poultry wherein a newly hatched chick develops a strong following response to a parental figure or siblings.
Latency	The time taken to initiate an action from some predesignated starting time.
Open-field activity	Activity of birds in a standardized test environment, usually a small arena.
Social dominance ability	The ability of a bird to cause another bird to submit or give way to it.
Tonic immobility	Immobility of a bird induced by fearfulness, often associated with being caught and handled; colloquially known as "playing dead."

7-C. GENETIC SELECTION TO MODIFY BEHAVIOR

There is growing evidence that a wide variety of behaviors of chickens can be modified by genetic selection (Table 7-2). Even when heritability (h^2) estimates based on sire family variation for behavior are not large, a response to selection may be demonstrated, e.g., cannibalism and measures of fearfulness. Although some behavioral traits may be influenced by non-additive genetic effects, e.g., a dominance effect in imprinting behavior, much of the genetic control of behavior appears to be additive in nature, meaning that quantitative changes in the performance of specific actions can be achieved through selection. Many behaviors which can affect production performance or well-being, such as ingestive behavior (eating and drinking), aspects of egg laying (posture, time of day), mating, cannibalism, fearfulness, social dominance, and pre-laying behavior, are at least potentially amenable to selection, indicating that genetic progress could be made using suitable selection programs to address specific aspects of behavior-performance or behavior-welfare relationships.

It is particularly encouraging that problems like cannibalism and feather pecking can be reduced through genetic selection (Table 7-2). The commercial egg and breeder industries have been heavily criticized for housing chickens in production systems which fail to accommodate the birds' behavioral needs, thus causing cannibalism and feather pecking. The industry is criticized still further for using beak trimming (a form of mutilation) to solve the problem of cannibalism. Time will tell if varieties of chickens will be developed which will not need to be beak-trimmed to avoid excessive plumage destruction, stress and mortality, and still be able to achieve high performance in commercial production environments. The potential for the adaptation of stocks in this manner, however, has been demonstrated.

Selection may not be able to change all behavior in the direction desired. The control of some types of behavior may be virtually fixed in the genetic makeup of the chicken, making them resistant to change by selection pressure. For instance, five generations of genetic selection did not reduce the responsiveness of chicks to an imprinting stimulus (Table 7-2). Intuitively, this result makes sense. Neither the chicken nor its ancestral type, the jungle fowl, gathers feed and brings it to its young. In the wild, any chick which failed to imprint on its dam and develop a following response would have little chance of survival. Responsiveness to an imprinting stimulus, therefore, may exemplify a behavioral trait that is genetically fixed, i.e., has very low heritability, because of its importance for survival. Actions which must be expressed in specific ways under specific circumstances by all birds, therefore, probably will be relatively unresponsive to selection.

Even if a trait has sufficient heritability, genetic selection may not provide a feasible solution for a behavioral problem because of conflicting

Table 7-2. Heritability Estimates (h^2) and Inheritance of Behavior in Chickens

Behavioral Trait	Selected References	Synopsis of Results
General Behavior	Hurnik, 1978	Heritabilities of the behavior of laying hens were estimated on the basis of sire family variation. Behavior — Estimated h^2 Eating — 0.28 Drinking — 0.82 Standing — 0.54 Resting — 0.51
Cannibalism	Craig and Muir, 1993	Family groups of White Leghorns were selected for hen-days without beak-inflicted injury from 16–40 weeks of age in 6-hen cages. Estimated h^2 — 0.05–0.17 Realized h^2 — 0.65 (2 generations of selection)
	Craig and Muir, 1996	Intact beaked hens from a stock selected 6 generations for family survival and egg production in 9- and 12-bird cages had much lower mortality due to beak-inflicted injury and had better feathering when housed in 12-bird cages than did intact beaked hens from a random-bred and a commercial stock.
Color Preference	Hurnik, et al., 1977	Two generations of Columbian Plymouth Rock chickens were selected for preference of chicks to move into areas dominated by specific colors. Color — Realized h^2 blue — 0.23 green — 0.23 yellow — 0.15 red — 0.03
Egg Laying (posture)	Carter, 1971	Using data from sire families of light-weight and medium-weight commercial layer strains and of a Brown Leghorn strain, the heritability of egg cracking incidence at egg lay was estimated to be 0.73, assuming no maternal effects. The author concluded on the basis of behavioral data that selection for low crack incidence at oviposition amounts to selection for reduced drop height and large-end-first emergence of the egg.
Egg Laying (time of day)	Lillpers, 1991	Estimated heritability of egg laying time for two lines of White Leghorns (WL) and one stock of Rhode Island Reds (RIR). Line — Estimated h^2 WL 1 — 0.38 WL 2 — 0.68 RIR — 0.78

Table 7-2. Continued

Behavioral Trait	Selected References	Synopsis of Results
Fearfulness	Gallup, 1974	One generation of chickens was bi-directionally selected for duration of tonic immobility at 3 weeks of age.
		Realized h^2: Short duration 0.59; Long duration 0.58
	Faure, 1980, 1981	Eight generations of bi-directional selection of Cornish chickens for open-field (OF) activity resulted in behaviorally distinct populations.
	Faure, 1977	Bi-directional selection of 2-day-old Cornish chicks for OF activity had a correlated influence on social dominance ability. The chicks with the highest OF activity in the active line (least fearful) or the lowest OF activity in the inactive line (most fearful) tended to have lower social dominance ability at 20 weeks of age than chicks which had intermediate OF activity.
	Shabalina and Malinova, 1988	Heritability estimates were calculated in regard to the performance of sire families in four breeds of chickens at 3 to 4 ages in three different tests of fearfulness.
	Craig and Muir, 1989	Heritability estimates were calculated based on the performance of White Leghorn hens in different sire families in regard to four different measures of fearfulness. These were tonic immobility duration, latency for one hen (Latency 1) and two hens (Latency 2) to feed near a ticking metronome, and Hansen's test (Hansen, 1976).

Gallup, 1974:

	Realized h^2
Short duration	0.59
Long duration	0.58

Faure, 1980, 1981:

Strain	Trait	Calculated h^2	Realized h^2
Active	OF activity	0.19	0.32
	Latency to move	0.15	0.35
Inactive	OF activity	0.12	0.27
	Latency to move	0.12	0.14

Shabalina and Malinova, 1988:

Breed	Test	Estimated h^2
White Cornish	Open field	0.02–0.36
	In Cage 1	0.00–0.47
	In Cage 2	0.03–0.24
White Rock	Open field	0.03–0.49
	In Cage 1	0.02–0.47
	In Cage 2	0.02–0.38
White Leghorn	Open field	0.02–0.39
	In Cage 1	0.05–0.58
	In Cage 2	0.01–0.33
Black Shoumen	Open field	0.01–0.12
	In Cage 1	0.01–0.10
	In Cage 2	0.05–0.25

Craig and Muir, 1989:

Measure	Estimated h^2
Tonic immobility	0.28
Latency 1	0.10
Latency 2	0.34
Hansen's score	0.08

<table>
<tr><td></td><td>Webster and Hurnik, 1989</td><td>Heritabilities were estimated for various open-field activities of White Leghorn pullets in sire families derived from two different sire parent stocks. Heritability estimates for actions having a significant sire component of variance in at least one stock are shown below.
<table>
<tr><th>Behavior</th><th>Measure</th><th>Stock</th><th>Estimated h^2</th></tr>
<tr><td>Neck extension</td><td>Latency</td><td>1</td><td>0.06</td></tr>
<tr><td></td><td></td><td>2</td><td>0.76</td></tr>
<tr><td>Stand</td><td>Latency</td><td>1</td><td>0.66</td></tr>
<tr><td></td><td></td><td>2</td><td>0.49</td></tr>
<tr><td>Number of actions</td><td>Number</td><td>1</td><td>0.18</td></tr>
<tr><td></td><td></td><td>2</td><td>0.65</td></tr>
</table></td></tr>
<tr><td>Mating</td><td>Dunnington and Siegel, 1983</td><td>23 generations of bi-directional selection of males for number of completed matings. Found that the frequency of forceful mating by high-mating-line males increased as selection progressed.
<table>
<tr><th>Line</th><th>Realized h^2</th></tr>
<tr><td>High mating frequency</td><td>0.18</td></tr>
<tr><td>Low mating frequency</td><td>0.90</td></tr>
</table></td></tr>
<tr><td></td><td>Shabalina and Malinova, 1988</td><td>Heritabilities were estimated on the basis of sire family variation for mating behavior at two ages.
<table>
<tr><th>Breed</th><th>Estimated h^2</th></tr>
<tr><td>White Cornish</td><td>0.20–0.43</td></tr>
<tr><td>White Rock</td><td>0.07–0.35</td></tr>
<tr><td>White Leghorn</td><td>0.02–0.24</td></tr>
<tr><td>Black Shoumen</td><td>0.09–0.23</td></tr>
</table></td></tr>
<tr><td>Imprinting</td><td>Graves and Siegel, 1968, 1969</td><td>Demonstrated both dominance effects and additive genetic variation for responsiveness to an imprinting stimulus in a standardized test. Bi-directional selection for five generations was able to increase, but not reduce, rate of responsiveness to the imprinting stimulus.</td></tr>
<tr><td>Pre-laying Behavior</td><td>Mills and Wood-Gush, 1983; Mills, et al., 1985a</td><td>Directional selection of a White Leghorn line for pre-laying pacing and a Rhode Island Red × Light Sussex line for pre-laying sitting for two generations resulted in respective realized heritabilities of 0.21 and 0.30.</td></tr>
<tr><td>Social Dominance/ Aggression</td><td>Komai, et al., 1959</td><td>In dam-daughter comparisons of social dominance rank for three strains of White Leghorns and one strain each of Black Australorps, Rhode Island Reds, and White Plymouth Rocks, the authors found no clear evidence of differences among strains in genetic variation for social aggressiveness. Mean intra-strain heritability estimates for social rank were 0.30–0.34.</td></tr>
<tr><td></td><td>Guhl, et al., 1960</td><td>Four generations of bi-directional selection of White Leghorns resulted in realized heritabilities of 0.18 and 0.22 for number of individuals dominated and percentages of confrontations won, respectively.</td></tr>
<tr><td></td><td>Craig, et al., 1965</td><td>Five generations of bi-directional selection of White Leghorn and Rhode Island Red males for social dominance ability over unselected control males.
<table>
<tr><th></th><th>Realized h^2</th></tr>
<tr><td>White Leghorn</td><td>0.16</td></tr>
<tr><td>Rhode Island Red</td><td>0.28</td></tr>
</table></td></tr>
</table>

Table 7-3. Genotype by Environment Interactions Involving the Behavior of Chickens

Behavioral Trait	Selected References	Synopsis of Results
Fearfulness	Phillips and Siegel, 1966	Method of rearing influenced the difference between two lines of White Rock chicks in development of fear-related responses to a sudden loud noise.
	Murphy, 1977; Murphy and Wood-Gush, 1978; Murphy and Duncan, 1978; Jones and Mills, 1983	A light (White Leghorn) strain (designated as flighty) and a medium strain (considered docile relative to the light strain) alternated in the nature of their responses to different fear inducing stimuli. The light strain was faster to approach and eat novel appearing food. The medium strain appeared to be made more fearful by sudden sounds, but recovered more quickly from being handled.
Feather Pecking / Cannibalism	Doyan and Zayan, 1984	Caged pairs of hens from a White Leghorn strain did little feather pecking at any floor space allowance, whereas feather pecking frequency by paired Rhode Island Red × White Plymouth was highest at 1,400 cm^2 per hen compared to 900 cm^2 and 1,900 cm^2 per hen.
	Craig and Lee, 1989, 1990; Craig and Muir, 1991; Craig, 1992	Feather loss, mortality and hen-housed production of a commercial White Leghorn variety of hens in multiple-bird cages were largely unaffected by beak-trimming because harmful pecking was low in this stock to begin with. Hens derived from other commercial White Leghorn varieties had better feathering, mortality and hen-housed production when beak-trimmed than when left with intact beaks. Beak trimming nullified the beneficial effects of selection of hens with intact beaks for high performance in multiple-bird cages by improving the survival of non-selected hens to equal that of hens of the selected line.
Pre-laying Behavior	Wood-Gush, 1972	While light strain (White Leghorn) hens consistently did more prelaying pacing than did medium strain hens (Rhode Island Red × Light Sussex in origin) in all circumstances, raising light intensity increased prelaying pacing in the medium strain but did not affect the light strain.
Sociality	Biswas and Craig, 1970	Selection of a strain of hens for high social dominance led to reduced production performance in multiple-bird cages but not in single-bird cages relative to the performance of a strain of hens selected for low social dominance.
	Webster and Hurnik, 1990b	In conventional cages, non-sibling pairs of hens derived from one male parental stock had poorer feed conversion per dozen eggs than sibling pairs. Social combination had no effect on feed conversion for hens derived from a second male parental stock.

objectives within a given type of chicken. For example, a propensity for high feed consumption is beneficial for growth in broilers but harmful for reproductive success and survival in broiler breeders, yet feed consumption in immature birds and in adults has proven thus far to be linked, preventing selection for reduced appetite in adult birds. Therefore, in some cases, perfect compatibility of stocks to environments may not be possible due to differing objectives within a given type of chicken.

7-D. GENOTYPE BY ENVIRONMENT INTERACTIONS

In a genotype by environment interaction, differences between stocks of chickens change depending on the environment in which the birds are kept. Genotype by environment interactions are not unusual for biological characteristics so it is not surprising that such interactions should be found for behavioral traits.

In most of the examples in Table 7-3, the environments acted differently on the stocks of chickens in ways which directly altered the birds' behavior. The genotype by environment interaction involving the effect of beak trimming on feather pecking/cannibalism noted in Table 7-3, however, may not be due entirely to changes in the behavioral characteristics of the stocks studied. Beak trimming minimizes beak inflicted injuries so that stocks which are predisposed to feather pecking and cannibalism actually survive and perform as well as stocks which have little pecking tendency. The interaction between beak trimming and feather pecking/cannibalism, therefore, would not require a change in pecking behavior, although such a change is conceivable. Beak trimming might minimize the tendency of some stocks of hens to peck other birds by reducing the likelihood of hens being positively reinforced by successful feather plucking or the drawing of blood. Incidentally, beak trimming as a management practice may have delayed the development of non-pecking stocks because its success in reducing harm from injurious pecking has nullified the performance benefit of selection for non-pecking.

The genotype by environment interaction for fearfulness in the studies involving novel stimuli, sudden sounds and handling, may only be apparent (Table 7-3). The differing reactions of the light and medium strains of hens in the different test situations may arise from the operation of behavioral systems specific to the stimuli being tested, and only be secondarily related to fearfulness.

Genotype by environment interactions for behavior complicate endeavors to produce genetic stocks of chickens which meet performance objectives and are well adapted to commercial environments. Choice of stocks and genetic selection programs would have to emphasize a specific match of genotype to environment.

8

Poultry Housing

by William D. Weaver, Jr.

Chickens, being warm-blooded (homeothermic) animals, have the ability to maintain a rather uniform internal body temperature (homeostasis). However, the mechanism for accomplishing this is efficient only when the ambient temperature is within certain limits; birds cannot adjust well to extremes. Therefore it is important that chickens be housed and cared for so as to provide an environment that will enable them to maintain their thermal balance (thermoneutral zone).

8-A. CONTROLLING BODY TEMPERATURE

The internal body temperature of birds shows more variability than mammals, and therefore there is no absolute body temperature. In the adult chicken the variability is between 105° and 107°F (40.6° and 41.7°C). Some variations within and outside of this range may be observed:

1. Body temperature of the newly hatched chick is about 103.5°F (39.7°C), and increases daily until it reaches a stable level at about 3 weeks of age.
2. Smaller breeds have a higher body temperature than larger breeds.
3. Male chickens have a slightly higher body temperature than females, probably the result of a higher metabolic rate and larger muscle mass.
4. Activity increases body temperature. For example, the temperature of birds on the floor is higher than that of birds kept in cages.

5. Molting birds have a higher temperature than those fully feathered.
6. Broody hens have a higher body temperature than non-broody hens.
7. After food enters the digestive tract and digestion begins, body temperature increases.
8. Body temperature of chickens is higher during periods of increased light intensity than during periods of lower light intensity or darkness.
9. There is a tendency for core body temperature to rise as ambient temperature goes above or below the thermoneutral zone (65 to 75°F/18 to 24°C for adult birds).

How Heat Is Lost from or Gained by the Body

Even though there are many contributing factors to a slight increase in deep body temperature, the rise will be excessive and potentially lethal if it is not possible for the bird to dissipate excess heat from the body. The chicken is continually producing heat through metabolic processes and muscular activity, and the heat lost from the body must equal the heat produced or body temperature will rise.

There are several principles of heat transfer:

Sensible heat. Heat that can be felt by the body. It is also the heat or energy that accompanies an actual change in temperature.

Radiation. When the temperature of the bird's surface is greater than an adjacent surface, heat is lost from the bird's body by radiation. Heat transfer ceases when the temperature of the surface of the adjacent object is similar to that of the surface of the bird's body.

Conduction. Loss of heat by conduction is caused when the surface of the bird comes in contact with the colder surface of any surrounding object, such as the floor or side walls. As the actual contact area is normally small, heat lost from the bird's body by conduction is generally very low.

Convection. When cool air comes in contact with the surface of the bird, the air is warmed. The heated air expands, rises, and heat is carried away as the warmer air moves away from the bird. When the speed of air moving over the body is increased, as with air from inlets or with tunnel ventilation, the amount of heat lost from the bird by convection increases. In many mammals, body heat is moisture-laden as sweat glands continually exude moisture, which evaporates, producing even greater cooling. Chickens have no sweat glands, therefore skin moisture is not normally a factor. As the ambient temperature rises, heat loss by convection decreases until ambient temperature reaches body temperature, when there is little or no loss by this method. Likewise, in still air, there is little heat loss.

Fecal Excretion. A small amount of heat leaves the body with fecal excretions.

Production of eggs. Loss of heat with the laying of eggs also occurs (eggs act as a heat sink), but is of minor importance.

Latent heat. Energy is defined as the heat required to change water from a solid to a liquid or from a liquid to a gas without changing its temperature. As a replacement for moisture lost through sweat glands in some mammals, the chicken uses a process of evaporative cooling by the vaporization of moisture from the damp lining of the respiratory tract (lungs and air sacs). Heat lost in this manner is the major way of eliminating heat from the body of birds when ambient temperature is high. The process is also known as the heat of evaporation.

Lethal Body Temperature

When the heat produced by birds is greater than that being dissipated through the various processes of elimination, deep (core) body temperature will rise. When it reaches a critical point, birds will die from heat prostration. This is called the upper lethal temperature and is about 116.8°F (47°C).

Mechanisms to Maintain Body Temperature

At 70°F (21°C), approximately 75% of the heat generated by birds is lost through radiation, conduction, and convection (sensible heat). However, as environmental temperature increases and approaches body temperature, sensible heat loss as a proportion of total heat loss lessens (Table 8-1).

The bird's ability to dissipate heat is influenced by skin temperature

Table 8-1. Sensible and Latent Heat Production as Influenced by Ambient Temperature (White Leghorn Hens)

Ambient Temperature		Sensible Heat (%)	Latent Heat (%)	Sensible Heat Production (BTU)		Latent Heat Production BTU	
°F	°C	%	%	Per lb	Per kg	Per lb	Per kg
40	4.4	80	20	8.8	19.4	2.2	4.8
60	15.6	75	25	8.9	19.6	3.0	6.6
80	26.7	60	40	7.1	15.6	4.7	10.3
100	37.8	10	90	4.0	8.8	36.0	79.2

Source: Ota and McNally, 1961

rather than by deep body temperature. As temperature of the air surrounding the bird decreases, blood vessels in and under the skin contract, thus reducing the flow of blood, which in turn acts to minimize the amount of heat lost from the body. When temperature of the surrounding air increases, the converse occurs as blood vessels dilate, increasing the flow of blood to the surface, thereby maximizing the amount of heat lost.

Panting necessary at high environmental temperatures. When heat cannot be adequately dissipated from the body by radiation, conduction, and convection, another mechanism is called upon. At this time, panting (more rapid and heavy breathing) occurs, bringing more outside air in contact with the membranes of the respiratory system. Heat is removed from the body by the incoming air itself, as well as the heat loss that occurs when water is evaporated from the moist surfaces of the respiratory tract. As mentioned previously, this latter principle is known as latent heat loss or heat of evaporation.

At a humidity of 50%, birds will begin panting when the ambient temperature reaches approximately 85°F (29.4°C). As the outside temperature increases above this level, so will the rate of respiration (rapidity of panting), allowing more heat to be eliminated from the body.

Panting and dehydration. The increase in respiration rate is accompanied by an increase in loss of moisture from the body. To compensate for this loss, birds drink more water to avoid dehydration. Many times birds drink more water than can be exhaled and the surplus is excreted in the droppings. The amount of moisture in the ambient air (humidity) also affects the panting rate; the higher the humidity, the more rapid is respiration.

High temperatures and high humidity. Chickens, regardless of their age, cannot withstand concurrent high temperatures and high humidity. When the surrounding air is moist, it cannot absorb as much moisture from the respiratory tract; consequently, the bird must pant more rapidly. Similarly, when high ambient temperature and humidity are present, birds may not be able to exchange enough air by panting to remove heat from the body. As a result, body temperature will rise and death may occur.

Heat production and feed consumption. As the process of digesting feed produces heat (heat of digestion), birds will reduce the amount of feed consumed during periods of hot weather. In turn, growth, egg production, and egg weight can be affected during hot periods. Removal of feed four to six hours prior to periods of high ambient temperature is a practice used primarily by broiler growers to reduce heat produced during digestion and consequently reduce death losses that may occur during hot weather. Another option when birds are not receiving continuous or almost continuous illumination is to provide feed during the cooler night time periods.

Bird activity. As ambient temperature changes, so does the activity of the birds. Birds move less during hot weather in an attempt to minimize heat production. Birds rest more, as evidenced by less eating and mating, and more sitting than standing or walking, with wings extended to expose more body surface for heat loss. Loss or molting of surface feathers may also occur. Conversely, when air temperatures are low, birds increase the production of body heat by increasing activity and feed consumption. During periods of extreme cold, birds can also conserve heat by fluffing their feathers and burying into the litter, which serves as an insulating material.

8-B. HEAT GAIN AND LOSS IN POULTRY HOUSES

Good poultry housing is designed to alleviate extremes in environmental conditions, and thus to assure that birds are comfortable and productive. A discussion of adequate housing must include an understanding of the contributing factors of heat and moisture production.

How heat is measured. Heat produced by the birds, brooders, and for that matter the sun, is measured in British thermal units (Btu). One Btu is the amount of heat required to raise the temperature of 1 lb of water 1 degree Fahrenheit, when at 59°F.

Heat production of birds. Data on the heat production of birds are highly variable and are influenced by a number of factors such as, type of bird (broilers, growing pullets, layers, etc.), age, caloric intake, ambient temperature, relative humidity, etc. But, inasmuch as heat production is discussed in the context of proper house designs and ventilation, tables have been developed to show average heat and moisture production by birds at different body weights and ambient temperatures (Tables 8-1, 8-2).

Data in Table 8-1 show that a 4-lb (1.8-kg) bird maintained at 80°F loses about 47 Btu of total heat (sensible and latent) per hour. On a unit of body weight basis, this is only about one-half as much as pro-

Table 8-2. Hourly Moisture Production of White Leghorn Hens—(1,000—4-lb hens)

Temperature		Respired		Defecated		Total*	
F	C	lb	kg	lb	kg	lb	kg
45	7.2	8.4	3.8	12.9	5.9	23.7	10.8
60	15.6	11.4	5.2	12.7	5.8	26.4	12.0
80	26.7	14.3	6.5	14.4	6.4	31.6	14.3
95	35.0	20.0	9.1	10.3	4.7	33.7	15.3

Source: Ota and McNally, 1961

* Includes wasted drinking water estimated at 10% of water consumed at 45 to 80°F (7.2 to 26.7°C) and 15% at 95°F (35°C)

Table 8-3. Sensible and Latent Heat Production for Broilers at Different Ages (82 to 87°F) (27.8 to 30.6°C)

Age (days)	Heat Production (Btu / lb)	
	Sensible	Total
25	11	24
32	8	20
39	7	18
46	5	17

Source: Reece and Deaton, 1970

duced by a 1-lb (0.45-kg) bird. This is partially illustrated in Table 8-3 which shows that 25-day-old broilers (approximately 2 lbs) produce 24 Btu of total heat, whereas 46-day-old broilers (approximately 5 lbs) produce only 17 Btu of total heat per pound. Thus, it must be kept in mind that heat produced per unit of body weight decreases as birds become heavier.

In ventilating a poultry house during hot weather the additional heat produced by birds must be removed from the building to reduce ambient temperature, but during colder weather a higher percentage of the generated heat must be kept in the house to help maintain temperature.

The following rules-of-thumb may be used for total heat production (sensible and latent) for chickens:

Standard Leghorn layer (3.5 lbs)	40 Btu (per hour) per bird
Brown-egg layer (4.5 lbs)	45 Btu (per hour) per bird
Meat-type breeder (7.0 lbs)	55 Btu (per hour) per bird
Broilers (4.5 lbs)	45 Btu (per hour) per bird

8-C. WATER PRODUCTION AND LOSS

The quantity of water consumed by birds depends on body weight, bird type, salt (sodium) levels in the diet, ambient temperature, and relative humidity. Water is eliminated from the body by the excretion of waste materials, of which about one-fourth is urine and three-fourths is moisture, from the intestinal tract. Water is also eliminated by respiration, and in the case of laying hens, the production of eggs. The amount lost by the birds, along with an estimate for the amount lost or spilled from the drinkers, must be determined before an adequate ventilating system can be designed for the poultry house.

At a normal ambient temperature and relative humidity [70°F (21°C) and 60% RH], moisture lost through respiration by a 4-lb (1.8-kg) bird approximately equals the amount lost through the feces; however, at lower body weights the proportion of water excreted in the feces is greater (see Table 8-2).

On a weight basis, the total amount of water eliminated through respiration and fecal discharge decreases as the size of the bird increases. For example, the quantity of water consumed is approximately one-half as much per unit of body weight for an 8-week-old broiler (6.6 lbs, 3 kg) versus a 1-week-old broiler (0.35 lbs, 0.16 kg) (National Research Council, 1994). This factor must be taken into consideration when ventilating a poultry house, as most ventilation systems are designed by using pounds (kg) of body weight and not the age of the chickens in the house.

Amount of water in the feces. This amount is highly variable as it is associated with ambient temperature and feed composition. Younger (smaller) birds generally have less moisture in their droppings than older (larger) birds. Most adult chickens in thermoneutral environments and consuming a standard commercial diet will produce feces containing 75 to 80% water.

Feed and water consumption affect water production. At 70°F (21°C) a chicken will normally consume two times as much water (by weight) as feed. But, as ambient temperature rises, feed consumption decreases while water intake increases. For example, with broilers, water consumption increases approximately 4% for each 1°F above 70°F (NCR, 1994).

Respiratory and fecal elimination of water. With a 4-lb (1.8-kg) bird at a temperature of 70°F (21°C), about 50% of moisture is lost through fecal excretions, while only about 25% is lost in this manner by a 1-lb (0.45 kg) bird. The remainder of moisture loss occurs through respiration.

8-D. THE ENVIRONMENTAL PROBLEM

Desirable housing, from an environmental standpoint, is necessary to meet requirements for bird growth, reduced stress, egg production, fertility, and the efficient utilization of feed. Briefly, housing must provide the flock an ambient temperature within the thermoneutral zone with good quality air—low levels of toxic gases and particulate matter—with adequate light and with the proper equipment to provide feed and water so that performance can be optimized.

Further, from a social or human perspective, housing for poultry must address concerns such as odors, dust, noise, flies, and other pests. Also houses should be constructed in a manner and located so as not to distract from the overall beauty of the surroundings. In many cases the planting of trees and other vegetation around the poultry house, and observing proper distances and setbacks from dwellings and other public places can minimize the negative impact of large poultry structures.

8-E. INSULATING THE POULTRY HOUSE

No matter what the climate, insulation in the ceiling or under the roof is essential. In colder climates, insulation in the side walls and end walls is also recommended. Increased levels of insulation become more economical as the difference increases between outside temperature and the desired inside temperature (Figure 8-1). During colder periods, insulation is beneficial for reducing the loss of heat from the building. Likewise during hot weather, insulation reduces the amount of heat allowed to enter the building.

Qualities of Insulating Materials and Their R-Value

For a material to qualify as a good insulator it must resist the transfer of heat. To accomplish this efficiently, a material must contain a large number of small, isolated dead air spaces. Therefore, the more small air spaces present in a cubic unit (inches, cm) of a material, the better insulator it

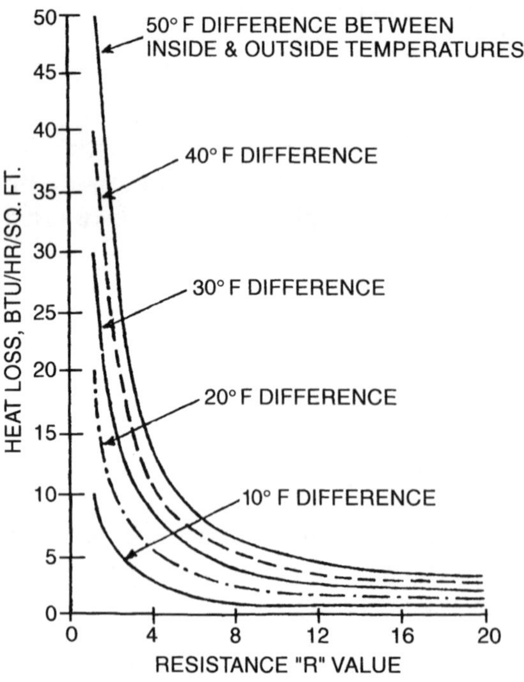

Figure 8-1. The effect of various amounts of insulation (R-value) in the transfer of heat (Btu's) at different ambient temperatures

becomes. The ability to resist the transfer of heat through a material can be measured and is referred to as the resistance, or R-value. The R-value of a number of common building materials are shown in Table 8-4. Remember, the higher the R-value, the greater its ability to restrict the transfer of heat.

Vapor Barrier

To be effective, insulating materials must remain dry, as moisture can act as a conductor, aiding in the transfer of heat. Certain insulating materials (polystyrenes, polyurethanes, vermiculite) do not absorb moisture and thus do not require a vapor barrier. However, a number of the other commonly used insulating materials such as cellulose, fiberglass, and various wool products will absorb moisture and therefore require a separate vapor barrier. Vapor barriers, by definition, must resist the movement of water vapor through them.

The ability of a material to allow or restrict this movement is known as its permeability (perm) value. A perm score of less than 0.5 is required for a material to be considered a desirable vapor barrier. Materials such as aluminum foil and various types of polyethylene films have perm values of less than 0.5 and are considered to be adequate vapor resisting materials. Other materials such as plywood and general framing lumber, bricks, concrete, and masonry blocks have perm values in excess of 0.5 and therefore are not considered as acceptable vapor barriers.

The vapor barrier must be installed on the inside (warm side) of the insulation to minimize the movement of water vapor into the material. This is critical because as air containing water vapor cools—reaching its dew point—condensation forms, wetting the insulation.

How Much Insulation?

While insulation is essential in hot climates to reduce the transfer of heat into the poultry building, generally the amount of insulation required is determined by how low outside temperatures become during the cold periods of the year. The following are recommended R-values for three types of climates:

	R-Value	
Climate Type	Ceiling	Side Walls
Hot ($\Delta t < 30°F$ or 17°C)*	9	6
Medium (Δt 30–50°F or 17–28°C)*	12	8
Cold ($\Delta t > 50°F$ or 28°C)*	20	14

* Δt = difference between inside and lowest outside temperatures

Table 8-4. R-Values of Various Building Materials

Item	Thickness		Resistance Rating
	inch	cm	
Insulation per 1 in (2.5 cm) of thickness			
Blanket bat	1	2.5	3.70
Balsam wool (wood fiber blanket)	1	2.5	4.00
Cellulose fiber	1	2.5	4.16
Expanded polystyrene, molded (bead board)	1	2.5	3.50
Expanded polystyrene, extruded (Styrofoam®)	1	2.5	5.00
Urethane foam	1	2.5	6.60
Fiberglass (glass wool)	1	2.5	3.70
Palco wool (redwood fiber)	1	2.5	3.84
Rock wool (machine blown)	1	2.5	3.33
Rock wool (blanket)	1	2.5	3.33
Foam glass	1	2.5	2.50
Glass fiber blanket	1	2.5	3.33
Mineral wool	1	2.5	3.33
Insulation board	1	2.5	2.37
Vermiculite (expanded)	1	2.5	2.05
Wood fiber	1	2.5	3.33
Sawdust or shavings (dry)	1	2.5	2.22
Straw	1	2.5	1.75
Materials (thickness as indicated)			
Air space, horizontal	0.75+	1.8+	2.33
Air space, vertical	0.75+	1.8+	0.91
Asbestos cement	0.12	0.3	0.03
Building paper			0.15
Concrete	8.00	20.3	0.61
Concrete block	8.00	20.3	1.11
Hardboard	0.25	0.6	0.18
Plywood	0.25	0.6	0.32
Plywood	0.50	1.2	0.63
Surface, inside			0.61
Surface, outside			0.17
Siding, drop	0.75	1.9	0.94
Sheathing	0.75	1.9	0.92
Metal siding			0.09
Glass, single			0.61
Shingles, asbestos			0.18
Shingles, wood			0.78
Roofing (roll, 55-lb)			0.15
Vapor barrier			0.15

Determining the R-Value of Walls and Roofs

As essentially all building materials have an R-value, the sum total of the R-values of the various materials used will give the total R-value for a wall or roof section. Using Table 8-4, an example of the resistance value of a wall section has been calculated below:

Wall Insulation Item	R-Value
Outside surface	0.17*
Metal siding	0.09
Vertical air space	0.91
$3\frac{1}{2}$-inch fiberglass (3.7 per 1 inch)	12.95
Vapor barrier	0.15
$\frac{1}{4}$-inch plywood	0.32
Inside surface	0.61*
Total resistance rating (R) of wall	15.20

* All exposed surfaces have an R-value. With no air movement (inside surface) the R-value is 0.61 and with a 15-mph wind (outside surface) is 0.17.

Determining Heat Loss From Buildings

Sensible heat loss (conduction, convection, and radiation) from the side walls and ceiling of poultry houses can be calculated by using the following equation:

$$Q = \frac{A \times \Delta t}{R}$$

where:

Q = Total heat loss in Btu's (per hour)
A = The area of outside wall and ceiling surfaces (ft^2)
Δt = Difference between inside and outside temperatures (°F)
R = R-values, or the resistance to the transfer of heat, of the various materials in the wall and ceiling sections.

Finally, heat loss can occur when ventilating with fans during the colder months while removing moisture from the poultry house (from the litter, as well as relative humidity in ambient air). Chapter 9 (*Fundamentals of Ventilation*) addresses the proper design and operation of mechanical ventilation systems so as to provide the proper environment while minimizing heat loss. Further, in properly constructed and insulated poultry houses, up to 75% of total heat loss can occur in this way. The following equation is used to calculate heat loss through ventilation:

$$Q = 0.018 \times CFM \times 60 \times \Delta t$$

where:

Q = Total heat loss in Btu's (per hour)
0.018 = A constant
CFM = The average cubic feet per minute of air being exhausted from the building by all fans
60 = Converting cubic feet per minute to cubit feet per hour
Δt = Difference between inside and outside temperature

9

Fundamentals of Ventilation

by William D. Weaver, Jr.

Ventilation can be best defined as a system that delivers fresh air throughout the poultry house, and in doing so, removes excess heat, moisture, and undesirable gases that may be present.

There are three general designs used for ventilation systems: positive pressure, negative pressure, and natural. The positive and negative pressure systems use mechanical fans to either direct air into the house (positive) or exhaust air from the house (negative) (Figure 9-1). With the positive pressure system, the exhaust or air outlet area is controlled, and with the negative system the air inlet area is controlled, which in turn increases the velocity (speed) of the air at that point and assists with the mixing of fresh incoming air with air in the house. Positive and negative turbo systems are deviations from the original positive and negative systems, and are used in many high rise layer houses (Figure 9-2). Natural ventilation systems generally consist of curtains or windows which are opened and closed automatically or manually with winches. The natural system can be an effective means of ventilation (although with less control than with fan systems) when higher levels of automation and control are not practical.

As mechanical, negative pressure systems are the most widely used in both broiler and layer houses, the remainder of the chapter will address its various principles and applications.

9-A. PSYCHROMETRICS

Psychrometrics is defined as the relationship between mixtures of air and water vapor at various temperatures. As ventilation deals with these

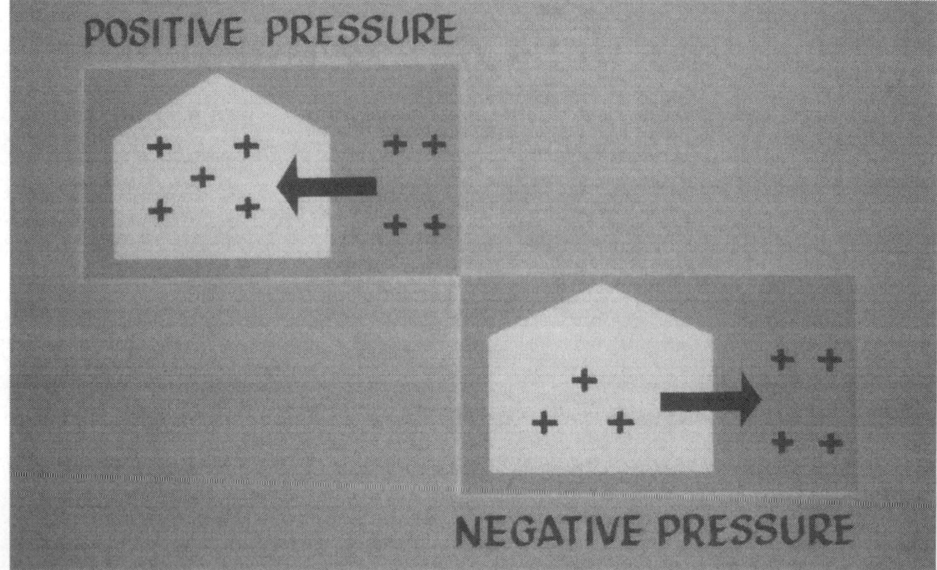

Figure 9-1. Diagram of Positive and Negative Static Pressures—Principles

relationships, some understanding of the components involved would be helpful. Following are definitions of some terms:

- *Dry Bulb (DB) Temperature:* Ambient temperature (temperature that you can feel).
- *Wet Bulb (WB) Temperature:* Temperature at saturation or 100% relative humidity.

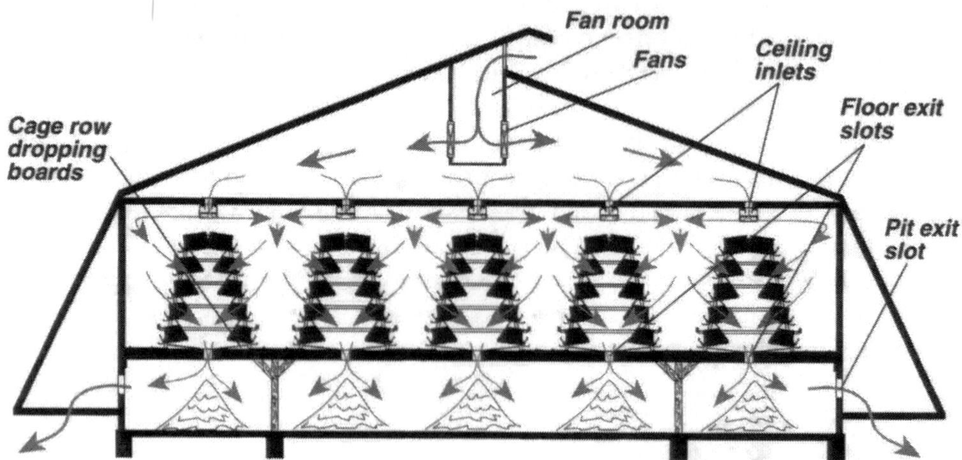

Figure 9-2. Diagram of Air Pathways in a Positive Pressure ''Turbo'' Ventilated Layer House
(courtesy of Chore-Time)

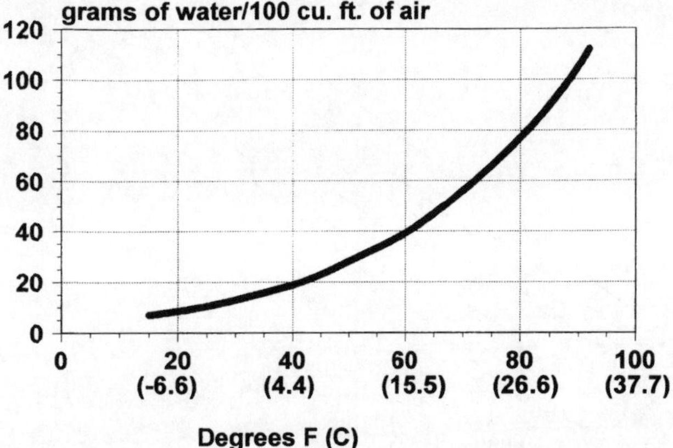

Figure 9-3. Moisture Holding Capacity of 100 Cubic Feet (2.8 cubic meters) of a Saturated Air at Various Temperatures

- *Relative Humidity (RH):* A ratio of the quantity of water vapor in the air compared with the total that can be held at a given temperature.
- *Dew Point (DP) Temperature:* Temperature at which water vapor is transformed back to a liquid (always lower than WB temperature).

While it is not the intent to provide a comprehensive discussion of psychrometrics and its relevance to poultry house ventilation, there is one point that does bear mentioning. That is; *warm air will hold significantly more water vapor than cold air* (Figure 9-3). The rule of thumb is that the water holding capacity of air approximately doubles with each 20°F (11°C) increase in temperature. For example, if outside air at 30°F (−1°C) and 100% RH is brought into the poultry house and allowed to warm up to 50°F (10°C), the RH of that air would drop to 50%. If outside air at 30°F (−1°C) and 100% RH is warmed to 70°F (21°C), representing a 40°F (22°C) increase in temperature, the RH would be decreased to approximately 25%. Therefore, during the colder months poultry producers with ventilation fans can bring relatively small quantities of air into the house—which may have a high RH—heat it to room temperature, and remove moisture and consequently ammonia from the house as the air is exhausted.

9-B. HOT VS. COLD WEATHER MANAGEMENT OF VENTILATION

Other than for the quantity of air required being greater in warm versus cold weather, the fundamental principles of ventilation systems used in the summer and winter months are quite similar. However, the reasons for ventilating during these two seasons are very different.

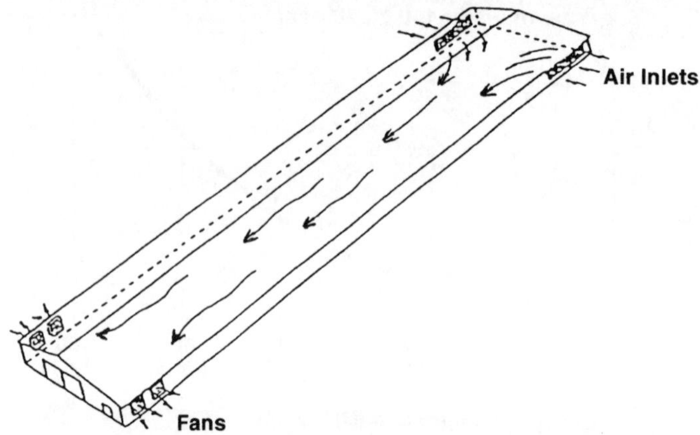

Figure 9-4. A Diagram Showing the Location of Air Inlets and Fans in a Typical Tunnel Ventilated Poultry House

Control Temperature

During the warmer months, the objective is to remove heat and therefore control temperature. This is generally accomplished by moving large quantities of air. More recently, in both broiler and layer houses, tunnel ventilation systems have been used to increase overall air speed (wind chill) and consequently promote convective cooling (Figure 9-4). These tunnel systems use fans located in one or both ends, or in the side walls in the center of the house, and large inlets (many times including foggers or evaporative pads) located in the opposite end(s) of the building.

Control Moisture and Ammonia

During the colder months the ventilation system must remove moisture and noxious gases, with the most critical being ammonia, while conserving heat. This is accomplished by using controllable air inlets (Figure 9-6) at the eaves—where the roof joins the side wall—on both sides of the house in combination with ceiling or side/end wall exhaust fans.

Following are definitions of several terms that will help explain some principles as well as describe components of the mechanical ventilation system:

- *CFM (cubic feet per minute) or CMS (cubic meters per second)*—Is generally used to describe the quantity or volume of air being moved by a fan or entering an air inlet.
- *Static Pressure*—The difference between inside and outside atmospheric pressure, which is expressed in inches (cm) of water column. Static pressure can be either negative or positive as determined by whether the fans ex-

Figure 9-5. Bank of Fans in a Broiler House

Figure 9-6. Intermittent Air Inlets in a Broiler House

haust air from the building (negative) or blow air into the building (positive).

- *Air Inlet*—A controllable opening, generally located at the eaves, allowing air to enter the house at the proper speed or velocity and discharging it in the proper direction. The size of the inlet determines the velocity of the air flow; large openings generally deliver air at slower velocities while small inlets deliver air at higher velocities.
- *Impingent Air Jet*—Air that is allowed to travel adjacent to a smooth surface, generally a ceiling or side wall. Such air jets will travel approximately 25% farther than a similar air jet be directed into open air.
- *Throw*—The distance an air jet will travel before its maximum speed (velocity) is decreased to 75 feet per minute (fpm, or 0.38 meters per second). The significance of throw in a negative pressure ventilation system is associated with the fact that air jets are used to mix fresh incoming air with moist, ammonia-laden air during colder periods, and hot air during warmer periods in the poultry house. Once the speed of an air jet is reduced to approximately 75 fpm, the momentum and consequently the air mixing ability of the jet is lost. At this point, the air will drift aimlessly toward the fans. Following is the equation for calculating throw:

$$X = \frac{K \times V_i \times b}{V_X}$$

where:

X = Throw—distance in feet from the air inlet (or side wall)
K = 10 (a constant)
V_i = Air velocity at the inlet (fpm)
b = Width or height/opening (feet) of the air inlet
V_X = Air velocity "X" feet from the air inlet, or 75 fpm*
(a constant)

* 75 fpm is defined as still air—incoming outside air that does not readily mix with air in the house.

Therefore, the objective in a house that is 50 feet (15 m) wide, with air inlets on both side walls, is to have throw (X) equal at least 25 feet (7.5 m), or one-half the width of the house. This will allow the air jets to reach the center of the house for the proper mixing of the incoming air with the inside air.

9-C. THE VENTILATION SYSTEM

A mechanical ventilation system has four distinct components. These are *fans, air inlets, controllers, and the producer or operator.* (Remember the systems are only mechanical and not automatic.)

1. *Fans*

Fans commonly used in poultry houses have a center shaft with propellers or blades, and are designed to exhaust air efficiently under a negative pressure of up to 0.15-inch (0.38-cm) water column. A good rule of thumb is that fans used in poultry houses should not lose more than 10% efficiency (capacity to exhaust air) when static pressure is increased from 0- to 0.10-inch (0.25-cm) water column.

Fans used in poultry houses can be either direct-drive or belt-driven. As either can be appropriate under typical situations, the important aspects to consider is whether the fan is rated to operate continuously under the desired static pressure, as well as the fan's energy efficiency (cfm/watt).

Fans used in poultry applications range in diameter from 24 to 60 inches (60 to 150 cm) and generally deliver from 4,000 to 25,000 cfm (110 to 710 cm). Further, in US applications, fans are generally mounted in the side or end walls, whereas in Europe they are typically located in the ceiling. Either location is acceptable as the more important aspect is where the air enters (air inlets) and not where it leaves (exhaust fans) the house. When possible, fans should exhaust air with prevailing winds and away from adjacent poultry houses. (Distances between houses should be a minimum of 1.5 times the width of the house.) Finally, other than in tunnel ventilation applications where all fans will be located in one or both ends of the house, fans or banks of fans should not be spaced more than 150 feet (45 m) apart. Greater spacings than these will contribute to larger differentials in temperature than is desirable in the poultry house.

Fan Requirements

Air volume requirements are normally based on the average body weight of the flock at market age (broilers) or maturity (layers). As egg laying breeds have smaller bodies and consequently lose more body heat per pound of weight than heavier meat producing (broiler) breeds, they normally require a higher rate of ventilation on a per-unit basis. Therefore, when considering a ventilation system that operates under negative pressure on a year-round basis, a minimum requirement of 1.5 cfm per pound (5.60 cmh/kg) should be provided for laying hens and 1.25 cfm per pound (4.67 cmh/kg) for broilers (Table 9-1). These levels are based on the needs

Figure 9-7. Evenly Spaced Fans on Wall of Layer House

of the flock for oxygen and the need to remove excess heat, moisture, and noxious gases. If birds are to be cooled, higher air volumes may be required.

When tunnel ventilation is used during hot weather to control temperature (see tunnel ventilation requirements, below), the poultry house will be operated by using dual systems. Consequently, with the dual system the goal of ventilation with fans in the side/end wall(s) (or ceiling) and air inlets at the eaves is to control moisture and ammonia during cold weather only. These fans are not used when the house is being operated under the tunnel mode. Therefore, as the tunnel system is used to remove excess heat during the warmer periods, the sidewall, cold weather system will be designed to provide only about 35% of the previously described

Table 9-1. Requirements for Ventilation at Different Ambient Temperatures (F) on a Cubic Feet Per Minute (CFM) Per Pound of Body Weight Basis (Broilers)

Ambient Temperature, F	C	CFM/lb of Body Weight
40	4	.48
60	16	.72
80	27	.96
100	38	1.20
110	43	1.32

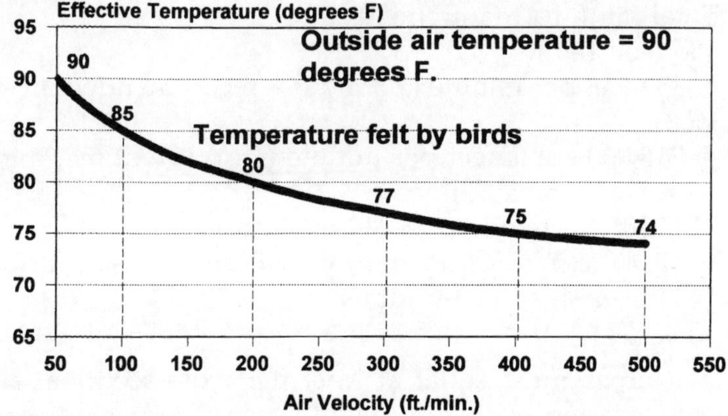

Figure 9-8. Actual and Effective Temperature Experienced by the Birds at Various Air Velocities in a Tunnel Ventilated Poultry House (wind chill effect)

volumes, or 0.5 and 0.4 cfm per pound (2.0 and 1.6 cmh/kg) for layers and broilers, respectively.

Calculations for cold weather fan requirements. Following is an example of how fan requirements are determined for a broiler house with 30,000 birds weighing 4.5 pounds (2.0 kg) at market age. The house is also equipped with tunnel ventilation, therefore, calculations for the eave inlet, cold weather system are based on only 0.4 cfm (1.6 cmh/kg).

30,000 birds @ 4.5 lbs per bird = 135,000 pounds (60,000 kg)

135,000 pounds × 0.4 cfm per pound = 54,000 cfm (92,000 cmh) total fan requirements

54,000 cfm ÷ 9,000 cfm per one 36-inch (0.91-m) fan = 6 − 36-inch (0.91-m) fans are required to remove the moisture and gases from the house during the colder months.

Tunnel ventilation requirements. Tunnel ventilation is a relatively new concept that places all fans in one end of the house, or in houses more than 600 feet (180 m) long, in both ends of the house. The long axis of the house is considered a "tunnel," and the air with the aid of exhaust fans travels the length of the house. When traveling at the recommended velocity of 450 feet per minute (2.3 m/sec) the wind chill that occurs as air passes over the bird's body is an effective means of cooling (Figure 9-8).

Calculations for Fan Requirements for Tunnel Ventilation

- Cross-section dimensions of the house:
  48 feet (15 m) width of house
  × 10 feet (3 m) average ceiling height (flat ceiling)
  = 480 feet^2 (45 m^2) cross-sectional area of the house

- Total fan (air) requirements:
  480 feet^2 (45 m^2)
  × 450 feet per minute (2.3 m/s)—recommended air velocity
  = 216,000 cubic feet per minute (cfm) (6,240 m^3/min)
- Number of fans required:
  216,000 cfm required (6,240 m^3/min)
  ÷ 19,000 cfm (538 m^3/min) per 48-inch (1.2-m) fan
  = 11 – 48-inch (1.2-m) fans

Tunnel inlet areas must equal at least the cross-sectional area of the house (480 ft^2), as a lesser opening would overly restrict air entering the building and consequently reduce the speed of air traveling down the axis of the house. The inlet can be located in the end wall(s), side walls, or a combination of both.

2. *Air Inlets*

Air inlets must be both properly designed and located to deliver fresh outside air into the house at the desired speed, or velocity, and in the proper direction.

Air speed (velocity). Air speed is normally measured in feet per minute (fpm) or meters per second (mps) (Figure 9-9). Because of the ease of measurement, it can also be expressed as static pressure (Table 9-2). Proper air velocity is necessary to create the proper mixing of the incoming air with air already in the building. Static pressures normally range from 0.04-inch (0.10-cm) water column (wc) during warm weather when larger volumes of air are needed, to 0.10-inch (0.25-cm)

Table 9-2. Relationship Between Static Pressure and Inlet Velocity

Static Pressure (in water column)	Inlet Velocity (ft/min)
.015	450
.025	625
.040	800
.055	925
.075	1,100
.100	1,250

(Adapted from "Ventilation for Poultry Houses," Cornell Extension Bulletin 1140, Cornell University, Ithaca, NY 14853)

Figure 9-9. Vane Anemometer for Measuring Air Velocity

wc under colder conditions requiring lesser amounts of air. Again, a producer must remember the goal is to have air jets reach at least the center of the room, therefore when air volume is decreased, the velocity must be increased by decreasing the inlet opening to assure proper mixing throughout the house.

Air direction. In many instances the direction of incoming air jets when leaving the air inlet may be such that it does not allow for proper mixing with the warm inside air (normally in the ceiling area) during cold weather. Likewise, in warm weather, air direction may not allow for the maximum cooling of the birds. Therefore, as a rule, air should be directed across the ceiling during cold weather and immediately over the birds during warm weather.

Location of air inlets. Air inlets are generally located at the eaves of the building, either in the ceiling or in the side wall, on both sides of the house. In wider broiler houses (greater than 70 feet or 21 m) and turbo-ventilated layer houses, inlets may be located in the ceiling away from the eaves. Inlets can be either continuous or intermittent. The decision as to which to install should be based upon whether a minimum of 1.5 inches (3.8 cm) of inlet opening can be maintained when fans are operating under minimum ventilation conditions. In practice, it has been found that when inlets are opened less than 1.5 inches it is difficult to have the incoming air jet reach the center of the building.

Controlling inlet openings. Inlets must be designed to open and close easily and completely. Although inlets can be controlled in series with hand operated winches, it is desirable to install a mechanical, manometer controlled winch that monitors differences between inside and outside atmospheric pressure (static pressure), and adjusts the inlet openings accordingly (Figure 9-10).

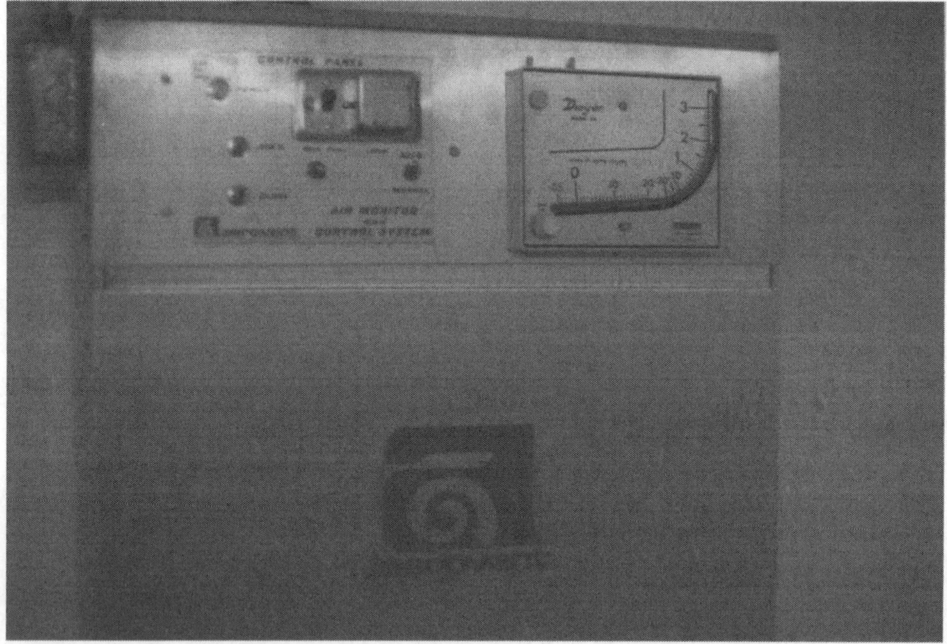

Figure 9-10. Automatic Air Inlet Using a Static Pressure Manometer

Calculating inlet requirements. Inlet requirements can be calculated once the number and volumes of fans have been determined (see below).

Calculations to Estimate Maximum Amount of Inlet Opening Needed

Assuming a requirement of 54,000 cfm
÷ 4 [one square inch of inlet opening is required for each 4 cfm (1.05 cmh) of fan capacity]
= Total area of inlet required: 13,500 in^2 (93.8 ft^2 or 8.7 m^2)

Determining size and number of air inlets. As mentioned earlier, inlets should be designed to open a minimum of 1.5 inches (3.75 cm) under minimum ventilation conditions. Therefore, in many instances in order to meet this condition, small, individual, intermittently spaced inlets *versus* continuous inlets must be installed. Generally these inlets are approximately 4 feet (1.3 cm) long with a 5 to 12 inch opening (13 to 30 cm). The inlet baffle should be constructed from materials that will not bend or warp or absorb moisture, and that has an insulating R-value >4. Polystyrene or polyurethane are examples of materials used for the construction of inlet baffles.

Calculations to Determine Size of Each Individual Inlet and the Number of Inlets Needed

46 in.* (117 cm) length × 6 inches (15 cm) width = 276 in^2 (1,755 cm^2)—area of each inlet

13,500 in^2 of total inlet space required (from above)
÷ 276 in^2 per inlet
= 49 total inlets measuring 46 inches (117 cm) by 6 inches (15 cm) required

* Normal 48 inch opening less structural (studs, etc.) wall members.

Again, inlets should be located at the eaves and spaced evenly on both side walls. Individual inlets should be located at least 10 feet (3 m) from exhaust fans, and to ensure adequate ventilation throughout the house, not more than 6 feet (1.8 m) from the ends of the building and all corners formed by internal partitions, i.e., brooding curtain(s).

3. *Controls*

Controls for operating fans generally consist of thermostats, many times with remote sensors, and proportion timers. While controlling temperature can be a problem during hot periods, thermostats with sensors have generally been found reliable for operating fans when temperature exceeds a predetermined set point. However, while both ammonia sensors and humidistats have been used experimentally to control ammonia and moisture in poultry houses, they have usually proven unreliable in commercial settings. Therefore, in most instances where supplemental heat is added, proportion or recycle timers have been used to operate fans during colder weather. The proportion of time a fan(s) may operate is influenced by the age of the birds, outside temperature and relative humidity, and inside litter moisture, relative humidity, and ammonia level.

More recently, manufacturers of climatic, computer-aided, controllers have marketed an array of devices to operate ventilation systems in poultry houses. Many of these units have the ability to monitor and compute average temperature readings from multiple locations in the house, and in some instances outside the house. In addition, they have the ability to be programmed to change set points over time. While these added features can significantly improve the performance, and in many instances the efficiency of the ventilation system, in reality they are simply monitoring temperatures and by doing so are causing something (fans, heaters, evaporative coolers, etc.) to be turned "on" or "off" based on a predetermined set point or range of set points. While several manufacturers of climatic controllers have offered humidistats for the control of moisture, most producers have found them to be less than reliable and therefore, have incor-

Figure 9-11. Principles of Evaporative Cooling

porated various types of proportion timers (recycle timers) into their units to control moisture and ammonia.

4. Operator

The operator or producer must first design and install a ventilation system that includes the specifications previously discussed, and then must operate the system (fans, air inlets, and controllers) in a way that will control temperature, moisture, and ammonia in the poultry house.

While actually controlling or maintaining temperature within reasonable limits in the poultry house may be difficult during periods of extremely hot weather, the actual recommended thermostat settings are quite simple. Under summer conditions fan thermostats should be set at the desired room temperature. In instances where a number of fans are controlled with multiple thermostats, the thermostats may be staged so that fans are uniformly turned on over a range of temperatures from 0 to

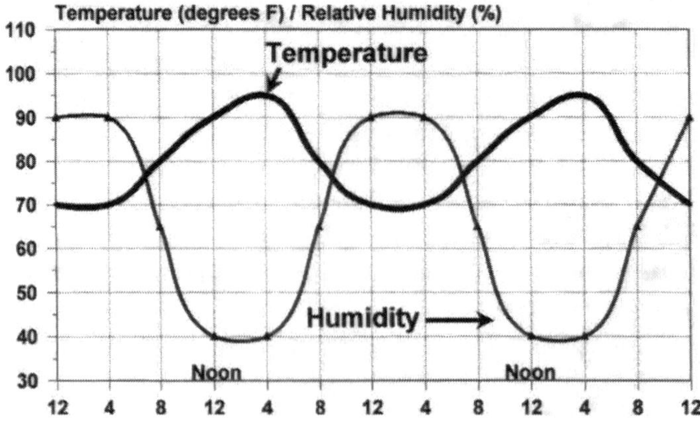

Figure 9-12. A Representation of the Inverse Relationship of Ambient Temperature and Relative Humidity Over a 48-hour Period During Hot Weather

Figure 9-13. High Pressure Foggers in Layer House

8°F (0 to 5°C) above the set point. In colder weather with young birds or when supplemental heat is required, thermostats should be set 1 to 3°F (0.6 to 1.7°C) above the desired room temperature, i.e., if the desired room temperature for brooding chicks is 90°F (32°C), fan thermostats should be set between 91 and 93°F (33 to 34°C).

Unfortunately, determining the proper timer settings to control moisture and ammonia during the colder periods is much more subjective. Normally, the manufacturer of the controls will provide some general recommendations for operation. However, ultimately the producers must observe litter moisture (recommendation is approximately 25%) and condition, and by using gas detection tubes, or by smell, estimate the ammonia level in the house (recommendation is less than 30 ppm). When either or both of these indicators are above the desired level, ventilation, by increasing time on the proportion or recycle timer, must be increased.

9-D. EVAPORATIVE COOLING

Evaporative cooling can be used even in environments with high relative humidity (RH) to cool the poultry house (Figure 9-11). As described earlier in the section on *psychrometrics*, warm air will hold more water vapor than cool air. Therefore, as temperature increases from early morning to late afternoon, RH decreases, providing capacity to hold additional water vapor (Figure 9-12). This principle allows for the use of evaporative cooling during the hottest periods of the day.

Several methods are used in poultry houses to aid in the conversion of liquid water to a vapor (evaporation). Evaporative pads, fogger pads, and medium (200 pounds per square inch, psi) and high pressure (more than 600 psi) fogger nozzles are the most commonly used systems in broiler, breeder, and layer houses (Figure 9-13).

As the heat required to convert water from a liquid to a vapor (evaporation) is known, the amount of water required to reduce air temperature by a given amount can be calculated.

Calculations for Evaporative Cooling

- 8,747-Btu required to convert 1 gal (3.79 liters) of water from liquid to a vapor. The equation used to calculate heat loss is similar to the one used to estimate ventilation heat loss.

$$Q = 0.018 \times \Delta t \times cfm \times 60$$

where:

Q = Heat production in Btu (per hour)
0.018 = A constant
Δt = Desired reduction in temperature (normally 8 to 10°F)
cfm = Quantity of air in cubic feet per minute being exhausted from the house. (In a tunnel ventilated house, this will equal the combined capacity of all fans.)
60 = Converting cfm to cubic feet per hour

Therefore: $Q = 0.018 \times 10°F \times 216{,}000$ cfm (refer to previous tunnel ventilation example) $\times$ 60.
Q = 2,332,800 Btu (per hour)

- Based on a desired reduction of 10°F (5.5°C) in incoming air temperature, the following amount of water must be evaporated:
  2,332,800 Btu ÷ 8,747 Btu/gal
  = 267 gallons (1,011 liters) of water per hour OR
  4.5 gallons (17 liters) per minute

Therefore, a producer exhausting 216,000 cfm (6,240 m^3/hr) of air from the poultry house using tunnel ventilation and evaporating 4.5 gallons (17 liters) of water per minute in the incoming air stream could expect to experience up to a 10°F (5.5°C) reduction in temperature in the house. In areas of low relative humidity (less than 25%) evaporative cooling can actually reduce in-house temperatures by 25°F (14°C) or more.

10

Fundamentals of Managing Light for Poultry

by Michael J. Wineland

10-A. PERCEPTION OF LIGHT

Controlling the light environment is a valuable tool for improving egg production and growth of poultry. Light can influence behavior, metabolic rate, physical activity, and physiological factors such as those involving the reproductive system. Light is typically supplied by a combination of natural and artificial sources; with the amount of each depending upon the season of the year and the distance from the equator. Visible white light is a composite of different colors that can be seen when sunlight passes through a prism. These colors represent specific regions of the light spectrum and the light energy produced represents specific wavelengths of the electromagnetic spectrum (Figure 10-1).

In most cases light is received through the eyes, but it can be received by extraretinal (not the eye) receptors in the brain. For instance, in addition to the eye, it has been demonstrated for reproductive purposes that light energy can also elicit its effect by penetrating the skin, feathers, and the skull to reach the extraretinal receptors in the brain (Figure 10-2). The pineal gland is considered an extraretinal receptor in some mammals, but not in poultry.

The ability of light to penetrate and reach the extraretinal receptors is believed to be a function of both intensity and wavelength. Thus, both natural and artificial light environments within the poultry house can significantly influence a bird's extraretinal reception. When the eyes perceive light, behavior and activity can be modified; this is important in egg laying and growing chickens for meat production. The chicken's eye perceives visible light similar to, but not exactly in the same way as the human eye. A bell shaped curve represents the amount of light energy perceived by

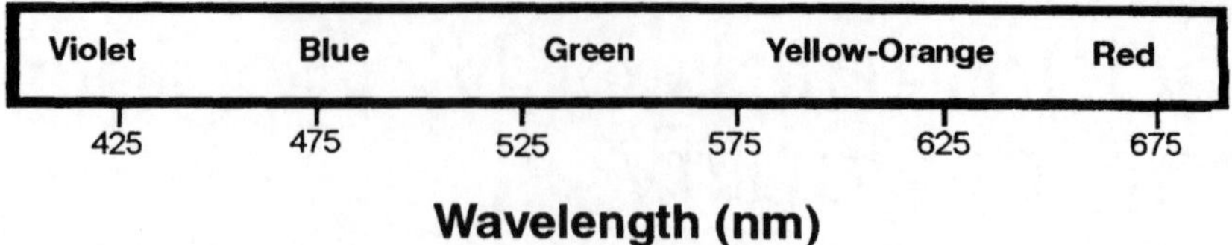

Figure 10-1. The Electromagnetic Spectrum of Visible Light. (The blue light represents the shorter wavelengths, while the longer wavelengths are represented by red light.)

the bird with a limited amount of visible light received at the spectral extremes (violet and red) and a maximal amount in the middle of the visible light spectrum (green) (Figure 10-3).

Additionally, it has been demonstrated that chickens can perceive a certain amount of ultraviolet light, however, ultraviolet bug killers have not been shown to impact reproduction when used in commercial egg-laying houses.

10-B. INFLUENCE OF LIGHT ON EGG PRODUCTION

1. *Day Length*

It has long been known that for some birds and mammals living in the temperate zones the changes in the daily hours of light is an important cue for the seasonal development of the reproductive system. Additionally, factors such as attaining a minimum body weight and age by time

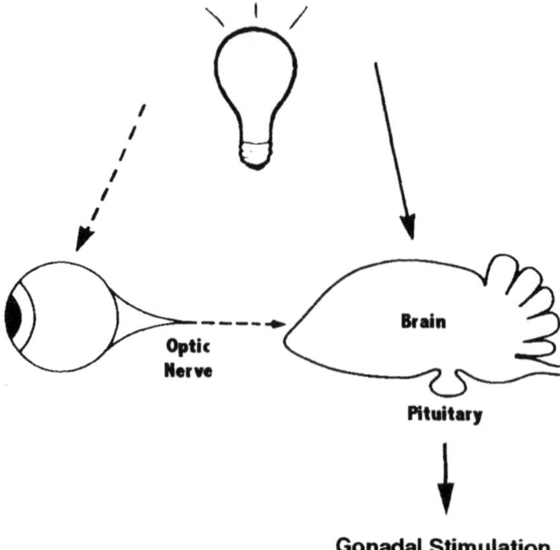

Figure 10-2. Pathways of Light Reception by Birds Effecting Gonadal Activity. (Solid lines indicate primary pathway and dashed lines the secondary pathway.)

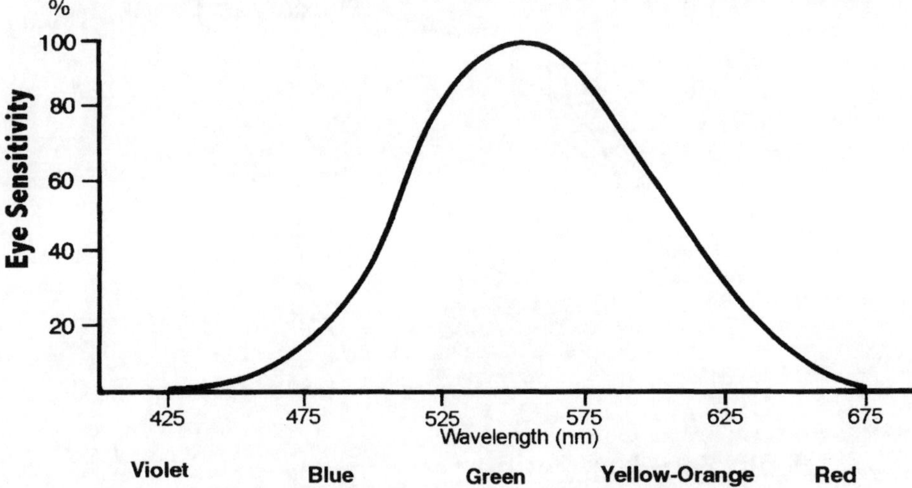

Figure 10-3. Approximate Spectral Sensitivity of a Light Meter. (All wavelengths of the visible light spectrum are perceived, but not equally. Note that less energy of the blue (short wavelengths) and red (longer wavelengths) colors of visible light spectrum are perceived).

of stimulatory light have been demonstrated to aid the bird to respond to stimulatory light and thus influence its reproductive status. It has also been shown that commercial egg layers depend less upon light programs to stimulate egg production than do heavy broiler breeders.

Lights are commonly used to stimulate poultry into egg production and maintain reproductive proficiency for extended periods of time. The importance of duration or "critical day length" is somewhat variable with the different types of poultry, but remains crucial. The need to be exposed to a critical day length is demonstrated elegantly with chickens used for egg production. Poultry have been shown to interpret day length by the occurrence, or the lack, of light during a "photosensitive period," which occurs approximately 11–16 hours after dawn in a 24-hour day (Figure 10-4). Birds perceive a stimulatory day (often referred to as "long day") if it perceives a "dawn" or "lights on" and then subsequently perceives light during the photosensitive period (11 to 16 hours later). If no light is perceived during the photosensitive period, then the bird will interpret the day as non-photostimulatory, often referred to as a "short day," similar to what is experienced during the winter season in the temperate zones of the world. In the equatorial regions of the world where day length is approximately 12 hours throughout the year, natural day length is marginally stimulatory. Chickens raised and brought into egg production in this region on natural daylight will demonstrate less uniformity with regards to onset and persistency of production.

There are specific requirements for day length (period of light) that must be met for poultry to become sexually mature. Broiler breeder hens should

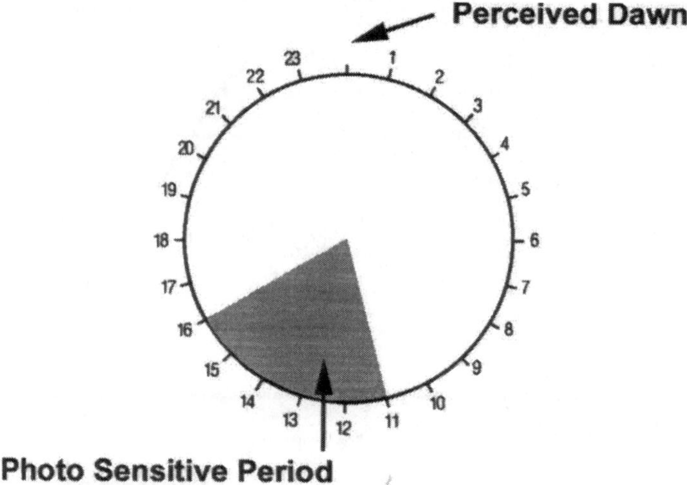

Figure 10-4. The Photosensitive Period During the Day. (Poultry have been shown to interpret day length by the occurrence or lack of occurrence of light during a "photosensitive period," 11–16 hours after dawn or first light in a 24-hour day.)

be exposed to short days before being exposed to the long stimulatory days. The exposure to nonstimulatory (short) day lengths allows the brain to become properly sensitized while disrupting the photorefractory condition. Photorefractoriness is a condition in which the bird is not capable of responding to long day lengths. It has been demonstrated that exposure to nonstimulatory day length for a minimum of 8 weeks prior to stimulatory lighting is necessary to properly attain sexual maturity and optimal production. The inability to properly provide for a nonstimulatory day length prior to sexual maturity in heavy breeder pullets reaching sexual maturity during the long summer days can prevent the females from attaining optimal production. In contrast, the male does not demonstrate the same need to be exposed to short day lengths prior to photostimulation to optimize sexual maturity. However, when females are photorefractory, simple exposure to long days is normally insufficient to stimulate the sexual maturation process.

2. *Photorefractoriness*

Long day lengths, greater than 11 hours (13 to 14 hours is the minimum commonly used) for properly sensitized females, will not only initiate the photostimulatory effect or photoperiodic drive but will also begin the photorefractory state in the female. Photostimulatory or photoperiodic drive refers to the "switching on" of egg production by light and photorefractoriness refers to the "switching off" of egg production by decreasing the ability of the bird to respond to stimulatory day lengths. Therefore, while exposure to photostimulatory day length is needed to stimulate egg production, it also begins the process in which the female becomes less and less responsive to the stimulatory nature of long days, thus decreasing egg production during the egg production period. Females that have been properly "sensitized" exhibit a much stronger photoperiodic drive than photorefractoriness. However, over time, photorefractoriness increases and causes a gradual decrease in egg production. There is evidence that shows the greater the stimulatory day length, the sooner and more pronounced the reduction in egg production will occur caused by photorefractoriness. Thus, stimulatory day lengths longer than 17 hours should not be used.

There is initial evidence that both photorefractoriness and photostimulation may require a different minimum number of hours of light (often times called a "critical day length") before they begin to initiate their effect. Work in turkeys has shown that the "critical day length" for photostimulatory effects changes depending upon the season. Additionally, research has demonstrated that the critical day length for the initiation of egg production is about 10 hours in Leghorns, with optimal egg production occurring at slightly longer day lengths (sometimes referred to as *saturation day length*) of approximately 12 hours. There are also indicators that different classes of poultry have slightly different critical day lengths. Further, normal biological variation within a flock, with regards to critical day lengths, necessitates providing a sufficient amount of light stimulation to optimize egg production, but creates the concern that excessive light stimulation will hasten photorefractoriness. Based on the information above, it is recommended that the photostimulatory period be 14 hours.

Raising pullets on short periods of daylight during the winter is advantageous in the temperate zones, while the long summer daylight hours pose a problem for pullets as they reach sexual maturity. Growing pullets on long daylight hours can cause a reduced peak production, indicative of poor uniformity at the onset of production, and reduced persistency of production. To counteract the effects of long daylight hours the poultryman should raise pullets in a house where daylight exposure can be controlled to provide shorter daylight hours, generally around 8 hours a day. If pullets must be grown in open or curtain sided houses where they reach sexual maturity during late spring and summer, a step-down or constant day length program should be used. A step-down lighting program provides 23 hours of light at one day of age and gradually decreases light until natural day length is reached at the time of photostimulatory lighting. This has been used successfully with commercial layers but has been less successful with broiler breeders. Another light program used at times provides a constant day length equal to the longest natural day length that the flock would be exposed to during the 10 weeks prior to photostimulatory lighting. It is important to remember that when pullets are being grown, the amount of daylight they receive should never increase until stimulation for sexual maturity. Additionally, hens that are in egg production should never experience a decrease in daylight.

Flocks grown using natural daylight in equatorial regions of the world, where daylight (at all times of the year) is close to 12 hours, have performed adequately. However, in these regions when stimulating pullets with light to bring them into egg production, an additional minimum of 2 to 3 hours of light should be provided.

3. *Unconventional Light Periods and Day Lengths*

Intermittent lighting programs are commonly used with table egg production. The programs, of which there are a number, provide alternating periods of light and darkness. As a minimum, these programs have a period of light (of varying length) which simulates dawn or "lights on," and then additional light periods over the next 14 to 16 hours, which may be used to service and care for the hens. This concept works when at least one of the periods of light is provided during the "photosensitive period" (11 to 16 hours after perceived dawn). While the total hours of light given the hen during the day is not typical of long day photostimulation, the periods of light are given in a way that the hen believes she is receiving a stimulatory photoperiod. This can be achieved with a typical intermittent lighting program for layers by using 15 minutes of light, followed by 45 minutes of darkness, and repeated for the duration of the normal lighting period of 14 to 16 hours. This type of program will provide the physiological requirements of the hen for light while reducing energy use and potential problems due to overactivity.

Ahemeral light programs, where a longer than 24-hour day is used, have been shown to slightly increase egg production, egg size, and egg quality when compared with conventional 24-hour programs. Ahemeral lighting must be used in environmental houses where strict light control can be maintained. The program provides a light stimulus at "dawn" and during the "photosensitive period," as with the more conventional programs. However, the day length is altered after the normal period of light by lengthening the period of darkness between the time lights are turned off and then turned back on for the start of a new day. The increase in egg size is primarily a result of increased albumen and increased shell quality resulting from increased time in the shell gland. However, there are inherent problems associated with using long ahemeral day lengths, as in some cases the period during which lights are on in the house occur during hours not normally considered regular employment hours for poultry workers.

4. *Light Intensity*

Intensity of light is important because a minimal level of light must be received to elicit a physiological response. The specific threshold intensity to elicit a response will vary among birds and species of birds; therefore, intensities normally recommended include a safety factor. For example, commercial layers appear to be more tolerant of lower light intensities than heavy breeders or turkeys. Common intensities for commercial layers are near 0.5 fc (5 lux), while for heavy broiler breeders 0.5 fc (5 lux) is not sufficient, and require 2 to 5 fc (20 to 50 lux) to optimize performance. There is research that demonstrates the ability of birds to interpret different intensities under different circumstances. In experiments where combinations of darkness, dim lights, and bright lights were used to elicit the onset of sexual maturity, contrast was important. Birds were capable of interpreting dim light as either a light or a dark period, depending upon which light intensity it was paired with. When paired with a totally dark period, the dim light was interpreted as light. However, when paired with a bright period of light it was interpreted as a period of darkness. Thus, contrast between the light:dark periods is important, and if the dark period is extremely dark, pullets may do very well on slightly lower intensity during the light periods.

5. *Light Color (Wavelength)*

Correct wavelength once was regarded as important to initiate and maintain proper photostimulation. Early work demonstrated that the longer wavelengths of visible light (toward the red end of the spectrum) were best for eliciting a sexual response. This is probably because the longer wavelengths are more capable of penetrating the skull and reaching the extraretinal receptors. When low light intensities are used, intensity may be marginal for providing a threshold response, and therefore, long spectral wavelengths may be important. But, when intensity is above the threshold needed by the birds, wavelength is less important. If light that provides less of the longer wavelengths is used, the additional energy from the higher intensity is generally capable of offsetting the shorter wavelengths.

10-C. INFLUENCE ON GROWTH

1. *Day Length*

Typically, chickens grown for commercial meat production use extended periods of light (23 to 24 hours) to encourage feed consumption and thus increased weight gain. Modern broilers, which have been intensely selected for fast growth, have also experienced increased metabolic problems such as ascites, flip over disease, sudden death syndrome, and skeletal (leg problems) disorders. Attempts to reduce the incidence of these problems by slowing down the early growth rate of broilers have been successful. This has been accomplished by physically restricting feed and by reducing the amount of light the birds receive during the first several weeks, therefore, reducing feeding time. Physiological effects of these early reduced lighting programs have shown increased testicular growth, but no effect on body composition (see *Broiler Management*, Chapter 43).

Intermittent light programs with alternating periods of light and darkness, such as one hour of light and two hours of darkness (1L:2D) throughout the 24-hour day, have been successful in improving feed conversion, increasing body weight, and decreasing the incidence of leg disorders. Broilers can anticipate changing from a period of light to a period of dark, and increase feed consumption activity just before lights go out. However, it is important as birds near market age to return to continuous lighting to ensure proper feed withdrawal prior to processing.

2. *Light Intensity*

Light intensity has also been found to influence growth in meat-type poultry. There is much conflicting data with regards to the effect of bright or dim lights, but most of it is the result of what the investigators termed *bright and dim*. Many investigators have demonstrated that reduced light intensity promotes heavier birds and, in some instances, improved feed conversions because of reduced physical activity, and thus, a decrease in energy expended. Increased light intensity at certain times can also be advantageous. Investigators have discovered that increased activity, such as wing stretching and other non-injurious activity, reduces the incidence of leg abnormalities such as tibial dyschondroplasia and enlarged hocks. Additionally, low light intensity 0.5 fc (5 lux) has been shown to increase fat pad size when compared to a higher intensity 15 fc (150 lux).

3. *Light Color*

Evidence supporting the importance of a particular wavelength or color of light is not clear. Part of this may be due to the interaction of wavelength and intensity with regards to the bird's spectral sensitivity, which must be kept in mind when evaluating research using different colors. It has been shown that broilers reared under blue, green, red, and white lights and subsequently given an opportunity to select a color, preferred first blue and then green light. The ability of broilers to discern intensity has also been demonstrated. When assessing the broiler's ability to recognize the intensity of either blue or red light, it has been shown that broilers require approximately 3 times greater intensity of blue than red. This may be due to the spectral sensitivity of the broiler for different wavelengths.

Blue light has been found to result in increased weight gain in turkeys and chickens, but how much of this is an intensity rather than a color effect is unknown. Where attempts to equalize intensity have occurred, red light has been demonstrated to increase activity with regards to wing stretching and other non-injurious activity. This increased activity has been shown to decrease leg problems in fast-growing birds. The timing of exposure to the red light has also been evaluated, when both red and blue lights were used for growing broilers. Red light provided during the first half of the growing period was superior to red light during the second half of the growing period; this is possibly due to increased activity during the early growing period which increased bone strength, thus reducing leg problems.

4. *Other Effects of Light*

The pineal gland, a small gland located near the top and center of the brain, produces a hormone called melatonin. Periods of light will inhibit melatonin production and periods of darkness stimulate it. Melatonin is an antioxidant that help cells remain healthy by destroying free radicals, and has been shown to enhance the immune response. Therefore, it is believed that intermittent light and dark periods are advantageous for growing poultry for reasons other than simply stimulating and controlling broiler activity.

10-D. LIGHT SOURCES AND INTENSITY

Chickens may be exposed to a variety of light sources and intensities during their lifetime, possibly from bright daylight, to dawn and dusk, and to artificial light. All commonly available artificial light sources can support egg production and growth; however, not all of them appear to be equivalent for producing equal egg numbers. This is possibly due to the interactions of intensity and wavelength. Light intensity may be a modifying factor in how lights with various wavelengths influence flock performance.

If adequate intensity is used in poultry facilities, the perception of day length will be the same when either natural or artificial light (any light source) is provided. Whether short or long daylight periods are provided, the light portion of the day must be of greater intensity than the minimum threshold level for the bird to adequately interpret it as light. Remember, light intensity must be sufficient to create contrasts between the light and dark period of the day. Insufficient light intensity may result from dirty lamps or reduced voltage to the light circuit. In blackout houses, light leakage from outside, when the lights are off, can create a brownout condition resulting in inadequate contrasts between the light and dark periods.

Table 10-1. Relative Light Intensity from Various Light Sources Measured as Footcandles Per Watt, then Compared to the Intensity from a 120 Volt Incandescent Lamp

Lamp Type	Relative Intensity[1]
Incandescent 120 volt	1.00
Incandescent 130 volt[2]	0.38
Vitalight Fluorescent	1.64
Warm White Fluorescent	2.97
Warm White Deluxe Fluorescent	2.03
Daylight Fluorescent	2.55
Cool White Fluorescent	2.69
Cool White Deluxe Fluorescent	1.96
Biaxial Fluorescent (compact, 2700°K)	3.74
High Pressure Sodium (with refractor)	2.20

[1] The values are relative to the light intensity observed from a 120-volt incandescent lamp on a per watt basis
[2] A 130-volt incandescent lamp, sometimes used in poultry houses because of its longer life expectancy, when used in a 120-volt circuit produces only 38% of light as measured in footcandles when compared with an equal wattage 120-volt incandescent lamp

Often, poultry manuals have suggested providing light on a basis of so many watts or lumens per square meter of floor space. This is not an adequate procedure because the light intensity perceived at the level of the chickens is dependent upon more than lumen output. It is also dependent on the height of the lamp above the chickens and the design and color of the interior of the house.

Different lamps have different abilities to emit visible light, even when evaluated on the amount of light per watt. Table 10-1 shows ten different light sources which have been evaluated on their ability to provide light on a per watt basis as measured with a light meter. These findings demonstrate a need for concern when replacing one lamp type with another.

The instrument most commonly used to measure light intensity is the conventional light meter or photometer. A common unit of measure for light intensity is the foot-candle (fc); however, some light meters also use the term lux, an international unit. A foot-candle is a unit of measure of illumination on a surface, and the intensity of light striking every point on the inside surface of an imaginary sphere having a one foot radius equipped with a one candlepower source of light at the center. Lux is the measure of light intensity equal to one lumen per square meter. One foot-candle is equivalent to 10.76 lux. A good light meter should be capable of measuring light intensity as low as 0.1 fc (1 lux) and at least as high as 10 to 20 fc (100 to 200 lux).

When measuring light intensity in floor operations the meter should be held at bird head height and the photoreceptor on the meter should be directed toward the light source(s). Light intensity in cage operations is best measured by holding the meter above the feed trough at bird head height. Since light can be emitted from more than one source in a poultry house, care must be taken when measuring light intensity not to position yourself between the meter and the light sources. The distance between the light source and the photometer is critical when measuring light intensity. When the distance from the light source is doubled, the intensity is reduced to one-fourth of the original value.

The photometer has a light receptor (sensor) which perceives light similar to the human eye (Figure 10-3), but slightly different than the bird's eye. Maximum reception of light energy by the photometer occurs at a wavelength of 555 nanometers (green light) and decreases to a minimum at the two ends (blue and red) of the visible light spectrum. This inability to detect all of the light energy present from a particular light source may not indicate the true amount of energy that is physiologically and behaviorally influencing the bird. Since the visible output of artificial light sources can vary significantly (Figures 10-5, 10-6, and 10-7) the sensitivity of the photometer to these light sources is not the same. While the photometer may detect a similar amount of light energy at the bird's eye, it does not measure all of the energy the bird is capable of receiving through its extraretinal receptors.

Light intensity measurements for different artificial light sources when using the photometer can be misleading, especially when measuring low light intensities near the physiological threshold of the bird. This is especially true with layers and breeders where adequate light intensity is critical. Since absorption of photons (unit of light energy) is required for any light induced effect to occur, the measurement of photons should be con-

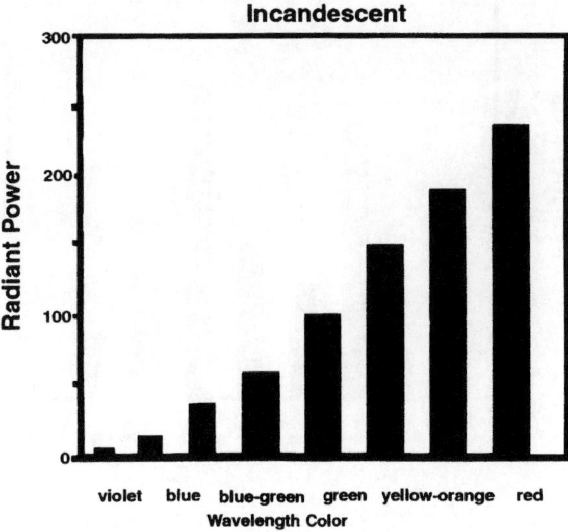

Figure 10-5. Visible Light Spectral Analysis for the Incandescent Lamp. (Note that while wavelengths representing all colors are present there is a predominance of yellow and red wavelengths plus considerable infrared energy beyond the visible light spectrum that produces considerable heat.)

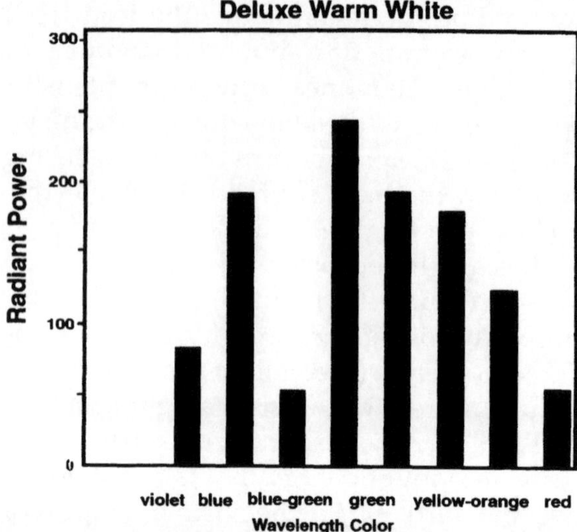

Figure 10-6. Visible Light Spectral Analysis for the Deluxe Warm White Fluorescent Lamp. (Note that wavelengths representing all colors are present but there is predominance of the longer wavelengths which give a slight yellow-white appearance.)

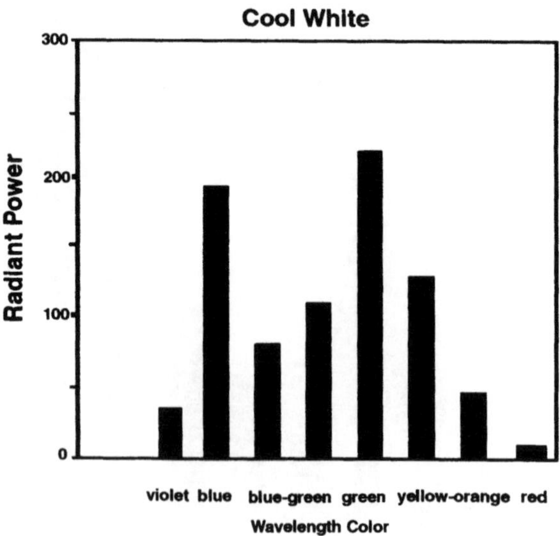

Figure 10-7. Visible Light Spectral Analysis for the Cool White Fluorescent Lamp. (Note that wavelengths representing all colors are present but there is less predominance of the blue to green wavelengths and thus a whiter light appears than in the deluxe warm white.)

Table 10-2. Correction Factors for Various Light Sources to Equalize the Number of Photons Per Footcandle

Light Source	Photons per Foot-Candle[1]
120 Volt Incandescent	0.215
130 Volt Incandescent	0.222
High Pressure Sodium	0.142
Cool White Fluorescent	0.154
Cool White Deluxe Fluorescent	0.202
Warm White Fluorescent	0.152
Warm White Deluxe Fluorescent	0.190
Day Light Fluorescent	0.167
Vita Light Fluorescent	0.196
Biaxial Compact Fluorescent (2700°K)	0.144

[1] These values may be used as correction factor (CF)

sidered. The energy of the light (the photon) which is received by the extra-retinal receptors is not necessarily measured by the photometer or light meter, as the light meter does not receive all of the energy across all wavelengths of light (Figure 10-3). To replace one particular type of light source with another while not changing the energy of light the bird receives, one would have to know the total amount of light energy emitted from the first light source using a radiophotometer and then replace with the second light source of sufficient wattage so as to have equal light energy. The radiophotometer is capable of measuring all of the light energy from a light source, while the photometer or conventional light meter is not. This radiophotometer is more expensive than the photometer and is impractical for most poultry operations. However, as the relationship between foot-candles and photons is proportional and linear, the relationship between light intensity photons ($\mu M/sec/m^2$) and foot-candles (fc) can be determined. Therefore, if a poultry producer is to achieve equal intensity with different light sources, the most sensible unit of measurement is the photon ($\mu M/sec/m^2$). Table 10-2 indicates correction factors for foot-candle readings from various light sources and can be used with the following equation to equalize intensity between two light sources when using a conventional light meter.

$$\text{Fc needed for NLS} = \frac{(\text{fc of CLS})\,(\text{CF of CLS})}{\text{CF of NLS}}$$

where:

NLS = New Light Source
CLS = Current Light Source
CF = Correction Factor (from Table 10-2)
Fc = Footcandle

10-E. LIGHT SOURCE AND ECONOMIC CONSIDERATIONS

Either natural or artificial light sources can be used to provide light for poultry, and are commonly combined to meet bird needs. Natural sunlight, while a common source for chickens, has variable duration, intensity, and wavelengths depending upon location, seasons of the year, and weather conditions. Daylight during the winter season in the temperate zones will provide the shortest duration of natural light and, conversely, during the summer season the longest days of the year (see Table 10-3). This is due to the earth revolving around the sun and the tilt of the earth on its axis in relation to the sun. The closer to the equator, the more uniform the day length, while increasing the distance from the equator toward the poles of the earth results in a more pronounced seasonal variation in day length. Additionally, because of the curvature of the earth, some light is experienced during dawn and dusk which is called civil twilight, the time just before the sun rises over the horizon and immediately after the sun sets below the horizon. Depending on the season, civil twilight typically will last 15 to 30 minutes at sunrise and sunset in the temperate zones.

The intensity of natural light varies considerably depending upon

Table 10-3. Approximate Natural Daylight at Latitudes of the Northern and *Southern* Temperate Zone (Does not include civil twilight)

Latitude	Approximate Location	December 21	March 21	June 21	September 21
15°	Manila, Philippines San Pedro Sula, Honduras	11.23 hours	12.0 hours	13.02 hours	12.0 hours
	Lima, Peru *Lusaka, Zambia*	*13.02 hours*	*12.0 hours*	*11.23 hours*	*12.0 hours*
25°	Miami, FL, USA Riyadh, Saudi Arabia	10.58 hours	12.0 hours	13.70 hours	12.0 hours
	Sao Paulo, Brazil *Pretoria, South Africa*	*13.70 hours*	*12.0 hours*	*10.58 hours*	*12.0 hours*
35°	Raleigh, NC, USA Tokyo, Japan	9.80 hours	12.0 hours	14.52 hours	12.0 hours
	Buenos Aires, Argentina *Cape Town, South Africa*	*14.52 hours*	*12.0 hours*	*9.80 hours*	*12.0 hours*
45°	Minneapolis, MN, USA Lyon, France	8.77 hours	12.0 hours	15.63 hours	12.0 hours
	Dunedin, New Zealand *Camarones, Argentina*	*15.63 hours*	*12.0 hours*	*8.77 hours*	*12.0 hours*
55°	Glasgow, Scotland Edmonton, Alberta, Canada	7.17 hours	12.0 hours	17.37 hours	12.0 hours
	Horn Island, Chile	*17.37 hours*	*12.0 hours*	*7.17 hours*	*12.0 hours*

weather conditions and the type and orientation of housing. If chickens are housed in curtain-sided buildings located in the warmer environments, light filters into the house and the birds can be exposed to varying intensities. Intensity in a curtain-sided house with the sun overhead can be as much as 200 to 400 foot-candles. Intensity increases considerably if sunlight directly enters the house. Cloudy and overcast conditions can significantly reduce light intensity.

Sunlight is a broad spectrum white light. Broad spectrum infers that the light contains all or most of the wavelengths (colors) of visible light. Sunlight, when passed through a prism, will exhibit its component colors, from the longer visible wavelengths of red through orange, yellow, green, to the shorter wavelengths of blue and violet. During the majority of the daylight hours, sunlight will be white (a combination of all colors). However, during the hours near dusk and dawn the sun will appear more red because of the low angle at which the light passes through the atmosphere, which allows only the longer and more penetrating wavelengths to reach the eye.

1. *Lamp Types*

Artificial lights (lamps) can be used as a sole source of illumination for chickens or as a supplement to natural daylight. Artificial lights are used in solid sidewall or curtain-sided houses to stimulate both egg production and growth. There are three points that are important when selecting an artificial light source for the chicken house:

- Artificial lights have different operating efficiencies with regards to electrical energy utilization
- Different warm up periods
- Different lamp life expectancies (Table 10-4).

These factors all influence the operating cost in the poultry house. Additionally, the color of light produced is a result of the length of the wavelength or various wavelengths produced by a particular lamp.

Incandescent lamps possess broad spectrum characteristics that predominantly display the longer wavelengths of the yellow and red end of the visible light spectrum as well as a considerable amount of infrared energy given off as heat (Figure 10-5). Since incandescent lamps are inefficient in converting electrical energy to visible light, they are expensive to operate. Additionally, incandescent lamps have a relatively short lamp life that requires frequent replacement. The advantage of incandescents is their low purchase price and minimal reduction in lumen output as the lamp ages.

Table 10-4. Characteristics of Various Lamps

Type of Lamp	Spectral Characteristics	Warm Up Time	Output (lumens / watt)	Average Life Expectancy
Incandescent	Broad spectrum, but predominately the longer wavelengths (yellow and red) along with considerable infrared energy	Negligible	10–18	1,000 hours or less
Standard Fluorescent	Broad spectrum, but predominately the blue and green wavelengths of visible light	Less than 5 seconds	45–72	9,000 to 16,000 hours
Compact Fluorescent	Broad spectrum, but predominately the blue and green wavelengths of visible light	Less than 5 seconds	35–70	10,000 hours
High Pressure Sodium	Broad spectrum, but predominately the yellow and orange wavelengths of visible light	Less than 5 minutes	52–105	24,000+ hours

Fluorescent lamps used in chicken houses are the tube and compact types (Figures 10-8 and 10-9). These lamps give variable spectral light output depending upon the phosphor used in their manufacture. Generally, the less expensive cool white or warm white lamps are used in chicken houses. While the spectral output of these fluorescent lamps are generally broad spectrum, they can show some predominance for the longer wavelengths as in the deluxe warm white or slightly more of the shorter wavelengths as seen in the cool white types (Figures 10-6 and 10-7). Also, the compact lamps may be rated by temperature with the 5500°K (Kelvin) compact lamp being similar to cool white fluorescent and the 2700°K compact lamp being similar to the warm white fluorescent.

2. *Economic Considerations*

Fluorescent lamps are considerably less expensive to operate and have a longer life expectancy than incandescent lamps. Most can also be dimmed if special equipment is installed in the circuit. Disadvantages are that fluorescent lamps, as well as other high intensity discharge lamps, experience significant decreases in light or lumen output as the lamp ages. Fluorescent lamps, depending upon type, will experience a 10 to 15% reduction in lumens by the half life of the lamp. Also some fluorescent lamps can have starting problems in cold temperatures. Additionally, fluorescent and other high intensity discharge lamps are designed to operate within a prescribed voltage range. If an area is prone to brownouts (reduced volt-

Figure 10-8. Tube Fluorescent Lighting in a Layer House

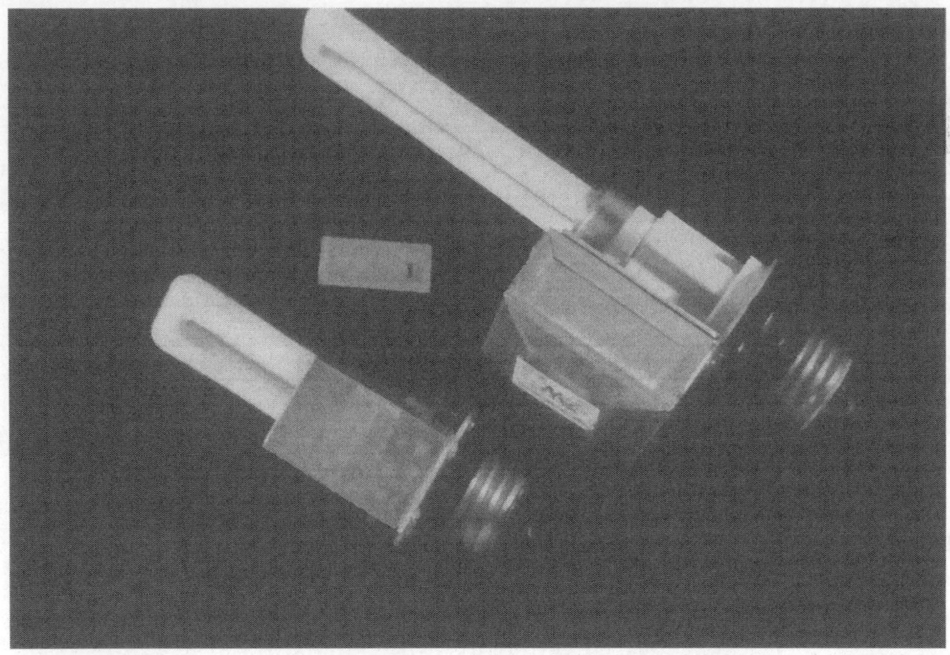

Figure 10-9. Compact Fluorescent Lamps

age) the life expectancy of the lamp is significantly reduced, resulting in increased replacement costs. Fluorescent lamps require a ballast and starter incorporated into the fixture, making them more expensive to purchase and install in a chicken house. The compact fluorescent is usually available as a two-piece unit with the plug-in lamp separate from the base containing the ballast. They are available for two types of installation, a screw-base retrofit for replacing a incandescent bulb (most common), and a box unit containing the ballast which is wired into the circuit, with the compact lamp plugged into the box. The retrofit type with the screw-base has been very popular in poultry houses and their costs have been quite competitive.

Other high intensity discharge lamps such as high pressure sodium have a narrower spectral output emitting predominately yellow-red wavelengths. They are used where higher intensity is desired and are less expensive to operate than incandescent or most fluorescent lamps. Also, because of the high lumen output, fewer high pressure sodium units are needed in a house. The disadvantage of high pressure sodium lights is that they are more expensive to purchase. Additionally, there is generally more variation in intensity of light in the house as the high intensity lamps are normally spaced greater distances (25 feet or 7.6 meters) apart. Sometimes mercury vapor lamps, commonly used as yard lights, have been used in chicken houses because of their low cost. However, they are considerably less efficient to operate than the high pressure sodium lights and have a greater reduction in lumens (approx. 20%) as they age when compared to other lamp types.

When a higher intensity artificial light is desired in a chicken house, the additional cost of the fluorescent or high pressure sodium lamp is usually justified because of its increased operating efficiency over the conventional incandescent lamp. However, if a specific minimum intensity is desired, a scheduled relamping program should be instituted because of the lumen reduction experienced with high intensity discharge lamps. Concern for a desired intensity requires attention to dust settling on the lamps and reducing the amount of light that the birds perceive. Also, lamps may be painted or manufactured with a color coating on the outside so as to emit a single color. These lamps are often used to invoke a particular behavior, but they do not provide as pure a wavelength as a specially manufactured lamp would.

Since lights are typically on for extended periods of time, life expectancy of the lamp is important. Light sources differ in their life expectancies (see Table 10-4). Incandescent lights have the shortest life expectancy, usually 1,000 hours or less. Fluorescent lights generally have a life expectancy of 10,000 to 20,000 hours and other high intensity lamps, such as high pressure sodium, have greater than 20,000 hours of life expectancy. The life expectancy of fluorescent and other high intensity discharge lamps is influenced by the frequency of turning the lamp on and off. Excessive on/

off incidences will decrease the rated life expectancy. Factors affecting intensity of artificial light sources in a poultry house include wattage, lamp type, number of lamps placed in the house, color of the inside of the house, whether light reflectors are used, cleanliness of lamps and the height of the lamps above the birds. Incandescent lamps and certain types of fluorescent lamps are capable of being dimmed so that intensity can be regulated to alter activity of the chickens.

3. *Location and Control of Lights*

Placement of lamps in the poultry house is critical for optimal performance. Typically, artificial lights in a cage layer house are low wattage incandescent or compact fluorescent fixtures located between the rows of cages at intervals that will provide uniform light intensity. Lights in a broiler house are in multiple rows of low wattage incandescent or compact fluorescent also spaced to provide a uniform intensity.

Lights in a typical breeder pullet house are incandescent or fluorescent and are normally spaced in two or three rows the length of the house to provide a uniform light intensity. The breeder house is commonly outfitted with incandescent or the energy-efficient fluorescent or high pressure sodium lights where a higher intensity is desired.

Computers, which are accurate to the minute, and conventional time clocks are used to regulate artificial light and provide the day length needed by the specific class of chickens. Remember if there has been a power outage, clocks generally must be reset, unless there is a backup system installed. Photoelectric cells can be used in open houses to turn lights off at sunrise and on at sunset. Additionally, they can turn the lights on whenever light intensity is below a present level on overcast days.

10-F. LIGHTING PROGRAMS

1. *Rearing: Broiler Breeders and Commercial Layers*

When rearing pullets exposed to natural daylight, the late spring and summer hatched birds (in season) reach sexual maturity as the days become shorter, which is advantageous. However, late fall and winter hatched pullets (out of season) will reach sexual maturity as the days become longer, therefore special precautions must be taken to ensure that birds do not reach sexual maturity too rapidly. Best results can be obtained when the duration of light (length of day) does not increase as pullets reach sexual maturity. Where long or increasing natural day lengths occur, good control of light duration can be accomplished by using light traps over the fans and air inlets to prevent sunlight from entering the house.

When restricting duration of light on broiler breeder pullets, it is critical to use short day lengths (8 hours), with all service and maintenance functions being performed during the times the lights are on. See *Managing the Breeding Flock,* Chapter 34 for specific recommendations for broiler breeder pullets, *Cage Management for Raising Replacement Pullets,* Chapter 51 for commercial egg laying pullets, and *Cage Management for Layers,* Chapter 52 for commercial layers.

2. *Egg Production: Broiler Breeders and Commercial Layers*

Stimulatory light, or an increase in day length, should be given when pullets have attained a suitable age and body weight. Once hens are provided the longer stimulatory day lengths it is important that day length not decrease throughout the production cycle. As natural daylight decreases, artificial light must be provided to maintain a constant day length. It is also critical that sufficient light intensity be provided, which in many instances will require the use of artificial light to supplement natural daylight on overcast days and at dawn and dusk.

3. *Rearing Broilers*

Light programs vary depending upon whether broilers are raised in curtain-sided or environmentally light-controlled houses. Light-controlled houses using only artificial lights provide the opportunity to utilize various lighting programs. See *Broiler Management,* Chapter 43 for specific lighting recommendations for broilers.

Summary

Light can have a variety of effects upon chickens by directly affecting the endocrine system, which in turn affects various organ systems, and indirectly by affecting bird activity. Both natural sunlight and several different types of artificial lights are used to satisfy the birds' need for light. Many different lighting programs have also been developed to help optimize production efficiency for breeders, layers, and growing birds. Research continues to be conducted, and in time, as more is learned, it may be possible to develop elaborate lighting programs for different seasons of the year, as well as for other specific applications. Currently, lighting programs help meet the needs of today's birds and new and different lighting programs may be needed in the future as birds change.

11

Waste Management

by Donald D. Bell

The wastes associated with poultry farming have an increased significance today as we become more aware of the harmful effects of polluting the environment. Today's modern poultry farms, because of their size, have enormous problems associated with the by-products of production. Manure is by far the number one waste problem, and its problems can be due to a number of different issues including disposal, odor, associated nuisances, and water and air pollution. Other wastes include hatchery residue, processing plant offal and waste water, eggshells, and dead birds.

11-A. POULTRY FARM POLLUTION PROBLEMS

All poultry farms have a problem with pollution as the term is defined today. Pressures will be made on farm owners to reduce their pollutants more and more each year. In many regions of the world and in individual states in the US, legislation has been enacted to restrict various types of operations as a direct result of past pollution problems. Much of this effort has focused on the very large intensively operated animal farm.

What constitutes pollution. Pollution is defined as the act of making something impure or unclean. There are various forms of poultry farm pollution:

1. Manure
2. Odors
3. Noise
4. Feathers
5. Contaminated air (dust, gases, and chemicals)

6. Water runoff
7. Insects and rodents
8. Dead birds
9. Hatchery debris
10. Dust from feed manufacturing plants
11. Processing plant wastes
12. Exhaust from internal combustion engines
13. Unsightliness
14. Toxic chemical residues in tissues and eggs
15. Lights

11-B. MANURE PRODUCTION AND DISPOSAL

The manure production from large poultry farms can create a problem of major proportions. In many instances, small farms can be well taken care of, but when tens or hundreds of thousands of birds are on a single site, manure disposal can oftentimes be a problem.

The production of manure, on either a weight or volume basis, has been reported to be from as little as 35% of feed consumption to as much as 145%. Obviously, these measurements are highly dependent upon when the manure was measured after defecation. Research by Ota and McNally (1961) indicated that White Leghorns produce between 0.31 and 0.43 pounds (140 to 195 g) of manure per day. Manure was collected in oil pans to prevent evaporation of water. This amount of manure was 1.45 times the amount of feed consumed. Bell (1971) measured 24-hour production of manure in 18 commercial flocks and found that the average hen produced 0.27 pounds per day (122 g), an amount almost equal to their feed consumption. Patterson and Lorenz (1996) measured manure production in high-rise layer houses and found that the amount of manure removed represented only 35% of the feed consumed, but this represented manure that had dried to 59% moisture and was partially decomposed. For purposes of comparison, the amount of manure produced on a daily basis is assumed to be equal to the amount of feed consumed.

Litter (manure + bedding material) production in meat bird or floor raised pullet flocks varies with the amount and type of bedding materials used per flock, the number of flocks per clean-out, the type and body weight of the birds being raised and water management practices. Patterson, et al. (1998) indicated that chicken meat flocks raised to 44 to 57 days produced between 49 and 57 pounds (22 to 26 kilos) of litter per day per 1,000 birds on an as-is basis. On a dry basis, this is equivalent to 0.71 to 1.23 dry tons per 1,000 birds to 44 and 57 days respectively.

In different regions, the problems of manure disposal vary because of climatic and usage conditions. In major regions of the US and the world, the spreading of manure as a fertilizer is not possible during the winter months because of snow and frozen ground. In other regions, crops are fertilized only in the spring or early summer. Because of this, poultry farms must either time their clean-outs to coincide with weather or crop usage periods or they must have storage facilities. In order to have adequate space for storage, the high-rise poultry house is commonly used, or some form of processing is used to reduce the quantity of manure to be stored.

Nutrient Management-Using Manure to Fertilize Crops

In many areas it is practical to dispose of poultry manure by spreading it on crop land or grassland. But, in other areas, the amount of available land may be limited. Nutrient management is a relatively new term that describes a program of balancing nitrogen and (possibly) phosphorus applications with the needs of a specific crop. It requires nutrient analysis of the soil, knowledge of crop needs, and an analysis of the manure to be applied. Manure applications to crop lands must be in quantities that meet but don't exceed the requirements for optimum production of the target crop. In some regions, disposal of manures must follow guidelines established by local or national governing bodies. In a sense, this means manure disposal "by prescription." Dumping of manure without considering the needs of the crop(s), is prohibited in many areas. Some data are given in Table 11-1 to show the estimated production of manure for different types of chickens.

The manure production estimates in Table 11-1 are based on the daily production of 0.225 lb per White Leghorn and 0.255 lb per brown egg layer (102 and 116 g, respectively) estimates per hen. As manure accumulates,

Table 11-1. Estimated Production of Manure for Table Egg Layer, Replacement Pullet and Broiler Flocks (fresh manure estimates are based upon feed consumption)

Birds (10,000)	Fresh– Av. Tons/day (2,000 pounds)	Fresh– Av. Tons/year* (2,000 pounds)	Dried– Tons/year** (2,000 pounds)
Table egg laying hens			
White egg type	1.13	410.6	136.7
Brown egg type	1.28	465.4	155.1
Replacement pullets (to 20 weeks)			
White egg type	0.54	179.4	59.8
Brown egg type	0.61	200.6	66.9
Broilers			
To 42 days	0.87	237.2	79.1
To 49 days	1.01	287.0	95.6
To 56 days	1.14	332.8	110.9

* Assume 2 weeks down time per flock (no litter)

** Dried to 25 to 35% moisture—note: multiply the figures in the chart by 1.1 for metric tons (2,200 pounds) depending on drying conditions, these weights and corresponding volumes will be lessened as moisture is lost and as decomposition occurs. While approximately 1.8 cubic feet of manure is produced for each hen during the year, natural drying and decomposition will reduce this to less than 1 cubic foot. Of the estimated 0.225 lb produced per day, less than 0.05 lb is dry matter.

11-C. THE IMPORTANCE OF REMOVING THE WATER FROM MANURE

The water in chicken manure is the source of most of its associated problems:

1. More conductive to fly breeding.
2. More expensive to transport because of added weight/volume.
3. Less value on a weight basis when used as a fertilizer.
4. Higher level of odor.

Fresh chicken manure from laying hens is defecated at 75 to 80% moisture. Individual flocks and birds within flocks vary from these standards. Additionally, moisture may be contributed from the drinking system and bird behavior while drinking. Moisture levels below 35% are recommended to minimize fly breeding. This can be achieved by maintaining normal bird densities, directing air movement onto the manure piles, increasing the height of the manure, and stirring the manure (see *External Parasites, Insects, and Rodents*, Chapter 12).

As illustrated in Table 11-1, 10,000 hens produce 410 tons of manure per year, if it is handled daily while still wet. By allowing it to dry to 25 to 35% moisure, only 137 tons will require transportation off the farm to its ultimate disposal site—a reduction to one-third of its original weight. In its original state, it would require 20 truck and trailer loads to remove this amount of manure, whereas in a dry state, this would be accomplished with no more than 7 loads.

Chicken manure has long been used by crop farmers as a fertilizer. It's an excellent source of both organic matter and various needed plant nutrients. In most major poultry areas, manure is available throughout the year in large quantities. In many areas it can be spread during most months of the year, however, in other areas its use is restricted due to cropping patterns and climatic conditions.

One of the principal complaints from users of manure is the uncertainty about the actual nutrient levels in individual deliveries. Manures may vary considerably from house to house and even from load to load making it difficult for the farmer to apply plant nutrients at precise levels. Nutrient

Table 11-2. The Loss of Weight As One Ton of Manure Dries

Percent moisture	80	70	60	50	40	30	20	10
Lbs of dry matter	400	400	400	400	400	400	400	400
Lbs of water	1,600	933	600	400	267	171	100	44
Total weight (lbs)	2,000	1,333	1,000	800	667	571	500	444
% of original weight	100	67	50	40	33	29	25	22
Total wt loss (lbs)	0	667	1,000	1,200	1,333	1,429	1,500	1,556
% water removed	0	42	63	75	83	89	94	97
Est. weight/cu ft	51.9	48.2	44.5	40.8	37.1	33.4	29.7	26.0
Est. cu ft	38.5	27.7	22.5	19.6	18.0	17.1	16.8	17.1
% of orig. volume	100	72	58	51	47	44	44	44

levels in manure or litter vary because of the type of chickens being raised, feed formulation, and the method of handling the manure.

One of the major reasons for this variation is the difference in water content. Fresh manure may contain more than 70% water. As the manure dries, the nutrients are not only concentrated on a weight basis, but also on a volume basis due to structural changes in the manure. Compared to fresh manure, manure with a moisture content of 30% or less has only 50% or less of the original volume.

The mathematics of water removal from manure is an important concept to understand. When fresh manure dries from 80% moisture to 70%, a ton is reduced to 1,333 pounds and to 72% of its original volume. If it is taken to a moisture level of 20%, it will be down to 25% of its original weight and 44% of its original volume. Table 11-2 illustrates these relationships.

Dehydration—Artificial and Natural

Some poultry producers use artificial dehydration to produce a higher quality product, to reduce the volume of manure, and to prevent bacterial activity that results in odor production. There are several types of dehydrators on the market; the temperature created in these varying from 700° to 1800°F (371° to 982°C). The length of the drying time is governed by the drying temperature, the moisture content of the incoming manure, the rate of flow, and the moisture content of the finished product. Most dehydrators will reduce the moisture content of manure from 70 to 10% in less than 10 minutes. The capacity of any dehydrator is rated by the number of pounds of moisture it will remove in 1 hour.

Natural drying (solar) is utilized in parts of the world where rainfall levels are low and drying conditions are suitable. In these regions, cage manure is commonly collected on a daily basis, spread thinly in a drying yard, and windrowed into piles when the drying process is complete. It is common to take manure with 75% moisture to less than 20% in one or two days. In general, the faster manure is dried, the higher the nitrogen content and consequently its value to the farmer.

Some cage systems have manure belts beneath cage rows to remove the manure. In some of these systems, air is directed over the droppings to promote drying on the belts.

Composting Manure

Composting is a natural aerobic, microbiological process in which carbon dioxide, water, and heat are released from organic wastes to produce a stable soil-like, humus-rich product. During composting, ammonia is typically released to the atmosphere, thereby, lowering the nitrogen level of the finished product and creating an odor problem. Composting can also be used to convert cage manure, litter, hatchery wastes, egg shells, and dead birds into a high-value by-product of the poultry and egg production industries.

The composting process consists of:

1. Properly mixing the waste with a carbon rich material (e.g., straw, wood shavings) in bins or in windrows. Carbon to nitrogen ratios of 20–25:1 are usually recommended. Pure manure can also be composted if all factors are carefully monitored.
2. Addition of air by periodic stirring.
3. Proper balancing of moisture levels (35 to 50% moisture).

Figure 11-1. Manure Composting

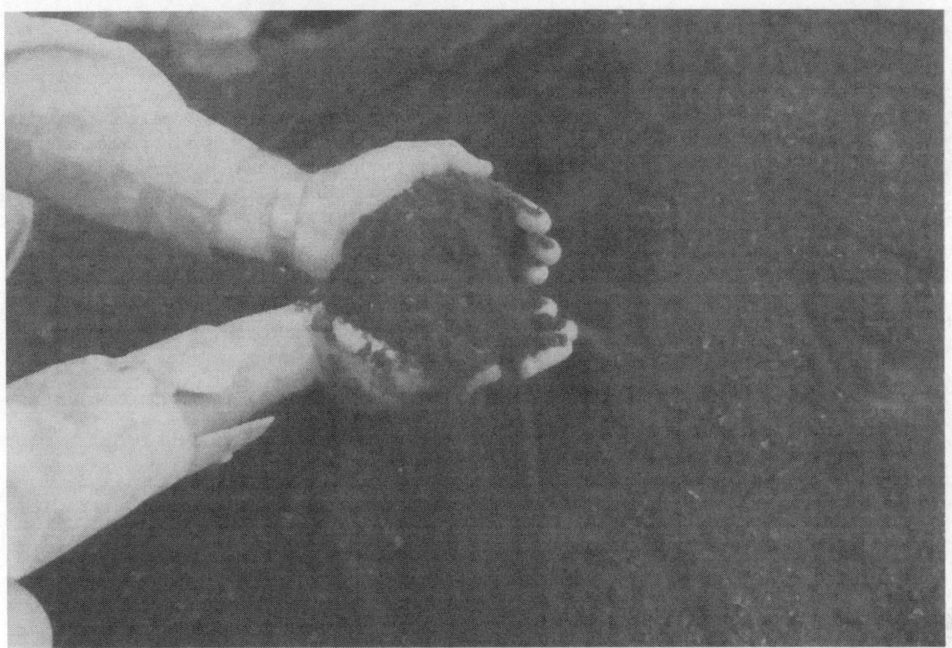

Figure 11-2. Composted Manure

4. Temperature monitoring to determine if composting conditions have occurred.

It should be noted that a true compost requires several weeks to process and that when finished, the material is stable with no further temperature buildups when used or stored.

Collecting poultry manure in pits under cages or slat or wire floors is a common, practical, and economical way to handle poultry wastes where some composting may take place. Other systems incorporate daily or twice weekly removal from the poultry house with belts delivering the droppings to composting units where the pure manure is turned daily for a 3- to 4-week period. In the high-rise system, manure may be allowed to accumulate for several years during which time a considerable amount of composting may occur. Such composting pits have been in operation for several years without manure removal. Afterward, the compost in the pit should be about 2 or 4 feet deep, depending on the stocking density of the hens and the number of years it is allowed to accumulate. The top foot is composed of fresh manure, the bottom foot is in an anaerobic condition, and the central portion is undergoing composting. Such systems produce both a true compost and wet or dry manure because the system is not completely controlled from top to bottom and only works when all conditions are right for composting.

The essential requirement for managing the deep pit house is to ensure that the fresh, wet material is adequately aerated to remove the moisture.

Figure 11-3. Hi-Rise House Manure Storage

To facilitate the composting process and to prevent odors, the pit must be tight so that outside water cannot enter. Care must be taken to prevent waterers from leaking or overflowing into the pit, for such overflow will normally increase moisture to a level that prevents proper bacterial action to occur in the manure. When managed correctly, there is little or no odor arising from the pits, and manure removal may be delayed for years. There is also practically no problem with flies.

Multiple-deck cage systems can also employ scraping devices or belts to remove manure on a daily basis. Some egg producers have combined these removal systems with a manure storage barn that incorporates hot air circulation systems and stirring to help dry as well as compost the manure.

11-D. THE NUTRIENT COMPOSITION OF CHICKEN MANURE

The chemical composition of manure is dependent upon several factors including:

1. the nutrient composition of the flock's diet
2. the flock's age and type
3. the manure collection and storage practices
4. the general house environment.

Table 11-3. Approximate Analysis of Air-dried Poultry Manure

Moisture (%)	Nitrogen (%)	Phosphorus*		Potassium*		Total Salts (%)
		P (%)	P_2O_5 (%)	K (%)	K_2O (%)	
75% (fresh)	1.13	0.74	1.70	0.63	0.76	3.86
35% (moist)	2.36	1.31	3.01	0.98	1.18	4.94
10% (dry)	3.84	2.01	4.62	1.42	1.70	6.18

Source: Bell (1971)
* $P_2O_5 = P \times 2.3$, $K_2O = K \times 1.2$

The quality of manure produced by a farm is highly dependent upon the rate of water removal. The faster this is accomplished, the higher the nutrient levels—especially nitrogen. The manure is simply more concentrated and less ammonia is lost.

For example, relatively dry manure (less than 35% water) would typically contain 65 pounds (30 kg) of nitrogen per ton, while moist manure (35 to 55% water) would contain about 44 pounds (20 kg) and wet manure (over 55% water) would contain only 27 pounds (12 kg) of nitrogen. A farmer who needs 65 pounds of nitrogen would have to purchase 2.4 tons of wet manure versus 1 ton of dry manure to acquire the same nutrient levels. Table 11-3 lists the approximate nutrient composition of caged layer manure with three different moisture levels naturally air-dried.

11-E. THE VALUE OF POULTRY MANURE

While generally poultry manure is considered a waste or by-product of poultry production, it is one that has considerable value as a fertilizer and feed nutrient for ruminants. To ensure the producer receives its value, its worth must be communicated to the buyer, it must be a consistent recognizable product, and it must not have any harmful ingredients or undesirable characteristics. A well-cared-for product can have a total nutrient value in excess of $25 per ton when used as a plant fertilizer and $50 per ton when used as a feed ingredient for cattle. Also, it is well known that when manure is used as a fertilizer, it has additional value associated with its organic properties.

Users complain that poultry manure is too wet, it's not a uniform product, it has too many feathers and weed seeds, it contains toxic chemicals, it gives unpredictable responses on their crops, it's too lumpy and does not apply evenly, it contains immature flies which may hatch and cause problems, it smells, it has the wrong balance of nutrients, and it burns plants. Even though the use of manures may be a very cost-effective way of fertilizing crops, the many negative factors associated with its use reduce its popularity and therefore must be addressed. Practically, all of these complaints can be solved with a good quality control program.

Manure As Poultry and Animal Feed

The fact that poultry manure contains many feed components that pass through the digestive tract without being digested and numerous by-products from metabolism, such as non-protein nitrogen for ruminants, suggests that it should have nutritional value if recycled through other animals, including poultry. Before it is used as an animal feed, care must be taken to determine whether or not this use is legal. Secondly, it must have the proper processing to standardize the product at its highest obtainable nutrient profile. Finally, contamination with pesticides, feed medications, herbicides, and foreign objects (metal, glass, soil) must be avoided.

Chemical analysis. Analyses of dried manure will vary with the age of the bird, the age of the sample, the condition of storage and handling, and the type of chicken involved. Typical cage layer manure (pure manure) should be similar to the following analysis:

Ash	26.9%
Crude fiber	13.7%
Crude protein	23.8%
True protein	10.6%
N-free extract	39.6%
Ether extract	2.1%
Calcium	7.8%
Moisture	7.4%

Source: Michigan State (1970)

Researchers have studied the nutritional value of dried poultry manure in various classes of livestock and poultry. Because of its relatively high crude protein values, dried poultry waste has become an alternative feedstuff for non-lactating ruminants. Experiments with monogastric animals, including chickens, have generally proven to be economically marginal because of the relatively low percentage of true protein and because of its high ash content.

11-F. MANURE HANDLING

Poultry manure on the farm is generally handled in one of the following ways:

1. Manure (without litter)—produced by birds in cages or under wire or slatted floors.
2. Litter—combined with various bedding materials.
3. Liquid—combined with large quantities of water.

Manure without litter is primarily a product of the table egg industry. Its estimated that 99+% of all laying hens in the US and 70 to 80% of the world's layer population are housed in cages along with more than 80% (US) of their replacements. Manure is handled in a variety of ways including daily or bi-weekly removal from the house by scraper or belt, periodic removal every 3 to 6 months, and annual or longer removal in high-rise houses.

In a 1991 survey of manure handling systems in the US table egg industry, it was projected that by the year 2000, an estimated 28% of the manure on commercial farms would be handled every week or oftener in a dry form (no water added), 51% would be handled infrequently in a dry form, and 21% would be handled in a liquid form (water added). Since the survey was completed, almost all new farms have adopted one of the dry manure systems—high-rise or manure belts. It was also predicted that by the year 2000, approximately 43% of the manure would generate income for the farm compared to only 28% in 1991. The manure handling programs on new farms are the result of public pressure to address manure handling issues before new farms or houses are approved. Equipment companies have responded by incorporating novel in-house manure drying and handling systems into new house designs.

Litter systems are used almost exclusively for the rearing of broilers, breeders, and some replacement pullets. The system allows birds free access to the litter. Litter is composed of wood shaving, rice hulls, chopped straw, or similar materials to provide absorbent bedding for the flock. Because of rising costs, many growers choose to replace litter only after several flocks have been grown instead of the more desirable once per flock system. Litter is removed completely or partially using tractors with front-end loaders or specialized equipment designed to remove only the wetter surface (caked) manure.

Liquid Systems Are of Two Principal Types

Scraping into a liquid manure storage tank. Some farms use a liquid holding tank for their manure. This system incorporates an underground tank for temporary storage and tanker trucks or tank trailers for delivery of the liquid manure to neighboring farms. Long-distance transportation is uneconomical because of the large amount of water added to the manure. In most cases, disposal from the poultry farm consists of application to nearby fields.

Wash-out systems into lagoons. Fresh poultry manure is collected in a concrete trough under the cages and is flushed into an open shallow pond known as a lagoon. Bacterial action reduces the quantity of waste material. As bacterial growth occurs only during the warm months, the use of lagoons is more common in warmer climates.

Figure 11-4. Cage Manure Belt System
(*courtesy of Chore-Time*)

Figure 11-5. Truck Loading from Manure Belt System

When aerobic action takes place, the lagoon produces very little odor; as the sludge builds up, anaerobic activity can take place and odors can become pronounced. Modern installations commonly recycle the water through the poultry houses on a daily basis.

11-G. OTHER PROBLEMS ASSOCIATED WITH MANURE

Wet droppings. High levels of protein and salt in the ration can be responsible for increased amounts of moisture in the droppings. Certain strains of chickens and high egg production rates are also associated with wet droppings. As environmental temperatures rise, birds drink more water, thereby increasing fecal moisture. Leaky and improperly installed foggers and watering devices can also contribute to excessive water in the dropping collection area. Some producers monitor their drinking systems with computers with built-in alarm systems sounding when water usage is more than planned.

Odor. Fumes from the ammonia in the droppings plus those created by bacterial action can be obnoxious. Further, odors are accentuated when droppings are wet. Some odors may be materially reduced by using various commercial products on the market. Regular removal of the manure below cage floors should be made a part of the management program. Farms with frequent manure removal systems gener-

Figure 11-6. Truck Being Loaded with Front-End Loader

ally have minimal odor problems. In pits, exposure to air and the use of fans will also help dry the droppings.

11-H. OTHER WASTE PROBLEMS

Most poultry farms are plagued with other waste problems and a management program must be established to care for these as well. Farms have to dispose of their dead birds each day, egg processing plants must have a waste water plan and a system for getting rid of rejected eggs and broken eggshells, and rain run-off from the roofs of buildings must be handled in a non-polluting manner.

The Disposal of Dead Birds

Dead bird disposal has become a major problem in the poultry industry as farms have become larger and as some of the older methods for disposing of birds are no longer environmentally acceptable. In addition to the need for disposal of normal mortality, occasionally, entire flocks may require disposal as a result of farm power failures, disease emergencies, or the inability to sell fowl through normal market channels. Traditional methods of disposal include burial, disposal pits, incineration, rendering, and composting.

General principals of dead bird management include:

1. Dead birds must be removed from the cages and the poultry house daily.
2. Keep the holding containers covered at all times to prevent contact with flies and other insects, dogs, cats, predatory animals, and free-flying birds.
3. Maintain the holding site in an isolated area of the farm.
4. After dead birds are handled or shipped, wash your hands, disinfect the facility and its equipment, and change to clean clothing.
5. Don't allow dead bird pick-up trucks or personnel to visit any area on the farm except the holding facility.

A dead bird disposal system must be able to process "normal" daily mortalities with some extra capacity for seasonal or flock-to-flock variations in mortality rates. The program must also be operable year-round. Alternative backup systems must be available for major flock losses due to disease, weather, or other emergency situations.

Table 11-4 lists the capacity requirements for different types of birds and different mortality rates. Multiply these numbers by the appropriate multiplier factor to determine the requirements for different farm sizes, e.g., a 100,000 bird White Leghorn flock dying at the rate of 0.2% per week would require a disposal system capable of handling 114 pounds (52 kilos) of dead birds per day.

Table 11-4. Daily Production of Dead Birds in Pounds/kg at Different Mortality Rates

	Rate of Mortality					
	0.10%/week		0.25%/week		0.50%/week	
Birds (10,000)	lb	kg	lb	kg	lb	kg
Table egg laying hens						
White egg type 4 pounds (1.8 kg)	5.7	2.6	14.3	6.5	28.5	13.0
Brown egg type 5 pounds (2.3 kg)	7.1	3.2	17.8	8.1	35.6	16.2
Replacement pullets (20 weeks)						
White egg type 3 pounds (1.4 kg)	4.3	2.0	10.7	4.9	21.4	9.7
Brown egg type 3.5 pounds (1.6 kg)	5.0	2.3	12.5	5.7	25.0	11.4
Broilers						
5 pounds (2.3 kg)	7.1	3.2	17.8	8.1	35.6	16.2
6 pounds (2.7 kg)	8.6	3.9	21.5	9.8	43.0	19.5

Burial is still a system commonly used worldwide, but the large size of modern commercial farms has made this system less feasible. Still, surveys in 1991 in the US showed that burial systems accounted for almost 50% of the systems in use at that time. Interestingly, this was expected to drop to about 30% by the year 2000. Society is concerned with possible contamination of ground water supplies, odor and nuisance insects. Shallow burial is generally unacceptable. Deep burial can be accomplished using rotary earth augers (commonly used to dig cesspools) to dig pits 30 to 48 inches (76 to 120 cm) in diameter and 30 to 40 feet (9 to 12 meters) in depth. Care must be taken with this type of burial system not to dig into the water table so as to avoid polluting the ground water supply.

Disposal pits are another type of burial system that are usually only 10 feet deep (3 meters). This system employs continuous bacterial action to break down the soft tissues of the dead birds. Decomposition will occur at a faster rate if the birds are chopped into smaller pieces before they are added. These pits and the deep burial pits are usually equipped with wood or concrete tops with fly-proof openings through which the carcasses can be dropped.

In general, disposal pits should be located on naturally high ground and at least 200 feet (60 meters) from dwellings and the nearest well, 300 feet (90 meters) from any flowing stream or public body of water, and 25 feet (8 meters) from the nearest poultry house. Pits should have a capacity of 50 cubic feet (1.4 cubic meters) for each one thousand bird mortality during the year.

Incineration is an excellent method of disposing of dead birds, but in most cases must be approved by local government because of the potential for air pollution. Incineration is a biologically secure system and one that does not create water pollution. The ash is easy to dispose of and it does not attract rodents or other pests. Its disadvantages include its slowness and cost of operation. If improperly located, unpleasant odors may result in complaints from downwind neighbors. In 1991, incineration accounted for 13% of the dead bird disposal and projections to the year 2000 estimated that only 7% of the layers would be incinerated.

Rendering is a commonly used method of dead bird disposal in some regions. It is estimated that approximately 35% of the mortality produced on commercial egg farms in the US are disposed of in this manner. Rendering is an excellent way to recycle dead birds by converting the carcasses into animal by-products, a feed ingredient. It is only feasible if there is a local rendering plant close enough for convenient and frequent pickup.

Rendering is probably one of the best methods of dead bird disposal because the entire system requires very little investment as its operation is done by an off-site service company—the renderer. With proper restrictions and sanitary precautions, it can also cause minimal risk. Some producers have built refrigerated holding rooms for dead birds and some rendering companies have provided freezers to the farm at no cost.

Composting is a very popular system employed mostly by the broiler industry. The size of farms is such that the use of a small on-site composter is a very practical method of dead bird disposal. Composting, as described earlier in this chapter, is environmentally sound, and if properly done, does not cause odors or water pollution. The final product is useful and the process is relatively inexpensive.

The key elements required for composting daily mortality include:

1. A proper mixture of smaller and larger particle sizes to obtain an optimum air exchange within the mixture and a buildup of temperature.
2. Moisture content of the composting pile should be approximately 60%. More than this may result in odor problems and less than this will reduce the efficiency of the composting process.
3. Carbon and nitrogen are vital nutrients for the growth and reproduction of bacteria and fungi. The carbon-to-nitrogen ratio must be in the range of 20:1 and 25:1 for proper composting. This is obtained by carefully balancing the dead bird and carbon sources.
4. The optimum temperature for composting is 130 to 150°F (54 to 66°C). If temperatures fall below 120°F (49°C) or rise above 180°F (83°C), the compost pile should be aerated or mixed immediately. Failure to do so will result in a poor compost.

The preparation of the mixture of dead birds, carbon source, and water involves a layering of ingredients (manure, then straw or another carbon source, then chickens, then manure—then repeat). Layers of material should be no more than 6 to 8 inches (15 to 20 cm) in depth. The dead birds should be layered only one bird deep. A recommended "recipe" is listed below:

Dead birds	1.0 parts by weight
Litter or manure	1.5
Straw	0.1
Water	0.2

Adding water may not be necessary, as too much water may result in an anaerobic condition, resulting in odors.

The principles of composting for dead bird disposal are not limited to

Figure 11-7. Dead Bird Composter—Broiler Farm

smaller farms. Large scale composting systems have been developed and used. Composting has also been used in emergencies where tens of thousands of birds require disposal. In such situations, windrow techniques are employed.

Detailed descriptions of these processes can be obtained from University Extension offices in most major poultry producing states.

Disposing of Reject Eggs and Egg Shells

Egg processing plants must reject all eggs with blood spots, adhering dirt or stains, leakers, and rots. These rejected eggs may represent as much as 2% of the entire production of a plant. A plant processing eggs from a one million hen complex will reject an estimated 16,000 eggs per day. This represents the disposal of 1,000 pounds (455 kilos) of eggs per day. In the US, if this product is to be accumulated, it must be first denatured and identified as an inedible product. Most commercial plants sell inedibles for use as pet food.

Eggshell disposal for egg breaking plants can be a major problem since 30% of all eggs produced in the US are broken out for use as products. In breaking plants, approximately 11% of the original product (the shell and membranes) remains as a waste. This is estimated to result in a 250 million pound (114 thousand metric ton) annual disposal problem in the US—or

1 pound (454 g) per average laying hen. Much of this waste is destroyed and buried in landfills, but some producers are converting it into high-calcium animal feed ingredient products.

Water Pollution Problems

A major problem associated with water pollution around poultry houses is with runoff from the site which has come in contact with manure, dead birds, or other contaminants. Simple rain water runoff can be managed with appropriate channeling, but if it comes in contact with animal wastes, it must not be allowed to flow into natural waterways.

To avoid this, manure must be protected from rain by proper cover. Field storage areas must have surrounding berms to prevent runoff. Manure piles should be stacked in such a way as to minimize exposure to the rain.

Processing plants, egg and poultry, can also be major sources of water pollution. High concentrations of phosphorus (from detergents) and organic compounds are present in egg wash waters and proper treatment and disposal is essential.

In today's society, all waste problems are potential community problems, and as such, are generally regulated by local or regional governments. Regulations vary between regions, and producers and processors must be fully aware of the restrictions and abide by them in every aspect of the operation. A full understanding of these regulations is necessary before a new facility is planned and a written agreement describing methods of compliance may be required. The technology outlined in this chapter can be utilized to minimize problems associated with poultry farm and processing plant pollution.

12

External Parasites, Insects, and Rodents

by Douglas R. Kuney

Each year external parasites, insects, and rodents cost the poultry industry millions of dollars, although the full extent of their economic impact is unknown. Some of these costs are direct, e.g., money spent by the industry for their control, while others are indirect, e.g., decreased flock performance or structural damage. All three types of pests have the potential for causing or transmitting diseases to poultry and some can present a threat to public health. Control of these pests, therefore, is not only important from a poultry health and production standpoint but also for public health reasons.

All pest control programs should have the following components:

- Pest identification
- Pest population monitoring
- Record of control actions taken
- Evaluation of control action effectiveness

Monitoring pest populations can help the poultry producer gain control over an infestation before populations become extreme, can result in reduced use of chemicals, and can save money. The ability to identify the pest and a thorough understanding of pest behavior are keys to good monitoring. Keeping records of actions taken and the effectiveness of those actions will help identify those methods that work efficiently, and may alert the producer to impending problems such as the development of pesticide resistance or pest behavior changes.

Most pest control programs include the use of toxic chemicals (pesticides) coupled with a variety of cultural and biological control methods. Cultural control refers to management practices (other than chemical or biological) such as manure stirring and frequent manure removal, that reduce or alter pest populations.

12-A. EXTERNAL PARASITES

External parasites are parasites that spend some or all of their time on the bird and feed on the surface of the chicken. Most cause little direct damage to the bird in low to moderate numbers, but if present in large enough populations may cause reduced performance (e.g., egg production, growth, and feed efficiency) or even death.

1. *Mites*

Mites are free-living external parasites belonging to the families Dermanyssidae and Macronyssidae. They are quite small, approximately (0.4–0.7 mm) in size. While on the bird, they feed by sucking blood. Heavy and chronic infestations may cause anemia in some chickens. Most mites can live for a few days to several weeks off their host, which makes their control difficult. There are several species of mites that can parasitize chickens. The three most common mites are:

- Red chicken mite *(Dermanyssus gallinae)*
- Northern fowl mite *(Ornithonyssus sylviarum)*
- Tropical fowl mite *(Ornithonyssus bursa)*

The *red chicken mite* is found worldwide and is a serious problem in warm temperate zones in older style poultry houses where roosts are used. The life cycle of the red chicken mite can be completed in as little as 7 days and they have been reported to survive as long as 34 weeks without a blood meal. The red chicken mite is most active during the summer and relatively inactive in cold houses during the winter. During the day, the red mite can be found hiding in cracks and crevices of the poultry environment and at night on the chicken where it takes it's blood meal. This mite has the capability of transmitting fowl cholera.

The *northern fowl mite* is the most important and common mite in caged layers in the US. This mite can complete its life cycle in less than one week on the chicken. Contrary to the red chicken mite, the northern fowl mite is usually most active during the winter and spends its entire life on the chicken. Mites do move off the chicken occasionally and can spread to different locations within the chicken house. Chickens caged singly are likely to have heavier mite infestations than groups of birds caged together.

The *tropical fowl mite* is most prevalent in warm regions of the world.

This mite closely resembles the northern fowl mite. Like the northern fowl mite, it can complete its entire life cycle on the chicken.

There are four main ways that mites can be introduced onto the farm.

1. Infested started pullets
2. Transport cages or racks used to carry infested birds
3. Personnel, crates, egg flats, or equipment
4. Wild birds

Control of Mite Infestations

Heavy infestations, once established, can be difficult to control. The best control programs, therefore, focus first on prevention. Replacement birds brought to the farm should be free of mites. Equipment used in transporting birds should be thoroughly washed as well as the poultry house where the birds will be housed. Whenever possible, wild birds should be excluded from the poultry facility.

Periodic monitoring for fowl mites is recommended to avoid populations becoming high enough to reduce production and become more difficult to control. Spot checking birds in different areas of a house is done by examining the vent area of several chickens in each area.

There are three classes of pesticides that can be used to control mites on birds and equipment.

1. Carbamates
2. Organophosphates
3. Pyrethroids

When applying pesticides to chickens to control mites, best results are achieved by spraying early before heavy infestations have occurred. Spraying of laying hens should be done after the last egg collection to avoid contamination of the eggs. When applying a pesticide to caged layers, a high-pressure liquid spray should be directed to the vent area of the birds, wetting the feathers to the point of runoff. Care must be taken not to contaminate feed, water, or eggs with pesticides. Floor birds can be provided dust boxes filled with a pesticide powder, which is approved for that purpose. Spraying of birds on the floor is oftentimes ineffective due to the difficulty of getting good coverage of each hen.

Modern layer cage configurations with multiple cage tiers and in some cases manure belts, may present difficulties when applying pesticides due to limited access to the cage floor with spray equipment. Systems of this type must rely heavily on preventive measures to control mites.

Mites can readily develop resistance to pesticides. Recent studies have found some mite populations have developed extremely high levels of resistance to certain pesticides. Frequent and indiscriminate use of pesticides can significantly contribute to the development of resistance as well as harm the environment.

2. *Ticks*

Ticks and mites belong to the same order, Acari. Unengorged adults range from 2 to 4 mm in size, while engorged adults can be more than 10 mm. Fowl ticks that live in poultry houses have soft bodies and belong to the family Argasidae.

Fowl ticks of the genus *Argas* are widely distributed and have been found on chickens and other fowl throughout the Americas, Europe, Africa, and Australia. Their distribution is thought to be nearly worldwide.

Ticks can cause damage directly to chickens by inducing anemia, which can sometimes be fatal, or at a minimum, cause slow growth and loss of production. They can transmit several diseases in chickens including spirochetosis *(Borrelia anserina)* and fowl cholera *(Pasteurella multocida).*

The complete life cycle of fowl ticks normally takes between 7 and 8 weeks. The life cycle includes an egg, larval, two nymph, and adult stages. Blood feeding occurs during the larval, nymph, and adult stages. Nymphs feed only at night, while the adults will feed both during the day and night. Ticks stay on the host for only a short period of time during their blood meal and then leave the host to hide. Adult ticks can survive for years without a blood meal while hiding in cracks and crevices of the poultry environment.

Control of Tick Infestations

Control requires treatment of the premises with an approved pesticide. Litter, floors, walls, and ceilings must be sprayed thoroughly forcing the pesticide into cracks and crevices. Ticks are rarely a problem in modern poultry houses with cages that are supported by metal materials, or in houses constructed of metal.

3. *Lice*

Lice are common external parasites of chickens and other birds. They have chewing mouth parts, antennae, flattened bodies, and no wings. There are several species of lice that have been reported on chickens. They appear as straw-colored insects crawling rapidly on the skin and feathers of the bird around the vent area, the undersides of the wings, the legs and the head. The adult louse is between 4 and 5 mm in length. More than one species of lice may be found on the same chicken.

Lice spend their entire life cycle (3 to 4 weeks) on the bird. Eggs take from 4 to 7 days to hatch and can be found in clusters attached to the feathers. The adults can live several weeks or perhaps months on the bird, but when off the bird will survive for only a few days.

Reports on the effect of lice on bird performance have been mixed. A few reports suggest that heavy infestations are related to loss of egg production. Reports from other studies have indicated that there is no effect on performance. There is clinical evidence that lice irritate the chicken's dermal nerve endings and therefore may interrupt sleep.

Control of Louse Infestations

Control for all species of lice is the same. The primary control method should be exclusion by preventing louse-infested birds from coming into the flock. Applying pesticide dusts to the litter that are approved for louse control is the best treatment for floor reared birds. Sprays are best for caged birds. Birds should be inspected for louse infestations during the fall and winter months when infestations are most common.

4. *Mosquitoes*

Although not a prominent parasite of chickens, mosquitoes have been included here because of their importance in transmitting fowl pox. Mosquitoes lay their eggs in standing water. Larval and pupal stages develop in water. Adults emerge and mate before they seek a host for their first blood meal. Their life cycle is completed in from 7 to 14 days during warm weather.

Control of Mosquitoes

Control is best achieved by preventing mosquito development, with particular emphasis on standing water. Draining undesirable swamps, ponds, or especially any containers that may collect standing water such as old discarded tires will eliminate developmental sites. These actions must usually be approved by local authorities. Tall vegetation near birds can harbor mosquitoes and should be mowed.

There are three species of mosquitoes that commonly transmit fowl pox to chickens:

- *Aedes stimulans*
- *Aedes aegypti*
- *Aedes vexans*

Vaccination programs for fowl pox should be considered in regions where heavy mosquito populations occur.

5. *Other External Parasites*

There are other external parasites that may be of economic importance to commercial chicken operations in certain regions of the world. These include several biting insects:

- Bedbugs
- Bird bugs
- Conenose bugs
- Fleas
- Biting midges
- Black flies

In general, biting insects have the potential for transmitting disease to chickens. However, some are not known to be involved in pathogen transmission (e.g., fleas and bedbugs).

12-B. NON-PARASITIC INSECTS

These insects, primarily flies and beetles, are important because they can vector disease, cause structural damage, or pose a nuisance to neighboring communities. Proper identification of these pests is important to their control. Most successful insect control programs rely on a combination of cultural, biological, and chemical control practices. Cultural control refers to management practices (other than chemical or biological) that reduce or alter fly breeding conditions in a way that minimizes or eliminates adult fly emergence and attraction. Biological and chemical control methods focus on the direct reduction of immature and adult stages through the action of the fly's natural enemies (mainly parasites and predators), and insecticides, respectively.

Control programs for flies and beetles depend mainly on manure and litter management. In general, good farm sanitation helps to reduce pest populations by reducing harborage, breeding sites, and food sources. In many cases, farm sanitation and manure management alone do not satisfactorily control these pests, and periodic pesticide applications may be necessary.

1. *Flies*

The potential impact of flies on human and animal health has always been a concern. Most flies are able to transmit causative agents of several

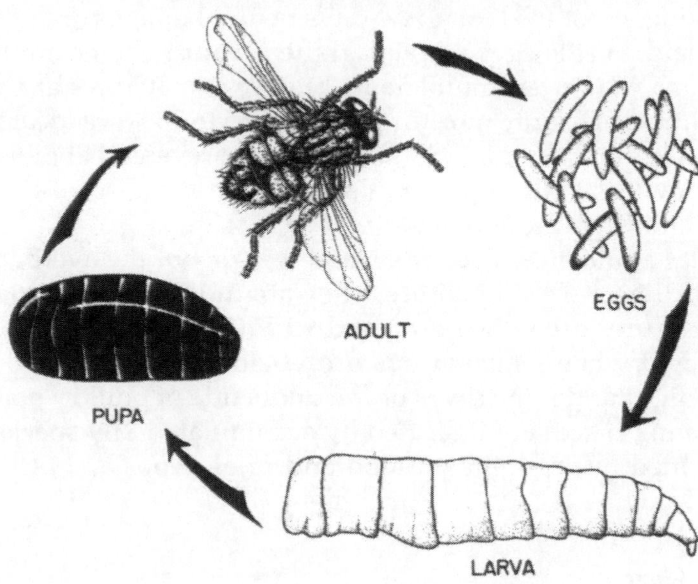

Figure 12-1. Life Cycle of the Fly

human and animal diseases under experimental conditions. However, in developed countries the actual role of flies as vectors of human disease is probably minor. Epidemiologic studies of the US during the velogenic viscerotropic Newcastle disease (VVND) outbreak in chickens in the early 1970's suggested that flies may have played a role in the transmission of the virus between flocks. Today, concern about flies is mainly due to their nuisance characteristics.

While there are several kinds of flies found in and around poultry houses, *Musca domestica* (common house fly) is the fly most commonly associated with nuisance complaints and is found worldwide. Studies have shown that *M. domestica* can transmit the causative agents of several poultry and human diseases and can serve as the intermediate host for the common tapeworm in chickens. Even though small numbers of flies may be capable of movement up to several miles from their breeding site, most are found no more than a few hundred yards from their source.

All flies pass through four life stages: egg, larva, pupa, and adult (Figure 12-1). Female flies deposit small, white elongated eggs on moist decaying organic material, where larvae develop. Mature larvae crawl out of this material and move to a drier place for pupation. Following pupation, the adult fly emerges from the pupal case and seeks a location to expand and dry its wings in order to fly. The mature adult then seeks food and proceeds to mate. On the poultry farm, adult flies feed on a wide range of materials including manure, decaying organic material from animal and plant sources, broken eggs, and spilled feed. They use this for their own nutritional needs and for the developing eggs. A newly emerged female fly, after mating, must feed for a few days before her eggs are mature and ready to be laid. Fertile eggs are laid in the manure, thus completing the life cycle. *M. domestica* can complete its life cycle in 10 days at a temperature of 85°F (29°C). Manure moisture levels of 65 to 75% are ideal for larval development, but development is hampered or prevented at moisture levels much below 60%. Managing manure and litter moisture is critical in the prevention of fly development.

M. domestica adults may live for several weeks, but the average fly usually lives less than a week in nature. They are most active during the day when temperatures are between 80 and 90°F (27 and 32°C), and become inactive at night when temperatures drop below about 50°F (10°C). Flies often rest at night under the eaves or the underside of poultry house roofs. Preferred resting places are indicated by accumulated "fly specks" which are spots formed by regurgitated food and fecal deposits.

Control of Flies

An integrated program can best achieve control by reducing conditions that attract flies or conditions favorable to fly breeding and development. The most effective and efficient programs integrate cultural control methods with biological and chemical control methods. In poultry systems, manure management is critical. It should be remembered, however, that nearly all moist organic materials can support fly development, including practically any decomposing, moist animal or plant material. All significant sources of fly breeding habitat must be managed for a program to be effective.

Cultural Control is centered mainly on manure management. Manure management is an important farm practice requiring as much attention as other routine management chores. The two general systems used for managing manure are frequent clean-out and buildup systems (see *Waste Management*, Chapter 11). The choice of system to use depends on housing design, ventilation, proximity of neighbors, availability of space and equipment, and the ultimate end use of the manure.

Frequent clean-out (from 1- to 7-day intervals) is a popular practice where rapid under-cage drying is impractical. New designs for laying houses, equipment (manure belts and scrapers), clean-out machines, liquid systems, and methods of manure processing make frequent clean-out economically feasible on many poultry farms. For effective fly control, manure should be cleaned out frequently enough to prevent adult fly emergence. If manure is to be solar dried on the farm following removal, it should be removed frequently enough to prevent late instar larval development and pupation. Late stage larvae, and especially pupae, are less susceptible to the lethal effects of sunlight and drying. *Musca domestica* achieves its highest populations during the warm summer period, and under ideal conditions, can reach late instar stages within 4 to 6 days following oviposition. Under these conditions, manure should be removed and immediately dried every 3 to 4 days.

Buildup systems are used to allow manure to accumulate in the poultry house for extended periods of time. Buildup systems rely on a combination of maintaining low adult fly populations, increasing the biological activity of fly predators and parasites, and ventilation to enhance manure drying to discourage fly oviposition and prevent the emergence of adult flies. A 4- to 8-inch deep (10- to 20-cm) pad of dry manure may be left under the cage rows following the eventual removal of the manure to encourage drying and coning of subsequent droppings and to aid in the reestablishment of beneficial mites and insects. The merits of this practice must be examined as it may affect the success of a cleaning and disinfectant program used to reduce disease exposure between successive flocks.

Biological Control, or the suppression of pests by their natural enemies, can be an important component of a complete management program for keeping nuisance flies at low levels. It is an attractive tool due to its lack of adverse environmental impact, its specificity for flies, its often low cost, and its potential for long-term effectiveness. Large populations of predators and parasites can have a suppressive effect on house fly populations, eliminating 90% or more of *M. domestica* before they emerge as adults. Efforts, therefore, should be made to conserve natural populations of beneficial insects that are present in the manure. To take advantage of biological control, operators should reduce or eliminate practices (such as the use of broad spectrum pesticides applied directly to the manure) that prevent the survival of beneficial insects. If pesticide applications are required, their impact on beneficial insects can be minimized by:

1. using fly-selective insecticides (growth regulator-type activity)
2. spot-treating wet places only
3. avoiding long-term and frequent use of larvicides.

Well-established populations of parasitic wasps can sometimes withstand occasional larvicide applications if part of the parasite population is developing within the protective fly pupal case at the time of the application. Other natural enemies, such as predaceous mites and beetles, are common in surface manure and are susceptible to even single applications of broad-spectrum larvicides. Applications of pesticides for control of northern fowl mites also may adversely affect natural enemies in the manure beneath the hens, if excess runoff occurs.

Chemical Control can complement the other components of an integrated program, and is sometimes necessary even in a well-managed poultry operation. Producers should monitor fly populations regularly in order to evaluate the fly management program, to decide when insecticide applications are required, and to determine which chemicals are most effective. Accurate records should be kept on formulations and rates of chemicals used, the name of the applicator, and the date of the application. The rotation of different insecticide types—carbamates, organophosphates, pyrethroids—can help delay or minimize the development of resistance. Most fly insecticides are toxic to predators and parasites and indiscriminate use may severely affect populations of non-target beneficial insects as well as contribute to fly resistance to the pesticide. However, applying insecticides carefully to fly resting sites or the use of insecticidal baits will directly target adult flies and spare their natural enemies.

There are several flies in addition to the common house fly that can be associated with poultry production and most if not all, can be controlled by good farm sanitation:

- Little house fly (two common species)
- False stable fly
- Blow fly (three common species)
- Flesh fly
- Black garbage fly
- Drone fly
- Soldier fly

2. *Beetles*

Several beetles (order Coleoptera) can be found in the litter and manure of poultry. Some are considered pests because they can cause substantial structural damage to wood structures and most insulation materials. Other beetles are considered beneficial since they feed on fly eggs and larvae. Beetles may also play a role in the transmission of avian disease. When poultry have access to their litter, they will feed on the beetles they find. Beetles (see *Diseases of the Chicken*, Chapter 27) can transmit several tapeworms and a few bacterial and viral diseases. In most poultry operations, beetles have been considered a serious nuisance pest because of their potential to cause structural damage to the poultry house, while in caged layer operations their ability to suppress fly emergence through their burrowing and foraging activities which aerate the manure is considered beneficial.

Four common beetles found on poultry operations are:

- Darkling beetle (*Alphitobius diaperinus*)
- Larder or hide beetle (*Dermestes lardarius* and *Dermestes maculotus*)
- Rove beetle (*Philonthus* spp.)
- Hister beetle (*Carcinops pumilio* and a few others)

The Rove and Hister beetles are predators, which primarily eat fly larvae and eggs, respectively. The Larder beetle feeds on decomposing organic matter including feathers but can become a structural pest. The Darkling beetle, also known as the lesser mealworm beetle, can be an important pest from the standpoint of structural damage and disease transmission. Studies have shown that the Darkling beetle has the potential for serving as a reservoir for bacterial (*Salmonella* spp.) and viral (Mareks's disease) pathogens; other studies have shown that its larva, the lesser mealworm, has the same potential.

Control of Beetles

Beetle control can be very difficult. General farm sanitation can help reduce beetle populations and includes the following:

- Cleaning up spilled feed
- Proper dead bird disposal
- Complete manure and litter removal

Applying pesticides to accumulating manure in the house should be avoided because of the toxic effects on beneficial predators and parasites. Pesticide application should be directed to the same areas where adult flies congregate such as fencing, outside walls and ends of buildings, and in weeded areas. Insecticide applications to walls and insulation may be of limited value, and require repetitive treatments.

12-C. RODENTS

The Norway rat (*Rattus norvegicus,* Figure 12-3), roof rat (*Rattus ratus*) and the house mouse (*Mus musculus,* Figure 12-2) are the most common and most important vertebrate pests from an economic and public health standpoint worldwide (Table 12-1). The annual economic loss to the poultry industry because of rats and mice runs into the millions of dollars. A single rat can eat 25 pounds of grain in a year, and in the process contaminate 10 times that amount of poultry feed by defecating and urinating.

Figure 12-2. House Mouse

These rodents cause extensive property damage as well. In an effort to keep their teeth from overgrowing, they gnaw on wood, cement blocks, insulation, and electrical wiring. Rodents also present a significant health hazard to poultry and humans. They can carry many pathogenic disease agents including salmonella, cholera, leptospirosis, coccidia, and rabies.

Rats and mice are primarily nocturnal and usually have two peaks of feeding activity during the night. Rats and mice have the ability to:

- Gnaw through most building materials
- Squeeze through openings the size of their skull
- Swim
- Jump vertically up to three feet
- Traverse wires
- Climb up vertical surfaces

Control of Rodents

Recognizing a rodent problem is the first step in control. Look for the following signs:

Figure 12-3. Norway Rat

Table 12-1. Characteristics of Various Rodent Species

Appearance	Norway Rat	Roof Rat	House Mouse
Eyes	Small	Large	Small
Ears	Short (do not reach eyes)	Large (can be pulled over eyes)	Large (can be pulled over eyes)
Head	Blunt muzzle	Pointed muzzle	Pointed muzzle
Size (without tail)	7–11 in (18–28 cm)	6–8 in (15–20 cm)	2.5–3.5 in (6.3–8.8 cm)
Tail	Shorter than head and body	At least as long as head and body	About as long as head and body

- Droppings and urine stains (fluoresce under a UV light source)
- Gnawed holes in feed sacks
- Damage to electrical wiring
- Gnawed wood and other building materials
- Oily appearing rub marks on wood structures
- Tail drag marks on dusty surfaces

Methods of control may include:

- Exclusion
- Trapping
- Glue boards

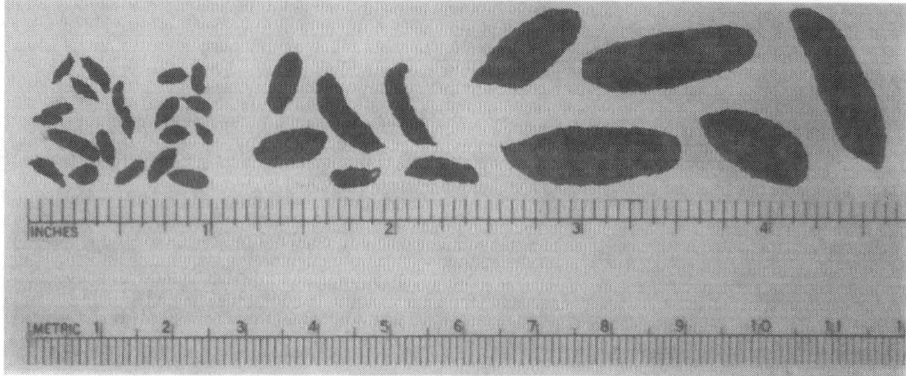

Figure 12-4. Rodent Droppings (House mouse, Roof rat, and Norway rat—left to right)

- Slow killing toxic baits
- Rapid killing toxic baits
- Tracking powder

1. Traps, glue boards, baits and tracking powders should be placed along frequently used runways, and in areas where the rodent feels unchallenged (secluded areas). Traps work best if placed for a period of time without setting them to avoid shyness. All of these methods require regular maintenance.
2. Toxic baits work best following a thorough clean up of the poultry facility. Usually this is most effectively done at the time of flock removal, when all feed and water sources have been eliminated.
3. Monitoring mouse and rat populations should be an integral part of the control program. Recognizing when populations start to increase tells the producer when to place more traps or baits. Observing a decrease in rodent populations indicates that the methods being used are working. Monitoring can be achieved by using traps or baits. Recording the number of animals trapped or the amount of bait that has disappeared over a period of time can give the producer a reliable indication of the overall rodent activity occurring in the poultry facility.

12-D. WILD BIRDS AND OTHER ANIMALS

Wild birds and other animals can carry disease-causing agents and can cause structural damage to the poultry premises. Some of the unwanted visitors that should be controlled or excluded include:

- Birds
- Skunks
- Squirrels
- Cats
- Dogs

Most control methods rely heavily on exclusion, although in some cases trapping and poison baits can be effective. Excluding these animals from poultry houses is most important. Openings into the house should be protected with a barrier that will prevent entry but not interfere with the performance of the house. All entry doors into the house should be kept closed. Open-type houses can be protected with poultry netting that is small enough to prevent animals from entering the house, yet wide enough so as not to restrict airflow into the house.

12-E. PESTICIDE RESISTANCE

Pesticide resistance is a change in susceptibility to specific pesticides within a pest population over time, which results in a decrease of pesticide effectiveness. Resistance is the result of genetic selection. Continued use of pesticides in the same chemical class can lead to resistance within a population of exposed pests. Pesticide resistance may develop within any species of organism and is common among flies and northern fowl mites.

Insects do not become resistant to pesticides within their lifetime. Instead, insects that are susceptible to a pesticide are killed by the chemical, leaving only the ones with some ability to survive the application available to reproduce. Their offspring are then more tolerant to the chemical, so a higher proportion of the population survives to pass along resistant genes. Because many pesticides (particularly carbamates and organophosphates) have the same mode of action, resistance to one member of either of these classes frequently confers resistance to both classes. For this reason, it is important to minimize pesticide use, to rotate the use of chemicals with different modes of action and to integrate non-chemical control methods.

12-F. PESTICIDE SAFETY

It is imperative that whenever chemicals are used, that all local regulations be strictly followed for each chemical applied. Restrictions, warnings and required safety instructions can be found on the label of the pesticide container. It must be emphasized that regulations concerning the use of specific pesticides differ between countries and even possibly between regions within a single country. Special consideration must be given to the following:

- Applicator health and safety
- Environmental protection
- Product safety
- Animal feed protection
- Non-target species

Recommended pesticides should be used at the times, in the amounts, and by the methods prescribed on the label to avoid excessive residues. Never allow pesticides to contaminate feed, eggs, or birds. In caged layer operations all eggs should be removed from the cage rollout trays before applying pesticides within the poultry house. Avoid spraying areas above the chickens in order to prevent contamination of feed and water.

Never store a pesticide in a food container or any container other than its own without the proper label attached.

Section II. Feeds and Nutrition

13

Feed and the Poultry Industry

by Paul W. Aho

The cost of feed is by far the most important factor in the competitiveness of a particular country, region, or chicken company. A number of factors including world availability of feedstuffs, local tariff rates, and the success of a hedging program, among others, influence the cost of feed. This chapter will investigate some of the issues related to grain and in particular those issues related to corn since 85% of the world's chickens use corn as their source of energy and corn represents approximately 70% by weight of chicken feed. The second most important ingredient for the world's chicken feed is soybean meal, the most widely used source of protein.

13-A. FEEDSTUFFS STOCKS LOW IN THE MIDDLE 1990'S

The most important issue for the chicken industry to consider in relation to feedstuffs is their availability and price. In that regard, the 1990's were a worrisome decade because of low worldwide stocks of feed grains in the middle of the decade. To illustrate the reduction in inventories, Figure 13-1 shows ending inventory of coarse grain (mostly corn and sorghum) as a percentage of total world use from crop year 1985–1986 through crop year 1998–1999. Note that stocks as a percentage of use dropped from a high of nearly 30% in 1985 to 11% in crop year 1995–1996. A similar reduction in inventories took place for soybeans. As can be seen on Figure 13-2, world soybean stocks fell from 17% in 1985–1986 to just 9% in 1995–1996. In the last two years, inventories of both grains have increased to higher levels.

Low inventories in the middle 1990's were caused primarily by agricul-

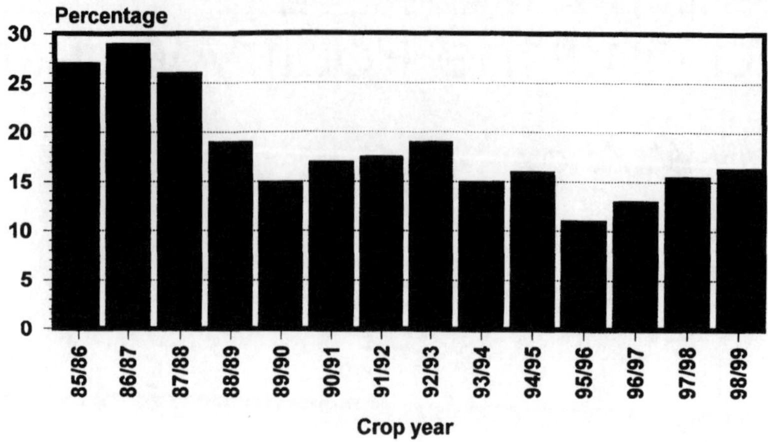

Source: USDA World Supply and Demand Estimates

Figure 13-1. World Coarse Grain Ending Stocks as a Percentage of Use, Crop Years 1985/1986 to 1998/1999

tural policy changes in the United States and the European Union. In both the European Union and in the United States, agricultural reform has reduced government subsidies and forced farmers to look more to the market and less to government for guidance in what to plant and grow. Reform has had the beneficial effect of reducing government expenditures and also should reduce the long-term average price of feed. In this respect, the reform is a significant long-term benefit to the world poultry industry.

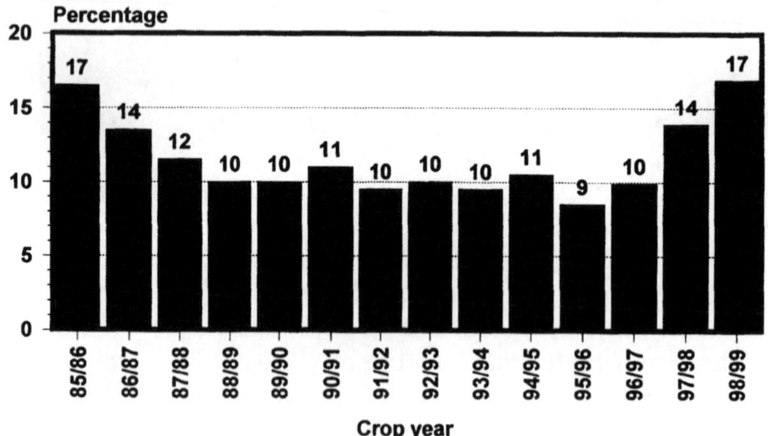

Source: USDA World Supply and Demand Estimates

Figure 13-2. World Soybean Ending Stocks as a Percentage of Use, Crop Years 1985/1986 to 1998/1999

However, the disadvantage of these reforms comes in increased volatility of prices. There will be increased volatility because governments are taking a smaller role in the storage of grain.

In earlier years, large quantities of government-owned grain were expensively stored in the United States, the European Union, Brazil, and elsewhere. In the new era of government frugality, the role of storing grain has now, for the most part, been privatized. Private companies do not store grain for charitable reasons and neither do private companies store unreasonable amounts of grain. As a result there is now less grain stored. With less grain in storage, the influence of the weather becomes magnified because grain use must be more severely rationed by price in the case of a poor harvest. A poor harvest such as the one in the United States in 1995, had an immediate and profound effect on grain prices all over the world. Although grain inventories have increased recently, inventory levels are still low by historical standards.

The Cause of Temporarily Lower Production

Weather conditions are often the cause of temporarily lower grain production. For example, the "El Niño" effect was blamed for the wet spring of 1995 that forced US farmers to plant corn a full month late resulting in the poor harvest. Another strong El Niño in 1997 did not affect harvests in the most important corn-producing areas. Needless to say, the ability to make accurate long-range predictions of how the weather will affect the success of the grain harvest is still relatively primitive. It is likely that drought in the US and/or China at some point in the next few years will cause production to be lower again.

The supply of grain is not the sole determinant of prices. The demand side of the equation is also important. There has been a strong and rapidly increasing worldwide demand for grain fueled in great part by the success of the world animal industries including the poultry industry.

Global Economic Trends and the Demand for Grain

From 1994 to 1997 the world economy grew at the unusually strong pace of 3 to 4% per year overall (Figure 13-3) and 6% in developing countries. Asia grew at an average fast pace of 8% in that period and the US economy grew at a healthy pace of about 3% per year. Even the former Soviet Union countries and Central Europe climbed out of their long economic depression in the middle of the 1990's that started with the fall of the Berlin Wall at the beginning of the decade. After the economic crisis in Asia in 1997, world growth slowed to just 2% in 1998 and 1999. However, world eco-

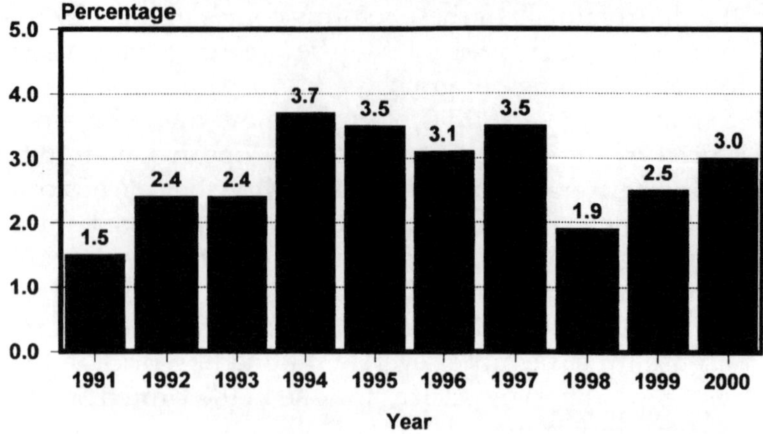

Source: International Monetary Fund, Aho

Figure 13-3. World Gross Domestic Product (GDP), 1991 to 2000

nomic growth is expected to return to a robust rate starting in the year 2000.

The strong overall world economy in the middle of the 1990's expanded the demand not only for poultry meat and eggs but also for all meats. The expansion of meat demand continued a long-term trend of increased meat consumption that has been in place for decades. Figures 13-4 and 13-5 illustrate the long-term trend of the use of corn for animal feeds. Corn fed

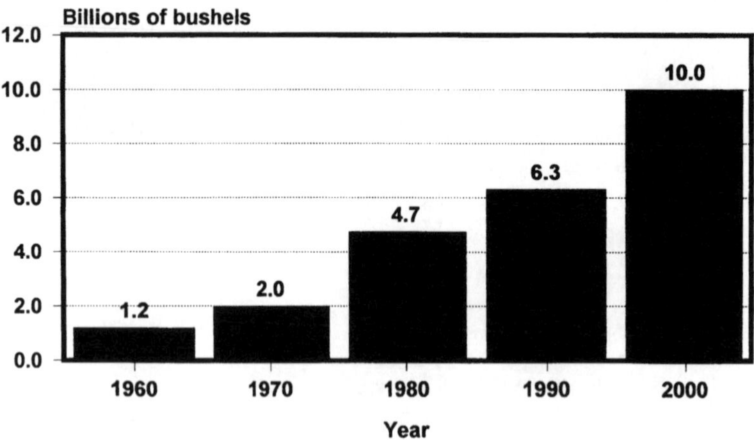

Source: Estimated by author (Aho)

Figure 13-4. Billions of Bushels of Corn Fed by the World Animal Industries, 1960 to 2000

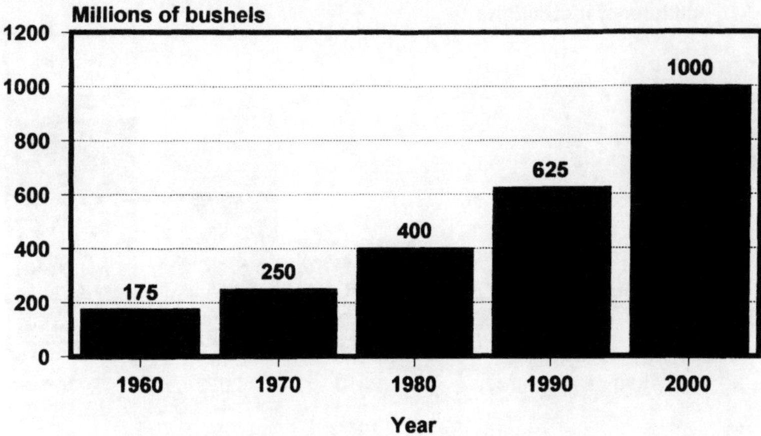

Source: Estimated by author (Aho)

Figure 13-5. Millions of Bushels of Corn Consumed by the US Broiler Industry, 1960 to 2000

to all world domestic animals has increased from just a billion bushels in 1960 to an estimated 10 billion bushels in the year 2000. In the 1990's alone, the amount of corn fed will increase by 4 billion bushels. In the US, corn fed to broilers has increased from 175 million bushels of corn in 1960 to an estimated one billion bushels by the year 2000.

To sum up the importance of the feed grain supply and demand issues:

1. The supply side is more volatile than before thanks to lower world grain reserves.
2. A poor harvest in a major grain-producing country like the United States and/or China will lead to sharply higher prices given lower world stocks
3. Demand for feed grains by the world's animal industries has increased rapidly along with world economic development
4. Increasing demand for feed grains lends support to feed grain prices

13-B. FUTURE AVAILABILITY OF FEED GRAINS

Observers of world grain production noted a worrisome trend in the first five years of the 1990's. Per capita production of grain was declining. Only in 1996 did this trend reverse itself. Does this mean that the world

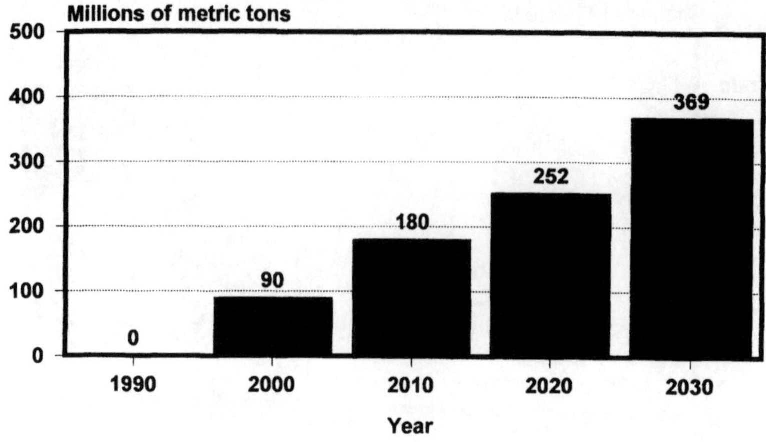

Source: Lester Brown, "Who will Feed China?"

Figure 13-6. Chinese Grain Imports Required, 1990 to 2030

is heading for a serious grain shortage in the future? There are two schools of thought on this matter, the alarmists and the *mainstream* forecasters.

The best known alarmist is Lester Brown[1] of the *Worldwatch Institute* in Washington, DC. He and others believe that the limits of agricultural land, water, and technology are now being reached. Lester Brown joins a long tradition of alarmists going back to Thomas Malthus (1766–1834) a British economist who published *An Essay on the Principle of Population* (1798).

According to Malthus, population tends to increase faster than the supply of food available for its needs and if population grows much faster than food production, growth is checked by famine, disease, and war. Malthusians (the followers of Thomas Malthus) have been waiting two hundred years to be proven correct (it is interesting to note that famines are almost always caused by war rather than vice-versa).

An important part of the disaster scenario painted by Mr. Brown is the enormous quantity of grain imports that he believes will be required by China in the future. He predicts that China will require 369 million metric tons (MMT) by the year 2030, an amount double the current total world grain trade. China will require huge imports of grain according to Brown because of its increased wealth leading to increased meat consumption combined with a large loss of agricultural land to urbanization and industrialization (Figure 13-6).

Mainstream forecasters look at the same data and see a very different scenario for the future. For example, the Food and Agriculture Organization of the United Nations (FAO) in Rome believes that the slowdown in grain

[1] "Who Will Feed China?" Worldwatch Institute, 1995.

production growth of the early 1990's was a natural response by major grain exporters to low prices and agricultural and trade policy reform. The world can and will increase agricultural production faster than population growth given the appropriate price signals from the market.

The rate of increase in food consumption will slow in China even if the economic growth rate remains high. A growing food and feed deficit in China would both stimulate production and curtail consumption in that country. There are opportunities to raise yields and production substantially in China by the more widespread use of technologies such as hybrid corn, a common technology in the rest of the world but not yet universally adopted in China.

Beyond China there are millions of acres that could be brought into grain production around the world. Some of those acres are found in the US set-aside program. Another area of great potential is Argentina. In that country there are millions of acres of some of the best grain-producing land in the world, the Pampas, that are now used for grazing livestock. The former breadbasket of Europe, Eastern Europe and the Ukraine, are also waiting in the wings to resume their natural role as a provider of grain.

In addition to the new land that could be placed into grain production is the potential future impact of biotechnology on crop yields. For example, in 1996, corn borer-resistant varieties of corn produced by genetic engineering became commercially available. Yields are higher and more consistent with corn varieties resistant to this pest. Other important research advances will come along in due time as long as agricultural research continues to be funded. The effect of adverse consumer reactions to genetically modified plants and animals is difficult to predict at this time.

The Price Signal

If the *mainstream* forecasters are right, all that is needed to stimulate production is the appropriate price signal from the market. The right price signal triggers increased production of all types of grain using new land and new technology. High prices in 1996 provided that signal and the world production of grain responded with the first increase of the decade, helped along by better weather.

End of the year inventories are important. When ending inventories are projected to be low, the market is more volatile and generally higher. Every bit of news about crop conditions can lead to a rapid increase or decrease in grain prices. When inventories are higher, prices are lower and less volatile. As an example of changing ending inventories, note the ending inventories on the following chart of US corn production in the middle 1990's. Note the lower domestic use and lower ending inventories in 1995–1996 and the resulting higher prices (Table 13-1).

Table 13-1. US Corn Production and Use 1994–1996

	US Corn Production and Use—Billions of Bushels—USDA Farm Price in Dollars Per Bushel		
	1994–1995	1995–1996	1996–1997
Beginning Stocks	0.8	1.5	0.4
+			
Production	10.1	7.4	9.3
=			
Total	10.9	8.9	9.7
−			
Exports	2.1	2.2	1.8
−			
Domestic Use	7.3	6.3	7.0
=			
Ending Stocks	1.5	0.4	0.9
Farm Price	$2.26	$3.20	$2.69

Figure 13-7. US Feed Mill with Rail Ingredient Delivery

13-C. GRAIN TARIFFS

A good world harvest of grain is not enough for the chicken industry. A chicken company must also be able to get the grain to its feed mills. Oddly enough the biggest obstacle for grain supply in the world are not mountains and oceans but rather tariff barriers.

Tariff barriers for the movement of grain from country to country are common in the world and act to reduce the competitiveness of local chicken industries. As a result of numerous tariff barriers to the purchase of grain, the most competitive countries are those that export grain. In addition, the rare country that has low tariff barriers to the importation of grain can also be competitive. In other words, a country where poultry producers pay no more than the world price for their grain results in the local poultry industry having the best chance to be competitive. Unfortunately there are numerous examples of countries where chicken producers are forced to pay far more than the international price of grain to feed their chickens, sometimes as much as 300% of the world price.

Table 13-2 shows some typical corn tariff rates as negotiated by the Uruguay round of the General Agreement on Trade and Tariff (GATT) negotiations that ended in the middle 1990's. High tariff rates allow domestic grain prices to exist at a higher than international market price. High grain tariffs result in high domestic feed grain costs and an uncompetitive poultry industry. High grain tariff countries have made the political decision that neither their poultry industry nor their feed grain industry should be

Table 13-2. Yellow Dent Corn Tariff—Uruguay Round of the GATT-Agreement

Country	Tariff
United States	None
Czech Republic	None
India	None
Malaysia	None
Egypt	5%
Costa Rica	15%
Tunisia	17%
Chile	25%
Hungary	32%
Mexico	37%
Indonesia	40%
Brazil	55%
Thailand	73%
European Union	94%
Poland	109%
Morocco	122%
Venezuela	122%
Pakistan	150%
Turkey	180%
Colombia	194%

globally competitive. The attitude expressed by French Agriculture Minister Louis Le Pensec is typical of this political stance when he said, "It is ludicrous to talk of cutting farm prices to compete on world markets."[1] It is interesting to note that this French restriction on competing in the world market does not include most industrial goods where the French, for the most part, do not consider it ludicrous to compete on the world market using world prices.

13-D. IDENTITY-PRESERVED CORN

An important point about grain in general and corn specifically is that the poultry industry will have a profound influence on the way corn is grown and marketed in the next century. That is likely to mean that there will be more identity-preserved corn that is tailor-made and delivered to the chicken industry around the world.

What Is Identity-Preserved Corn?

Historically, the world grain system has focused on volume and cost considerations to move as much grain as cheaply as possible. As a result, grades and standards for corn resulted in a commodity that was relatively homogeneous for a limited number of traits such as test weight and moisture content. Grades and standards have not included traits that can have a substantial impact on the value of the corn to poultry firms such as protein content and oil (energy) content. These non-grade nutritional attributes can vary widely within a given grade.

In the future it is likely that the measurement of these non-grade standards will become an increasingly important part of the production and marketing of corn and other feedstuffs. As these nutritional attributes become better known, the world grain system will preserve more and more of the identity of grain through the marketing chain. Not only will these nutritional attributes become transparent to buyers but they will also be actively manipulated by seed genetics firms. In the next several years the second wave of Ag biotech will begin to affect seed genetics. The first wave emphasized input traits such as insect resistance and herbicide resistance. The second wave are the improvements in oil content, amino acid composition, and other grain quality attributes of interest to a feed end-user.

There is, of course, a cost associated with improving nutritional attributes and maintaining the identity of grain through the marketing channel. However, chicken producers will gladly pay an additional cost for

[1] Feedstuffs, April 27, 1998, page 3.

these services because the chicken industry is no longer interested in just "cheap" corn. The idea of "cheap" corn belongs to the old concept of agriculture. The new concept of agriculture looks for corn that will contribute the most to the bottom line. Just as the chicken industry is not just looking for "cheap" genetics or "cheap" vaccines it is also not looking just for "cheap" corn.

The ideal corn for the chicken industry is one that supplies low cost energy and arrives at the feed mill with a consistently high and known quality. The new corn will probably have a name, a pedigree, and a history. In other words, it will in many cases be an identity preserved corn. An early example of an identity preserved corn is Optimum 80 High Oil Corn® from the DuPont/Pioneer joint venture. There will surely be additional examples in the future.

13-E. GRAIN HEDGING

As a consequence of the increased volatility of grain prices, managers will be exposed to the risk of rapidly changing prices for grain. One way to manage that risk is through the use of hedging.

Hedging is the purchase of futures or options on a commodity exchange to help manage the risk of fluctuating commodity prices. In the case of the chicken industry, the commodities of interest are corn and soybean meal. The easiest way to think about the futures market is to imagine a high stakes poker game where people bet on the price of grain and other commodities. With rare exceptions, only pieces of paper are traded in the futures market. Buyers of futures contracts do not take delivery of a commodity, they buy one piece of paper at the beginning of the poker game and trade it for another piece of paper at the end of the poker game. At that point they are either richer or poorer.

Futures are derivatives, they are derived from the underlying cash market for grain. Futures represent the right to purchase a certain grain at a certain price at a certain time in the future. There are two groups of traders attracted to this casino, speculators and hedgers. Speculators are true gamblers willing to make a bet on the future price of anything. Hedgers are different. Hedgers include all of those businesses that are either short or long on a commodity on the cash market and wish to fix their costs.

For example, all poultry companies that use corn and soybean meal are "short" corn and soybean meal because they *need* corn and soybean meal all the time. A company that is "short" corn and soybean meal is continually exposed to the risk of higher prices. In effect, a gamble has been taken because of the nature of the business. That gamble can be hedged by taking the opposite gamble in the futures market from the one already taken in the everyday cash market. If a company is short corn and soybean meal by the nature of their business, they can hedge by going long (buying) futures. The net result of hedging is to lock in the price of corn and soybean meal.

Locking in the price of feedstuffs does not guarantee that a company will pay the lowest price it merely locks in a price. If a company establishes a hedge on corn and then corn prices fall, then the company will pay more for corn than non-hedging companies because of losing money on their futures positions. Hedging is a tool that can be used to make the cost of something more predictable and is particularly effective if the price of the end product (chicken meat) can also be made more predictable. Each company needs to examine its situation to determine whether or not the hedging tool is appropriate. It is important to remember that ignoring the issue of hedging is the same as making the conscious decision not to hedge.

A hedging program should be established and utilized every year. Hedging only when the manager thinks that hedging will make money is a dangerous strategy. A company that hedges every once and awhile is likely to end up hedging when hedging loses money and neglecting to hedge when hedging would bring a tremendous profit.

Another concept related to hedging that is important to mention is that of "basis." Basis is the difference between the price of a grain at a given location (a local feed mill) and the price at a given market (Chicago). If corn is $2.50 in Chicago and $3.00 at the local feed mill, then the basis is $0.50. Basis can fluctuate. At times the basis may be $0.30 and at other times $0.70, for example. Therefore, in addition to the risk of a change in market price there is also the risk of a change in basis. Hedging can reduce market price risk but does not reduce the risk of a change in basis.

Conclusion

The world is not facing a serious feed ingredient shortage of either corn or soybean meal anytime soon. Nevertheless, volatile prices can be expected due to the reduction in world inventories. There is likely to be a wide range of prices year to year and within each year. Poultry producers will have to learn to protect themselves from these wide swings in prices by using tools such as hedging. The greatest barriers to the competitiveness of poultry companies in most countries are tariff barriers on feed grain that prevent access to the world price of feedstuffs. The use of identity-preserved grain can help reduce the uncertainty of feed quality.

14

Digestion and Metabolism

by Craig N. Coon

The information discussed in this chapter refers to the mechanisms within the chicken to convert feed into useful nutrients and energy for maintenance, growth, and egg production. The chapter emphasizes the digestion and metabolism of carbohydrates, lipids, and proteins.

14-A. BASIC NUTRITIONAL COMPONENTS

All animals, including birds, require certain basic nutritional constituents to sustain life, grow, and reproduce. The list includes:

carbohydrates	minerals
lipids (fats)	vitamins
proteins	water

The digestion of these dietary components varies greatly, and each section of the digestive tract is responsible for certain parts of the process. (see *Anatomy of the Chicken,* Chapter 4).

14-B. WHY A CHICKEN EATS?

The lack of satiety (fullness) in certain sections of the alimentary tract induces the primary need for feed. Chickens are continuous nibblers compared with most animals that resort to eating a meal, then resting while the meal digests. They fill their crop and gizzard to capacity, then wait until some feed leaves these organs before they eat again. The process will be repeated many times a day if feed is present.

Many things affect feed consumption. A few of the important ones are breed and strain of the birds, environmental temperature, body weight, sex, age, degree of egg production, egg size, feather cover, activity, type of housing, feed palatability, energy content of the feed, quality of feed ingredients, water consumption, body fat content, and degree of stress.

14-C. DIGESTION

A large percentage of the feed ingredients consumed by a chicken is in a form that necessitates chemical and other reactions before it can be utilized by the bird. The alimentary canal is a long tube through which the food passes while these reactions take place. Therefore, *digestion* refers to those changes that occur in the alimentary canal to make it possible for the feed to be absorbed through the intestinal wall and enter the bloodstream.

Within certain sections of the digestive tract, proteins are produced to facilitate the digestive process. These are known as *enzymes,* and each of the several types has a specific function in producing the necessary chemical reaction. Enzymes are catalysts produced by living cells to aid certain chemical reactions without entering into them. Today, commercial enzyme preparations are being added to poultry feed to enhance the digestion of poorly digested carbohydrates, proteins, and phytate phosphorus in many feed ingredients.

Other chemicals are secreted to alter the acidity or alkalinity of the tract so that the chemical reactions may be expedited. Bacteria also play an important role in the digestion of foods. All-in-all, the digestive process is quick, continuous, and constant.

Mouth

Secreted in the mouth of the chicken is a fluid known as *saliva.* It is very slightly alkaline and contains the enzyme *ptyalin,* which has the capacity to hydrolyze starch, converting it to sugar. However, food is held just a short time in the mouth of the chicken and the hydrolysis in this area is minor.

Crop

After leaving the mouth, food continues down the esophagus to the *crop,* a reservoir for storage. The food material remains here for varying lengths of time depending on its particle size, on the amount consumed, and on the quantity of material in the gizzard. In the crop, the feed particles are softened, and *ptyalin* from the mouth continues to hydrolyze the starches. No enzymes are produced in the crop.

Proventriculus

The *proventriculus* is a bulbous organ situated just before the gizzard, and is sometimes known as the *glandular stomach.* It is here that the gastric enzyme, *pepsin,* and hydrochloric acid are produced. The pepsin acts to break down the complex protein molecules; the hydrochloric acid changes the contents of the digestive tract from alkaline to acidic and thus aids in protein digestion.

The proventriculus is small and holds little food, and therefore food passes quickly through it to the gizzard. Because food is held in the proventriculus for such a short time, little or no actual digestion takes place here.

Gizzard

The *gizzard* is a highly muscular portion of the alimentary tract and is capable of exerting pressures of up to several hundred pounds per square inch. It is here that large particles of feed material undergo mechanical grinding, sometimes in the presence of "grit" in the form of sand, granite, or other abrasives to help facilitate the process (grit is rarely used in commercial diets today because finer grinds of feed are used). Although highly variable, the contents comprise about 50% water when in the gizzard. No enzymes are secreted in the gizzard, but digestion continues as the result of the secretions of the proventriculus.

Small Intestine

The foremost portion of the small intestine is known as the *duodenum.* It takes the form of a loop known as the *duodenal loop;* imbedded within the loop is the *pancreas,* a gland that empties its secretions into the intestine. The pancreas produces *pancreatic juice* that contains the enzymes *amylase, lipase,* and *trypsin.* These, along with other enzymes from the pancreas, continue the process of digestion in the duodenum, although most of the absorption takes place in the next section of the small intestine, the *jejunum.* The small intestine *mucosal* cells also contain specific carbohydrate and peptide hydrolysis enzymes that continue the digestion of feed into final products. The third section of the small intestine is the *ileum* which is the location for final absorption of nutrients.

Bile is secreted by the liver and flows into the duodenum as a thick green material. It does not contain enzymes, but helps emulsify the fats and plays a part in other digestive processes. When the feed contents leave the gizzard they are slightly acid as the result of the hydrochloric acid secreted in the proventriculus, but the contents become alkaline as they pass through the jejunum and the ileum. Relatively speaking, little digestion takes place until the food reaches the small intestine. Here, most of it is completed.

Large Intestine

Some of the processes of digestion may continue in the large intestine, although no enzymes are secreted here, any digestion is merely a continuation of processes initiated in the small intestine. Water moves in and out of the large intestine, but outward transfer predominates to bring the intestinal contents into a more solid state. This movement of water is related to conditions associated with dehydration and edema of the tissues. *Dehydration* is a condition produced as the result of a loss of sodium or potassium from the muscle cells. Retention of water produces *edema*, a condition arising when too much salt is consumed, and the body tries to dilute the salt in the cells of the tissues and in the space between the cells by osmosis. Both dehydration and edema of the tissues affect the transfer of water through the walls of the large intestine.

Ceca

At the juncture of the small and large intestines are two blind pouches called the *ceca*. Fermentation and some digestion take place here. Fermentation is instrumental in digesting the very small quantity of crude fiber the chicken is able to utilize.

14-D. GENERAL METABOLISM

Metabolism is a term used to describe those chemical changes in food components that occur after digestion and absorption. Since the various components of feed (protein, carbohydrates, fats, vitamins, and minerals) have been converted to compounds capable of absorption during digestion, they must be restructured before they can become usable by the bird. For the tissues of the body to be able to utilize the simpler compounds carried to them by the blood system, further chemical reactions must take place. By these additional processes energy is developed, fat is stored, heat is liberated, and many end products not of value to the bird are eliminated through the kidneys.

How Food Material Is Utilized in the Body

The body has almost an hourly need for certain food materials in order to carry out its normal physiological processes. These materials perform the following general functions:

1. Maintenance of life
2. Growth
3. Production of feathers
4. Egg production
5. Deposition of fat

To carry out these functions, food must be metabolized. As the processes in metabolism are both involved and complex, the following sections will attempt to give only an overview of how carbohydrates, lipids, and proteins are transformed after digestion for use by the bird.

14-E. CARBOHYDRATES: DIGESTION AND METABOLISM

A major proportion of poultry diets consist of cereal grains, and the main energy component of poultry feeds is the starch that is contained in these grains. Oil seed protein feedstuffs, such as soybean meal, have less than 1% starch content; however, legumes such as field beans and peas utilized in poultry feeds in many countries outside the United States have starch content ranging from 20 to 58% of the dry matter content. Poultry have the ability to digest starch with their endogenous production of *pancreatic amylase.* The amylase will break the starch into shorter polymers called *dextrin* and the carbohydrate can be further hydrolyzed to *maltose, isomaltose,* and *glucose* units. The maltose and isomaltose can be hydrolyzed from intestinal secretions producing the enzymes *maltase* and *isomaltase* that will further hydrolyze the carbohydrates into *glucose* units. The glucose can be actively absorbed in the intestine of poultry. Carbohydrates that are absorbed can be metabolized

1. to produce chemical energy that can be used by the animal for productive purposes
2. to synthesize a reserve glucose source called glycogen that can be used in acute stressful situations
3. to form structural components such as chondroitin sulfate in cartilage
4. to form lipids.

Cereal grains contain two forms of starch which are described as *amylopectin* and *amylose.* Amylopectin starch is the predominant form normally representing approximately 70 to 90% of the grain starch. Amylopectin is also the main form of starch in legumes; wrinkled peas contain as much as 65% amylose starch. Generally, an increased amylopectin percentage of the total starch content of feedstuffs increases the solubility characteristics of the starch.

In recent years, geneticist have altered the quantity of starch types found in some new cultivar grains. A "waxy" corn or barley means the starch content is approximately 100% amylopectin. Starch is stored in the endosperm of cereal grains as starch granules with the cell wall of some cereals containing both different amounts and types of non-starch polysaccharides. Poultry have been shown to digest approximately 99% of the isolated starch from corn and wheat, whereas the digestibility of pea and bean starch has been shown to be 98.0% and 94.5% for adult cockerels, and 94.2 and 78.2% for young broilers, respectively.

Water Soluble Non-starch Polysaccharides (NSPs)

There is a tremendous variation in the amount of metabolizable energy (ME) found in cereal grains. Key reasons for the variation in ME have been attributed to a decreased digestion of protein, fat, and starch caused by the type and amount of NSP surrounding the starch granules of the grains. For instance, barley and oats contain relatively high levels of *β-glucan*, an NSP, which accounts for their lower digestibility.

There is some confusion regarding the detrimental aspects of isolated wheat *arabinoxylans* (water-soluble and non–water-soluble NSPs) on wheat starch digestibility. The water-soluble arabinoxylans are thought to be the antinutritive factor in wheat because of their ability to change intestinal fluid viscosity; however, research has shown isolated water-insoluble arabinoxylans slightly decrease the starch digestibility of sorghum diets, and the isolated water-soluble *pentosans* from wheat caused no change in starch digestibility of sorghum. The cell wall structure with the different types of NSPs may also restrict or slow down the digestion of the starch specifically.

The addition of commercially prepared microbial enzymes (*β1–4 endoglucanase* and *β1–4 endoxylanase*) in poultry feeds containing large amounts of wheat and barley have greatly increased the ME value, feed conversions, and weight gain of growing poultry. In European countries and in Canada where large amounts of wheat and barley are used in poultry diets, the economic value of adding enzymes may be justified. The value of adding feed enzymes to layer feed containing these grains has not been as effective. The adult bird may have less trouble with viscosity in the digestive tract, possibly because of a more mature microbial population that may help hydrolyze the NSP carbohydrates.

Poultry have the ability to digest *sucrose* in feeds because birds have the *sucrase* enzyme that is included in the intestinal juices that are secreted during digestion. Sucrose is hydrolyzed to *fructose* and *glucose* which can be absorbed in the intestine. Sucrose is not found in high levels in many cereal grains, but is found in levels between 5 and 6% in soybean meal. The enzyme lactase is not produced in birds, therefore, there is limited digestion of lactose or milk sugar.

Protein feedstuffs, such as canola and soybean meal, primarily contain

Table 14-1. Carbohydrate Content of Selected Feedstuffs (% Dry Matter)

			Canola		
Component	Wheat	Barley	Brown	Yellow	Soybean
Glucose and fructose	0.2	0.3	0.5	0.6	0.5
Sucrose	1.1	1.4	7.7	9.8	6.9
Galactooligosaccharides[1]	0.5	0.3	2.5	2.4	5.3
Fructosans[2]	0.9	0.6	Tr.	Tr.	—
Starch	68.0	60.4	2.5	2.6	0.7
Non-starch polysaccharides	10.6	16.8	17.9	21.4	20.3
(Cell wall components)					
–Cellulose	2.1	5.0	4.6	6.0	5.5
–Hemicellulose / Pectin	—	—	13.3	15.4	14.8
–β-glucan	0.8	4.6	—	—	—
–Arabinoxylan	6.5	6.7	—	—	—
Lignin (not a carbohydrate)	0.8	2.3	8.0[3]	3.2[3]	1.0
Total carbohydrates	81.3	79.8	31.1	36.8	33.7
Total fiber	11.5	20.5	30.1	27.3	24.1

[1] Includes raffinose and stachyose
[2] Includes ketose, isoketose, and neoketose
[3] Includes lignin and polyphenols
Source: Slominski, 1991

oligosaccharide carbohydrates and sucrose instead of starch as the reserve carbohydrate source (Table 14-1). The primary oligosaccharides found in the protein feeds are first *stachyose* and then *raffinose.*

Water Insoluble Non-starch Polysaccharides

The type of carbohydrates included in this group include cell wall carbohydrates such as *cellulose, hemicellulose, pectic substances,* and *lignin.* The major hemicellulose polymers in cell walls are xylan-related, such as *arabinoglucuronoxylan* found in wheat bran and *glucuronoxylan* in soybean hulls. Pectic substances are also a major component of the cell wall carbohydrates of legumes. The neutral detergent fiber (NDF) value used to indicate fiber content of feeds does not include pectic substances, therefore, the NDF value for legumes may be misleading as an expression for fiber content. Research indicates there is no significant digestion in poultry of cellulose, water insoluble polysaccharides, or pectic substances in either adult or young poultry. The fiber content of feedstuffs may be a fairly good predictor of the feeding value because of the non-digestible diluting effect of the cell wall material and because of the negative effect of fiber on digestibility of crude protein and fat.

14-F. LIPIDS: DIGESTION AND METABOLISM

Lipids are classified as simple, compound, or derived. The lipids that are the most important to poultry dietary formulations are simple lipids containing fats and oils which are esters of fatty acids with glycerol. Lipids are a tremendous value in feeds as an energy source with a gross energy value of 9.45 kcal compared to only 4.1 kcal for a typical carbohydrate. The metabolizable energy (ME) content of feedstuffs are greatly affected by the lipid content of the feed. The chain length and degree of unsaturation of the fatty acids making up the lipids are important factors determining ME values as well as physical and chemical properties. Lipids in feeds are also important for the absorption of vitamins A, D_3, E, and K, because fat-soluble vitamins utilize lipid micelles in the intestinal tract as a carrier. *Lipid micelles* are small oil droplets that contain free fatty acids, monoglycerides, bile salts, fat soluble vitamins, cholesterol, and phospholipids.

Poultry nutritionists must also provide both growing birds and adults with essential fatty acids that are needed for membrane integrity, prostaglandin formation, fertility, and hatchability. A 1% level of *linoleic acid* is required for both growing and adult birds. If linoleic levels are deficient, poor feathering and reduced growth in chicks may be noticed. If the linoleic acid content is too low in breeder diets, there is a significant drop in hatchability of fertile eggs. The deficiency of essential fatty acids in adult birds may be hard to induce if birds are fed linoleic acid during the rearing period, because they can mobilize stored abdominal fat. Research has also shown that egg weights can be significantly increased when levels higher than 1% linoleic acid are fed to layers.

Digestion of Lipids

Lipids are digested in the intestine by forming lipid droplets with the help of bile salts serving as emulsifying agents. The lipids are hydrolyzed by pancreatic lipase in the intestinal lumen and dispersed into small mixed *micelles*. The mixed micelle crosses the microvilli of the intestine by diffusion. The fatty acids are reesterified into triglycerides in the mucosa cell and combined with phospholipids, proteins, and cholesterol into a *lipoprotein* structure that is transported by the portal blood to the liver. The digestibility and absorption of fat is the key factor influencing the ME value of the fat. Digestibility of fats will be affected by the ratio of unsaturated and saturated fatty acids, amount of glycerol, length of carbon chains, number of double bonds, and age of the birds (Table 14-2). The digestibility of fats are lower for younger birds. The reason the ME value of corn oil or soybean oil is higher than pig lard or beef tallow is because of their higher level of unsaturated fatty acids compared to saturated fatty acids. However, the digestibility of animal fats are also very good and when fed with cereal grains, such as corn, there seems to be a synergistic effect improving the absorption of both types of lipids.

Dietary lipids that are natural constituents of feed, and added fats or oils, have become a more important component of broiler and layer diets

Table 14-2. Absorbability Values of Various Fatty Acids, Monoglycerides, Triglycerides, and Hydrolyzed Triglycerides as Determined in the Chicken

		Absorption, percent	
		Chicks 3–4 wks	Chickens over 8 wks
Fatty acids			
Lauric	12:0	65	—
Myristic	14:0	25	29
Palmitic	16:0	2	12
Stearic	18:0	0	4
Oleic	18:1	88	94
Linoleic	18:2	91	95
*Monoglycerides**			
Monocaprylic	8:0		100
Monocarpin	10:0		93
Monolaurin	12:0		89
Monomyristin	14:0		67
Monopalmitin	16:0		55
Monostearin	18:0		41
Monoelaidin	18:1 (trans)		93
Monoolein	18:1 (cis)		98
Monolinolein	18:2		96
Triglycerides			
Soybean Oil		96	96
Corn Oil		94	95
Lard		92	93
Beef tallow		70	76
Menhaden oil		88	—
Hydrolyzed triglycerides			
Soybean oil fatty acids		88	93
Corn oil fatty acids		90	92
Lard fatty acids		82	83
Beef tallow fatty acids		61	67

* When monoglycerides were fed, the pancreatic ducts were ligated to eliminate pancreatic lipase from the lumen
Source: Scott, et al., 1982

in the last twenty years. Poultry are growing at a faster rate and are producing more egg mass with less feed. The addition of feed grade fat to poultry diets allows the energy concentration to be increased without using a large amount of dietary space. Feed mills producing pelleted feed for broilers that contains added fat above 2% (maximum for pellet quality) must spray the additional fat on the hot pellets. Dietary lipids have an advantage in hot climates because the bird can use a portion of the fat directly in eggs or store in tissue without it being metabolized. Therefore, dietary lipids produce a lower heat increment and potentially lower body temperature in poultry than other forms of dietary energy. The ability of poultry to maintain a lower body temperature in a hot environment will allow the bird to increase feed intake and therefore consume higher levels of calories and other nutrients needed for optimum gains and egg production. Nutritionists should increase the ME concentration of the diet in hot environments by adding feed grade fats and oils.

Extra Caloric Effect

Dietary fats produce an "extra caloric effect" when fed to poultry. The *apparent metabolizable energy-nitrogen* (AMEn) corrected value of feed grade fat has been shown to underestimate the net energy or effective energy of fat for both layers and growing broilers in energy balance studies. It has been suggested that the "extra caloric effect" is increased from 10 to 20% from fats for maintenance and production compared to the efficiency of utilization of AMEn from carbohydrates. Dietary fats have also been shown to have an "extra metabolic effect" caused by slowing down the passage rate of feed in layers thus increasing the ME of other feed constituents. The practice of increasing the ME concentration of feed when keeping poultry in a colder climate is not the most efficient method of providing calories for maintenance and egg production. Unless ambient temperatures are extremely cold, in which the birds cannot consume adequate energy to maintain body temperature, the most efficient method would be to provide a more consistent diurnal thermoneutral environmental temperature with improved poultry housing. Cooler temperatures will increase feed intake, and if the diets contain high ME levels, layers fed free choice will tend to overconsume calories, gain extra body weight, and produce larger eggs than needed.

Fat Quality

Nutritionists purchasing feed grade fats and oils must be familiar with terminology describing fat quality. Free fatty acids in fats and oils occur from oxidation and may indicate rancidity. Acidulated soapstocks also contain large amounts of free fatty acids which may cause corrosion of metal storage and handling systems in feed mills. Moisture content of fats has no nutritional value and may cause rust and corrosion of equipment. Impurities (such as small particles of fiber, hair, hide, bone, and soil) can be found in purchased fats. Impurities cause clogging of filters and a buildup of sludge in the storage tanks at the feed mill.

Unsaponifiable lipids are sterols, hydrocarbons, pigments, fatty alcohols, and vitamins that are not hydrolyzed by alkaline saponification. The ME value of unsaponifiables will be variable depending upon its components. The MIU value of a commercial fat is the combination of moisture, impurities, and unsaponifiable lipids. There is no ME value for moisture and impurities, thus a quality commercial fat should have an MIU value of

2% or less. The initial peroxide value (IPV) and the 20-hour active oxygen method (AOM) peroxide value describes the rancidity state of the fat and its ability to resist oxidation. A low IPV value (<5) means the fat or oil has a low peroxide content and is not rancid. An AOM value greater than 25 means the fat or oil was not stabilized with an antioxidant. Nutritionists must be sure that fat manufacturers are adding antioxidants to their feed grade fats and oils. A common practice is to also add an antioxidant such as *ethoxyquin* to the poultry feed to also prevent oxidation during transportation and storage. A fat or oil containing higher levels of unsaturated fatty acids will go rancid more rapidly than a more saturated fatty acid. It has been shown that unstabilized fat without antioxidants that has been oxidized will cause damage to the villi of the small intestine of poultry.

14-G. PROTEINS: DIGESTION AND METABOLISM

Poultry do not have a specific requirement for dietary crude protein, but the dietary protein level must be adequate to provide essential amino acids and amino acid nitrogen for non-essential amino acid synthesis. Dietary protein is normally added to supply all requirements of the essential amino acids except *methionine, lysine,* and *threonine.* Feed grade synthetic methionine, methionine hydroxy analogue, lysine hydrochloride, and threonine can be added to poultry feeds to supply the methionine, lysine, and threonine requirements at an economical cost and to minimize overfeeding protein. Methionine and lysine are usually first- and second-limiting amino acids for both growing birds and layers for all types of cereal grains and oilseed protein type diets.

Protein digestion or denaturation begins in the proventriculus with the production of pepsin and hydrochloric acid. The acid and pepsin begin the initial process of breaking down some of the peptide bonds to help expose the central core of the protein to further digestion in the small intestine. The majority of protein digestion takes place in the small intestine when the undigested feedstuff causes the pancreas to release pancreatic juices containing key protease digestive enzymes in an inactive form. The trypsin from the pancreatic juice is activated into an active form in the small intestine by a proteolytic reaction and then the trypsin activates other protease digestive enzymes such as *chymotrypsin, carboxypeptidase A and B,* and *elastase* in a similar manner. The digestive enzymes that attack the peptide bonds of protein are very specific for amino acids in the chain. Some of the protease enzymes hydrolyze peptide bonds inside the chain containing basic amino acids, aromatic amino acids, and aliphatic amino acids whereas some of the carboxypeptidase enzymes hydrolyze peptides with specific amino acids on the ends of peptide chains.

There are also specific protease enzymes in the intestinal mucosa that hydrolyze short and longer chain peptides. The proteins need to be completely digested into free amino acids for active absorption from five or more transport pathways or into short chain peptides that are absorbed and catabolized inside the intestinal cells. The amino acids are metabolized primarily in the liver and kidney of poultry into key intermediates such as *creatine phosphate* needed for muscle energy, *glutathione* needed for key oxidation reduction systems in the body or into blood proteins that are necessary for homeostatic mechanisms. The amino acids that pass through to muscle tissue can be utilized to synthesize proteins if all of the amino acids are available. Amino acids are also catabolized in the liver, kidney, and muscle tissue of poultry into metabolites that can be converted to glucose or ketones for tissue energy and the nitrogen is eliminated as uric acid in the excreta. A key problem with amino acids is a lack of body storage which provides a short amount of time that dietary amino acids will be useful for forming key intermediates or making muscle because the amino acids will be metabolized.

An Ideal Protein

An "Ideal Protein" profile of a feedstuff or diet is one that contains the correct proportion and quantity of the essential and non-essential amino acids without having an excess. An "Ideal Protein" may be different for maintenance than for growth or egg production. Larger birds or layers in the same flock with smaller birds will have different maintenance needs which will affect the overall optimum "Ideal Protein" needed. Realistically, there is no such thing as an "Ideal Protein" diet because the only way an excess of some amino acids could be avoided during formulations with natural ingredients is to feed only the crude protein needed to provide the least limiting amino acid and then add synthetic amino acids for the remaining amino acid requirements. Presently, the price of synthetic amino acids other than methionine, lysine, and threonine are too high to use in feed formulations.

Essential Amino Acids

Essential amino acids are those that cannot be synthesized by the animal. For all species of poultry, they include *methionine, arginine, tryptophan, threonine, histidine, isoleucine, leucine, lysine, valine,* and *phenylalanine. Glycine, serine,* and *proline* are considered as semi-essential amino acids for growing poultry because the bird can synthesize these amino acids, but cannot produce enough for optimum growth. The requirement for glycine and serine is considered together because either amino acid can be used to synthesize the other amino acid. Proline can be synthesized in poultry from glutamic acid. Likewise, *tyrosine* and *cystine* are not considered essential because the bird can synthesize tyrosine and cystine from phenylalanine and methionine, respectively. The amount of dietary tyrosine and cystine in feeds is also important because phenylalanine and methionine do not have to be catabolized to synthesize these amino acids.

Non-essential amino acids are also required by poultry for growth and egg production, however the bird has the ability to synthesize these amino acids. The bird primarily needs a form of nitrogen such as amino acid nitrogen or diammonium citrate that can be converted to ammonia. In practical feed formulations, specific non-essential amino acids will be synthesized in growing birds or layers from dietary non-essential amino acids and catabolized essential amino acids from the protein feedstuffs. An optimum ratio of dietary essential and non-essential amino acids for broilers for optimum gain, feed conversion, and carcass composition has been reported to be 55:45.

The amino acid concentration needed in poultry feed for optimum performance may be affected by many factors. Research suggests the response of poultry to amino acid levels is affected by environmental temperature, sex, species, immunological stress, age, ME concentration, ionophore coccidiostats in the diet, and amino acid balance. Researchers have explained these effects by suggesting the variables are changing feed intake therefore altering the amount of amino acid consumed. Amino acid antagonism caused by a structurally similar amino acids such as arginine and lysine, or the branch chain amino acids (valine, isoleucine, and leucine) actually change the efficiency of utilization of the amino acids.

Amino Acid Digestibility, Absorption, and Metabolism

There are many types of feedstuffs that are not equal when based on digestibility in the intestine of chickens. Some feedstuffs contain antinutritional factors that affect endogenous protease activity, passage rate, or absorption of the amino acids by the villi in the small intestine. Some feedstuffs also contain non-digestible carbohydrates that may bind the amino acids. Feeds that need to be heated to destroy inhibitors or potential pathogens also provide opportunities for overcooking or undercooking the feedstuffs.

Total vs Digestible Amino Acid Values

Formulating poultry feed based only on total amino acid values may cause (1) wasted protein and amino acids because their margin of safety is too large or (2) underfeeding of certain amino acids because the protein source is less digestible. The formulation of feeds using digestible amino acids allows the nutritionist to formulate closer to the actual requirement

Table 14-3. Mean and (Range) Digestibility and/or Availability Estimates (%) of Some Amino Acids in Various Feedstuffs Summarized from Studies with Poultry

Feedstuff	Lysine	Methionine	Cystine	Arginine
Corn	88 (84–91)	94 (93–95)	93 (86–100)	91 (88–92)
Wheat	92 (88–95)	97 (93–99)	92 (83–98)	93 (91–95)
Sorghum	67 (42–92)	64 (45–98)	56 (10–98)	67 (35–96)
Soybean meal	91 (68–100)	92 (64–100)	87 (58–100)	90 (66–92)
Canola meal	84 (68–94)	84 (72–98)	83 (74–96)	86 (66–92)
Corn Gluten meal (60%)	94 (92–97)	96 (89–99)	92 (89–96)	98 (97–99)
Fish meal	86 (69–98)	91 (79–102)	90 (86–92)	84 (74–96)
Mean and Bone Meal	86 (73–104)	78 (34–98)	65 (59–66)	88 (84–90)
Cottonseed Meal	68 (48–89)	74 (56–93)	—	90 (84–96)
Feather Meal	70 (5–95)	80 (58–95)	76 (39–97)	83 (55–97)
Poultry By-Product meal	95 (88–98)	97 (94–99)	95 (93–97)	90 (82–98)
Bloodmeal	82 (55–101)	92	88	90 (89–92)
Mean	84	87	83	88

Parsons, 1985

with less margin of safety and less digested and non-digested protein and amino acid nitrogen lost in excreta. The excess nitrogen in poultry waste from poorly digested proteins and from overfeeding protein and amino acids will increase environmental problems caused from contamination of the fresh water supply. The ability to formulate diets using digestible amino acid values will also provide an economic assessment of lower quality protein sources.

A range of digestion percentages for methionine, cystine, lysine, and arginine found in poultry feeds are shown in Table 14-3. The large variation in digestibility of amino acids from the feedstuffs emphasizes the importance of formulating poultry diets with known digestible amino acid values when possible. Many of the companies that market synthetic amino acids have conducted extensive research and have complete listings of average digestible amino acids from large numbers of samples for a wide variety of feedstuffs. Nutritionists using high quality protein sources will need less space in the diet to provide the needed amino acids, and therefore will have more space available for energy and other components (Table 14-3).

14-H. TIME REQUIRED FOR FOOD TO PASS THROUGH THE ALIMENTARY TRACT

Many factors affect the flow of food through the alimentary tract. The signal from the gizzard for more food will determine the length of time that feed remains in the crop. Sometimes it may be only minutes; at others it may be several hours. If the feed is in fine form it can pass the gizzard in a very short period of time, but if it is coarse it must first be broken down into small particles before it can enter the intestines. Some feed may leave the gizzard after a few minutes; in other cases, as with whole grains, the grinding action may take hours. Actually, the entire process of digestion is rapid. If the alimentary tract is empty, feed will pass through it in about 3.5 hours. When feeding is more or less continuous, the entire process of transfer will take about 12 hours.

14-I. CHRONOLOGICAL AGE EFFECTS ON DIGESTION AND ABSORPTION CHANGES

Young broiler chicks cannot utilize some sources of nutrients as effectively as older birds, thus the young birds gain less metabolizable energy from these feeds. Research shows that digestive enzymes, bile salt secretion, and absorptive efficiency of the gastrointestinal tract for young chicks increase at different rates during the first two to three weeks of age. Nutritionists should always use the appropriate digestibility and ME values of feedstuffs when formulating feeds for different ages and types of poultry.

15

Major Feed Ingredients: Feed Management and Analysis

by Craig N. Coon

Commercial poultry rations today are known as *complete rations;* that is, they contain all the essential ingredients for the bird to perform well, whether it be in growth, feather renewal, egg production, or the production of meat. For the most part, the bird, being closely confined to its quarters, has no other source of food material. Therefore, its nutrients requirements must be gotten from the feed it is given each day.

Certain components of the feed come from the common and major feed ingredients such as cereal grains, protein and fat supplements, certain mill by-products, and the major minerals. But in most cases, a mixture of these ingredients would not satisfy the bird's total nutritional requirement, nor would it be economical. Certain vitamins, minerals, by-products, and other ingredients must be added to "balance" the diet. This chapter deals with the major feed components; Chapter 20 includes the minor components including the vitamins, minerals, and trace ingredients.

15-A. CARBOHYDRATES

Weight per Bushel

Cereal grains and soybeans are measured either in 100 lb (cwt) or bushels. The major ingredients and their bushel weights are as follows:

Weight per bushel

Grain	*Lbs.*	*Kilo*
Barley	48	21.8
Corn, shelled	56	25.4
Sorghums	56	25.4
Oats	32	14.5
Rice, rough	45	20.4
Soybeans	60	27.2
Wheat	60	27.2

Barley

(Note: Refer to Tables 15-1 and 15-2 for analyses of feedstuffs.)

Barley is produced abundantly in some areas and is used in many poultry feeds as a fine-ground ingredient. Compared with corn, it contains about 75% as much energy and three times as much fiber. Therefore, its use is limited, especially in feed mixtures that must be high in energy and low in fiber. Although the fiber of barley is practically indigestible, the grain may be soaked at high temperatures or treated with enzymes to improve its qualities. The cost of energy in normal barley must be considered when it is substituted for a high-energy cereal such as corn. In many areas it would be uneconomical to use.

Cassava

Cassava or cassava root is produced in abundance in many tropical areas under a variety of names: *mandioca, manioca, tapioca, yucca,* and *manioc.* By enzymic action, the roots release a poisonous compound, *prussic acid.* Special washing is necessary to make the root edible. In ground form, cassava root may replace up to half the cereal grains in a ration if its low levels of methionine and protein are provided for.

Corn (Maize)

In most areas, corn is the predominant source of energy in poultry feeds, mainly because of its availability, price, and high digestibility. Corn is, however, a variable cereal grain, and in many countries is sold by "grade," which gives an indication of its moisture content, weight, kernel composition, and the presence of foreign material. Corn also has a variable protein content, ranging from 8 to over 11%. Most corn is now the result of hybrid breeding in an endeavor to produce plants adaptable to certain climates, rainfall, and soil composition. Corn is a good source of *linoleic acid,* an essential fatty acid.

Yellow corn. Yellow corn contains an abundant quantity of carotenoid pigments called *xanthophylls,* which impart yellow pigment to the fat deposits of chickens and to egg yolk. Yellow corn is a fair source of vitamin A activity, but storage can reduce its content by as much as 30%.

White corn. White corn is similar to yellow corn in most respects except that it contains little or no xanthophyll and has practically no vitamin A activity.

High-lysine corn. A special hybrid corn has been developed that is high in the amino acid, lysine, but costs more to produce because of lower yields. The hybrid is specifically known as *Opaque-2,* after the gene responsible. The corn contains about 11% total protein, about 30% more than normal dent corn. The amount of lysine is about 50% greater than the lysine content of normal corn. It is questionable whether the hybrid can be economically fed to chickens at current price levels for synthetic lysine and conventional corn.

High-oil corn. The selection of corn to increase the oil content has been occurring for over a century. The oil content has been shown to increase, but the yields were much lower than standard commercial varieties. Recently, with a technique patented by DuPont, scientists have been able to produce equivalent yields of high oil corn varieties comparable to regular commercial dent corn varieties. The scientists use male-sterile female hybrids that contain disease resistance, root and stalk strengths, and yield potential and then pollinate the elite hybrids with male pollinators with the high oil gene. The oil content averages 6.8% and crude protein content is approximately 8.7%. The economics of using high oil corn in poultry rations will depend upon the costs or regular corn and the costs of feed fat available in a given area. The high level of oil in the corn will increase the *True Metabolizable Energy* (TMEn) of the grain. Regular corn with 3.5% oil has been shown to contain 3,408 kcal TMEn/kg compared to high oil corn with 6.8% oil and 3,615 kcal TMEn/kg.

Molasses

Usually, molasses is a by-product of the cane sugar and beet sugar industries. Beet molasses contains about 6% protein; cane molasses, about 3%. Although both are relatively high in energy, molasses is primarily used in poultry feeds to prevent dustiness. Care must be exercised in mixing to prevent balling of small molasses particles.

Oats

Although an excellent feed for chickens, oats are limited in their use. They contain a large amount of fiber because of their husk, and are therefore low in energy. With a fiber content of about 12% compared with 2% for corn, oats contain only about 75% as much energy as corn. In most instances, the energy from corn is more economical than from oats. Because of this, oats cannot be used in quantity in a high-energy broiler ration; their value lies in growing, laying, and breeding feeds. Because oats vary in weight, their protein content is highly variable. When incorporated in a mash, oats should be finely ground in order to pulverize the hulls thoroughly.

Oats have also been incorporated in pullet developer feeds with good success. Growers report pullets with better developed digestive systems when oats are used. This is thought to be attributable to the higher fiber levels in oats-based feeds. Oats have also been used with good results as the only feed used during the first four weeks of an induced molt in layers, in place of the usual feed removal practice.

Rice

Rice is second to wheat in worldwide production. However, only where it is produced in abundance is any incorporated in poultry feeds, and then the use of only inferior grades and broken kernels is common. New varieties of rice have materially increased the yield of rice by several times. They are short-strawed, lodging-resistant, respond better to nitrogen as a fertilizer, and have a much shorter growing period.

Sorghums

There are several sorghum grains, but kafir and milo are the two generally used in poultry rations. Sorghums are grown extensively in many areas and make up an important part of many poultry feeds. Nutritionists need to be aware of variable analysis of sorghums dependent upon environmental conditions and genetic variety. Although somewhat unpalatable in ground form, they may be used effectively to replace two-thirds of the cereal grain portion of most rations. If the feed is pelleted, the percentage can be higher. Kafir and milo are quite comparable to yellow corn in feeding value except that they have no vitamin A activity or any pigmenting xanthophylls.

Bird-resistant sorghums. Special strains of sorghums, high in *tannins,* have been developed to prevent wild birds from eating the grain in the fields. In general, darker-colored sorghums contain more tannin than lighter-colored sorghums. Tannins are known to cause growth depression in chickens, and egg mottling in the yolks of eggs produced by layers. Bird-resistant sorghums should not replace over 40% of the cereal grain portion of the ration.

High-lysine sorghums. These variants have been shown to produce results superior to normal sorghums because of their higher protein content.

Wheat

Whole wheat has an energy relationship analogous to corn and contains a higher percentage of protein. The protein may vary between 10 and 17%, depending on the type of wheat and the area where grown. However, wheat is practical in poultry diets only when it is available in quantity and will provide an economical source of energy. Because of its great use in human diets, it generally carries a higher price.

Wheat is *gelatinous,* and when ground and used at high percentages, it has a tendency to "paste" on the beaks of birds. The pasting may sometimes produce *beak necrosis.* If the wheat incorporated in a poultry mash is coarsely ground, or if the feed is pelleted, most of the difficulty is overcome. Wheat has no vitamin A activity or pigmenting properties.

15-B. MILL BY-PRODUCTS

Rice Bran

Rice bran is composed mainly of the pericarp and germ of rice as a by-product of the milling of raw rice to produce an edible product. It contains about 13% protein, slightly less than wheat bran, and about 90% as much energy as corn. The high fat content of rice bran (13–15%) makes it a fairly good poultry feed.

Wheat By-products

Wheat bran. Wheat bran is composed of the outer layer of the wheat kernel. It is one of the by-products of wheat milling and contains about 15.6% protein and 510 kcal ME/lb (1,322 kcal ME/kg).

Figure 15-1. Grain Elevator

Wheat middlings, shorts. These are mixtures of milling by-products including the finer particles of bran, germ, flour, etc. Wheat shorts contain about 16% protein and 890 kcal of ME/lb (1,958 kcal ME/kg).

15-C. FATS AND OILS

Although the fat content of a feed is usually calculated as the percentage that will dissolve in ether, known as lipids, fats are better identified with only pure fatty acid esters of glycerol, called *triglycerides*. Fats are solid, while oils are liquid.

Fatty acids contain carbon, oxygen, and hydrogen and are classified as *saturated, monosaturated,* or *polyunsaturated.* A saturated fatty acid contains all the hydrogen it can hold; a monosaturated fatty acid has room for two additional hydrogen atoms per molecule; and polyunsaturated fatty acids have room for four or more hydrogen atoms.

It has long been known that when the vegetable oil content of the diet is increased, egg size is larger, even when the total calories in the ration remain the same. Most of the effect is due to the increases of readily absorbable fatty acids including *linoleic* and *oleic* acid in the vegetable oil. Even when large amounts of yellow corn are used in the diet with no added fat, some rations may be marginal for these fatty acids. The problem may become acute when milo, barley, or oats are substituted for corn. Most commercial feed ingredients are low in linoleic acid. Consequently, fats and oils, such as certain stabilized vegetable oils, should be added to many rations in order to prevent a deficiency of linoleic acid, which has a requirement of about 1.0% of the ration.

Types of Fats and Oils

Because of their high energy content, relatively large amounts of fats or oils are added to some poultry rations, particularly in broiler diets. As a added benefit, they reduce the dustiness of the mixed feed and improve its palatability. Up to 8% of a commercial diet can be added fat; however, chickens can tolerate over twice this amount. In many instances the practical use of fats or oils is determined by the price relationship between their energy and the energy derived from corn, milo, wheat, and rice. When fat energy is inexpensive compared with the energy of any of these grains, it is economical to use more fat or oil. There are several feed grades of fats:

1. *Hard fats.* Most of these are solid at room temperature and come from slaughtered cattle; they are known as *tallow* and *lard.* Their melting point is above 104°F (40°C).
2. *Soft fats.* These are semisolid, and are termed *greases.* Their melting point is below 104°F (40°C).
3. *Hydrolyzed animal fats.* These are by-products, mostly from the manufacture of soaps, and are sold as *hydrolyzed animal fat* or *hydrolyzed vegetable fat.* They must contain no less than 85% total fatty acids.
4. *Vegetable oils.* Oils in this group come from plants such as coconut oil, corn oil, soybean oil, palm oil, canola oil and so forth, and are used as an energy source in poultry feeds.

The following table shows the comparison between corn and several fats in regard to their metabolizable energy content and the utilization of the energy by chickens:

	Approximate Kcal of ME per		*Energy Utilization*
	Lb	*Kilo*	*%*
Corn	1,530	3,366	70
Lard	4,000	8,800	80
Hydrolyzed animal and vegetable fat	3,400	7,480	72
Grease (yellow)	3,400	7,480	84

Tallow (beef, feed grade)	3,130	6,886	80

Omega-3, -6 Fatty Acids

In a review written by Miles and Jacob (1998), the authors stated that current research findings have given us a better understanding of the growing role of nutrition in both the development and prevention of chronic disease. Animal tissues require the *omega-3* and *omega-6* fatty acids for their proper functioning and good health. The omega-3 and omega-6 families of compounds are derived from *linolenic* and *linoleic* acid, respectively. The omega-3 polyunsaturated fatty acids have been found to protect against heart disease and some cancers. Animals cannot synthesize either linolenic or linoleic acids, so these essential fatty acids must be supplied by the diet.

The review by Miles and Jacobs also describes how the practical application of omega-3 research in the egg industry has resulted in the development of "designer eggs" which are high in omega-3 fatty acids. The development and marketing of omega-3 rich eggs meets the shifting demand of consumers away from the traditional food choices toward those offering more health-protecting and therapeutic benefits.

Antioxidants for Fats

Fats, particularly the unsaturated fatty acids, are subject to *oxidative rancidity.* To prevent oxidation, antioxidants are usually added, particularly if the fats are to be stored. Several commercial products are available including ethoxyquin and BHT.

15-D. PROTEINS OF ANIMAL ORIGIN

Dried Blood

This protein supplement, composed of ground dried blood, contains about 80% crude protein and is an excellent source of the amino acid lysine, of which about 80% is available to the bird. Blood meal contains very high levels of leucine and a disproportionate low level of the amino acid isoleucine, therefore rendering it a protein of poor quality, and only token amounts should be included in the ration if maximum growth and egg production responses are to be realized.

Meat By-products

Two meat by-products are of value in poultry feed formulation. Although for breeding birds, vegetable protein supplements have essentially replaced meat by-products in today's rations. Meat products are often excluded from poultry diets because of *Salmonella* contamination concerns. Even though the rendering process destroys the organism, the meal is often subject to recontamination. Poultry breeders and others commonly avoid using animal products in their rations for this reason.

Meat scrap. This is a dry-rendered product made from animal flesh and tissues and contains about 50 to 55% protein. It must be guaranteed low in phosphorus indicating that little or no bone was incorporated. It is high in lysine, but low in methionine, cystine, and tryptophan. Where meat scrap is used in poultry rations, it is normally limited to 5 to 10% of the ration.

Meat and bone meal (scrap). This product is more readily available than meat scrap and is a good supplement with 47 to 50% protein. It contains a high percentage of ground bone, making it a source of both calcium and phosphorus. Up to 10% may be used in the ration.

Poultry By-product Meal

This product consists of ground dry-rendered poultry offal including the heads, intestines, and other organs, but excluding the feathers. It contains 55 to 60% protein, and unless extracted, about 12% fat. It is considered as an excellent source of protein for both meat- and egg-type chickens.

Poultry Feather Meal (Hydrolyzed)

Hydrolyzed poultry feather meal contains at least 70% protein, of which 75% is digestible. However, the protein is high in cystine and deficient in the amino acids methionine, tryptophan, histidine, and lysine. Feather meal must be used sparingly in the ration, with thought given to its deficiencies. It should not replace more than 10% of the soybean oil meal in the ration.

15-E. PROTEINS OF FISH ORIGIN

There are many types of protein supplements derived from fish, with variations arising from the many types of fish and the part of the fish used in producing the meals. The number of different products is also increased because of the four different methods of processing: (1) sun-dried, (2) vacuum-dried, (3) steam-dried, and (4) flame-dried. Of the four processing methods, only vacuum-dried and steam-dried have any commercial significance, as sun-dried meal is usually of low quality and little flame-dried product is available today.

Most fish meals are a source of good-quality protein for poultry diets because of their well-balanced amino acid profile. But all fish meals are not similar in their makeup of amino acids or in their digestibility. Broadly, fish meals may be grouped into two categories.

1. *Whitefish meals.* These are processed from the non-edible portions of tuna, cod, halibut, and other fish, and are low in fat.
2. *Dark fish meals.* These come from such fish as sardine, herring, menhaden, anchovies, etc., and are usually high in fat.

Fish meals vary in their content of crude protein from 55 to 75%. For instance, herring meal is high, menhaden and sardine meals are medium, and tuna meal is low in protein.

Antioxidant in Manufacture

Antioxidants are added to many fish meal products to prevent oxidation. This materially improves the value of the meals.

Salt in Fish Meals

The salt content of fish meals must be carefully monitored when certain processing practices are used and for different sources of fish meal. As salt produces a laxative effect in the chicken, the salt content of the various fish meals should be carefully and routinely determined. Meals should contain less than 3% salt for best results, but legally may contain up to 7%. Formulation of diets which include fish meal should be based on *tested levels* of sodium and chloride rather than on book values because of product variability.

Pricing Fish Meals

Because of their variability in protein content, fish meals are usually priced on the units of protein for the particular product.

Amount of Fish Meal in the Diet

Because of their relatively high costs coupled with a usual shortage of supply, fish meals are included at about 5% in broiler rations and about 2% in other poultry rations. However, levels up to 8% will usually show productive improvement. Fish products are considered by some to include *unidentified growth factors* (UGF) and therefore are commonly used to satisfy this nebulous "need."

Fish Flavor in Meat and Eggs

The oil from fish carries a definite "fishy" taste and odor that can be transferred to the poultry meat and eggs when the diet contains more than 6 to 10% fish meal or 1% fish oil, depending on how much fat is in the meal.

Fish Solubles

The wet processing procedure of producing fish meal leaves a water byproduct, known as *stick*, that may be condensed or dried. The value of these products lies not in the fish protein, but in vitamin B_{12} and certain UGF. Fish solubles can also act as a laxative, even more effectively than dried skim milk or dried buttermilk.

15-F. PROTEINS OF VEGETABLE ORIGIN

After cereal grains, protein supplements of vegetable origin comprise the largest component of most poultry rations. Soybean oil meal is most commonly used because of its available supply, good nutritional value, and relative economy. The objective of most US nutritionists is to build a diet composed of corn and soybean oil meal, adding other ingredients only to compensate for their deficiencies. Nutritionists in other countries have the same objectives, but with local or easily obtained feedstuffs.

Many of the vegetable protein supplements are derived from seeds that have had their oil extracted and have been further processed, as in their raw form they cannot be utilized efficiently by chickens. The seeds must undergo heat or other treatment to eliminate certain toxic factors. Such treatment thereby increases the nutritional value of the seeds.

Corn Gluten

Corn gluten comes in two forms:

1. *Corn gluten feed.* This is the part of the corn remaining after extraction of most of the starch and germ when making corn starch and syrup. It contains about 22% protein.

2. *Corn gluten meal.* The meal is similar to corn gluten feed, except that the bran portion of the corn kernel has been removed. Although a good source of vegetable protein (60% protein), its main value lies in its ability to provide yellow pigment to the skin of chickens and to egg yolks.

Coconut (Copra) Meal

Coconut meal is the result of grinding the portion of the coconut remaining after the oil has been extracted. Its average protein content (solvent product) is about 22%. Ten percent of the diet seems the limit for optimum coconut meal usage. The meal has a low energy value because of its high fiber content (14%) and is also low in methionine and lysine.

Cottonseed Meal

This meal is generally available in many areas and is the product remaining after oil is extracted from cottonseed. The expeller process was first used, but in many instances has been replaced by the solvent extraction process, which removes more oil from the seed leaving less in the meal.

Although cottonseed meal is a vegetable protein of good quality with about 41% protein, it is inferior to soybean meal. Dehulled cottonseed meal will contain 45% protein. Neither should be used as the only vegetable protein source in the ration because of the presence of *gossypol* and their low levels of lysine.

Gossypol content. Cottonseed oil contains gossypol in minute quantities, with the amount left in cottonseed meal after oil extraction being adequate to cause the production of eggs with pink to dark mottled yolks. Free gossypol is toxic and reduces growth and egg production.

Linseed (Flax) Meal

This product is unpalatable and is generally not suitable for poultry feeding, but in the absence of good vegetable protein supplements a modest amount could be incorporated in the ration. Linseed oil meal contains large amounts of omega-3 fatty acids and is presently being used to increase omega-3 fatty acids in special marketed eggs.

Peanut (Groundnut) Meal

Peanut meal is a good vegetable protein supplement and, where available, large amounts may be used in the ration. It contains 42 to 50% protein depending on how it is processed. Although peanuts contain a trypsin inhibitor, it is destroyed in the heating process. Peanut meal should not replace more than 10% of the soybean oil meal in the ration. Care must be taken when feeding peanut meal as it is susceptible to the production of *mycotoxins*. Peanut meal is also very low in the essential amino acid lysine.

Rapeseed Meal (Canola Meal)

While rapeseed oil meal has a good amino acid balance, containing 38% protein, it should be fed cautiously as it tends to irritate the digestive system. Meal made from the older varieties of rapeseed should not be used at levels above 10% of the diet, and preferably not above 5%. Such meals, when fed in excess, cause liver degeneration, thyroid hypertrophy, reduced feed efficiency, and loss in egg production as they contain high levels of *glucosinolate* and *erucic acid*.

New rapeseed varieties (canolas) have been developed in recent years that are low in glucosinolate, and erucic acid is neglibible with the meals produced from these varieties capable of replacing up to 75% of the soybean oil meal in the ration for most poultry. Canola should be restricted for brown-egg layers because it contains 1.5% *sinapine* that will produce fishy-flavored eggs from certain breeds/strains of brown-egg layers.

Safflower Meal

Decorticated safflower meal has historically been used in moderate amounts in poultry rations. Since it is low in lysine, it should not be fed at levels above 5% during the first 5 weeks of a chick's life, and 15% thereafter. Much more may be fed if adequately supplemented with lysine, however, the lower energy value of the meal, because of the 14% fiber, still restricts the economical use of large amounts in the feed.

The following two products are generally available:

1. 27% protein product
2. 32% protein product (dehulled)

Soybean Meal

Historically, soybean meal was considered a by-product of oil extraction, but today, because of its widespread use and value, the meal may be of equal economic importance. The abundance of soybean production in different regions of the world and the high nutritional value of the processed bean have made it possible to use high percentages of the meal in most poultry rations. To properly balance the amino acids in soybean

Figure 15-2. Soybeans—Ready to Harvest

meal, supplementation with methionine and lysine from animal or fish meals, or synthetic amino acids is usually required. Raw soybeans should not be fed. They contain a *trypsin inhibitor* (trypsin is an enzyme associated with protein digestion) that must be destroyed by heat treatment. Soybean meal contains from 43 to 50% protein, depending on the method of processing:

1. *Expeller soybean meal.* This process does not remove as much valuable oil as solvent extraction, although the meals are nutritionally comparable. It has 43% protein.
2. *Solvent soybean meal.* Solvent extraction of the oil from soybeans is predominantly in use today. The resulting meal is of excellent quality, although lower in fat than those resulting from expeller processing. The protein content is 44%.
3. *Dehulled (solvent) soybean meal.* A meal higher in protein (47–50%), lower in crude fiber (3.3%), and higher in energy can be produced by removing the hull from the soybean prior to the oil being extracted. When higher energy diets are required, such as with broilers, dehulled meal is recommended.

Full-Fat Soybeans (Roasted or Extruded)

The use of full-fat soybeans in certain countries may be a practical way to increase the metabolizable energy level of their rations. The use of vegetable oils in poultry rations is often too expensive because of its use in human foods. Some countries also have limited access to animal fats because of religious and other reasons. In such cases, full fat soybeans may be economically added to supply high energy lipid calories to the poultry diets. Full fat soybeans are prepared by cooking, roasting, or by extruding. The soybeans have to be heat treated just like soybean meal to destroy the antinutritional components. The average crude protein content of full fat soybean meal is 38%. The ME is reportedly 1,523 kcal/lb (3,350 kcal/kg) and is dependent upon the processing method utilized.

Sunflower Seed Meal

While low in lysine, dehulled sunflower meal contains from 38 to 44% protein. It may be substituted for 50% of the soybean oil meal in the ration, and up to 100% if lysine is added. Sunflower seed meal is sticky when being consumed and may cause necrosis of the beak at higher levels. Pelleting a feed containing sunflower seed meal will prevent the stickiness on the beak. The product is becoming more available because of increases in sunflower production.

15-G. GREEN LEAFY PRODUCTS

The tops from many grasses and legumes may be dried and fed to chickens as a source of β-carotene, xanthophyll, and for the UGF. Some green leafy products are good sources of vitamin K. Most common are from alfalfa products.

Alfalfa Products

There are several alfalfa products resulting from different methods of curing hay and the portion of the plant used to make the meal.

1. *Sun-cured alfalfa meal.* Originally, alfalfa hay was sun-cured and ground, but the product was highly variable. This was probably due to the variations in moisture content and the indefinite time required for drying.
2. *Dehydrated alfalfa meal.* Alfalfa hays are now properly and uniformly dried artificially by heat, and then ground.
3. *Dehydrated alfalfa leaf meal.* This is a product made from only the leaves of the alfalfa plant.

Vitamin A Activity

Dehydrated alfalfa products are much higher in β-carotene than sun-cured products, however, the β-carotene in the meal is easily lost through oxidation. To prevent this loss, an antioxidant is added to the ground alfalfa, or the meal is pelleted to reduce air exposure. When pellets are used, they are ground prior to feed mixing.

Alfalfa meals are measured by their vitamin A activity instead of their β-carotene content. A meal of high quality should contain no less than 100,000 units of vitamin A activity per lb (454 g).

The use of dehydrated alfalfa meal in poultry diets has decreased significantly because current vitamin premixes now have competitively priced synthetic vitamin K and xanthophylls, and because of alfalfa's relatively low ME values.

15-H. MACROMINERALS

This section is devoted to sources of the four major minerals, calcium, phosphorus, sodium, and chloride, commonly used in quantity in poultry diets.

Dicalcium Phosphate

Dicalcium phosphate comes from rock phosphate or bone after chemical processing. Dicalcium phosphate derived from rock phosphate may contain an appreciable amount of *fluorine,* most of which must first be removed before the product is acceptable for poultry feeding. Dicalcium phosphate contains approximately 18% phosphorus and 22% calcium.

Rock Phosphate

Much ground phosphate rock is so high in fluorine that the raw rock must be defluorinated (by exposure to very high temperatures) before it is fed. Such a product is sold as *defluorinated rock phosphate,* containing no more than one part fluorine to 100 parts of phosphorus. Defluorinated rock phosphate contains about 32% calcium and 8% phosphorus.

Steamed Bone Meal

A source of phosphorus that comes from bones of animals, steamed bone meal contains an appreciable amount of calcium. Most products contain about 30% calcium and 12.5% phosphorus.

Limestone

Used as a source of feed calcium, limestone contains 38% calcium. Care should be taken to not use a limestone that contains appreciable amounts of magnesium (up to 10% magnesium), sometimes known as *Dolomite limestone.*

Oyster Shell

Oyster shell is an excellent source of supplemental calcium, especially for egg-producing birds, because of its availability to the bird and its particle size. Oyster shell is composed of about 94% calcium carbonate (38% calcium).

Salt (NaCl)

Salt is a source of sodium and chlorine. Although necessary in small quantities, large amounts in the diet increase water consumption and have a laxative effect. Generally, no more than 0.25% of free salt is added to the poultry ration. Deficiencies and excesses of salt are both very harmful to poultry and problems of this nature are quite common.

15-I. ANALYSIS OF FEEDSTUFFS

In order to formulate poultry rations it is necessary to have tables showing the analyses of the various feedstuffs. These values are necessary to build formulas that are properly balanced for the type and age of birds involved and for the environment under which they are kept. Many practicing nutritionists prefer to develop their own tables based upon current ingredient analyses, as local conditions may require changes in nutrient standards. The ability to formulate a nutritionally sound and economical diet is the result of both experience and training in the field of poultry nutrition.

There are many combinations of feedstuffs that could provide the calculated requirements for growth and reproduction, yet many would not be good diets. Many of the feedstuffs could have deleterious effects when given at percentages greater than the optimum, as they could be unpalatable, toxic, or otherwise impractical.

But once the specified amounts of a feedstuff are included in a formula, the analysis tables provide a basis for determining whether the minimum nutritive requirements for carbohydrates, fats, proteins, minerals, vitamins, and so forth have been met.

15-J. EXPRESSION OF NUTRITIVE REQUIREMENTS

There is no consistent manner in which the component parts of a ration or the nutritional requirements are expressed. Some of these variations are due to the fact that all countries and all scientists do not use the same units of measure. Some of these variations are given below.

Major Feed Ingredients

Usually these are expressed in percentages by weight.

Minor Feed Ingredients

Vitamin A is most often given as *International units* (IU) per pound or kilo. Vitamin D_3 is expressed as *International chick units* (ICU) per pound or kilo. In the case of vitamin E, IU or milligrams per pound or kilo are used. Most other vitamins and trace minerals are expressed as milligrams, while amino acids are given as a percentage.

Useful Conversion Factors

See *Appendix* for useful conversion factors associated with feed formulation.

Energy Terms

Small calorie (cal). A small calorie is the amount of heat required to raise the temperature of 1 g of water 1-degree C and is designated by the small letter "c." The small calorie is seldom used as a measure of heat or energy in animal nutrition.

Large calorie (Cal). The large Calorie is the amount of heat required to raise the temperature of 1,000 g of water 1-degree C. Thus, 1 (large) Calorie is equal to 1,000 small calories. The large Calorie is often spoken of as a kilocalorie (kcal), meaning 1,000 small calories. Often the energy value of a ration is given only as calories, meaning large calories. It is conventionally capitalized when denoting a large Calorie.

Therm. One million small calories or 1,000 large Calories equal 1 therm. One therm also equals one Megacalorie (Mcal) which can also be used to describe a quantitative amount of energy.

Joule. In some European countries the use of joule is becoming more common as the measurement for a unit of energy, mainly because it was adopted by the *International Union of Pure and Applied Chemistry.* One small calorie is equal to 4.184 joules or one kilocalorie is equal to 4,184 joules or 4.184 kilojoules. One joule is equal to 0.239 calories.

15-K. ENERGY

The dietary energy concentration of poultry feed is the main factor regulating the optimum intake of all nutrients. Poultry are thought to regulate their feed intake based on their daily energy requirement. There is current research suggesting that the modern broiler may be more affected by their *fill* instead of strictly calorie consumption. Scientists are finding that broilers often consume as much feed of a high energy diet compared to low energy diets. Presently, it is thought if a nutritionist formulates a lower energy concentration diet, the bird will try to consume additional feed to obtain the calories needed. The percentages of other nutrients would need to be adjusted downward to correlate with increased daily intake. Depending on the quantity of energy concentration decrease, birds can adjust to the new feed if the bulk of the diet does not become a limiting factor. In certain situations if the bird density in cages or on the floor is high in proportion to feeder space, feeding low energy diets may not provide adequate caloric intake. The utilization of a high energy diet usually requires adding a feedstuff with a higher fat or oil content or directly adding feed grade fat to the feed. In theory, higher energy concentrations should decrease poultry feed intake but often the increased fat calories and energy concentration produces an increased output of gain or egg mass, thus only small decreases or no change in feed consumption occurs. A major concern for today's poultry nutritionist is being able to provide adequate caloric intake in various types of environments and stressful conditions at an economical price.

Determination of Feed Energy

The energy value of a feedstuff may be measured in several ways. First, there is the total or *gross energy* (GE), the energy released as heat by burning the feedstuff in a bomb calorimeter. But all of the GE consumed by the chickens is not used for productive purposes, as a considerable amount of energy is excreted in the feces. Metabolizable energy in feedstuffs is easily measured because the excreta contains both the undigested energy from the digestive tract and the non-retained metabolic energy from the urine. The metabolizable energy of a feedstuff is determined by subtracting the GE of excreta (feces and urine) from the GE of the feed. The gaseous products from digestion are not measured because they are not considered of major significance. The metabolizable energy (MEn) of a feedstuff should be corrected for nitrogen retention in order to compare metabolizable energy studies for different ages and types of poultry. The nitrogen correction is based on 8.22 kcal/g nitrogen retained. The MEn values of ingredients are used in this book and are primarily used in feed formulations.

The MEn values of practically all feed ingredients have been determined (see Table 15-1 for some of the practical ingredients). Notice that fats are the highest in MEn with 3,130 to 3,720 kcal per pound. Alfalfa products are the lowest with 500 to 640 kcal of ME per pound.

True Metabolizable Energy

The use of *True Metabolizable Energy* (TME) as a description of feed energy has created some confusion. The original method of Anderson et al. (1958) using a substitution method of replacing glucose with a test ingredient was used to determine *Apparent Metabolizable Energy* (AME). The method, however, provided an inherent correction for fecal and urinary energy (NRC, 1994). In 1978, Sibbald developed a 48-hour, force feeding system to determine AME and discovered that the AME caloric values were in error if not adjusted for metabolic fecal and urinary energy. When these adjustments were made, the TME could be calculated.

Since knowledge of the dietary energy concentration is critical for feed formulation, a nutritionist needs to continually evaluate feedstuff energy values used in their formulations. Nutritionists primarily use either the TME method with nitrogen correction or the AME method corrected for nitrogen. The TMEn method is based on force feeding test ingredients and collecting the excreta during a 48 to 72-hour time period depending on the type of feedstuff being evaluated. The assay also includes the determination of *endogenous energy lost* (EEL) during the assay and the EEL is subtracted from the excreta energy. The evaluation of energy by this system provides an energy value for the feed only, and will increase the energy value of the feed compared to not subtracting the EEL with this assay procedure. The AMEn system is based on feeding a balanced basal diet and then replacing part of the basal diet with the test feedstuff. The difference in energy balance between the basal diet and the basal diet with test ingredient is used to determine the energy value of the test feed. Free choice feeding is used in determining the AMEn of feeds, whereas specific levels of the individual feedstuff is used for the TMEn system. Scientists suggest the difference between TMEn and the AMEn for feedstuffs in practical feed formulations may actually be very small because the effect of free choice feeding larger amounts of test feedstuff reduces the effect of EEL on the AMEn value (Figure 15-3). The MEn values for ingredients are

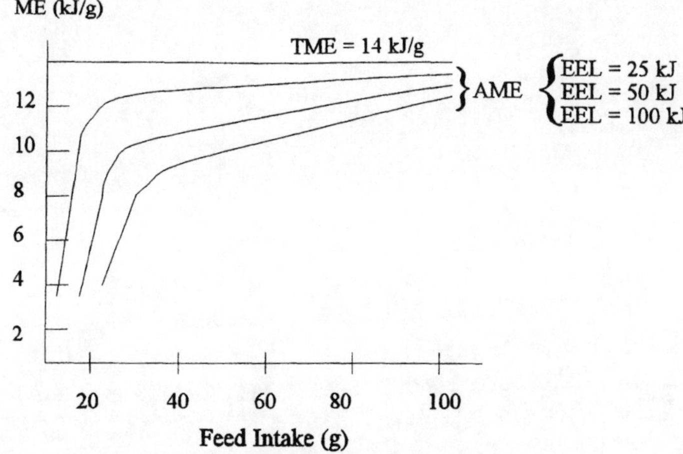

Figure 15-3. Relationship Between Apparent (AME) and True Metabolizable Energy Values (TME) with Different Endogenous Losses (EEL)

similar to TMEn values for most feedstuffs when feed intake is ad libitum for the MEn studies (NRC, 1994).

15-L. INGREDIENT ANALYSES

The analyses of some common poultry feedstuffs are given in Tables 15-1 and 15-2.

15-M. FEED MANAGEMENT

Most early poultry rations were used to supplement locally produced cereal grains and other feeds grown on the farm. When commercialism entered the poultry business chicken farms increased in size, birds were closely confined to houses, and the knowledge of poultry feeding increased.

Complete Feed

Little by little it became possible to formulate poultry rations that would include all the known nutrients. These were *complete feeds,* requiring no supplementation. However, feed formulation did not become static; new nutritional discoveries were made every year and the cost of ingredients changed, making formula substitutions necessary. Changes in the genetic

Table 15-1. Poultry Feed Ingredient Analysis

Ingredient	ME kcal/lb	Percent						
		Dry Matter	Protein	Fat	Fiber	Calcium	Total Phosphorus	Non-Phytate Phosphorus
Alfalfa meal (20% protein)	741	92	20.0	3.6	20.2	1.67	0.28	
Alfalfa meal (17% protein)	545	92	17.5	2.5	24.1	1.44	0.22	0.22
Barley, ground	1,200	89	11.0	1.8	5.5	0.03	0.36	0.17
Canola	909	93	38.0	3.8	12.0	0.68	1.17	0.30
Copra meal	693	92	19.2	2.1	14.4	0.17	0.65	
Corn, yellow, ground	1,523	89	8.5	3.8	2.2	0.02	0.28	0.08
Corn, gluten feed	795	90	21.0	2.5	8.0	0.40	0.80	
Corn gluten meal (60% protein)	1,691	90	62.0	2.5	1.3		0.50	0.14
Cottonseed meal	1,091	91	41.4	0.5	13.6	0.15	0.97	0.22
Defluorinated rock phosphate						32.00	18.00	
Dicalcium phosphate						22.00	18.70	
Dried bakery product	1,755	92	8.0	10.5	1.2	0.13	0.24	
Fat, stabilized								
Animal tallow	3,409	98						
Fish oil	3,841	99						
Hydrolyzed animal & vegetable fat	3,686	99						
Poultry oil	3,920	99						
Fish meal								
Herring (72% protein)	1,450	93	72.3	10.0	0.7	2.29	1.70	
Menhaden (58–65% protein)	1,282	92	60.0	9.4	0.7	5.11	2.88	
Anchovy	1,173	92	61.2	5.0	1.0	3.73	2.43	
Fish solubles, condensed	664	51	31.5	7.8	0.2	0.30	0.76	
Grain sorghums, Milo	1,494	87	8.8	2.9	2.3	0.04	0.30	
Hydrolyzed poultry feathers	1,073	93	81.0	7.0	1.0	0.33	0.55	
Limestone, ground (38% calcium)						38.00		
Meat and bone meal (50% protein)	977	93	50.4	10.0	2.8	10.30	5.10	
Oats, ground	1,159	89	11.4	4.2	10.8	0.06	0.27	0.05
Oyster shells, ground						38.00		
Peanut meal	1,136	90	42.0	7.3	12.0	0.16	0.56	
Poultry by-product meal	1,341	93	60.0	13.0	1.5	3.00	1.70	
Soybean meal (dehulled)	1,109	90	48.5	1.0	3.9	0.27	0.62	0.22
Soybean meal (44% protein)	1,014	89	44.0	0.8	7.0	0.29	0.65	0.27
Wheat, ground	1,318	87	14.1	2.5	3.0	0.05	0.39	0.13
Wheat bran	591	89	15.7	3.0	11.0	0.14	1.15	0.20
Wheat middlings	909	88	15.0	3.0	7.5	0.12	0.85	0.30

Source: National Research Council, 1994

Table 15-2. Poultry Feed Ingredient Analysis of Minor Nutrients

	mg per pound				%				
Ingredient	Riboflavin	Pantothenic Acid	Choline	Niacin	Arginine	Lysine	Methionine	Cystine	Tryptophan
Alfalfa meal (20% protein)	6.91	15.5	645	18.2	0.92	0.87	0.31	0.25	0.33
Alfalfa meal (17% protein)	6.18	11.4	637	17.3	0.69	0.73	0.24	0.19	0.23
Barley, ground	0.82	3.64	450	25.0	0.52	0.40	0.18	0.24	0.14
Canola	1.68	4.32	3045	72.7	2.08	1.94	0.71	0.87	0.44
Copra meal	1.59	2.95	495	10.8	1.97	0.50	0.28	0.28	0.12
Corn, yellow, ground	0.45	1.82	282	4.1	0.38	0.26	0.18	0.18	0.06
Corn, gluten feed	1.09	7.73	690	30.0	1.01	0.63	0.45	0.51	0.10
Corn gluten meal (60% protein)	1.00	1.36	150	25.0	1.82	1.03	1.49	1.10	0.36
Cottonseed meal	1.82	3.18	133	18.2	4.66	1.76	0.51	0.62	0.52
Defluorinated rock phosphate									
Dicalcium phosphate									
Dried bakery product	0.64	3.77	420	11.8	0.47	0.31	0.17	0.17	0.10
Fat, stabilized									
Animal tallow, feed grade									
Fish oil									
Hydrolyzed animal & veg. fat									
Fish meal									
Herring (72% protein)	4.50	7.73	412	42.3	4.21	5.47	2.16	0.72	0.83
Menhaden (58–65% protein)	2.23	4.09	138	25.0	3.68	4.51	1.63	0.57	0.49
Anchovy	3.23	6.82	300	5.5	3.81	5.07	1.95	0.65	0.78
Fish solubles, condensed	6.64	15.9	600	76.8	1.61	1.73	0.50	0.30	0.31
Grain sorghums, Milo	0.59	5.64	304	18.6	0.35	0.21	0.16	0.17	0.08
Hydrolyzed poultry feathers	0.95	4.54	405	12.2	5.57	2.28	0.57	4.34	0.55
Limestone, ground (38% calcium)									
Meat and bone meal (50% protein)	2.00	1.86	907	0.91	3.28	2.61	0.69	0.69	0.27
Oats, ground	0.50	3.54	430	5.5	0.79	0.50	0.18	0.22	0.16
Oyster shells, ground									
Peanut meal	2.36	21.3	752	5.5	4.35	1.26	0.45	0.52	0.39
Poultry by-product meal	5.00	5.59	270	18.1	3.94	3.10	0.99	0.98	0.37
Soybean meal (dehulled)	1.32	6.82	124	10.0	3.48	2.96	0.97	0.72	0.74
Soybean meal (44% protein)	1.32	7.27	1270	13.2	3.14	2.69	0.62	0.66	0.74
Wheat, ground	0.64	4.50	495	21.8	0.60	0.37	0.21	0.30	0.16
Wheat bran	2.09	14.1	560	84.5	1.02	0.61	0.23	0.32	0.23
Wheat middlings	1.00	5.91	654	44.5	1.15	0.69	0.21	0.32	0.20

Source: National Research Council, 1994

make-up of modern birds and improvement in the management of chickens require that changes be made in feed formulas as an on-going process.

Self-Feeding vs Controlled Feeding

Within limits, the chicken has the ability to control its feed intake according to its nutritional needs. In the early days chickens were self-fed, that is, a complete mash was kept before them at all times and they could consume all they needed.

Later, it was found that chickens did overeat and got too heavy or did not produce eggs or meat economically. Today, market broilers and commercial white and brown layer egg strains are self-fed, while the heavier broiler breeders must have their feed intake controlled (restricted) from three to four weeks of age through egg production (see *Feeding Broiler Breeders*, Chapter 19).

15-N. FORM OF FEED

Particle Size and Its Effect

Particle size of the mash mixture affects water consumption, as the coarser the texture, the less water the birds drink. Particle size also has a major effect on the extent of self-selection by the birds and on separation of feed components during transport down the feed trough.

Bulkiness Affects Water Consumption

The bulkier the diet (more fiber) the more water consumed, and, therefore, more water is voided in the feces. As high-energy diets are less bulky, the birds drink and excrete less water than when lower density diets are consumed.

Different forms of feed. Most poultry rations are available in either mash, crumble, or pellet forms:

1. Leghorns and brown-egg varieties: mash
2. Broilers: mash or crumbles for 3 weeks, then pellets

Mash Form

Many feed ingredients are purchased in a ground form; others, such as the whole grains, must be ground prior to mixing the ration. Mashes of complete feeds composed of cereal grains and oilseed proteins of medium particle size improve the bird's ability to eat them readily because finely ground mashes are usually too dry, sticky, and less palatable. Chickens have a tendency to pick out the larger cereal grain particles from the mash first, leaving the finer material until last. In the past, this was a problem with certain types of mechanical feeding systems, especially when installed in excessively long houses. Most of the problems have been corrected by changing the design of the delivery system and by moving the feed at faster rates.

Pellet Form

The mash may be compressed by running it through specialized equipment to form pellets of various sizes. The pelleting machine has a die with hundreds of holes of a specific diameter through which the feed is forced under pressure to form pellets.

With pellets, chickens cannot pick out certain parts of the feed, but must eat an entire pellet. This is particularly advantageous with young chicks as they are consuming such a small amount of feed and all nutrients must be obtained in each day's food intake.

Producing firmer pellets. Steam is added to the mash during pelleting to produce a firmer pellet. This moisture, plus the heat generated during pelleting, increases gelatinization of the mixture, thereby forming a firmer pellet. When excess fat (>3%) is added to the feed mix, the pellets tend to crumble because the fat acts as a lubricant rather than an adhesive. Fat may be added in larger quantities to pelleted feed by spraying it on after the pellet is made.

Physical Makeup of Pellets

Size of pellets. The size of pellets is determined by their diameter and length. A knife cuts the material extruded from the die into pellets of varying lengths; however, pellets are merchandised according to their diameter rather than their length. Broiler chicks are usually started on mash or crumbles, then changed to pellets at 3 to 4 weeks of age.

Fine material not a disadvantage. Although pellets look better if there is no fine material, experimental work has shown that small amounts of "fines" are not detrimental to feed consumption or feed conversion, but large amounts increase feed waste and reduce growth, particularly in broilers.

Pelleting alters nutritive value. Pelleting improves palatability and increases nutrient availability of certain feedstuffs by denaturing protein and gelatinizing carbohydrates. Pelleting a mixed feed containing a poor source of soybean meal that has been undercooked may improve the protein digestion of the diet by destroying some of the protease inhibitors. The heat generated during the pelleting process will destroy some of the carotene (provitamin A) in feedstuffs, and it is wise to increase the *minimum* allowances by 10 to 20% to compensate for this. However, pelleting destroys some trypsin growth inhibitors found in soybean meal, which is an offsetting advantage.

Advantages and Disadvantages of Pellets

The production of pellets is an expensive procedure. If the costs are to be regained, the advantages of pelleting must outweigh the disadvantages.

Advantages of Pellets

1. Wind loss is less with pellets than with mash.
2. Feed dustiness is reduced.
3. When handling feeds, there is minimal separation of ingredients, except for the fines fraction.
4. Pelleting destroys some bacteria in the feed (e.g., *Salmonella*).
5. Pelleting increases feed density and birds can consume more low-energy (high-fiber) feeds.
6. Certain feed ingredients are unacceptable to chickens (e.g., rye, buckwheat, barley), but when feeds are pelleted, consumption is markedly increased
7. There is less feed waste from the feeders.

Disadvantages of Pellets

1. There is the added cost of pelleting the mash.
2. Pellets increase water consumption.
3. The droppings are wetter with pellets than with mash.
4. Pellets may increase the incidence and severity of cannibalism.

Crumble Form

When pellets are coarsely ground, or preferably run through special cracking rolls, a type of product midway between mash and pellets is formed. It has most of the advantages and disadvantages of pellets, but because of the smaller size, it may be fed to younger chicks. Often crumbles are used for the first 3 to 4 weeks.

Coarseness of Crumbles

The texture of crumbles should be intermediate in size, neither too coarse nor too fine. In fact, crumbles are best prepared by leaving some finer material in them. This enables younger chicks to eat more rapidly, and prevents some of the cannibalism which results from compressing all the feed particles.

16

Broiler Nutrition

by Craig N. Coon

In all probability, more is known about the nutrition of the broiler than any other type of chicken. In the clamor for rapid growth and superior feed conversion, scientists have spent countless hours developing feed formulas that will produce rapid and economical gains in the broiler house.

The problem with feeding broilers today is not the knowledge of optimum nutrients to use for maximum gains and feed efficiency but how to align the growth of broilers to minimize mortality and skeletal disorders to produce more saleable meat after processing. Geneticists have developed breeding stocks that will produce broilers that grow at a rapid rate, mainly because of the birds insatiable appetite. The modern broiler has the genetic potential to grow at such a rapid rate that the bird will gain more body mass than its heart, lungs, or bones can support. When modern broilers are provided conditions to achieve maximum growth potential in the early portion of the growth curve with high nutrient density diets and 23-hour lighting, unacceptable levels of mortality caused by ascites, flip-over, and leg weaknesses occur. In general, the industry uses different management practices such as intermittent feeding and reduced lighting in the earlier feed periods to reduce the daily feed consumption to control mortality and skeletal disorders caused by rapid growth. The broilers are allowed to increase feed consumption later in the growing period after the earlier problems associated with rapid growth have passed. See *Broiler Management*, Chapter 43, for specific feeding and lighting programs.

The main things that have changed in the US with broiler diets during the past decade have been the lower Metabolizable Energy levels in all diets, but especially in the starter period for broilers being fed to larger weights for further processing. Since breast meat production in modern broiler strains has become a major economic product, dietary methionine and lysine levels tend to be higher in many diets because of the known

Figure 16-1. Delivery of Feed by Truck

relationship of these amino acids for improving breast meat yield. Many companies are also reducing diet cost as much as possible by removing a large portion of vitamins, minerals, and some amino acids during the feed withdrawal period at the end of the grow-out period.

16-A. BROILER FEEDING

Today, there are many different feeding programs utilized in the industry. Some broiler companies use as many as six diets or as few as three to cover the production cycle. For most companies, fewer feeds and diet changes are made when birds are marketed at younger ages and more feeds with larger (roaster) broilers. A four-feed program for light straight-run broilers (approximately 4 lbs live weight) and a five-feed program for heavy straight-run broilers are as follows:

	Average Time Period of Feeding	
Feed Name	*Four-Feed Program Days*	*Five-Feed Program Days*
Starter	1–18	1–18
Grower	19–30	19–30
Finisher	31–	31–35
Initial withdrawal		36–
Final withdrawal	last 5	last 5

Figure 16-2. Broilers feeding

Drug withdrawal period alters feeding program. In order that poultry meat will be void of drugs at the time the birds are slaughtered, many drugs used in broiler feeds have a withdrawal period of 3 to 5 or more days prior to marketing. Often the finisher feed may also be used as the withdrawal feed by eliminating these drugs from the formula and altering the feeding period to coincide with the withdrawal period. In some cases, a final withdrawal feed is used where, in addition to withdrawing the drugs, the nutritionist also significantly reduces other feed ingredients such as vitamins, added amino acids, and minerals.

16-B. ENERGY IN BROILER RATIONS

The primary sources of energy in broiler feed are carbohydrates and fats. However, when protein is fed in excess, it too may become a source of energy. But to feed protein for energy is uneconomical; the balance between carbohydrates, fats, and protein in the diet must be carefully constructed.

Metabolizable Energy (MEn) Content of Broiler Rations

Following are recommended MEn contents of broiler rations for males, females, and straight-run birds given a three-feed program (Table 16-1).

Table 16-1. Metabolizable Energy Level for Rearing Sexes Separate and for Straight-Run Broilers

	Males			Females			Straight-Run		
	Amt.	Kcal ME		Amt.	Kcal ME		Amt.	Kcal ME	
Feed Type	Feed (g/bird)	Per lb	Per kg	Feed (g/bird)	Per lb	Per kg	Feed (g/bird)	Per lb	Per kg
Starter	250	1,395	3,070	250	1,395	3,070	250	1,395	3070
Grower	1,000	1,439	3,166	1,000	1,396	3,071	1,000	1,439	3,166
Withdraw	to mrkt	1,494	3,286	to mrkt	1,439	3,166	to mrkt	1,466	3,226

Source: U.K. Cobb-Vantress Broiler Management Guide, 1998 (40–45 days)

These recommendations will differ with changes in ambient temperature. The dietary energy should be increased 100 kcal per pound in hot environmental temperatures by increasing calories derived from fat.

The percentage of fat calories in broiler diets should be increased under high ambient temperature conditions in order to decrease broiler heat production. An increase in fat calories without simultaneously reducing significantly equal quantities of carbohydrate and protein calories, generally from cereal grains, will cause the metabolizable energy of the diet to increase. The protein that will be lost from cereal grains during the displacement with dietary fat may be partially replaced with the addition of feed grade amino acids and a source of highly digestible amino acids such as dehulled soybean meal. In general, the energy level of the diet may be increased by 50 kcal MEn/pound (23 kcal/kg) with fat calories by adding 2.5% tallow.

Effect of Energy Value on Growth and Feed Conversion

Increasing the dietary energy level has been shown to increase gain and improve feed conversions. There are two schools of thought regarding the effect of dietary energy on broiler feed intake. In the past broilers were thought to regulate their feed consumption to obtain needed energy for growth and maintenance. Research (Waldroup, 1996) has recently shown that feeding broiler diets ranging in energy from 3,023 to 3,383 kcal MEn/kg (1,374 to 1,538 kcal MEn/lb) caused broilers to consume slightly less feed with increasing energy levels but the reduction in feed intake was not in proportion to the increased nutrient consumption with the higher energy diets. The calorie:protein ratio was kept the same for all of the energy diets associated with each growing period (starter feed, 0–21 days; grower feed, 21–42 days; and finisher feed, 42–63 days). The researchers showed that broilers gained more weight with increased energy levels and had significantly improved feed conversions. The research indicated that the response to increasing dietary energy plateaued for 21- and 42-day-old broilers at 3,267 kcal MEn/kg (1,485 kcal MEn/kg) with 6% added poultry fat, and plateaued at 3,304 kcal MEn/kg (1,502 kcal MEn/kg) with 7% added fat for heavier weight broilers fed to 50 or 63 days of age.

Many researchers now believe that since the modern broiler has been selected for appetite, the feed intake regulation mechanism may be slightly different than previously thought. The increased intake of energy with the high energy diets may also help explain why there is an improvement in gain with high density diets compared to lower density diets.

Table 16-2 has been constructed from Waldroup's research. The research shows:

1. Decreasing the feed energy reduces the 42- and 63-day body weight.

Table 16-2. Response of Two Ages of Male Broilers Fed Diets with Increasing Energy Levels

ME in Ration		Body Weight (g)		Total Feed Consumed Per Broiler (g)		Feed Conversion, g feed/g gain		Total ME Consumption, Mcal (Therm)	
kcal/lb	kcal/kg	42 d	63 d	42 d	63 d	42 d	63 d	42 d	63 d
1,374	3,023	2,119	3,589	3,864	8,033	1.823	2.237	11.652	24.509
1,395	3,069	2,114	3,521	3,829	7,796	1.811	2.213	11.699	24.238
1,413	3,109	2,179	3,651	3,909	7,933	1.793	2.172	12.091	25.101
1,431	3,148	2,158	3,663	3,825	7,927	1.771	2.163	11.977	25.315
1,449	3,188	2,187	3,786	3,832	8,174	1.751	2.159	12.151	25.870
1,467	3,227	2,201	3,717	3,841	7,937	1.744	2.135	12.332	25.811
1,485	3,267	2,224	3,778	3,884	8,062	1.746	2.134	12.623	26.635
1,502	3,304	2,204	3,772	3,788	7,941	1.718	2.106	12.459	26.487
1,520	3,344	2,210	3,624	3,719	7,672	1.683	2.117	12.385	26.024
1,538	3,383	2,200	3,628	3,727	7,584	1.694	2.091	12.553	26.114

Source: Waldroup, 1996

2. Increasing the dietary energy does not affect total feed consumption as long as broilers are positively responding to energy.
3. Total calorie consumption increases with increasing dietary energy concentration until broilers stop positively responding to energy.
4. Decreasing the feed energy results in poorer feed conversion.

Feed Density

The subject is partially discussed in *Major Feed Ingredients,* Chapter 15. Results have shown that density of broiler rations, as measured by weight per cubic foot of feed, had an effect on broiler growth. A comparison of results with mashes of many densities showed that broilers fed less dense feeds grew more slowly. The same trend was found with pelleted diets. Broilers fed pellets grew faster and reached a weight of 3.89 lb (1,766 g) about 3 days earlier than those fed mash.

Ambient Temperature, Growth, and Feed Conversion

Bird growth and feed conversion are better during moderate temperatures than during hot or cold. Of special importance is that feed consumption drops off drastically as house temperatures increase. The effect of housing temperatures on body weights and feed conversions is discussed in *Broiler Management,* Chapter 43.

16-C. FAT IN BROILER RATIONS

Adequate fat deposition in market broilers is necessary to give a pleasing appearance to the dressed carcass and to improve the quality of the flesh, but too much fat is a detriment. *Triglyceride* is the major type of fat deposited in the tissues of the chicken. About 95% of the triglycerides come from the diet; 5% are synthesized. Dietary fats are delivered to the fat cells in the body as *lipoproteins,* and therefore they represent the limiting factor in fat deposition. Fats may leave the fat cells to re-enter the blood system and be delivered to other regions of the body when the need arises.

How Much Fat in Broiler Rations

The gross energy value of fat is approximately 2.25 times that of most carbohydrates (starch); therefore, fat is usually added to broiler rations in order to increase the ME value of the ration to the high levels necessary.

When fats are included in broiler rations the utilization of all consumed energy is also improved, so the value of added fat is twofold. The added fat will slow down the transit time of digestion going through the intestine which will allow a more efficient efficient utilization of nutrients from other feedstuffs. Up to 8% of fat may be added to broiler feeds if a portion of the fat is sprayed on after pelleting, with more being added to diets when used after 4 weeks of age than prior to this age. The usual added fat percentage is 2 to 4%.

The availability of fat in the diet is highly variable. Not only do fats themselves differ but age of the bird, strain, type of diet, level of fat in the diet, fat composition including free fatty acid content, and degree of saturation and fat purity produce variability in availability (see Table 14-2 in *Digestion and Metabolism,* Chapter 14).

Abdominal Fat in Broilers

The deposition of fat in specific locations is related to age and the growth curve of the broiler. Abdominal fat is laid down primarily during the early stages of growth. Most quantitative differences in abdominal fat are the result of differences in growth rate. There is an inherent increase in abdominal fat with increased broiler weight. The dietary calorie:protein (amino acid) ratio of broiler diets has a significant affect on the percentage of abdominal fat/live weight or carcass weight. Feeding higher energy diets will not increase the percentage of abdominal fat if the protein and amino acid levels are also increased to maintain the same ratio of ME:protein (amino acids). High density diets will often increase weight gain and feed efficiency but the percentage of abdominal fat does not necessarily increase. However, reducing dietary protein and amino acids with the same level of dietary energy will increase the percentage of abdominal fat (Table 16-3). Likewise, increasing the dietary energy level while maintaining the same dietary protein and amino acid levels will also increases the abdominal fat percentage (Table 16-4). In both Tables 16-3 and 16-4, female broilers are shown to have a higher percentage carcass fat and abdominal fat compared to male broilers at both 42 and 50 days of age.

Increasing body fat is usually associated with poorer feed conversions because it requires more feed to produce a unit of fat than a unit of meat.

High-fat Diets During Hot Weather

Broilers consume less feed during hot weather than during cool, thereby reducing the consumption of the daily amount of protein and other feed constituents. The often-used procedure of removing fat from the dietary ration in order to cause the birds to consume more feed so as to meet the

Table 16-3. Effect of Dietary *Protein* Levels on Performance and Carcass Parameters of Broilers at 42 and 50 Days of Age[1,2]

Days	Parameter	Diets		Protein (%)		
	Starter	17.4	19.3	21.2	23.0	24.9
	Grower	16.3	18.2	20.1	22.0	23.8
	Finisher	14.5	16.4	18.3	20.2	22.0
Female						
42	Body weight (g)	1,643[a]	1,700[ab]	1,720[ab]	1,827[bc]	1,984[c]
	Abdominal fat (%)	4.32[d]	4.21[d]	3.56[c]	3.07[b]	2.09[a]
	Carcass protein (%)	41.58[a]	43.93[ab]	45.76[bc]	45.86[bc]	47.48[b]
	Carcass fat (%)	48.92[d]	46.76[c]	43.78[b]	41.14[a]	39.83[a]
50	Body weight (g)	1,851[a]	2,027[ab]	2,080[ab]	2,207[bc]	2,463[c]
	Abdominal fat (%)	3.96[c]	3.81[bc]	3.63[b]	3.62[b]	3.38[a]
	Carcass protein (%)	37.12[a]	38.89[ab]	41.11[ab]	42.02[bc]	43.62[c]
	Carcass fat (%)	55.19[b]	52.92[ab]	50.11[ab]	48.61[a]	47.88[a]
	Feed efficiency	2.29[c]	2.25[bc]	2.23[c]	2.06[ab]	2.01[a]
Male						
42	Body weight (g)	1,893[a]	1,933[ab]	2,039[b]	2,119[bc]	2,293[c]
	Abdominal fat (g)	2.66[d]	2.16[c]	1.90[bc]	1.73[b]	1.11[a]
	Carcass protein (%)	44.60[a]	47.01[b]	48.32[b]	48.40[b]	48.71[b]
	Carcass fat (%)	45.84[c]	43.18[b]	42.02[b]	39.97[ab]	38.18[a]
50	Body weight (g)	1,993[a]	2,200[b]	2,344[bc]	2,453[cd]	2,633[d]
	Abdominal fat (%)	3.26[d]	3.18[cd]	3.01[bc]	2.88[ab]	2.76[a]
	Carcass protein (%)	40.60[a]	41.23[a]	42.96[ab]	44.68[b]	47.55[c]
	Carcass fat (%)	47.92[b]	47.50[b]	47.34[b]	45.09[b]	43.83[a]
	Feed efficiency	2.12[a]	2.11[a]	2.03[a]	1.94[a]	1.88[a]

[a,b,c,d] Means within a row that are followed by the same letter are not significantly different ($P < 0.05$)

[1] Carcass composition determined on whole carcass including feathers, blood, and fat pad. Abdominal fat expressed as percent of live weight. Carcass protein and carcass fat determined on a dry matter basis

[2] Dietary MEn for starter, grower, and finisher / withdrawal diets was 3,113, 3,212, and 3,289 kcal / kg, respectively

Source: Zollitsch, et al., 1995

critical amino acid requirements has been shown to produce a negative growth effect.

Dietary fat has a lower heat increment, requiring less energy to be utilized by the bird compared with carbohydrates and protein. Research results have shown that fat should not be withdrawn from the ration during hot weather, but perhaps should be increased in order to allow the birds to consume sufficient calories.

Reducing Fat in Broilers

Throughout the years of commercial broiler production there has been a never-ending endeavor to grow a larger bird in a shorter amount of time with a better feed conversion. But with these increases the birds generally will carry more fat. In the past, the abdominal fat pad (leaf fat) was mar-

Table 16-4. Effect of Dietary *Energy* Levels on Performance and Carcass Parameters of Broilers at 42 and 50 Days of Age[1,2]

Days	Parameter	Diets		ME (kcal/kg)		
	Starter	3,025	3,069	3,113	3,157	3,201
	Grower	3,124	3,168	3,212	3,256	3,300
	Finisher	3,201	3,459	3,289	3,333	3,377
Female						
42	Body weight (g)	1,478[a]	1,507[a]	1,559[a]	1,620[ab]	1,693[b]
	Abdominal fat (%)	2.33[a]	2.49[ab]	2.99[bc]	3.17[c]	3.44[d]
	Carcass protein (%)	45.00[a]	43.67[b]	43.04[b]	41.67[c]	40.79[c]
	Carcass fat (%)	47.71[a]	48.76[b]	49.22[b]	50.23[c]	51.47[c]
50	Body weight (g)	1,726[a]	1,808[a]	1,976[b]	2,035[b]	2,052[b]
	Weight gain (g/day)	38.13[a]	40.10[ab]	41.81[bc]	42.57[bc]	44.21[c]
	Abdominal fat (%)	2.85[a]	3.23[ab]	3.41[b]	3.60[c]	3.82[c]
	Carcass protein (%)	42.65[a]	41.29[b]	29.97[c]	39.24[d]	38.21[c]
	Carcass fat (%)	50.24[a]	52.33[b]	53.71[c]	55.21[d]	55.74[d]
	Feed efficiency	2.33[a]	2.17[ab]	2.13[bc]	1.96[bc]	1.82[c]
Male						
42	Body weight (g)	1,693[a]	1,791[ab]	1,860[abc]	1,934[abc]	2,013[c]
	Abdominal fat (g)	2.23[a]	2.43[ab]	2.56[b]	2.61[b]	2.72[c]
	Carcass protein (%)	48.75[a]	47.74[b]	46.66[c]	46.63[c]	45.03[b]
	Carcass fat (%)	42.18[a]	44.24[b]	45.57[c]	46.70[d]	48.00[e]
50	Body weight (g)	1,966[a]	2,095[b]	2,216[c]	2,308[cd]	2,404[d]
	Weight gain (g/day)	47.87[a]	49.31[ab]	50.73[bc]	50.93[bc]	52.33[c]
	Abdominal fat (%)	1.85[a]	2.26[b]	2.61[bc]	2.89[c]	3.04[c]
	Carcass protein (%)	45.38[a]	44.05[b]	43.47[b]	41.80[c]	40.83[d]
	Carcass fat (%)	45.53[a]	49.36[b]	50.34[c]	51.58[d]	53.10[e]
	Feed efficiency	2.11[a]	2.00[ab]	1.98[ab]	1.89[b]	1.70[c]

[a,b,c,d] Means within a row that are followed by the same letter are not significantly different ($P < 0.05$)
[1] Carcass composition determined on whole carcass including feathers, blood, and fat pad. Abdominal fat expressed as percent of live weight. Carcass protein and carcass fat determined on a dry matter basis
[2] Dietary protein for starter, grower, and finisher/withdrawal diets was 21, 20, and 18%, respectively
Source: Zollitsch, et al., 1995

keted with whole broilers along with the neck and giblets. Today, a large portion of broiler meat is sold as cut-up parts and further processed meat, therefore there is less opportunity to market abdominal fat. Consumers are also very health conscience and they are consuming less fat in their daily diets. Since poultry fat is no longer marketed with the whole broiler, the fat pad is presently being removed at the processing plant and a percentage of the fat is then sent to renderers. The poultry fat is rendered into a highly digestible fat product that can be re-used for making high energy diets utilized in poultry feeds.

Greasy Broilers

On occasion, certain deposited fats and oils remain fluid in processed chilled broilers. These cause a condition known as *greasy broilers.* As such fluid fat encompasses most of the body of the broiler, the parts are difficult to coat with batter just prior to cooking. Reducing the quantity of unsaturated fatty acids in the diet or changing the source of any added dietary fat will be of some benefit.

16-D. PROTEIN IN BROILER RATIONS

It is not the broiler's requirement for *total* protein that is important but the daily need for the *individual amino acids*. The National Research Council (1994) states that broilers do not have a protein requirement per se. The dietary protein should be sufficient to provide adequate amino acid nitrogen for the synthesis of non-essential amino acids. The age and sex of the broiler also alters the protein requirement. The kcal of ME per pound (kilo) of ration affects the protein requirement. The higher the ME, the greater the protein percentage required.

Age and Sex as They Affect Dietary Protein

Theoretically, the diet of the broiler should contain about 21 to 22% protein in a 3,058 to 3,135 kcal MEn/kg (1,390 to 1,425 kcal MEn/lb) diet during the first 2 weeks, and protein should gradually decrease thereafter. However, it has not been considered practical to make a large number of different diets because of the problems associated with extra hauling expenses, increased need for additional feeding bins, and the increased opportunity for making mistakes. While many different practices have been used by the industry, most integrators feed a specific quantity of starter feed to the broilers and then feed the remaining diets based on a set number of days. In the US, a typical distribution of feed usage might be starter—-12%, grower—33%, finisher—25%, and withdrawal—30%.

A three to five feed program equalizes the necessary protein requirement during the starting, growing, and finisher/withdrawal periods, and matches the feeding schedule involved with the MEn program (Table 16-5). An MEn program consists of a series of diets that are fed for a specific response by an integrator. The program matches the energy levels

Table 16-5. Dietary Protein Levels for Rearing Sexes Separate and for Straight-Run Broilers

	Males		Females		Straight-Run	
Feed	Amount (g/bird)	Protein (%)	Amount (g/bird)	Protein (%)	Amount (g/bird)	Protein (%)
Starter	250	23.0	250	23.0	250	23.0
Grower	1,000	23.0	1,000	21.0	1,000	22.0
Withdrawal	to market	20.0	to market	19.0	to market	19.0

Source: U.K. Cobb-Vantress Broiler Management Guide, 1998 (40–45 days)

Table 16-6. Recommended Practical Broiler Nutrient Levels

Market Wt.		% of Feed Fed		
kg	lb	Starter	Grower	Withdrawal*
1.75	3.85	25	42	33
2.00	4.40	24	42	34
2.25	4.95	21	45	34
2.50	5.50	17	48	35
2.75	6.05	15	48	37
		Starter	Grower	Withdrawal*
Protein %		21.50	20.25	18.00
Calories / lb (kcal, MEn)		1,400	1,450	1,475
Calories / kg (kcal, MEn)		3,080	3,190	3,245

* The withdrawal feed schedule will depend upon the desired market weight. The program presented is based on an average broiler body weight of 4.15 to 4.25 lbs (1,885–1,930 kg)
Source: U.S. Cobb-Vantress Broiler Management Guide, 1998

of the diets with the protein levels that are fed on a length of time basis or by quantity of feed. The overall program will depend upon the sex of the flock, strain, product needs, marketing age, and feed costs involved.

The protein levels in Table 16-5 correspond with the dietary MEn levels in Table 16-1. The male rations have higher levels of nutrients because of their increased requirement for skeletal growth. The same breeder in the US recommends a different level of protein and energy for straight-run broilers (Table 16-6). The protein levels in broiler diets in the US tend to be lower and the energy levels may be slightly higher than in other international locations. Recommendations are also given for feeding different percentages of starter, grower, and withdrawal diets depending upon the needed market weight instead of just feeding the diets a specific amount of time.

Requirements Expressed on a Calorie/Protein Basis

There is a definite relationship between the kcal of MEn and the protein (amino acid) requirement of the growing broiler. This relationship is known as the calorie / protein ratio and is calculated as follows:

1. Per pound basis
   kcal MEn/lb ration ÷ % protein = Calorie / Protein Ratio
   Example: If the kcal ME / lb of ration is 1,400, and the protein is 22%, the ratio would be 63.6

2. Per kilo basis
   kcal MEn/kg ration ÷ % protein = Calorie / Protein Ratio
   Example: If the kcal MEn / kg of ration is 3,080, and the protein is 22%, the ratio would be 140.0

Variations in the Calorie: Protein Ratios

The calorie / protein ratio increases with age as older broilers require higher energy levels and less protein in their diet than younger birds. Some countries or markets may require faster growth because of less available housing, thus may feed higher levels of protein in all feeds. An example of this is the use of less protein and additional energy levels in the US diets (Table 16-6) compared to the UK diets (Tables 16-1, 16-5). The U.K. diets have a lower calorie / protein ratio.

Important. If either the kcal of MEn or the percentage of protein (amino acids) of the feed formula is altered, then the remaining one must be adjusted so the calorie / protein ratio remains the same. If this adjustment is not made, then either calories or protein (amino acids) will be wasted.

Today, the importance of formulating feed for the proper calorie / protein ratio receives less attention because nutritionists are more concerned with the relationship between amino acids and calories. The best system for using a ratio with amino acids is to invert the ratio and express it as an amino acid / calorie ratio. Amino acids can be converted to quantitative amounts like mg of amino acid per Therm (Megacalorie) of dietary energy or percent amino acid per Megacalorie. Using the percent amino acid per Megacalorie system allows a tabular set of values to be used for each amino acid. Individual diets can be formulated by multiplying the tabular values in Table 16-7 times the Megacalories / lb of the diets to be made.

16-E. AMINO ACID REQUIREMENTS FOR BROILERS

The amino acid requirements for straight-run broilers for three different ages are given in Table 16-8. It is strongly recommended that the requirements for amino acids be formulated on a digestible basis when digestible amino acid values for ingredients are available. Amino acid digestibility in broiler diets can vary when using ingredients that have known anti-nutritional factors (tannins, goitrogens, alkaloids, gossypol) that are not sensitive to heat treatment, variable processing conditions for ingredients that need to be heat-treated to denature anti-nutritional factors or for microbial degradation (oil seed proteins and animal / fish proteins), ingredients that contain lower digestible amino acids because of location within

Table 16-7. Suggested Amino Acid Recommendations for Broiler Diets in Relationship to the Energy Content of the Diet[1,2]

	0–21 days		22–42 days		43–53 days	
Nutrient	Male	Female	Male	Female	Male	Female
Arginine	0.88	0.83	0.77	0.73	0.66	0.59
Lysine	0.81	0.76	0.70	0.67	0.53	0.50
Methionine	0.34	0.31	0.32	0.29	0.25	0.22
Methionine + cystine	0.60	0.60	0.56	0.50	0.46	0.41
Tryptophan	0.16	0.15	0.12	0.11	0.11	0.10
Histidine	0.24	0.22	0.22	0.20	0.20	0.18
Leucine	0.84	0.64	0.81	0.76	0.76	0.72
Isoleucine	0.54	0.51	0.52	0.48	0.45	0.40
Isoleucine	0.56	0.54	0.55	0.51	0.47	0.42
Phenylalanine	0.59	0.53	0.51	0.47	0.42	0.38
Phenylalanine + tyrosine	1.04	0.98	0.95	0.87	0.78	0.70
Threonine	0.53	0.49	0.43	0.41	0.42	0.38
Valine	0.66	0.63	0.62	0.57	0.51	0.46
Glycine + serine	1.00	0.94	0.80	0.74	0.70	0.63
Protein[3]	15.25	15.25	13.75	13.75	12.25	12.25

[1] All nutrients are expressed as percent per metabolizable megacalorie per pound of feed. To determine the actual percentage of the nutrient required in the diet, multiply the value in the table by the metabolizable megacalories per pound of diet to be formulated, e.g., 0.88 (arginine for males in starter diet) × 1.400 (megacalories MEn per lb in starter diet) = 1.23% (level of arginine in starter diet). Caution: calculations are only valid on a per lb basis

[2] Amino acid recommendations are based on data published by Thomas et al. (1992)

[3] The broiler does not have a protein requirement per se. However, there should be a sufficient amount of crude protein to ensure an adequate nitrogen pool for synthesis of nonessential amino acids. The values suggested are typical of corn-soybean meal-based diets and may be reduced if amino acid supplements are used to provide essential amino acids and if a good balance of high-quality protein is used

Source: Waldroup, 1999

the cereal grain (germ, aleurone layer), or just the quantitative amount of fiber in the overall diet.

Ideal Amino Acid Profiles for Broilers

The ideal amino acid profile for broilers is shown in Table 16-9. Since one set of amino acid requirements cannot apply to all broilers, the use of an amino acid profile allows nutritionists to adjust the overall level of the "ideal protein" to compensate for multiple dietary, environmental, sex, body composition, and genetic differences that could affect amino acid requirement of broilers. The amino acid profile has been correlated to lysine levels in the diet because lysine is relatively easy to analyze, it is not a precursor to substrate production, and research indicates there is a lower maintenance requirement for lysine compared to other amino acids. The negative side of correlating other amino acids to lysine is the potential for

Table 16-8. Amino Acid Requirements of Broilers as Percentages of Diet (90 percent dry matter)

Nutrient	Unit	0 to 3 Weeks[a]	3 to 6 Weeks[a]	6 to 8 Weeks[a]
Protein and amino acids				
Crude protein	%	23.00	20.00	18.00
Arginine	%	1.25	1.10	1.00
Glycine + serine	%	1.25	1.14	0.97
Histidine	%	0.35	0.32	0.27
Isoleucine	%	0.80	0.73	0.62
Leucine	%	1.20	1.09	0.93
Lysine	%	1.10	1.00	0.85
Methionine	%	0.50	0.38	0.32
Methionine + cystine	%	0.90	0.72	0.60
Phenylalaline	%	0.72	0.65	0.56
Phenylalanine + tyrosine	%	1.34	1.22	1.04
Proline	%	0.60	0.55	0.46
Threonine	%	0.80	0.74	0.68
Tryptophan	%	0.20	0.18	0.16
Valine	%	0.90	0.82	0.70

[a] The 0-to-3, 3-to-6, and 6-to-8-week intervals for nutrient requirements are based on chronology for which research data were available; however, these nutrient requirements are often implemented at younger age intervals or on a weight-of-feed consumed basis
[b] These are typical dietary energy concentrations, expressed in kcal ME_n/kg diet. Different energy values may be appropriate depending on local ingredient prices and availability
Source: National Research Council, 1994

Table 16-9. Ideal Amino Acid Profiles for Broilers

	Baker, 1993, 1996		Hruby and Coon, 1991[1]		Mack, 1999[2]
Amino Acid	0–21 d	22–42 d	0–21 d	22–43 d	20–40 d
Lysine	100	100	100	100	100
Methionine	36	36	35	36	ND*
Methionine + cystine	72	75	71	69	75
Threonine	67	70	63	65	63
Arginine	105	108	97	101	112
Valine	77	80	67	75	81
Isoleucine	67	69	64	62	71
Leucine	109	109	94	117	ND*
Tryptophan	16	17	16	18	19
Histidine	32	32	29	31	ND*

* Not determined
[1] The levels of digestible lysine were 1.07% and 0.96% for the 3,150 kcal MEn/kg starter and 3,302 kcal MEn/kg grower diets, respectively, i.e., in a starter diet with a lysine requirement of 1.10%, the methionine requirement would be 0.385% ($1.10 \times 0.35 = 0.385$)
[2] The digestible lysine was 1.15% for a 3,158 kcal/kg grower diet
Source: Hruby et al., (1999)

overfeeding other amino acids, especially if the amino acid profiles were developed with lysine requirements based on feed efficiency instead of nitrogen or protein gain, as researchers have shown that broiler requirements for lysine and methionine are higher for feed efficiency compared to gain. The requirements are also higher for breast meat yield. Other amino acids do not show similar differences for different parameters.

The principal advantage of using amino acid profiling is that all essential and non-essential amino acids are equally limiting and would provide maximum nitrogen retention. This is going to become more critical in the future because of the need to decrease nitrogen in animal wastes in order to improve the environment. The practicality of feeding an ideal amino acid profile at the present time is not realistic, because to do so would require feeding a 15% protein diet supplemented with all essential amino acids. At the present time, only feed grade levels of methionine or methionine analogues, lysine, and threonine are economically available for formulating commercial broiler diets.

16-F. VITAMIN REQUIREMENTS OF BROILERS

The vitamin requirements of broiler rations are given in Table 16-10 (see also *Vitamins, Minerals, and Trace Ingredients,* Chapter 20).

16-G. MINERAL REQUIREMENTS OF BROILERS

The mineral requirements of broiler rations are given in Table 16-11.

16-H. OTHER DIETARY SUPPLEMENTS

Following are some other dietary supplements used for broiler feeding (see also *Vitamins, Minerals, and Trace Ingredients,* Chapter 20).

Antibiotics. Although most antibiotics are used for disease prevention and treatment, there are several that have growth-promoting properties when fed continuously at low levels. Feed levels should be those recommended by the manufacturer. Many countries may have different guidelines regarding the use of antimicrobials in feeds and it will be necessary to comply with the regulations in effect for each location.

Coccidiostat. A coccidiostat is usually added to broiler rations (see *Diseases of the Chicken,* Chapter 27).

Antioxidant. Ethoxyquin, BHT, or other antioxidants may be added to broiler diets to help prevent the appearance of *encephalomalacia.* Antioxidants must be added according to the manufacturer's directions. Antioxidants are very beneficial when adding fat in poultry rations

Table 16-10. Vitamin Requirements of Broilers as Units per Kilogram of Diet (90 percent dry matter) (3,200 kcal MEn per kg diet)

Nutrient	Unit	0 to 3 Weeks[a]	3 to 6 Weeks[a]	6 to 8 Weeks[a]
Fat soluble vitamins				
A	IU	1,500	1,500	1,500
D_3	ICU	200	200	200
E	IU	10	10	10
K	mg	0.5	0.5	0.5
Water soluble vitamins				
B_{12}	mg	0.01	0.01	0.007
Biotin	mg	0.15	0.15	0.12
Choline	mg	1,300	1,000	750
Folacin	mg	0.55	0.55	0.50
Niacin	mg	35	30	25
Pantothenic acid	mg	10	10	10
Pyridoxine	mg	3.5	3.5	3.0
Riboflavin	mg	3.6	3.6	3.0
Thiamin	mg	1.8	1.8	1.8

[a] The 0-to-3, 3-to-6, and 6-to-8-week intervals for nutrient requirements are based on chronology for which research data were available; however, these nutrient requirements are often implemented at younger age intervals or on a weight-of-feed consumed basis
Source: National Research Council, 1994

to prevent oxidation (turning rancid) of fats and to preserve fat soluble vitamins.

Xanthophylls. Although natural feedstuffs may contain xanthophylls in a quantity ample to properly color the skin and shanks of broilers; on occasion, it may be necessary to add other sources of these pigmenters to the ration (see *Feeding for Broiler Skin Color,* section 16-J).

Table 16-11. Mineral Requirements of Broilers as Percentages or Units per Kilogram of Diet (90 percent dry matter) (3,200 kcal MEn per kg diet)

Nutrient	Unit	0 to 3 Weeks	3 to 6 Weeks	6 to 8 Weeks
Macrominerals				
Calcium[1]	%	1.00	0.90	0.80
Chlorine	%	0.20	0.15	0.12
Magnesium	mg	600	600	600
Nonphytate phosphorus	%	0.45	0.35	0.30
Potassium	%	0.30	.030	0.30
Sodium	%	0.20	0.15	0.12
Trace minerals				
Copper	mg	8	8	8
Iodine	mg	0.35	0.35	0.35
Iron	mg	80	80	80
Manganese	mg	60	60	60
Selenium	mg	0.15	0.15	0.15
Zinc	mg	40	40	40

[1] The calcium requirement may be increased when diets contain high levels of phytate phosphorus (Nelson, 1984)
Source: National Research Council, 1994

Table 16-12. Sample Broiler Rations

		Starter (lb)	Grower (lb)	Finisher (lb)	Withdrawal (lb)
Ingredient					
Ground yellow corn		1,257	1,303	1,412	1,507
Soybean meal (dehulled, 47.5%)		548	505	400	300
Meat and bone meal (50%)		125	129	122	123
Fat, animal-veg blend or equivalent		38	37	44	49
Limestone		10	8	4	7
Salt		5.5	8	8	8
DL-Methionine or equivalent		5	5.4	5.8	4.4
Defluorinated phosphorus		4			
L-Lysine, 78.8%		2.3			0.1
Broiler Vitamins		1.5	1	1	0.5
Coccidiostat		1.5	1.8	1.8	
Liquid choline (70%)		1.3	1.3	0.9	
Growth promotant		1	0.5	0.5	
Trace minerals		1	1	1	0.5
Vitamin and mineral supplements					
Vitamin A	(IU)	15,000,000	10,000,000	10,000,000	5,000,000
Vitamin D	(IU)	3,750,000	2,500,000	2,500,000	1,250,000
Vitamin E	(IU)	15,000	10,000	10,000	5,000
Vitamin K	(mg)	3,000	2,000	2,000	1,000
Vitamin B_{12}	(mg)	22	15	15	7.5
Riboflavin	(mg)	9,000	6,000	6,000	3,000
Niacin	(mg)	45,000	30,000	30,000	15,000
Calcium pantothenate	(mg)	22,500	15,000	15,000	7,500
Biotin	(g)	112.5	75	75	37.5
Folic acid	(g)	900	600	600	300
Pyridoxine	(g)	22,500	15,000	15,000	7,500

Copper	(g)	1.5	1.5	1.5	0.75
Iodine	(g)	1	1	1	0.5
Iron	(g)	14	14	14	7
Zinc	(g)	45.4	45.4	45.4	22.7
Manganese	(g)	54.48	54.48	54.48	27.24
Selenium	(mg)	136.2	136.2	136.2	68.1
Totals	(lb)	2,001.1	2,001	2,001	2,000
Calculated nutrients					
Metabolizable energy	(kcal/lb)	1,420.5	1,430.48	1,465.1	1,490.9
Protein	(%)	21.09	20.25	18	15.98
Lysine	(%)	1.268	1.12	0.96	0.823
Methionine	(%)	0.598	0.608	0.6	0.501
TSAA	(%)	0.951	0.951	0.91	0.789
ME:protein ratio	(%)	67.3	70.64	81.39	93.3
Fat	(%)	4.8	4.85	5.32	5.72
Fiber	(%)	2.61	2.61	2.59	2.56
Calcium	(%)	0.927	0.835	0.71	0.756
Total phosphorus	(%)	0.651	0.617	0.585	0.57
Non-phytate phosphorus	(%)	0.429	0.401	0.381	0.379
Vitamins					
Vitamin A activity	(IU/lb)	8,438	5,974	6,056	3,630
Vitamin D_3 (added)	(IU/lb)	1,874	1,249.4	1,249.4	625
Vitamin E	(IU/lb)	10.8	8.38	8.63	6.36
Vitamin K	(mg/lb)	1.5	1	1	0.5
Riboflavin	(mg/lb)	5.2	3.68	3.63	2.09
Niacin, total	(mg/lb)	26.57	18.9	18.3	10.32
Pantothenic acid	(mg/lb)	13.72	9.95	9.87	6.05
Choline	(mg/lb)	725.2	705.6	595	420.6
Xanthophyll	(mg/lb)	5.29	5.46	5.84	6.18

Source: Richard Arnold, Southwest Nutrition Services, 1999

16-I. BROILER RATIONS

Typical broiler starter, grower, finisher, and withdrawal rations are given in Table 16-12.

16-J. FEEDING FOR BROILER SKIN COLOR

The color of yellow-skin chickens is due almost entirely to a group of chemicals known as *xanthophylls,* substances closely related to the carotenoids (see *Feeding Commercial Egg-Type Layers,* Chapter 18, for a discussion of yolk pigmentation). Not only are xanthophylls easily oxidized from natural feed ingredients but the fat and skin of the chicken also lose its yellow color in this manner. To maintain any skin color, the quantity of xanthophylls consumed must approximate that lost from the bird.

There are many xanthophylls capable of imparting a yellow-orange color to the skin of chickens, all classified as *hydroxycarotenoid* pigmenting compounds. Alfalfa leaf meal is a source of *lutein.* Yellow corn is an excellent source of several xanthopylls, but corn contains mostly *zeaxanthin.* The petals of a marigold species, *Tagetes erecta,* are a very potent source of xanthophylls. Another is algea, mostly *Spongiococum excentricum.* A synthetic carotenoid, *beta-apo-8′-carotenal,* produces a skin color similar to the xanthophylls.

Measuring Skin Colors

There are several methods for measuring skin colors, as follows:

Roche® color fan. The simplest procedure is to visually match the skin color with those on the Roche color fan.

IDL color-eye. This is a type of photometer used for determining skin colors.

NEPA standard (National Egg and Poultry Association). Graded solutions of potassium dichromate are used and compared to the ether extract of a small piece of skin. The scores range from 0 to 5, as follows:

NEPA Number	*Approximate Skin Color*
0	Very pale
1	Light yellow
2	Dark yellow
3	Normal orange
4	Dark orange
5	Very dark orange

Table 16-13. Variations in Surface Skin Color of Broilers

Area	Color Index
Web of toe	100 (base)
Foot pad	108
Shank	122
Back	132
Breast feather tract	163

Source: North and Bell, 1990

Skin-color Variations

Skin color is not the same in all sections of the bird as shown in Table 16-13, which represents a comparative guide, using the color of the toe web as a base of 100.

Xanthophyll Content of Feedstuffs

The total xanthophyll content of some feedstuffs is shown in Table 16-14.

Color Transmission of Xanthophylls

Xanthophylls differ in color transmission. The ability of the xanthophylls in alfalfa leaves, yellow corn, and corn gluten meal to increase the density of yellow-orange color in the skin of the chicken is not equal. For example, per unit of xanthophyll, alfalfa leaves will impart only 75% the density of yellow corn.

Time necessary to color broilers. It takes about 3 weeks to produce the desired color in a broiler. The older the broiler, the higher the percentage of xanthophylls transferred from the feed to the skin, but oxida-

Table 16-14. Total Xanthophyll Content of Feedstuffs

	Approximate Total Xanthophyll	
Feedstuff	mg per lb	mg per kg
Marigold petal meal	3,182	7,000
Algae meal	909	2,000
Alfalfa meal (17% protein)	100	220
Alfalfa meal (22% protein)	150	330
Corn gluten meal (60% protein)	132	290
Yellow corn	8	17
Alfalfa protein concentrate (40% crude protein)	364	800

Source: National Research Council, 1994

Table 16-15. Dietary Mixed Xanthophylls Necessary to Produce NEPA Scores in Broilers

NEPA Number	Approximate Xanthophyll Quantity Necessary per Unit of Ration	
	mg/lb	mg/kg
1	5	11.0
2	10	22.0
3	16	35.2
4	23	50.6
5	30	66.0

Source: North and Bell, 1990

tion of xanthophylls is also greater in older birds. Of the two xanthophylls (lutein and zeaxanthin, found in alfalfa and yellow corn, respectively), zeaxanthin is far superior in producing a darker pigment, and it will increase the density more in the breast than in the shank.

Xanthophylls necessary to produce various skin colors. Amounts of mixed xanthophylls to produce various NEPA skin numbers are found in Table 16-15.

16-K. LEAST-COST FEEDING OF BROILERS

Least-cost formulation of broiler diets is discussed in *Feed Formulation and the Computer,* Chapter 21.

16-L. FEEDING "ROCK-CORNISH GAME" (SQUAB) BROILERS

Rock-Cornish game broilers are marketed at a very young age, at about 4½ to 5 weeks and at about 2.2 to 2.6 pounds (1.0 to 1.2 kg) live weight. This produces a dressed weight of between 1.5 and 2.0 pounds (0.7 and 0.9 kg).

A broiler starter diet containing 23 to 24% protein and 1,386 kcal of ME per pound (3,190 kcal MEn/kg) of ration is fed to young broilers for the first 18–21 days and then fed a grower/withdrawal diet to market weight.

An example feeding program for Rock-Cornish Game broilers is shown in Table 16-16.

16-M. FEEDING HEAVY WEIGHT MALES FOR FURTHER PROCESSING

A larger percentage of broilers are being fed for further processing. Males are normally used for producing the heavy weights because they

Table 16-16. Sample of Nutrient Specifications Needed for Producing Rock-Cornish Game Broilers with Live Weight of 2.2 lb (1 kg) at 28 to 32 Days

	Starter	Grower/Withdrawal
Crude protein, %	24.00	23.50
ME (kcal/lb)	1,386	1,455
(Kcal/kg)	3,050	3,200
Fat, %	4–7	4–7
Fiber, %	2–3	2–3
Lysine, %	1.30	1.25
Methionine, %	0.62	0.60
Methionine + cystine, %	1.10	1.00
Tryptophan, %	0.24	0.24
Calcium, %	0.90	0.90
Available phosphorus, %	0.45	0.45
Salt, %	0.34	0.34
Sodium, %	0.18	0.18
Chloride, %	0.16	0.16

Source: Ross Breeders Broiler Management Guide, 1999. Broiler Nutrition Summary

are better converters of feed to meat, and also the length of time to reach the heavy weights is shorter for males than for females. Presently, males are being fed to approximately 8 pounds, or 3.64 kg, in 63 days. Feed is the main item of cost in producing the heavier males.

Heavy male starter diets are lower in both dietary energy and protein to reduce early growth. Heavy male grower and withdrawal rations are similar to traditional broiler grower and withdrawal diets with increases in dietary energy and a reduction in protein. Ultrarapid growth is not always advantageous when raising heavy males because of extreme stress on the skeletal system. Furthermore, when broilers are grown to heavier weights there is likely to be a high incidence of breast blisters and skeletal deformi-

Table 16-17. Heavy Male Feeding Program

	Starter 0–18 d	Grower #1 19–36 d	Grower #2 37–50 d	Finisher 51–63 d
CP, %	18–19	20	18–19	17–18
ME (kcal/kg)	2,850	3,100	3,150	3,150
(kcal/lb)	1,295	1,400	1,430	1,430
Calcium, %	0.95	0.95	0.90	0.85
Available phosphorus, %	0.42	0.42	0.40	0.38
Sodium, %	0.16	0.18	0.18	0.18
Methionine, %	0.40	0.55	0.50	0.45
Methionine + cystine, %	0.72	0.95	0.90	0.85
Cystine, %				
Lysine, %	0.95	1.20	1.05	0.95
Anticoccidial	+	+	+	–
Growth promoter	±	+	+	±

Source: Hubbard, Hi-Y Broiler Management Guide, 1996

ties. In addition, metabolic disorders such as ascites and flip-over syndrome are more common.

Feeding Program for Heavy Males

A recommended feeding program for heavy males is given in Table 16-17.

17

Feeding Egg-Type Replacement Pullets

by Craig N. Coon

The period represented in this discussion is from 1 day of age until sexual maturity, which is approximately 18 to 20 weeks.

Growing and developing a good pullet is one of the most important items in the operation of an egg-production enterprise. Undoubtedly, the quality of the bird at the time her production cycle begins will greatly determine how profitable she will be during her period of lay. Therefore, special emphasis must be placed on feeding the growing bird so that she may develop into a healthy productive individual and one that can fulfill her genetic potential. Mistakes made during the growing phase cannot be corrected during the laying cycle.

The modern layer is capable of producing close to 300 eggs per hen in a normal one-year laying cycle because they are sexually mature at an earlier age than ever before. The increase in eggs obtained from the modern layer is mainly the additional 10 to 15 extra eggs that are produced at the beginning of the cycle (Figure 17-1). The modern egg-type pullet, depending upon the individual strain, is capable of being light stimulated to reach sexual maturity from 16 to 18 weeks of age compared to pullets a decade ago that were sexually mature at 20 weeks of age. Today's flocks will also reach peak egg production in only 4 to 6 weeks compared to past flocks reaching peak production in 7 to 8 weeks. The early sexually maturing pullet creates new demands on proper feeding and management.

17-A. FACTORS AFFECTING PULLET DEVELOPMENT

There are several factors of importance in pullet development. But the major rule that applies is in two parts, and each part is extremely important. Each group of egg-type pullets must reach sexual maturity:

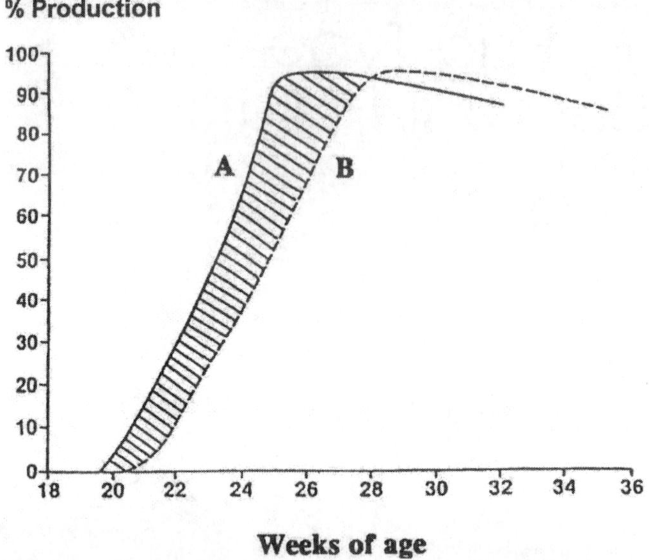

Figure 17-1. Comparison of Modern Earlier Sexually Maturing Pullet (A) with Slower Sexually Maturing Pullet of the Past (B)
Source: Summers, 1993, Maryland Nutrition Conference

1. At the correct weight for that particular strain.
2. At an age that is optimum to produce eggs economically during her laying year.

From a feeding standpoint, the following have a bearing on the above rules.

Genetics

The size (weight) of a strain of layers at sexual maturity is a derivation of genetics and each strain will have a different recommended weight. Although there is little evidence that the basic feed formula need be altered according to the strain involved, feed management during the growing period is important so the birds may reach their desired weight and body condition when they lay their first eggs.

Season of Hatch

Two problems that confront the poultry producer when chicks are started at different months of the year, *with only natural daylight* and full feeding, are as follows, and are shown in Table 17-1.

Table 17-1. Effect of Date of Hatch on Age at Sexual Maturity and Egg Size

Month Hatched	Days to 25% Egg Production	12 Months' Production: Large Eggs %	12 Months' Production: Eggs under 22 oz per doz %
January	164	87.1	10.5
February	172	86.5	10.6
March	184	89.0	7.9
April	187	93.1	4.8
May	189	94.1	3.9
June	195	93.8	3.7
July	190	86.4	8.9
August	202	93.6	4.1
September	200	93.4	4.1
October	179	80.2	14.3
November	150	78.5	16.6
December	147	72.3	22.6

Source: Skoglund, Univ. of Delaware

a. Those pullets raised during decreasing light days reach sexual maturity at an older age while birds raised under increasing day lengths reach sexual maturity at an earlier age.
b. The younger the bird at sexual maturity, the smaller the size of her first eggs and, conversely, the older the bird at sexual maturity the larger the size of her first eggs.

These normal variations in flock performance must be controlled in order to secure profitable laying house performance. The requirements for profitable laying house performance will, naturally, vary seasonally—year-to-year and country-to-country—depending upon the prevailing market conditions and demand for egg size.

Light Stimulation

Table 17-1 clearly indicates the importance of having a light-control program during the growing period. Feed control (restriction) will also delay the onset of egg production, but its effects are most pronounced with flocks maturing in open-sided houses during days of increasing day length.

Research in 1980 to 1982 with 68 commercial flocks of White Leghorn chickens showed a somewhat different picture under commercial conditions with artificial light programs. The season in which the birds were grown significantly affected body weight and subsequent weight throughout the life of the flocks. These flocks were raised under commercial lighting programs and California temperature extremes were probably greater

Table 17-2. Effect of Season of Hatch on Performance in Commercial Leghorn Flocks

Month of Hatch	18-wk Body Weight		Egg Weight				Egg Production	
			24 wk		60 wk		24 wk	60 wk
	lb	kg	oz/doz	g/egg	oz/doz	g/egg	%	%
Jan through Mar	2.62	1.188	20.4	48.2	26.8	63.4	54.1	71.7
Apr through June	2.62	1.188	20.7	48.8	26.4	62.3	44.7	70.4
July through Sept	2.60	1.179	21.0	49.5	25.9	61.2	63.0	72.1
Oct through Dec	2.76	1.252	21.1	49.9	27.2	64.2	62.8	70.9
Avg	2.65	1.202	20.8	49.1	26.6	62.8	56.1	71.3

Source: University of California, 1982

than in the Delaware research (see Table 17-2). Early egg production was affected (decreased) the most in spring-hatched flocks when egg production commenced during the hottest months of the year. Twenty-four week egg size was also the lowest for winter- and spring-hatched flocks reflecting both body weight differences and the effects of hot temperature at the beginning of lay. Fall-hatched flocks had larger eggs at both 24 and 60 weeks of age. (See *Egg Production and Egg Weight Standards for Table-Egg Layers,* Chapter 55.)

Stress

Chickens are subjected to stresses during the growing stage; many of these are man-made. Most stresses cause a reduction in the daily feed consumption, and therefore some method of increasing the nutritional intake to offset the stress becomes the responsibility of the poultry caretaker.

One of the most stressful events in the life of a growing pullet is beak trimming. Feed consumption is depressed following trimming and body weight is often reduced. Table 17-3 illustrates the effects of beak trimming at 12 weeks of age on daily feed consumption and body weight during the following week. Another stress that is becoming of increasing importance is the vaccination schedule that is imposed on many flocks. Because of the handling and injections, vaccination of birds with inactivated or killed vaccines appears to affect the birds more than the use of a live vaccine that may be administered by spray or in the water.

Management Practices

Many management practices require changes in feed formulation and feeding method. Whether birds are to be raised on a litter floor or in cages

Table 17-3. Effects of Beak Trimming at 12 Weeks of Age on Feed Consumption and Body Weight

Days Post Trim	Feed Consumption				Body Weight			
	Controls		Trimmed		Controls		Trimmed	
	lb	g	lb	g	lb	g	lb	g
0					1.99	903	2.07	939
1	0.14	63	0.07	30	2.00	907	1.98	898
2	0.15	67	0.01	5	2.04	925	1.91	866
3	0.15	69	0.01	4	2.07	939	1.81	821
4	0.15	69	0.01	5	2.09	948	1.75	794
5	0.15	68	0.01	4	2.13	966	1.71	776
6	0.15	68	0.05	22	2.16	980	1.72	780
7	0.15	70	0.09	41	2.18	989	1.75	794

Source: University of California, 1988

with wire floors is a typical example. As birds exercise less in cages than on the floor, their daily energy requirement is lower. Floor birds, on the other hand, have further to travel to obtain feed and water and this can affect dietary energy needs. Environmental conditions are also often not as uniform in floor rearing systems as can be achieved in cages.

Feed Management

When full-fed, each strain of layer has an inherent ability to grow at a certain rate and to reach sexual maturity with a certain body size. Today's pullet is capable of being sexually mature earlier than before and so the challenge is to get the pullet weight and frame size ready for egg production in a shorter amount of time. Rarely will pullets housed under commercial conditions be overweight. Although the bird has some control over her caloric intake to meet her demands, this mechanism is far from perfect. She cannot compensate for all the environmental variations, stresses, etc., with which she may come in contact.

If 6-week pullet body weights are less than recommended in the *Breeder Management Guide*, the main nutritional adjustment required is to feed the starter diet for a longer period. A pullet producer should feed the starter diet until the pullets are back on target weight for their age instead of automatically changing diets at 6 weeks. The grower diets can also be increased in nutrient density, if body weights are still below target values, by increasing the energy and protein (amino acid) concentration of the diet.

Growth During the First 6 Weeks

When based on percentage increases in the body weight at the end of the previous week, the most rapid gains are made when chicks are young.

Table 17-4. Increases in Weekly Weight Through 6 Weeks

Age		Increase in Weight over Previous Week (%)	
weeks	days	White Egg Pullets	Brown Egg Pullets
1	7	60	60
2	14	67	58
3	21	65	68
4	28	42	56
5	35	28	42
6	42	18	30

Source: DEKALB Delta, 1999; DEKALB Gold, 1995 Management Guides

Consequently, it is at this time that the bird is most efficient at converting feed consumed to body weight gain. As the chick grows older, the weekly increments of weight increases become less, as shown in Table 17-4. *As a rule of thumb,* Leghorn pullets generally reach 1 lb (0.45 kg) of body weight in 6 weeks, 2 lb (0.90 kg) by 12 weeks, and 3 lb (1.35 kg) by 18 weeks. Strains and breeds vary, however, and therefore standards must be based upon the breeder's recommendations.

Feed Consumption

Weekly feed consumption during the first few weeks will vary according to the strain of birds, the energy content of the ration, temperature, chick size and quality, and a vast number of other conditions. Guidelines for white and brown egg-type growing pullets are shown in Table 17-5.

Table 17-5. Daily Feed Consumption per 100 Pullets During the First 6 Weeks

	Feed Consumption Per Day			
	White Egg Pullets		Brown Egg Pullets	
Week	lb	kg	lb	kg
1	2.64	1.2	2.64	1.2
2	3.30	1.5	3.96	1.8
3	4.40	2.0	5.28	2.4
4	5.50	2.5	6.60	3.0
5	6.71	3.1	7.42	3.6
6	7.92	3.6	9.02	4.1

Source: DEKALB Delta, 1999; DEKALB Gold, 1995 Management Guides

17-B. FEEDING DURING THE FIRST 6 WEEKS— THE "STARTER PERIOD"

During the first 6 weeks of the life of the chick, a well-balanced, high density diet is of particular importance. In some instances, the starter is fed for more than 6 weeks, especially if body weights are below standard for their age; however, any period longer than 8 weeks is usually uneconomical. The starter diet may also be fed in a crumble form to improve feed intake at this time, when achieving proper body weights is a concern.

Energy Requirements for Starter Diets

Starting and growing feeds are formulated to contain a prescribed amount of energy, usually measured in terms of kilocalories per pound

Figure 17-2. Young Replacement Pullets Feeding on the Floor

Figure 17-3. Young Replacement Pullets Feeding in Cages

or per kilo of ration. Chick starting diets used during the first 5 to 6 weeks of the growing period should contain 1,300 to 1,350 kcal of ME per lb (2,860 to 2,970 kcal / kg) of ration. Seldom is the energy in the chick starting ration altered much from these figures even during periods of higher or lower environmental temperature.

Protein Requirement for Starting Diets

The protein requirement of the growing chick is based on its need for various amino acids in the proper balance. Therefore, *quality protein*, defined as the proper balance of amino acids, is a requisite of proper feed formulation and affects the level of protein required in the diet. As most poultry rations are modifications of a corn-soybean meal base, certain amino acids are likely to be deficient, with methionine generally being the most limiting. Any amino acid deficiencies in natural feedstuffs can be overcome by adding other protein supplements or synthetic amino acids to the ration.

An analysis of a feed mixture or feed formulation must include the amounts of certain essential amino acids as well as total protein. These requirements are given in Table 17-6.

In practical rations, most nutritionists formulate for a margin of safety of between 5 and 15% to compensate for bird-to-bird differences in feed consumption and the variability found in ingredients and mixing. Most

Table 17-6. Protein and Amino Acid Requirements of Young Egg-Type Chicks

	Starting Chickens 0–42 Days	
	White Egg %	Brown Egg %
Protein	18.00	17.00
Arginine	1.00	0.94
Glycine + serine	0.70	0.66
Lysine	0.85	0.80
Methionine	0.30	0.28
Methionine + cystine	0.62	0.59
Tryptophan	0.17	0.16

Source: Nutrient Requirements of Poultry, 1994

poultry producers would rather pay a little more for feed than risk growth problems due to marginal formulations. Care must be taken, though, to not adjust levels of individual nutrients arbitrarily as the requirements of some are closely associated with requirements for others, and therefore imbalances can occur.

Mineral Requirements of Young Chickens

The mineral requirements of young growing chickens are given in Table 17-7. The list contains macrominerals and some trace minerals that may not be adequately available in natural feedstuffs.

Table 17-7. Mineral Requirements of Young Egg-Type Chickens

	Starting Chickens 0–42 Days					
	White Egg			Brown Egg		
	%	mg/lb	mg/kg	%	mg/lb	mg/kg
Calcium	0.90			0.90		
Phosphorus (non-phytate)	0.40			0.40		
Sodium	0.15			0.15		
Potassium	0.25			0.25		
Manganese		40.9	90		25.5	56
Magnesium		273	600		263	579
Iron		36.4	80		34.1	75
Copper		2.3	5.0		2.3	5.0
Zinc		18.2	40		17.3	38
Selenium		.068	0.15		.064	0.14

Source: Nutrient Requirements of Poultry, 1994

Table 17-8. Vitamin Requirements of Young Egg-Type Chickens

	Starting Chickens 0–42 Days			
	White Egg		Brown Egg	
Vitamin	Per lb	Per kg	Per lb	Per kg
Vitamin A (IU)	682	1,500	645	1,420
Vitamin D_3 (ICU)	91	200	86	190
Vitamin E (IU)	4.6	10	4.3	9.5
Vitamin K (mg)	0.22	0.50	0.21	0.47
Thiamin (mg)	0.46	1.00	0.46	1.00
Riboflavin (mg)	1.6	3.6	1.5	3.4
Pantothenic acid (mg)	4.6	10.0	4.3	9.4
Niacin (mg)	12.3	27.0	11.8	26.0
Pyridoxine (mg)	1.4	3.0	1.3	2.8
Biotin (mg)	0.067	0.150	0.064	0.140
Choline (mg)	591	1,300	557	1,225
Vitamin B_{12} (mg)	0.004	0.009	0.004	0.009

Source: Nutrient Requirements of Poultry, 1994

Vitamin Requirements of Young Chickens

Vitamin requirements for young chickens are given in Table 17-8. These values include no margin of safety, and feed formulators will often prescribe much larger amounts, particularly for the fat soluble vitamins that are subject to oxidation.

17-C. FEEDING FROM 6 THROUGH 20 WEEKS—THE "GROW PERIOD"

This is a critical period in the development of an egg-type pullet, for how well she is grown will have an important bearing on her productivity during the laying period. A pullet must develop at a rate appropriate for the strain and reach sexual maturity at an opportune and economical age. The basic size of the chicken is established during this period.

The nutritional requirements during the growing phase are vastly different from those during the starting period, with the primary difference involving the amount of protein in the growing diet. Compared with the starting diet, protein may be reduced materially not only to satisfy the bird's requirement but to produce a pullet at the lowest cost possible. Guidelines for feed consumption from 7 to 18 weeks of age are given in Table 17-9.

Dietary Energy for Egg-Type Growing Pullets

The amount of energy in the growing ration should be from 1,250 to 1,318 kcal ME per pound (2,750 to 2,900 kcal/kg) of ration. However, this

Table 17-9. Daily Feed Consumption per 100 Pullets Between 7 and 20 Weeks of Age

	Feed Consumption Per Day			
	White Egg		Brown Egg	
Week	lb	kg	lb	kg
7	9.2	4.2	10.1	4.6
8	10.3	4.7	11.2	5.1
9	11.2	5.1	12.1	5.5
10	11.9	5.4	13.0	5.9
11	12.5	5.7	13.6	6.2
12	13.1	6.0	14.3	6.5
13	13.6	6.2	15.0	6.8
14	14.1	6.4	15.6	7.1
15	14.5	6.6	16.3	7.4
16	15.0	6.8	16.9	7.7
17	15.4	7.0	18.0	8.2
18	15.8	7.2	18.0	8.2

Source: DEKALB Delta, 1999; DEKALB Gold, 1995 Management Guides

diet may not prove optimal under all conditions. A problem often arises during hot weather because the birds will not eat enough feed and thus body weights will suffer. During cold weather, pullets housed in open-sided housing or in a low population density environment, may consume additional feed to provide energy for an increased maintenance requirement. Note: The consumption of additional feed energy to meet an increased maintenance requirement is normally not as economical as reducing heat loss by decreasing ventilation rate and providing more insulation in the pullet house. It must be cautioned, though, that poor air quality can not be tolerated as an outcome of reduced ventilation rates. It is not common for today's early maturing pullets to gain weight too rapidly in commercial conditions. Energy intake is the main nutrient controlling body weight during the growing period, whereas protein is much more important during the starting period.

From 8 through 16 weeks, body fat layers develop in the bird, but too much or too little body fat development is to be avoided. Excessive fat accumulations can be determined by palpation of the abdominal area of the bird. To help control the development of excessive body fat depositions, energy in the growing ration should be about 1,318 kcal ME per pound (2,900 kcal/kg) from 6 to 14 weeks of age, and reduced to about 1,250 kcal ME per pound (2,750 kcal/kg) after 14 weeks. Each pullet grower will need to assess this problem as it may apply to his/her own conditions. Exceedingly fat pullets usually suffer an increased rate of prolapse. Insufficient energy consumption, on the other hand, will result in poor laying house performance with birds having insufficient energy reserves to support and maintain early egg production. Typically under such

Table 17-10. Percentage Change in Feed Consumption for Each 1°F Change in House Temperature at Various Temperatures

Average Daytime House Temperature Between		Change in Feed Consumption for Each 1°F (0.6°C) Change in Temperature
°F	°C	%
90–100	32.2–37.8	3.14
80–90	26.7–32.2	1.99
70–80	21.1–26.7	1.32
60–70	15.6–21.1	0.87
50–60	10.0–15.6	0.55
40–50	4.4–10.0	0.30

circumstances, a dip in egg production just after peak production will occur. If pullets are chronologically old enough but not physiologically ready for sexual maturity, it is much better to delay increasing the lighting to stimulate sexual maturity until the pullets are at target weight.

Temperature and Feed Consumption

Table 17-10 shows that there is a change in feed consumption as house temperatures increase or decrease, but the relationship is not constant at various house temperatures. The percentage changes in feed consumption are much larger during hot weather than during cold weather.

Tables 17-11 and 17-12 compare the feed consumption changes in Table 17-10 on a different basis. Table 17-11 shows how feed consumption decreases as temperatures increase; Table 17-12 shows how feed consumption increases as temperatures decrease.

Table 17-11. Percentage Decrease in Feed Consumption as Average Daytime House Temperatures Increase

% Decrease in Feed Consumption Between Two Temperatures as Temperatures Increase							
From		To					
°F	°C	50°F (10.0°C)	60°F (15.6°C)	70°F (21.1°C)	80°F (26.7°C)	90°F (32.2°C)	100°F (37.8°C)
40	4.4	3	8	16	27	42	60
50	10.0		6	14	25	40	59
60	15.6			9	21	37	56
70	21.1				13	31	52
80	26.7					20	45
90	32.2						31

Table 17-12. Percentage Increase in Feed Consumption as Average Daytime House Temperatures Decrease

% Increase in Feed Consumption Between Two Temperatures as Temperatures Decrease

From		To					
°F	°C	90°F (32.2°C)	80°F (26.7°C)	70°F (21.1°C)	60°F (15.6°C)	50°F (10.0°C)	40°F (4.4°C)
100	37.8	46	82	110	130	143	151
90	32.2		25	44	58	67	72
80	26.7			10	26	34	38
70	21.1				10	16	20
60	15.6					6	9
50	10.0						3

Dietary Protein and Amino Acids for Egg-Type Growing Pullets

The percentage of protein in the ration for the growing pullet should be reduced as body weight increases, as long as body weight standards are met. As the daily protein requirement of the growing bird is relatively constant, and because she is eating more feed each day, the percentage in the ration must be reduced. Any protein that is consumed above the daily requirement is wasted and simply adds to the cost of maturing the pullet.

The growing bird's requirement for protein is better indicated by body weight than by age. Normally, the total protein in the ration should be reduced by about 1% per week after the sixth week until it reaches 13% at about 14 weeks of age. However, weekly adjustments may not be practical. Changes, therefore, are usually made twice during the growing period: 6 to 11–12 weeks (first period) and 11–12 to 17–18 weeks (second period).

The second growing period is often called the *developer period.* On occasion, the growing period may be divided into three phases. But regardless of the breakdown, there may also be additional formula adjustments needed for changes in environmental temperatures and other factors. Although some nutrition guidelines list the requirements for a pre-lay diet from 16–18 weeks of age to first egg, some breeders believe it is more suitable to skip the pre-lay diet and initiate feeding the layer feed at the time of lighting.

Amino Acid Requirement of Leghorn Growing Pullets

Feed formulations must include the minimum requirement of all essential amino acids. The requirements of the most limited amino acids in normally used ingredients are given in Table 17-13. Note: The amino acid requirements do not include safety margins.

Table 17-13. Protein and Amino Acid Requirements of Egg-Type Growing Pullets

	Amount in Ration					
	White Egg %			Brown Egg %		
Item	6–12 wk	12–18 wk	18 wk, 1st egg	6–12 wk	12–18 wk	18 wk, 1st egg
Protein	16.00	15.00	17.00	15.00	14.00	16.00
Arginine	0.83	0.67	0.75	0.78	0.62	0.72
Glycine + serine	0.58	0.47	0.53	0.54	0.44	0.50
Lysine	0.60	0.45	0.52	0.56	0.42	0.49
Methionine	0.25	0.20	0.22	0.23	0.19	0.21
Methionine + cystine	0.52	0.42	0.47	0.49	0.39	0.44
Tryptophan	0.14	0.11	0.12	0.13	0.10	0.11

Source: Nutrient Requirements of Poultry, 1994

Mineral Requirements from 6 Through 20 Weeks

The mineral requirements of egg-type growing pullets are given in Table 17-14 (A & B).

Vitamin Requirements from 6 Through 20 Weeks

The minimum vitamin requirements for egg-type growing pullets are given in Table 17-15 (A & B).

Table 17-14(A). Mineral Requirements of White-Egg Growing Pullets

	Feed Requirement								
	6–12 wk			12–18 wk			18 wk, 1st egg		
	%	mg/lb	mg/kg	%	mg/lb	mg/kg	%	mg/lb	mg/kg
Calcium	0.80			0.80			2.00		
Phosphorus, non-phytate	0.35			0.30			0.32		
Sodium	0.15			0.15			0.15		
Potassium	0.25			0.25			0.25		
Manganese		13.6	30		13.6	30		13.6	30
Magnesium		227	500		182	400		182	400
Iron		27.3	60		27.3	60		27.3	60
Copper		2.7	4		2.7	4		2.7	4
Zinc		15.9	35		15.9	35		15.9	35
Selenium		0.045	0.100		0.045	0.100		0.045	0.100

Source: Nutrient Requirements of Poultry, 1994

Table 17-14(B). Mineral Requirements of Brown-Egg Growing Pullets

	Feed Requirement								
	6–12 wk			12–18 wk			18 wk, 1st egg		
	%	mg/lb	mg/kg	%	mg/lb	mg/kg	%	mg/lb	mg/kg
Calcium	0.80			0.80			2.00		
Phosphorus, non-phytate	0.35			0.30			0.32		
Sodium	0.15			0.15			0.15		
Potassium	0.25			0.25			0.25		
Manganese		12.7	28		12.7	28		12.7	2.8
Magnesium		214	470		168	370		168	370
Iron		25.4	56		25.4	56		25.4	56
Copper		1.8	4		1.8	4		1.8	4
Zinc		15.0	33		15.0	33		15.0	33
Selenium		0.045	0.100		0.045	0.100		0.045	0.100

Source: Nutrient Requirements of Poultry, 1994

17-D. ATTAINING OPTIMAL BODY WEIGHT AT SEXUAL MATURITY

The importance of correct body weight during the growing period and at sexual maturity cannot be overstressed. Changes must be made in the feeding program, and often in the ration, so that pullets will mature not only at an optimal body weight, but at an optimal age. Remember that changes in the rearing lighting program will also affect the timing of the onset of sexual maturity, and the feeding program must match this for optimal performance.

Table 17-15(A). Vitamin Requirements of White-Egg Growing Pullets

	Feed Requirement					
	6–12 wk		12–18 wk		18 wk, 1st egg	
Vitamin	mg/lb	mg/kg	mg/lb	mg/kg	mg/lb	mg/kg
Vitamin A (IU)	682	1,500	682	1,500	682	1,500
Vitamin D_3 (ICU)	91	200	91	200	91	300
Vitamin E (ICU)	2.3	5.0	2.3	5.0	2.3	5.0
Vitamin K (mg)	0.22	0.50	0.22	0.50	0.22	0.50
Thiamin (mg)	0.45	1.00	0.36	0.80	0.36	0.80
Riboflavin (mg)	0.82	1.80	0.82	1.80	1.00	2.20
Pantothenic acid (mg)	4.54	10.0	4.54	10.0	4.54	10.0
Niacin (mg)	5.0	11.0	5.0	11.0	5.0	11.0
Pyridoxine (mg)	1.36	3.00	1.36	3.00	1.36	3.00
Biotin (mg)	0.045	0.100	0.045	0.100	0.045	0.100
Choline (mg)	409	900	227	500	227	500
Vitamin B_{12} (mg)	0.0014	0.003	0.0014	0.003	.0014	0.004

Source: Nutrient Requirements of Poultry, 1994

Table 17-15(B). Vitamin Requirements of Brown-Egg Growing Pullets

	Feed Requirement					
	6–12 wk		12–18 wk		18 wk, 1st egg	
Vitamin	mg/lb	mg/kg	mg/lb	mg/kg	mg/lb	mg/kg
Vitamin A (IU)	645	1,420	645	1,420	645	1,420
Vitamin D_3 (ICU)	86.4	190	86.4	190	127	280
Vitamin E (ICU)	2.14	4.7	2.14	4.7	2.17	4.70
Vitamin K (mg)	0.214	0.47	0.214	0.47	0.214	0.470
Thiamin (mg)	0.454	1.0	0.364	0.8	0.364	0.800
Riboflavin (mg)	0.77	1.70	0.77	1.70	0.77	1.70
Pantothenic acid (mg)	4.27	9.4	4.27	9.4	4.27	9.40
Niacin (mg)	4.68	10.30	4.68	10.30	4.68	10.30
Pyridoxine (mg)	1.27	2.80	1.27	2.80	1.27	2.80
Biotin (mg)	0.041	0.09	0.041	0.09	0.014	0.090
Choline (mg)	386	850	214	470	214	470
Vitamin B_{12} (mg)	0.0014	0.0030	0.0014	0.0030	0.0014	0.0030

Source: Nutrient Requirements of Poultry, 1994

Optimal Mature Body Weight

The weight of the sexually mature egg-type pullet varies with the strain of bird. Added to this is the variability of individual birds within the flock; some mature earlier than others and some have heavier weights at maturity. As with all chickens, flock averages must be used in making feeding recommendations, and the assumption is that most strains of Leghorns will reach sexual maturity at about 18 weeks of age, weighing about 2.9 lb (1.3 kg). Medium-size pullets for the production of brown-shelled eggs will mature at about the same age, with a weight of approximately 3.5 lb (1.6 kg).

Most Egg-Type Strains Should Be Full Fed

When energy and protein levels in the ration are properly adjusted for age of the flock and house temperature during different seasons of the year, most breeders feel that egg-type pullets should be full fed, with no restriction being practiced. One of the problems associated with feed restriction is the difficulty in maintaining flock uniformity of body weight, as is often seen with heavy-type birds such as broiler breeders. Therefore, whenever possible, an adjustment in nutrient density is preferable over feed restriction, thus allowing the birds to continue on full feed.

Feeding Pullets with Automatic Feeding Equipment

When birds are hand-fed, feed should be kept before them at all times. To keep a low level of feed in the troughs, feed must be added to the feeders once or twice a day. But when automatic feeders are used, the feeders should be operated intermittently. Run feeders until the feed is distributed equally to all cages, then allow them to remain idle for an hour or two, then run again, etc. The length of the house and the amount of feed in the troughs or pans at the end of the period when the feeders are not operating will determine the exact on-and-off times. Some feed should be in the troughs or pans at all times. As automatic feeding equipment is operated with a time clock, settings may be made for any time interval.

17-E. FEED CONTROL AND OPTIMAL MATURE WEIGHT

Obtaining Correct Body Weights

The growing pullet offers few indications of how rapidly she is advancing toward sexual maturity. Growing body weight seems to be the best criterion, and is the only one available to the poultry producer.

Today's egg-type pullets rarely become too heavy because of the high stocking densities in grower houses. A larger problem is obtaining adequate weight necessary to stimulate sexual maturity with increased lighting at the suggested age of the breeder. This becomes a real problem in summer conditions when pullet weights are usually too low and first eggs are often too small. If pullets are too light at 6 weeks, it is highly advisable to continue feeding the starter diet containing a higher level of protein. Then switch to the grower feed when pullets have obtained target body weights as recommended by the breeder.

The main factor in rearing pullets is to not vary too much from the recommended breeder body weights during the 7- to 18-week period. If pullets are not gaining adequately, dietary energy and protein levels may need to be increased in grower diets. In hot climate, increase airflow to help move body heat away from the pullet. Adjustments in time and number of feedings with automated feeding systems may also help increase feed intake. It is necessary to monitor body weights on a weekly basis during the grower period. One cannot wait until near the end of the growing period to adjust body weights in an attempt to match the breeder's recommendation.

Weighing Growing Birds

Average flock body weights must be established once each week during the growing period. These weights are of the utmost importance if a good pullet is to be raised. *Individually* weigh a sample of birds in each house. Then calculate the average weight of the birds and their uniformity. Weigh a sample of birds from several areas within the house. Be sure to note the location of each sample so that you can return to the same location in subsequent weighings. Weigh each bird in the sample to be sure that smaller birds are not being overlooked. Uniformity of flock body weight is important to ensure that the majority of the flock commences egg production at a similar body weight and that problems with prolapse and mortality do not occur. (See *Cage Management for Raising Replacement Pullets,* Chapter 51.)

Feeding During Stress

Stresses created by vaccination, beak trimming, disease, high and low temperatures, and moving must be compensated for in the feeding program. When stresses occur, pullets may need additional nutrients to help them reach target body weight. Body weight should be at or above standards prior to beak trimming. Because of the severity of this particular stress, body weight will usually drop well below the standard for several weeks, resulting in a temporary depression of the growth curve.

Stress and Low-Protein Rations

Diets too low in protein will delay the recovery rate when birds are subjected to severe and prolonged stresses. The difficulty is that under such conditions feed consumption is drastically reduced. Therefore, in such circumstances, it is advisable to feed a ration higher in protein.

17-F. GROWING FEED FORMULA VARIATIONS

Hot Vs Cold Weather

It must be realized that during hot weather, birds of all types require less energy to maintain body temperature than in cold weather. If the caloric content of the diet remains the same, birds will eat less feed when the environmental temperature rises, resulting in a reduction in protein intake and slower growth rates. Therefore, the nutrient density of the diet must be increased to compensate for the reduction in feed intake caused by high environmental temperatures.

Sample Pullet Rations

Typical starter, grower, and developer diets for White Leghorn pullets are shown in Table 17-16.

Table 17-16. Sample Rations for Commercial Leghorn Pullets[1]

Ingredient	Starter lb	Grower lb	Developer lb
Ground yellow corn	1,266.5	1,471.2	1,560.9
Soybean meal (dehulled, 47.5%)	607	445	337
Meat and bone meal (50%)		11	40
Fat, animal-veg blend or equivalent	40.1		
Limestone	24	30	26
Salt	10	7	7
DL-Methionine or equivalent	3.25	1.89	1.86
Dicalcium phosphate (18.5%)	42.6	29.9	23.9
Vitamin premix	2	2	2
Vitamin E supplement	2		
Choline chloride (60%)	2	1.5	0.5
Trace minerals	1	1	1
Vitamin and mineral supplements			
Vitamin A (IU)	6,000,000	6,000,000	6,000,000
Vitamin D_3 (IU)	2,600,000	2,600,000	2,600,000
Vitamin E (IU)	43,760	3,760	3,760
Vitamin K (mg)	2,200	2,200	2,200
Vitamin B_{12} (mg)	5	5	5
Riboflavin (mg)	3,760	3,760	3,760
Niacin (mg)	20,000	20,000	20,000
Calcium pantothenate (mg)	6,000	6,000	6,000
Zinc (g)	86.018	85.482	85.236
Manganese (g)	90	88.327	87.518
Selenium (mg)	136.2	136.2	136.2
Calculated nutrients			
Metabolizable energy (kcal/lb)	1,375	1,375	1,400.9
Protein (%)	19.33	16.5	15
Lysine (%)	1.05	0.85	0.74
Methionine (%)	0.47	0.37	0.35
TSAA (%)	0.82	0.68	0.63
Fat (%)	4.24	2.52	2.74
Fiber (%)	2.81	2.75	2.68
Calcium (%)	1	1	1
Total phosphorus (%)	0.73	0.62	0.61
Non-phytate phosphorus (%)	0.6	0.5	0.5
Vitamins and other ingredients			
Vitamin A activity (IU/lb)	3,691	3,802	3,851
Vitamin D_3 (IU/lb)	1,300	1,300	1,300
Vitamin E (IU/lb)	27.62	8.44	8.78
Vitamin K (mg/lb)	1.1	1.1	1.1
Riboflavin (mg/lb)	2.68	2.64	2.62
Niacin, total (mg/lb)	14.25	13.13	12.4
Pantothenic acid (mg/lb)	6.37	6.1	5.88
Choline (mg/lb)	766.4	646	487
Xanthophyll (mg/lb)	6.3	7.36	7.8

Source: Sandy Gretebeck, Bios Unlimited, 1999

18

Feeding Commercial Egg-Type Layers

by Craig N. Coon

Feeding the laying bird is only a continuation of feeding the growing pullet by supplying the necessary ingredients in the correct proportions so that the bird can produce an abundant number of eggs. The nutrient demand for caged layers for egg production in modern strains of chickens is tremendous. The eggs produced by a pullet during a layer year can weigh ten times as much as she weighs, and she will also increase her body size by one-third. To do this, she will eat nearly 20 times her body weight in feed.

18-A. FEED CHANGES AT SEXUAL MATURITY

Just prior to the time the laying flock begins to produce eggs, the first of several management and feed changes must be initiated:

1. The total length of day must be increased.
2. The growing ration must be replaced with a pre-lay or laying ration.
3. Calcium consumption must be increased.

Feed Consumption Pattern Changes

The amount of feed consumed by the *individual* pullet just prior to and after the production of her first eggs has a remarkable pattern, greatly different than that shown by flock averages. During the month prior to her first egg, the individual Leghorn pullet consumes an almost constant

Figure 18-1. Traveling Hopper Cage Feeding System (courtesy of Salmet Poultry Systems)

daily amount of feed, about 16.5 lb (7.5 kg) per 100 pullets. About 4 days prior to her first egg, daily feed consumption decreases by about 20% and remains at this low level until the first egg is produced. This is followed by a rapid daily increase in feed intake during the first 4 days of egg production, and a moderate increase thereafter until 4 weeks later, after which time it increases very slowly. Individual birds tend to eat what they require. *Average* flock feed consumption prior to peak production is not representative of the consumption of the higher-producing individual birds.

Body Weight Increases

On an individual bird basis, a sudden increase in body weight occurs 2 to 3 weeks prior to, and 1 week after, the production of her first egg.

Figure 18-2. Fixed Feeder System for Cages

The weight gain during this period for a Leghorn pullet will be between 0.50 and 0.75 lb (227 and 340 g) and for a brown-egg layer between 0.75 and 1.00 lb (341 and 454 g). The weight gain that occurs prior to egg production is related to the development of the reproductive tract and increase in ova weight. During the following 10 to 12 weeks, the young pullet gains weight very slowly. In fact, many birds may lose weight during this period. Depending on the body weight and conditioning of the flock when the pullets are stimulated with increased light to lay, the pullets may not have adequate weight to sustain egg production. Therefore, the layers may lose weight because feed intake is not adequate to provide for maintenance and egg production.

Calcium Needs Increase

The pullet's requirement for calcium is relatively low during the growing period, but, when the first eggs are produced, the need is at least four times as great, with practically all of the increase being used for the production of eggshells.

It is a natural procedure for large amounts of calcium to be deposited in the long bones of the body just prior to the time the bird lays her first egg. This process has been shown to take place during the 2 weeks prior to the day the pullet lays her first egg, and not before. During this period, and for a week after, it is essential that there be ample calcium consumed if high egg production is to be attained.

The best recommendation that can be made is not to increase the calcium intake until 10 days before the flock is expected to produce its first eggs. Certainly the calcium stored in the bones of the first pullets to lay will be somewhat inadequate, but the last pullets to lay will not suffer as much from earlier added calcium feeding. Today, most breeders recommend that layer calcium levels be fed when pullets are moved to the layer house and light is increased to stimulate sexual maturity.

18-B. BASIC NUTRITIONAL REQUIREMENTS OF LAYING HENS

Feed is necessary for the following reasons:

1. *Body maintenance.* The amount of feed necessary for body maintenance varies with the weight of the bird and the environment.
2. *Body growth.* A Leghorn pullet should gain from 0.75 to 1.00 lb (350 to 454 g) during her laying year. A medium-size layer (producing brown-shelled eggs) should gain from 1.00 to 1.25 lb (454 to 570 g).
3. *Feather production.* This includes the growing of new feathers to replace those molted or pulled out.
4. *Egg production.* The feed required for the production of eggs is determined by both the number and size of eggs laid (egg mass).

18-C. ENERGY REQUIREMENT FOR MAINTENANCE

The weight of the layer will, in part, determine the daily energy requirement for maintenance (without egg production or growth). Maintenance requirement values are shown in Table 18-1.

Ambient temperature also influences the energy requirement for maintenance. Table 18-2 shows the effect of temperature on the average daily maintenance energy requirement of the laying Leghorn.

The metabolizable energy (ME) requirement is also greatly affected by the interaction of temperature and feather cover (Table 18-3). When layers are housed in cold temperatures and have less than normal feather cover the hens will need to consume additional energy to supply their needs for maintaining body temperatures. Layers housed in a hot environment need less dietary energy to maintain body temperatures. Layers with less than 100% feather cover in hot temperatures will consume additional calories because of the increased ability to eliminate body heat.

Table 18-1. Feed Requirement of Laying Hens for Maintenance[1]

Weight of Hen			Feed Required for Maintenance		
			Feed per Hen per Day		
lb	kg	Feed per Unit of Body Weight	lb	g	kcal of ME per Hen per Day
2.86	1.3	0.037	0.105	47.8	134
3.08	1.4	0.036	0.111	50.4	141
3.30	1.5	0.036	0.117	53.2	149
3.52	1.6	0.035	0.123	55.7	156
3.74	1.7	0.034	0.129	58.6	164
3.96	1.8	0.034	0.135	61.1	171
4.18	1.9	0.034	0.140	63.6	178
4.40	2.0	0.033	0.146	66.1	185
4.62	2.1	0.033	0.151	68.6	192

[1] 1,300 kcal/lb (2,800 kcal/kg); 70°F (21°C)
Source: Zhang and Coon, 1998

Table 18-2. Effect of Ambient Temperature on the Maintenance Requirement for Energy

Temperature		Maintenance Requirement in kcal ME per Laying Hen per Day	
°F	°C	White Egg[1]	Brown Egg[2]
50.0	10.0	182	223
60.0	15.6	169	207
70.0	21.1	156	191
80.0	26.7	143	176
90.0	32.2	131	160
100.0	37.8	118	144

[1] White egg hen weighs 1.6 kg
[2] Brown egg hen weighs 2.1 kg
Source: Zhang and Coon, 1998

Table 18-3. Mean Feed Intake (g/hen/day) of Hens with 0, 50, and 100% Feather Coverage for a 6-Week Period at 55, 75, and 93°F

		Feather Coverage		
Temperature °F	Temperature °C	100%	50%	0%
55	12.8	108[cd]	128[b]	147[a]
75	23.9	105[d]	112[c]	128[b]
93	33.9	82[g]	91[f]	99[e]

[a–g] Means with different superscripts differ significantly ($P < 0.05$)
Source: Perguri and Coon, 1993

Example. Data in Table 18-2 show that a White Leghorn pullet would have to consume 38 kcal more ME per day when the ambient temperature is 60°F (15.6°C) than when it is 90°F (32.2°C) to maintain the same rate of egg production. This amount would be equivalent to 2.93 lb (1.33 kg) of additional feed per 100 pullets per day based on a 2,850 kcal MEn / kg laying ration.

Example. Data in Table 18-3 show layers consuming 40 g more feed / day when housed at 55°F (12.8°C) with no feather cover compared to fully feathered hens. The diet contained approximately 2,900 kcal ME per kg suggesting hens needed at least 116 kcal of additional ME to maintain body temperature. The cold temperature situation is very uneconomical without feather cover, whereas hens consuming 10 to 20 g additional feed in a hot temperature, because of their ability to eliminate excess body heat (without feathers), may consume enough extra nutrients to produce more egg mass and profit.

18-D. ENERGY REQUIREMENTS FOR EGG PRODUCTION

Leghorn and medium-size layers producing brown-shelled eggs are usually fed a diet with about 1,300 kcal of ME per lb (2,860 kcal / kg) of feed. The optimum energy level of a layer diet should be based on least cost computer formulating that provides for adequate daily intake of all essential nutrients as affected by environment, age, economics, and performance. Layers on the floor should consume slightly more MEn than layers kept on wire floors (cages) to allow for increased activity.

The daily energy requirement of the laying bird is highly variable. Some reasons for this are:

1. Variation in body weight of pullets
2. Environmental temperature
3. Amount of bird activity
4. Variations in egg production
5. Differences in egg size
6. Prevalence of stress
7. Age of the bird
8. Amount of feather cover
9. Strain of layer

The only compensating fact in overcoming the above variations is that each bird is usually able to govern her feed intake according to her energy needs. But whether the governing mechanism is efficient is a question. The factors listed above that make the energy requirements for layers variable are the same factors that affect the feed consumption of a flock. The MEn content of the diet is also an important factor that determines the amount of feed consumed. Good feeding requires changes in the feed formula or in the method of feeding to help the bird regulate her energy intake.

Layer Weight Gain

The daily feed consumption for layers is far from consistent throughout the egg production period. Not only does the weight of the layer influence consumption but birds must also gain weight, and this gain in weight is not uniform. Studies have shown that practically all individual birds have periods of weight gain followed by intervals when they gain no weight. From a flock standpoint, however, there should be some slight weekly increase in average body weight during the laying period.

Energy Requirement

The MEn requirement of a 3.5 lb (1.6 kg) layer, kept at a moderate temperature of 70°F (21° C), and gaining 2 grams of body weight and producing approximately 55-g egg mass per day is about 305 kcal per day (Table 18-4). The requirement will increase in cold weather and decrease in hot weather.

The amount of energy in the diet will positively influence feed consumption. The relationship is shown in Table 18-4, which gives feed consumption necessary per day as the caloric content of the ration varies. Actually under field conditions (especially in hot temperatures), rations higher in energy, because of added fat, will usually be more economically efficient than those having lower energy.

Note. The values in Table 18-4 are calculated for specific temperature and performance criteria. White Leghorns will require less energy compared to brown egg layers.

Table 18-4. Dietary Energy in the Feed and Daily Feed Requirement for a 1.6-kg Hen at 70°F (21°C)

kcal of ME per		Feed Required per Day per 100 Hens to Supply 305 kcal ME per Hen		Feed per Dozen Eggs Produced[1]	
lb of Ration	kg of Ration	lb	kg	lb	kg
1,200	2,640	25.5	11.6	3.41	1.55
1,250	2,750	24.5	11.1	3.41	1.55
1,300	2,860	23.5	10.7	3.15	1.43
1,350	2,970	22.7	10.3	3.01	1.37
1,400	3,080	21.9	10.0	2.93	1.33
1,450	3,190	21.1	9.6	2.82	1.28

[1] 90% hen-day egg production, 55 g egg mass, 2 g gain/day
Source: Zhang and Coon, 1998

Environmental Temperature and Feed Consumption

As the hen's requirement for energy is higher in cold weather than in hot weather, there are differences in the amount of feed hens will consume under these conditions. These variations in feed consumption are smaller for each degree change in temperature when the weather is cool than when it is hot.

Example. Between 44° and 57°F, each degree change alters feed consumption about 0.69%, while between 87° and 95°F each degree change alters consumption about 5.79%. The determination of percentage change in feed intake is based on a 45 g change per day over the entire range from 44°F to 95°F. The relationships between temperature and feed intake, caloric intake, and feed conversion are shown in Tables 18-5, 18-6, and 18-7. If feed consumption appears to be too high, particularly toward the end of the laying cycle, the environmental temperature of a layer house may be kept slightly higher to cause the flock to decrease its feed intake and control body weight gain. The effect of dietary energy concentration and temperature on ME consumption is more meaningful than simply evaluating the effects on feed consumption. Many parts of the world utilize significantly lower dietary energy concentrations than in the US, therefore the temperature effects on layer feed intake would be less useful. The optimum environmental temperature for layers in Table 18-7 for optimum feed conversions was approximately 30.5°C or 87°F. The layers at this temperature required less feed for maintenance and could provide more nutrients for egg mass production.

Dietary Caloric Content and Feed Consumption

As the energy content of the feed increases, hens will eat less, and vice versa (Table 18-5).

Rule of thumb. Feed consumption will be reduced by 2.5% for each increase of 50 kcal of ME per pound (454 g) of ration, or vice versa.

Summer and Winter Caloric Requirement

The consumption of feed and consequently energy (calories) changes with increasing age and with temperature changes. Table 18-8 illustrates these relationships in a 1978 California study of 100 commercial Leghorn table-egg flocks.

Directions for Table 18-5. To use Table 18-5, first determine the MEn of the ration being fed, then find the feed consumption of the flock of birds for the appropriate average daytime ambient temperature. Estimated feed consumption per 100 pullets per day for other temperatures will be found in the same row. Energy intake models usually consider body weight, temperature, egg mass, and sometimes feathering.

Table 18-5. Effect of Feed Energy and Temperature on Feed Intake (g/day) of White Leghorn Layers from 20 to 36 Weeks of Age

	Temperature F° (C°)											
Feed Metabolizable Energy, kcal/kg	45 (7)	48 (9)	57 (14)	61 (16)	66 (19)	71 (22)	76 (25)	82 (28)	87 (31)	92 (33)	95 (35)	Av.
	Feed Intake (g/day)											
2,645	118	119	113	112	111	105	105	100	91	70	72	102
2,755	113	115	111	114	108	103	102	102	91	68	67	99
2,865	114	111	105	109	111	104	102	88	88	66	66	97
2,976	104	109	110	106	108	99	96	84	84	61	65	94
*	ab	a	c	bc	c	d	d	e	f	g	g	
Means	112	113	108	110	109	103	101	98	88	66	67	

* a–g: Means between columns with no common letter differ ($P < 0.05$)
Source: Peguri and Coon, 1991

Table 18-6. Effect of Feed Energy and Temperature on Metabolizable Energy Intake of White Leghorn Layers from 20 to 36 Weeks of Age

	Temperature °F (°C)											
Feed Metabolizable Energy, kcal/kg	45 (7)	48 (9)	57 (14)	61 (16)	66 (19)	71 (22)	76 (25)	82 (28)	87 (31)	92 (33)	95 (35)	Av.
	Metabolizable Energy Intake (kcal/kg)											
2,645	312	315	300	296	294	278	278	264	241	186	189	268
2,755	312	316	306	313	297	284	280	280	251	187	184	274
2,865	326	317	301	312	317	299	291	273	251	188	188	279
2,976	311	324	309	315	320	295	284	287	249	183	194	279
*	a	ab	c	bc	b	d	d	e	f	g	g	
Means	315	318	304	309	307	289	283	276	248	186	189	

* a–g: Means between columns with no common letter differ ($P < 0.05$)
Source: Peguri and Coon, 1991

Table 18-7. Effect of Feed Energy and Temperature on Feed Conversion (g feed/g egg mass) for White Leghorn Layers from 20 to 36 Weeks of Age

	Temperature °F (C°)											
Feed Metabolizable Energy, kcal/kg	45 (7)	48 (9)	57 (14)	61 (16)	66 (19)	71 (22)	76 (25)	82 (28)	87 (31)	92 (33)	95 (35)	Av.
	Feed Conversion (g feed/g egg mass)											
2,645	2.84	2.72	2.62	2.52	2.50	2.38	2.42	2.34	2.15	2.47	2.23	2.47
2,755	2.82	2.69	2.51	2.53	2.34	2.32	2.33	2.26	2.13	2.25	2.15	2.39
2,865	2.60	2.59	2.40	2.40	2.35	2.28	2.25	2.20	2.03	2.21	2.08	2.31
2,976	2.51	2.47	2.37	2.36	2.33	2.18	2.16	2.16	2.00	2.25	1.94	2.25
*	a	b	c	c	d	de	e	e	f	de	f	
Means	2.69	2.61	2.47	2.45	2.36	2.29	2.29	2.24	2.07	2.30	2.10	

* a–g: Means between columns with no common letter difference ($P < 0.05$)
Source: Peguri and Coon, 1991

Table 18-8. Consumption of Feed and Calories by Leghorn Layers in Relationship to Their Age and the Season

	Age in Weeks										
	21/24	25/28	29/32	33/36	37/40	41/44	45/48	49/52	53/56	57/60	Av.
						lb					
Daily Feed Consumption											
Summer	0.164	0.199	0.213	0.226	0.227	0.232	0.231	0.231	0.221	0.217	0.216
Winter	0.179	0.214	0.233	0.240	0.242	0.246	0.245	0.246	0.251	0.240	0.233
Year	0.171	0.205	0.229	0.236	0.236	0.239	0.238	0.238	0.239	0.235	0.227
						g					
Summer	74	90	97	103	103	105	105	105	100	98	98
Winter	81	97	106	109	110	112	111	112	114	109	106
Year	78	93	104	107	107	108	108	108	108	107	103
						kcal ME					
Daily Energy Consumption											
Summer	212	252	268	286	290	294	293	291	278	273	274
Winter	229	272	296	302	306	313	313	315	321	306	297
Year	220	260	291	300	300	304	303	302	304	299	288

Source: University of California, 1978

Rate of Egg Production and Feed Intake

A whole egg contains approximately 1.6 kcal / g of egg and the efficiency of converting dietary ME to egg energy is approximately 63%. A 60-g egg will therefore require approximately 150 kcal of dietary ME to produce.

Rule of thumb. If egg size remains the same, each 10% change in hen-day egg production will alter the feed requirement by 4 to 5%.

Egg size and feed requirement. As larger eggs contain more calories of energy than smaller eggs, the dietary energy required to produce them is greater.

Rule of thumb. A hen needs 1.8% more energy intake for each 1 oz/doz (2.4 g / egg) increase in egg size.

Body Weight Alters Feed Intake

The larger the hen, the greater the feed requirement for maintenance; therefore, feed consumption increases as body weight increases during the egg production year.

Rule of thumb. For each 0.1 lb (45 g) increase in body weight, a laying hen requires 1.2% more energy intake.

Predicting Energy Consumption with Mechanistic Models

The ability to measure feed consumption of a flock is often difficult. Knowledge of feed consumption is critical when formulating diets on a daily feed intake basis. Since layers tend to consume feed based on their daily needs for energy, feed consumption of layers can be predicted by determining the amount of dietary energy that is needed to maintain their performance in specific environmental conditions and when the level of ME in the layer feed is known. In Table 18-9, the consumption of ME was

Table 18-9. Actual ME Intake (kcal/d) and Predicted ME Intake (kcal/d) Using Prediction Models

Actual				Predicted			
Temp. (°F)	Temp. (°C)	Strain	ME Intake	Zhang and Coon (1998)	Emmans (1974)	NRC (1994)	Pesti (1992)
50.0	10.0		324	324	357	338	364
65.0	18.3		297	311	333	322	333
80.0	26.7		282	294	305	301	301
95.0	35.0		223	215	214	219	328
		B	277	278	290	286	338
		CV	296	302	326	313	335
		DD	291	293	209	302	293
		HW	264	273	287	282	362

Source: Zhang and Coon, 1998

measured over a three-month period at four different environmental temperatures with four different strains of white-shell egg layers. The four strains were known to consume different amounts of feed and produce large differences in egg mass. The model that is used in the research was developed by Zhang and Coon (1998). The model is as follows:

$$MEI = ((W^{0.75}) \times (143.7 - 1.612T)) + (5\Delta W) + (EM \times EEC/0.63)$$

where:

- MEI = Predicted metabolizable energy daily intake
- $W^{0.75}$ = Metabolic body weight (kg)
- T = Degrees in Centigrade
- ΔW = Change in daily body weight (g)
- EM = Egg mass (g/day)
- EEC = Egg energy concentration (energy concentration of whole eggs) (kcal/g)

The weight of the layer is converted to metabolic body weight ($W^{0.75}$), which is used to better compare different sizes and weights of animals. The metabolic weight of the animal is in kg. Temperature has a large impact on energy requirement for maintenance, so in order to adjust for environmental temperature in the model, the mean daily temperature in Centigrade is multiplied by 1.612. The final component in the parenthesis (143.7 − 1.612T) is multiplied by the metabolic body weight of the layers to determine the ME requirement for maintenance. A 1.6-kg layer housed in a 21.1°C (70°F) environment would have a maintenance requirement of 156 kcal. The change in daily body weight is multiplied by 5 kcal. A mature layer is primarily gaining fat after the reproductive tract is developed, thus a layer gaining one gram of weight would be actually gaining 0.40–0.45 g of fat and 0.55–0.60 g of water, with the kcal of gross energy in fat being approximately 9.1 kcal/g. The efficiency coefficient for fat retention into body weight is thought to be 0.8; therefore, the ME needed for one gram of body weight = 0.44 g fat × 9.1 kcal/g of fat = 4.004 kcal/g gain/0.8 = 5 kcal. The last portion of the energy equation relates to the egg mass produced. The equation does not include a specific component for EEC which is the egg energy concentration. The layers tested for the model in Table 18-8 each produced eggs that had a specific energy concentration that was different and ranged between 1.6 to 1.67 kcal/g of whole egg (including shells). If layers are producing more yolk compared to albumen or other components such as egg shell, the energy concentration would be higher. An average value of 1.6 kcal (EEC) per gram of egg mass can be used to determine kcal of gross energy produced. The egg mass produced per day times 1.6 kcal is then divided by the determined average coefficient for retaining energy (0.63) within eggs. A flock laying at a rate of 92% with average eggs weighing 60 grams would be generating 55 grams of egg mass daily. Therefore the flock would require 140 kcal of ME per day for egg mass production. The total ME required per day for a layer housed at 21.1°C, weighing 1.6 kg, gaining 2 grams of weight per day, and producing 55 grams of egg mass would be 306 kcal ME.

The Relationship of Metabolizable Energy, Egg Weight, and Egg Production

The ability of layers to partition dietary energy into maintenance, egg weight, and egg numbers is primarily related to the environmental temperature in which the layers are housed. Research with layers housed in thermoneutral temperatures indicated layers primarily used ME intake to produce egg numbers and were not partitioning energy for increasing egg weight. Recent research with layers housed in both a 70°F (21.1°C) thermoneutral temperature and also a cycling warmer temperature from 95 to 80°F (35.6 to 26.7°C) has shown layers housed at 70°F (21.1°C) only increased egg numbers with increasing ME intake while layers housed in the warmer environment partitioned energy to egg numbers and egg weight (Figure 18-3). The layers housed in the warmer temperature had

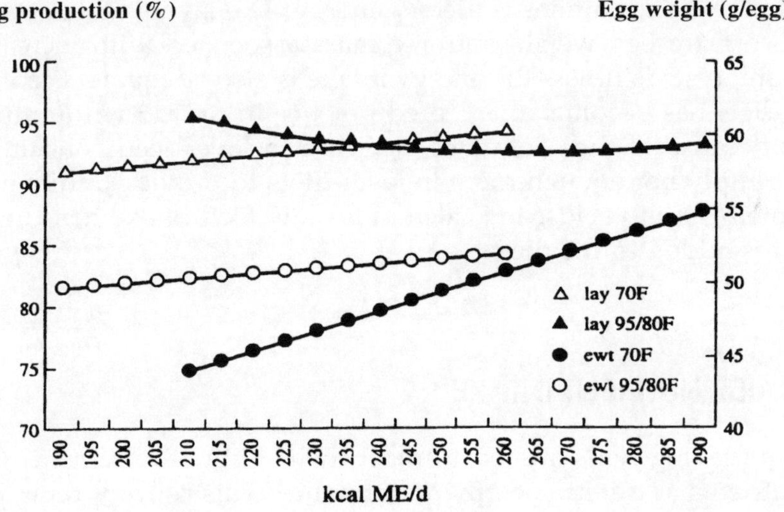

Equations for the four responses are: egg weight (70F) $y = 114.63 - 0.423x + 0.0008x^2$; egg weight (95/80F) $y = 49.81 + 00396x$; egg number (70F) $y = 40.80 + 0.162x$; egg number (95/80F) $y = 74.13 + 0.0398x$.
Source: Zhang and Coon, 1994.

Figure 18-3. Effect of ME and Environmental Temperature on Egg Production and Weight

a significantly lower maintenance requirement and therefore had more energy to direct to the production of additional egg mass.

18-E. FAT IN THE LAYER RATION

Fats are primarily used in poultry feeds to supply a concentrated source of energy. Most feed grade fats have more than two times the ME of feed grains. In addition, the utilization of energy in the diet is enhanced when energy derived from fats is substituted for carbohydrate or protein sources of energy. The fatty acid composition of eggs can also be manipulated by incorporating high levels of corn or sunflower oils in the diet. Eggs for specialty markets have been produced containing additional linolenic acid, an essential fatty acid normally found in fish, by feeding flax (linseed) meal to birds.

Modern strains of layers are consuming significantly less feed/day when compared to layers in the past—oftentimes consumption is less than 90 g per day. In high density cage systems, individual layers from strains known to consume less feed may not consume adequate daily energy unless the diet contains added fat. The first limiting nutrient for high producing hens in these conditions is dietary energy. The layers will not be able to produce extra egg weight and egg mass associated with increases in dietary amino acids unless the energy intake is also adequate. Feeding fat in layer diets has become an accepted practice in the US within the past 10 years because of the continuing decrease in layer feed consumption. There is simply not enough room in layer diets to provide additional energy, protein, amino acids, and calcium for low feed intake hens unless 1 to 3% fat is added to the diets.

18-F. DAILY NUTRIENT INTAKE

Egg-type laying hens are fed to meet their daily requirements for all major nutrients as determined by performance. This concept replaces the older concept "percentage of the diet," which is still quoted throughout the industry and is used in this book for reference purposes. The nutritionist's concern is not with the percentage in the feed as much as with the daily intake of nutrients as *affected by feed consumption* and *percentage of each nutrient in the feed.*

The essential elements of the daily intake feeding program are (1) the amount of feed each flock is eating, (2) the requirements for each production level, and (3) the nutrient content of the diet. Its success is dependent on the accuracy with which feed consumption can be measured, the validity of the standards used in establishing the requirements, and the precision achieved in feed formulation and manufacture.

Nutrient intake is calculated by multiplying the measured feed consumption per hen per day by the percentage of the nutrient in the diet. For example, if a flock eats 100 g of feed per hen per day, and it is formulated to contain 0.35% methionine, the average daily intake of methionine is 350 mg. Many primary breeders provide tables containing the recommended nutrient requirements for their strain of bird at a particular stage of its laying cycle. This may serve as a guide for designing a feeding program.

Projecting feed consumption for next week's usage requires knowledge of present consumption and reliable estimates of whether or not a change is likely to occur because of changes in ambient temperature or other factors. Present consumption is usually obtained by estimating feed disappearance from the feed tank during the present week or by actually weighing the amount of feed used. Several systems are available that will electronically monitor the quantity of feed in a feed tank and consumption on an hourly, or more often, basis.

18-G. PROTEIN REQUIREMENTS FOR EGG PRODUCTION

The protein requirement of laying birds is closely associated with the rate of egg production and egg size. When egg production reaches its peak, the requirement may be as high as 17 to 19%. At the end of the production cycle, it may drop to as low as 14%. However, in most countries there has been a trend toward feeding so-called *amino acid equivalent* diets. In adopting this practice the nutritionist formulates the diets on the basis of the bird's requirement for amino acids rather than on an absolute protein basis. In this way, the diet may contain 2 to 3% less crude protein but be formulated with sufficient added essential amino acids to be equivalent to a higher protein diet. A reduction in the dietary level of protein in this manner reduces feed costs while supplying the bird's requirements for amino acids.

Amino Acids

To speak of the protein requirement for egg production is to speak of the amino acid requirement. Proteins must be well balanced and of high quality for a hen to lay her maximum number of eggs, and to produce them economically. Of the amino acids, methionine is often the most deficient in the laying ration. The laying hen's *digestible amino acid requirements*

Table 18-10. Protein[1] and Digestible Amino Acid[2] Requirements of Layers

	Daily Intake		
	NRC	Minnesota	
Nutrient	(mg/day)	(mg/day)	Egg Mass (mg/g)
Protein	15,000	—	—
Arginine	602	880	17.4
Lysine	593	675	13.2
Methionine	258	329	6.4
Methionine + cystine	499	547	10.6
Tryptophan	138	132	2.7
Isoleucine	559	579	11.4
Valine	602	689	13.3

Source: [1]National Research Council, 1994 (a 100-g daily feed intake and a 0.86 coefficient was used for converting total amino acid requirements to digestible amino acids)
[2] Coon and Zhang, 1999 (2,900 kcal ME/kg ration, layers weighed approximately 1.5 kg, 50-g egg mass/day)
Note: Layers also require the essential amino acids threonine, leucine, phenylalanine, and histidine. Actually, when protein levels are maintained at 13 to 15% with commonly used ingredients, the amino acids listed in Table 18-10 tend to be the most limiting and in some instances will not be supplied at adequate levels

are given in Table 18-10. The amino acid requirements are expressed as minimum values with no margin of safety added.

Researchers using modern layer strains have reported the digestible amino acid requirement of layers on a daily intake basis. The layers weighed approximately 1.5 kg and were producing a minimum of 50-g egg mass per day. Assuming amino acid values reported by NRC (1994) are from research using corn-soy diets and a 0.86 coefficient for converting total amino acids to digestible amino acids is used, the digestible amino acid values in Table 18-10 are still higher, mainly because of higher egg mass production from modern layers.

The feeding of molted flocks should be no more difficult than feeding first cycle layers. The layers will be producing larger eggs immediately after coming back into lay and may be slightly heavier in body weight than beginning first cycle pullets. The flock manager must watch egg size and body weight of the recycled flock and regulate amino acid levels to produce the most economical size egg for their markets. A molted flock may start producing maximum egg mass almost immediately after peaking because of increased egg size. If a nutritionist is feeding high dietary energy levels by adding fat, the energy concentration should be reduced more quickly in the second than the first cycle to keep hens from becoming too fat. The persistency of lay will also drop off more rapidly in the second cycle thus leaving additional calories for gaining body weight.

The commercial application of poultry nutrient requirements requires knowledge of the quality of local feedstuffs, allowances for milling or feed separation problems, unexpected changes in feed consumption, and an assessment of the losses (production and otherwise) which could result from insufficient nutrient intake. For these reasons, practical diets are always formulated to higher specifications by providing a 5 to 15% margin of safety.

Daily Amino Acid Requirement of Leghorn-Type Chickens

One of the most controversial points surrounding the nutrition of laying hens is the minimum daily requirement of amino acids. There have been countless experiments and much has been written, but there is still confusion. As dietary protein and their corresponding amino acids are expensive, every feed formulator wants to be sure there are adequate but not excessive levels in the diet.

Impact of flock performance on requirement. To understand amino acid requirements, one must understand the variables involved. In the first place, nothing is constant during the laying year. Birds increase in body weight, egg production rises rapidly then falls gradually, and egg size increases. Then to complicate matters, individual birds within the flock are not uniform. Some start to lay at an early age, while others lay later. In addition, body weight varies, as does egg size—all of which influence the amount of amino acids necessary.

When body weight increases, more amino acids are required. During the year, feather growth decreases, necessitating feeding less amino acids for this component. Egg production is highly variable, and the dietary amino acids necessary to produce egg protein therefore is also highly variable. As egg size increases, more amino acids are needed to deposit the additional protein into the larger eggs. To add to these variables, dietary amino acids are not well utilized in the production of eggs. The efficiency of dietary nitrogen from amino acids being utilized for producing egg nitrogen for laying hens is approximately 40 to 45%. The conversion of dietary nitrogen to egg nitrogen can be observed in the 14-day balance study shown in Table 18-11. Feeding 14% protein diets supplemented with amino acid levels, to provide ideal amino acid profiles, allowed the hens to retain 46% dietary nitrogen in eggs and carcass. Essentially all of the nitrogen need is for egg production, as the nitrogen gain in body tissue is very minimal in layers.

Determining the amino acids needed with factorial models. Table 18-12 shows factorial coefficients used for different amino acids for maintenance and production as reported by Hurwitz and Bornstein (1973). Table 18-13 summarizes the predicted critical amino acid require-

Table 18-11. Layer Fourteen-day Nitrogen Balance Study

Diets	N Intake* (g)	Egg N (g)	Change Carcass (N) (g)	N Loss (g)	Nitrogen Retention (%)
1. 18CP	38.14[a]	14.57[a]	0.205[a]	23.37[a]	38.89[b]
2. 16CP	34.33[b]	13.60[ab]	−0.216[a]	19.86[bc]	42.37[ab]
3. 16CP+Lys	34.40[b]	14.18[a]	0.019[a]	20.20[b]	41.48[b]
4. 14CP+Met+Lys +Trp+Arg	30.25[c]	12.43[b]	−0.652[a]	18.33[bc]	39.56[b]
5. Diet 4 +Ile+Val	30.74[c]	13.73[a]	0.59[a]	17.17[c]	45.94[a]
Mean	33.54	13.70	−0.01	19.87	41.57
S.D.	4.29	1.47	1.55	3.57	5.16
S.E.	0.62	0.21	0.22	0.53	0.76
Crit. LSD value (0.05)	2.94	1.19	N.A.	2.69	4.26
P value	0.0001	0.011	0.53	0.0007	0.022

* a–c: Means within columns with no common letter difference ($P < 0.05$)
Source: Zhang and Coon, 1998

ments for different body weights and egg mass production by using the factorial equation. After reviewing the table one can see there is a great potential for variation in the daily amino acid requirement depending upon flock performance. The table shows the impact of two different body weights on daily amino acid requirements producing three different quantities of egg mass. All combinations within the table would not apply during the laying year. For example, all large eggs would not be produced at the peak of lay, nor would body weight and egg weight be low at the end of the laying cycle.

Most Leghorn flocks average about 2.75–3.00 lb (1.25–1.40 kg) in body weight at sexual maturity and lay a small number of small eggs. Body weight will steadily increase until adult weights of 3.6 lb (1.6 kg) to 4.0 lb (1.8 kg) are reached. Egg mass will reach its maximum level at 35 to 40 weeks of age. A slight increase in daily amino acid intake may be required at the lower body weights for young flocks because some growth may still be occurring.

Amino acid consumption varies with individual layers. The example requirements in Table 18-13 are average factorial values and do not show the variation of requirements within the flock. Assuming hens are at peak lay, many hens in the flock would be laying at a rate of 100%—an egg a day—and this amount of amino acid may not be adequate for 100% egg production. How then will this amount of dietary amino acid maintain a high rate of production with many birds laying at 100%? In other words, if the flock average production is 92%, some hens are laying at 100% while others are laying at 84% or lower, consuming less feed and less dietary amino acids daily. When the flock peaks from 4 to 6 weeks after the first hens start to lay, some birds are just starting to produce eggs. These birds would be eating less

Table 18-12. Estimated Amino Acid Needs for Maintenance, Weight Gain, and Egg Production for White Leghorn Laying Hens

Amino Acid	Maintenance mg/kg of body weight/day	Body Weight Gain mg/g/day	Egg Mass mg/g/day
Isoleucine	76.2	8.7	10.5
Lysine	31.6	15.9	11.1
Methionine	76.2	3.8	5.5
Tryptophan	20.4	1.7	2.2
Valine	65.1	14.2	13.0

Source: Zhang and Coon, 1994; Hurwitz and Bornstein, 1973, factorial model

Table 18-13. Estimated Digestible Amino Acid Needs of Laying Hens[a]

	Egg Mass g/hen/day	Body Weight kg/hen (lb)	Amino Acid Needed mg/hen/day
Lysine	50	1.5 (3.3)	682
		2.0 (4.4)	698
	55	1.5 (3.3)	737
		2.0 (4.4)	753
	60	1.5 (3.3)	793
		2.0 (4.4)	809
Methionine	50	1.5 (3.3)	353
		2.0 (4.4)	391
	55	1.5 (3.3)	436
		2.0 (4.4)	474
	60	1.5 (3.3)	463
		2.0 (4.4)	501
Isoleucine	50	1.5 (3.3)	683
		2.0 (4.4)	721
	55	1.5 (3.3)	735
		2.0 (4.4)	773
	60	1.5 (3.3)	788
		2.0 (4.4)	826
Tryptophan	50	1.5 (3.3)	149
		2.0 (4.4)	159
	55	1.5 (3.3)	160
		2.0 (4.4)	170
	60	1.5 (3.3)	171
		2.0 (4.4)	181
Valine	50	1.5 (3.3)	819
		2.0 (4.4)	851
	55	1.5 (3.3)	884
		2.0 (4.4)	916
	60	1.5 (3.3)	949
		2.0 (4.4)	981

[a] Body weight gain of 5 g/hen/day is assumed in the calculations
Source: Zhang and Coon, 1994; Hurwitz and Bornstein, 1973, factorial model B, assume 0.85 digestion coefficient for all protein when converting to digestible amino acids

because their energy requirement is less. Likewise, as a result of consuming less feed, these birds will be consuming a lower level of all nutrients than the average of the flock.

The Reading model. This model was developed in Reading, England to predict the amino acid requirement of a layer flock that takes into consideration the variation of egg size in the flock, variation of body weight, and utilizes an economic ratio consisting of cost of amino acid / value of egg mass in deciding how much amino acid to add.

Daily Amino Acid Requirement of Brown-Shell Layers

The daily amino acid requirement for medium-size layers producing brown-shelled eggs is higher than the needs shown for white-shell egg producing Leghorns. The brown-shell layer requires higher daily intake of amino acids for maintenance because they are slightly larger in body weight and require additional intake for increased daily egg mass. The factorial coefficients utilized to predict amino acid requirements for layers in Table 18-12 could be used to estimate the daily intake requirement for brown-shell layers weighing approximately 2 kg and generating 55–60-g egg mass daily.

Larger Hens Get More Amino Acids

Individual bird weights within the flock vary, with larger birds consuming more feed to meet their increased need for energy. In doing so, they also consume a higher level of amino acids each day. Egg size is greater in the larger birds, but dietary conversion of protein and amino acids to egg amino acids is generally poor (40 to 45%). Therefore, the more uniform body weight, the easier it is to formulate for flock needs.

Environmental Temperature and Amino Acid Requirement

In hot temperatures, feed consumption will be lower and unless nutrient density is increased there can be inadequate amino acid intake. Table 18-14 shows the advantage of formulating diets for layers housed in hot temperatures by increasing concentrations of amino acids to provide similar daily intakes of amino acids when compared to layers housed in cooler temperatures. The egg mass production was similar for layers housed in each of the three temperatures, while feed conversion of layers in the 29.9°C environment was less than 2 grams of feed per gram of egg mass because of lower energy requirements for maintenance. The best way to add additional amino acids to rations for layers housed in hot

Table 18-14. Performance of Hens from 37 to 65 Weeks of Age Housed at Different Temperatures and Fed Different Levels of Protein and Amino Acids

	Temperature °F (°C)		
	65 (18.3)	75 (23.9)	85 (29.9)
Hen-Day Egg Production (%)	83.0	84.7	84.5
Egg Weight (g)	58.7	58.3	58.5
Egg Mass (g)	48.7	49.4	49.4
Feed Consumption			
—lb / 100 hens / day	25.2	23.4	21.5
—g / hen / day	112.8	106.9	97.9
Feed Conversion			
—lb feed / lb egg	3.64	3.31	3.05
—g feed / g egg mass	2.32	2.17	1.98
Body Weight (lb) at 65 wks	3.49	3.47	3.39

Source: Peguri and Coon, 1991; each group of hens consumed a similar daily intake of amino acids by increasing the dietary concentration of protein and amino acids for hens in warmer temperatures

temperatures is to increase synthetic amino acid levels (methionine, lysine, and threonine) as much as practical with minimum increases in dietary protein. When layers are severely heat stressed and feed intake has dropped sharply, the decline in energy intake will eliminate the value of adding dietary amino acids. Layers housed in hot temperatures will have a tremendous advantage when fed with rations formulated on a digestible basis. The digestible amino acid formulations will emphasize the value of protein quality because of the need to keep dietary protein and excess amino acids to a minimum under hot conditions. The number one limiting nutrient for layers in hot temperatures is ME. Higher heat increments and heat production is created in layers when synthesizing uric acid from excess nitrogen. The additional dietary energy needed to produce uric acid in hot conditions could be used to make egg mass if diets were properly formulated.

Protein and Egg Size

Although the size of the egg yolk has a greater influence on egg size than the amount of albumen, the amount of the latter is also important. The solids in egg albumen are almost entirely protein. Because the egg's demand for protein and amino acids is great, any lack of dietary protein results in a decrease in the amount of albumen, and consequently egg size even though the quantity of yolk may be similar. Increasing the protein and amino acid content of the diet has a marked effect on increasing egg size, particularly when there are small eggs. Excessive protein and amino acid consumption may increase egg size too much, while too little may result in an excessive number of medium eggs.

Smaller eggs during the summer months are commonly the result of lower energy intake as egg producers usually adjust feed formulas to maintain uniform protein and amino acid intakes throughout the year. Methionine is the first limiting amino acid and is often used to control egg size in layer operations. Producers often increase and decrease dietary methionine to increase and decrease egg weights, respectively. Many nutritionists also believe that methionine intake can be decreased somewhat without affecting egg numbers. This may be true up to a point but Morris and Gous (1988) have shown that decreasing the limiting amino acids or protein in the diet will affect both egg numbers and egg weight, depending on the level of supplementation (Figure 18-4). When the rate of lay is maintained at a level that is close to its potential, a slight decrease in the methionine intake affects mainly egg weight. However, when a

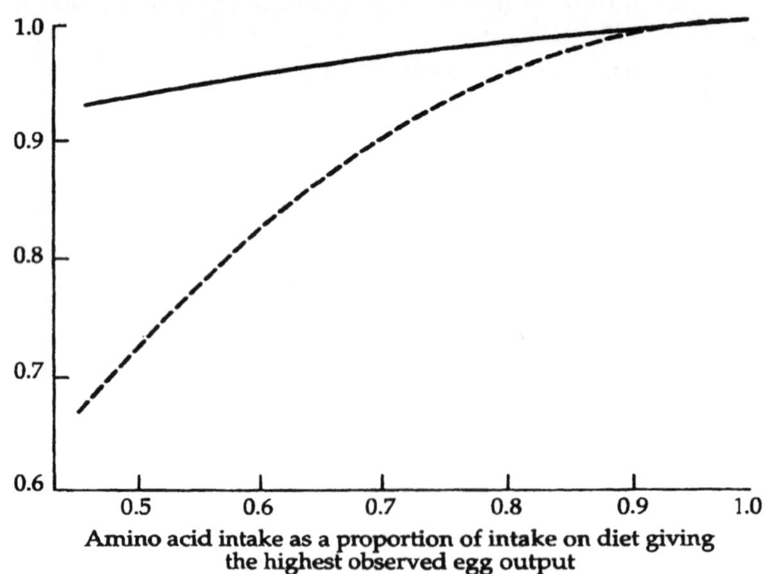

Rate of lay (- - -). Egg weight (——)
Equations for the two responses are: relative egg weight = $1 - 0.07353x - 0.1042x^2$; relative rate of lay = $1 - 0.03734x - 1.02927x^2$.
Source: Morris and Gous, 1988

Figure 18-4. The Relationship Between Intake of a Limiting Amino Acid and Rate of Lay

more severe amino acid deficiency develops, the rate of lay may also be reduced.

18-H. MINERAL REQUIREMENTS FOR EGG PRODUCTION

Mineral requirements for laying rations are shown in Table 18-15. These do not include any margin for safety.

Calcium Requirement of Layers

Once egg production begins, the need for calcium is much greater than during the pullet growing period because of eggshell formation (see *Feeding Egg-Type Replacement Pullets*, Chapter 17). Of particular importance is that too much calcium during egg production is detrimental, as it depresses appetite. It is also uneconomical because it takes up valuable space in the ration and surpluses are excreted in the feces.

Only a portion of the calcium consumed by the laying hen is retained, with the balance being excreted. The retention is about 55% for young laying hens, and 40% for older layers. Of equal importance is that the calcium be increased before the first egg is laid. In the past, nutritionists would feed a pre-lay diet containing 2% calcium for a 2-week period prior to the anticipated onset of egg production. However, under such a regimen many layers were developing rickets with rubbery like keels during the early part of the production period because of an inability to mobilize adequate calcium.

Today, most Leghorn breeders recommend that layers be fed a layer diet containing a minimum of 3.25% calcium when pullets are moved to the layer house and lighting is increased to stimulate sexual maturity. Half of the calcium should be supplied in a coarse particle size so that the earlier

Table 18-15. Average Mineral Requirements of Leghorn Laying Hens[1,2]

Mineral	19–40 Wk of Age
Calcium (%)	3.25
Phosphorus, non-phytate (%)	0.25
Sodium (%)	0.15
Chloride (%)	0.13
Manganese (mg/kg)	20
Selenium (mg/kg)	0.06
Zinc (mg/kg)	35

Source: [1]National Research Council, 1994
Source: [2]Zhang and Coon, 1995. Layers over 40 wk of age should consume between 3.5 and 4.0 g Ca per day

maturing pullets can self-select to meet their calcium needs. Also, the larger particles are an advantage for eggshell formation. An optimum method for feeding calcium is to start feeding the layer level (3.25% or more) slightly before the first egg, because after hens start laying there is a negative calcium balance when consuming the pre-lay calcium level of 2.0%.

Dietary Calcium Variations

The amount of calcium necessary in the laying ration is determined by several major factors, all of which may require alteration of the diet.

1. *Rate of lay*. (The higher the rate, the more calcium needed.)
2. *Size of bird*. (Larger birds consume more feed therefore the calcium percent of the diet may be decreased to provide the same daily calcium intake levels of smaller hens. The calcium level in the feed must be based upon both feed consumption and egg mass.)
3. *Age of birds*. (Those past 40 weeks of age require more dietary calcium.)
4. *ME content of the ration*. (The higher the figure, the less feed consumed.)
5. *House temperature*. (Birds eat less when temperatures are high; therefore, the ration should contain more calcium.)

Table 18-15 shows the average percentage of calcium needed in the ration, but because of the above factors actual percentages needed can vary substantially.

Decline in Eggshell Quality as Hen Ages

Although eggshells are poorer in quality (shell thickness and texture), as the laying cycle progresses and as the flock ages, no one has been able to determine the exact cause. Shell thickness is not just a simple relationship to age, as eggs of a similar size from molted flocks are significantly thicker. The association with shell quality, therefore, is with the *length* of the laying period. One hypothesis is that the hen is capable of generating a fairly uniform daily quantity of eggshell material throughout her life, and as eggs get progressively larger, the shell material must be spread over a larger area, and thus is thinner. Some researchers have reported that efforts to reduce egg size during the latter stages of production by

Table 18-16. Effect of Egg Weight on Various Eggshell Characteristics (Three Leghorn Strains)

	Shell Characteristics			
	Weight			
Egg Weight (g)	(g)	(%)	Thickness (Microns)	Specific Gravity
44 wk of age				
<60	5.7	9.9	363	1.0870
60–64	6.1	9.8	368	1.0864
65–69	6.5	9.7	373	1.0860
>70	6.7	9.3	384	1.0854
Avg	6.1	9.8	368	1.0863
56 wk of age				
<60	5.4	9.4	356	1.0815
60–64	5.8	9.4	366	1.0816
65–69	6.1	9.2	363	1.0808
>70	6.6	9.1	371	1.0815
Avg	5.9	9.3	363	1.0813

University of California, 1987

limiting protein and amino acid consumption have resulted in improved eggshells. This method should be applied with caution, though, because as previously discussed, egg numbers may be affected as well.

Larger eggs within a sample at a given age tend to have thicker shells and consistently greater shell mass, reflecting the individual bird's ability to add more shell to eggs of increasing weight. See Table 18-16.

Eggshell Quality Is Poorer During the Summer

During hot weather, feed consumption is reduced and the daily intake of critical minerals may be less than optimum. Even though it is a common practice to offset lower feed consumption with higher concentrations of calcium and phosphorus, eggshell thickness generally decreases during the summer months. Table 18-17 illustrates this problem. Poorer egg shell quality in the warmer months occur due to blood acid/base imbalances resulting when birds attempt to increase heat loss by panting. The use of

Table 18-17. Effects of Season on Eggshell Thickness

Season	Age (wk)	Shell Thickness Micron	Shells Less Than 356 Microns in Thickness (%)
Winter	50	365	30
Summer	50	355	43
Winter	60	369	26
Summer	60	352	47

University of California, 1982

sodium bicarbonate to replace part of the salt in the ration may help to reduce this problem.

Feeding Coarse-Particle Oyster Shell or Limestone

The formation of the eggshell usually occurs during the nighttime hours, when the hen is not eating. If the source of dietary calcium is finely ground, the calcium passes through the gizzard quickly; little is available to the bird when the eggshell is being deposited.

To improve the situation, one-half to two-thirds of the dietary calcium supplement should be in the form of large-size flaked oyster shell or coarse limestone (>1.0 mm; average of 2.5 mm diameter). This material leaves the gizzard more slowly with a larger amount passing through the gastrointestinal (GI) tract during the dark hours when the eggshell is being formed. An in vitro solubility assay using a dilute hydrochloric acid solution is often used to determine how quickly or slowly a calcium source will become soluble in the GI tract of layers (Cheng and Coon, 1990, Zhang and Coon, 1998). While there is often a more noticeable benefit of feeding a large particle calcium source during the latter part of the laying cycle, hens fed large particles during the entire cycle will usually have better egg shells and bones at the end of the laying period. Research to support these field observations has indicated that layers fed small particle calcium supplements tend to mobilize more bone calcium throughout the laying cycle to help make eggshells. Contrastingly, layers fed large particle calcium from the beginning of lay have been shown to have significantly stronger bones at the end of a laying cycle.

Phosphorus Requirement of Layers

Much of the phosphorus in plant ingredients is in the form of phytin phosphorus, an organic compound not well-utilized by the chicken. It is thought that only about 30 to 40% of total phosphorus is available from plant ingredients. Phosphorus recommendations are presently based on non-phytate phosphorus with the assumption that non-phytate phosphorus is completely available.

The laying hen's need for phosphorus is low, mainly because there is little phosphorus in the eggshell. However, too little or too much phosphorus will prevent proper shell calcification. One of the chief causes of poor eggshell quality and strength is an excess of phosphorus in the diet; however, rations low in total phosphorus can increase flock mortality. The recommended daily intake of non-phytate phosphorus in the laying ration is somewhat controversial, but levels of 350 to 450 mg per hen per day are considered adequate. Typically, the dietary level of available phosphorus is reduced as the bird ages because of economics similar to decreasing dietary levels of protein and amino acids. This generally results in the layer being provided a larger margin of safety of available phosphorus early in the production cycle and an adequate level at the end of the production cycle.

Due to environmental concerns associated with phosphorus levels in the manure and potential pollution problems by it, there has been a move within the industry to add a phytase enzyme preparation to the feed which makes a greater quantity of the phytate phosphorus found in grains and oilseed protein available to the bird. When added to the feed, phytase reduces the need for much of the inorganic phosphorus currently used and reduces the level of total phosphorus in the diet and, consequently, in the manure. Research indicates that a phytase enzyme with good activity is capable of providing the equivalent of 0.1% non-phytate or available phosphorus. For example, if an egg producer wanted to provide a daily intake of 350 mg of available phosphorus and the hens were consuming approximately 100 g of feed daily, the layer formulation may contain 0.25% non-phytate phosphorus and adequate phytase activity to provide the additional 0.10% non-phytate phosphorus.

Trace Minerals Requirement of Layers

The requirement of the laying bird for trace minerals is very indefinite. Except for manganese and zinc (Table 18-15), natural feedstuffs seem to supply a large portion of these needed minerals. Many layer rations, however, include supplementary mineral premixes that contain manganese, zinc, iodine, copper, and iron. Some include selenium. Supplementation with selenium is contingent upon the soil levels of selenium in the region where the feed grains are grown.

18-I. VITAMIN REQUIREMENTS FOR EGG PRODUCTION

The dietary vitamin requirements of laying hens are given in Table 18-18. Also see *Vitamins, Minerals, and Trace Ingredients,* Chapter 20. The vitamins most often added to a laying ration include:

A	riboflavin	B_{12}
D_3	pantothenic acid	
E	niacin	
K	choline	

18-J. XANTHOPHYLLS AND EGG-YOLK COLOR

The xanthophylls in feed are the main contributors to yolk color. In the United States, consumer preference is for yolks that are pale to deep yellow

Table 18-18. Vitamin Requirements of Laying Hens (White Leghorn) Consuming 100 g of Feed per Day (Add 10% for Brown-egg Layers)

	Amount per Unit of Feed	
Vitamin	Per lb	Per kg
Vitamin A activity (IU)	1,364	3,000
Vitamin D_3 (ICU)	136.4	300.0
Vitamin E (IU)	2.27	5.00
Vitamin K (mg)	0.227	0.500
Thiamin (mg)	0.318	0.700
Riboflavin (mg)	1.14	2.50
Pantothenic acid (mg)	0.91	2.00
Niacin (mg)	4.54	10.00
Pyridoxine (mg)	1.14	2.50
Biotin (mg)	0.045	0.100
Choline (mg)	477.3	1,050
Vitamin B_{12} (mg)	0.0018	0.0040

Source: National Research Council, 1994

rather than darker shades, but this is not the preference in some other countries. Huge quantities of egg yolks are used in the preparation of noodles, cake mixes, and many other bakery products, where deep orange-colored yolk is normally preferred.

There are many xanthophylls and they represent a group known as *hydroxy-carotenoids.* These compounds are absorbed from the intestinal tract of the chicken and deposited in the egg yolks and fatty tissues in the body in the same form as they are consumed. The xanthophylls can also impart yellow color to the skin and shanks of layers.

In the United States, diets containing yellow corn are rarely associated with yolk color problems. Diets that depend more on grain sorghum (milo) or wheat will usually have very pale yolks unless supplemented with xanthophyll. Some of these ingredients are listed in Table 18-19.

Table 18-19. Mixed Xanthophyll Content of Various Feedstuffs

	Total Xanthophyll Content	
Feedstuff	mg per lb	mg per kg
Marigold petal meal	3,182	7,000
Algae (common, dried)	909	2,000
Alfalfa meal (20% protein)	150	330
Alfalfa meal (17% protein)	100	220
Corn gluten meal (60% protein)	132	290
Yellow corn	8	17

Source: National Research Council, 1994

Causes of Yolk-Color Variations

The quantity and type of dietary xanthophylls are not the only causes of variation in yolk color. Some others are as follows:

Strain and breed difference. These can cause as much as 14% variation in intensity of yolk color.

Individual bird variation. The genetic capability to absorb and deposit xanthophylls in egg yolk varies between hens within a single strain.

Cages. Hens kept in cages are able to make better use of yolk pigmenters than hens kept on litter floors.

Morbidity. Disease reduces the bird's ability to absorb xanthophylls from the intestinal tract. This is particularly true with certain coccidial strains of *Eimeria.*

Stress. Any stress reduces the transport of xanthophylls to the ovary.

Fat in the diet. There is an increase in xanthophyll absorption as the level of fat in the diet is increased.

Oxidation of the xanthophylls. Xanthophylls are easily oxidized in their pure state or in mixed feeds, thereby reducing their effectiveness. When possible, an antioxidant should be used with xanthophylls.

Certain ingredients. On occasion meat scraps, soybean oil meal, charcoal, and sulfur have been shown to reduce egg-yolk color, probably because of lowered intestinal absorption of the xanthophylls.

Egg/feed ratio. Rate of egg production is a cause of variability in yolk color. As flock egg production increases, the dietary xanthophylls are spread over more egg yolks with a corresponding decrease in yolk color, and vice versa. Rations for flocks laying at higher rates should contain more xanthophylls than those laying at lower rates.

18-K. FEED REQUIREMENT

As previously discussed, the daily feed requirement for egg production is based on energy and protein (amino acid) requirements. Furthermore, birds vary their feed intake according to their caloric needs, thus affecting the amount of protein (amino acid) consumed. Therefore, to provide a recommended feed intake for all conditions to which flocks may be subjected, and for all strains, is an impossibility. Only average consumption values are given in Tables 18-20 (for white-shell layers) and 18-21 (for brown-shell layers) for moderate weather with a diet containing 1,275 kcal of ME per lb (2,805 ME/kg). Computer software is available to fine-tune feed consumption projections which include factors such as body weight, feathering, ambient temperature, ME content of the feed, and egg mass.

Assumptions when using Tables 18-20 and 18-21 are:

1. Feed consumption is based on breeder standard body weights at the start of egg production.

Table 18-20. Feed Consumption of White-Egg Layers

Age (wk)	Feed Consumed Per 100 hens/day (lb)	(kg)	Cumulative per Hen housed (lb)	(kg)	Age (wk)	Feed Consumed Per 100 hens/day (lb)	(kg)	Cumulative per Hen housed (lb)	(kg)
18	15.9	7.23	1.11	0.50	48	22.0	10.0	44.81	20.37
19	17.0	7.73	2.30	1.05	49	22.0	10.0	46.29	21.04
20	18.0	8.18	3.55	1.61	50	22.0	10.0	47.78	21.72
21	18.7	8.50	4.86	2.21	51	22.1	10.0	49.26	22.39
22	19.5	8.86	6.22	2.83	52	22.1	10.0	50.75	23.07
23	20.1	9.14	7.62	3.46	53	22.1	10.0	52.23	23.74
24	20.8	9.45	9.06	4.12	54	22.1	10.0	53.71	24.41
25	21.1	9.59	10.52	4.78	55	22.1	10.0	55.19	25.09
26	21.3	9.68	12.00	5.45	56	22.1	10.0	56.67	25.76
27	21.4	9.73	13.49	6.13	57	22.1	10.0	58.15	26.43
28	21.5	9.77	14.98	6.81	58	22.2	10.1	59.63	27.10
29	21.6	9.82	16.47	7.49	59	22.2	10.1	61.11	27.78
30	21.6	9.82	17.96	8.16	60	22.2	10.1	62.58	28.44
31	21.6	9.82	19.45	8.84	61	22.2	10.1	64.06	29.12
32	21.7	9.86	20.94	9.52	62	22.2	10.1	65.53	29.79
33	21.7	9.86	22.44	10.20	63	22.2	10.1	67.01	30.46
34	21.7	9.86	23.93	10.88	64	22.2	10.1	68.48	31.13
35	21.8	9.91	25.42	11.55	65	22.2	10.1	69.95	31.80
36	21.8	9.91	26.92	12.24	66	22.3	10.1	71.42	32.46
37	21.8	9.91	28.41	12.91	67	22.3	10.1	72.89	33.13
38	21.8	9.91	29.90	13.59	68	22.3	10.1	74.36	33.80
39	21.9	9.95	31.40	14.27	69	22.3	10.1	75.83	34.47
40	21.9	9.95	32.89	14.95	70	22.3	10.1	77.30	35.14
41	21.9	9.95	34.38	15.63	71	22.3	10.1	78.76	35.80
42	21.9	9.95	35.87	16.30	72	22.3	10.1	80.23	36.47
43	21.9	9.95	37.36	16.98	73	22.3	10.1	81.69	37.13
44	21.9	9.95	38.85	17.66	74	22.3	10.1	83.15	37.80
45	22.0	10.0	40.34	18.34	75	22.4	10.1	84.61	38.46
46	22.0	10.0	41.83	19.01	76	22.4	10.2	86.07	39.12
47	22.0	10.0	43.32	19.69	77	22.4	10.2	87.53	39.79
					78	22.4	10.2	88.99	40.45

Source: Dekalb Poultry Research, Inc., 1998/1999

2. No allowances are made for temperature or seasonal variations.
3. The values are based on the number of birds present with zero mortality.

Egg production and feed consumption results for the data in Tables 18-20 and 18-21 are listed on the next page.

Table 18-21. Feed Consumption of Brown-Egg Layers

	Feed Consumed					Feed Consumed			
	Per 100 hens/day		Cumulative per Hen housed			Per 100 hens/day		Cumulative per Hen housed	
Age (wk)	(lb)	(kg)	(lb)	(kg)	Age (wk)	(lb)	(kg)	(lb)	(kg)
18	18.7	8.50	2.6	1.18	48	24.0	10.91	50.7	23.04
19	19.4	8.82	3.9	1.77	49	24.0	10.91	52.3	23.77
20	20.3	9.23	5.3	2.41	50	24.0	10.91	53.9	24.50
21	21.2	9.64	6.8	3.09	51	24.0	10.91	55.6	25.27
22	21.8	9.91	8.3	3.77	52	24.0	10.91	57.2	26.00
23	22.3	10.14	9.9	4.50	53	24.0	10.91	58.8	26.73
24	22.7	10.32	11.5	5.23	54	24.0	10.91	60.5	27.50
25	22.9	10.41	13.1	5.95	55	24.0	10.91	62.1	28.23
26	23.4	10.64	14.7	6.68	56	24.1	10.95	63.7	28.95
27	23.4	10.64	16.3	7.41	57	24.1	10.95	65.3	29.68
28	23.5	10.68	17.9	8.14	58	24.1	10.95	67.0	30.45
29	23.5	10.68	19.6	8.91	59	24.1	10.95	68.6	31.18
30	23.5	10.68	21.2	9.64	60	24.1	10.95	70.2	31.91
31	23.6	10.73	22.8	10.36	61	24.1	10.95	71.9	32.68
32	23.6	10.73	24.5	11.14	62	24.1	10.95	73.5	33.41
33	23.7	10.77	26.1	11.86	63	24.1	10.95	75.1	34.14
34	23.7	10.77	27.7	12.59	64	24.1	10.95	76.7	34.86
35	23.7	10.77	29.4	13.36	65	24.1	10.95	78.4	35.64
36	23.8	10.82	31.0	14.09	66	24.1	10.95	80.0	36.36
37	23.8	10.82	32.7	14.56	67	24.1	10.95	81.6	37.09
38	23.9	10.86	34.3	15.59	68	24.1	10.95	83.2	37.82
39	23.9	10.86	35.9	16.32	69	24.1	10.95	84.8	38.54
40	23.9	10.86	37.6	17.09	70	24.2	11.00	86.4	39.27
41	23.9	10.86	39.2	17.82	71	24.2	11.00	88.1	40.04
42	23.9	10.86	40.9	18.59	72	24.2	11.00	89.7	40.77
43	23.9	10.86	42.5	19.32	73	24.2	11.00	91.3	41.50
44	23.9	10.86	44.1	20.04	74	24.2	11.00	92.9	42.23
45	23.9	10.86	45.8	20.82	75	24.2	11.00	94.5	42.95
46	23.9	10.86	47.4	21.54	76	24.2	11.00	96.1	43.68
47	23.9	10.86	49.0	22.27	77	24.2	11.00	97.7	44.41
					78	24.2	11.00	99.3	45.14

Source: Dekalb Poultry Research, Inc, 1995

	White-Egg Layer[1]	*Brown-Egg Layer*[1]
Eggs per hen	333	334
Feed consumed per hen (lb)	88.9	99.3
Feed consumed per hen (kg)	40.4	45.1
Feed consumed per doz eggs (lb)	3.21	3.56
Feed consumed per doz eggs (kg)	1.46	1.62

[1] Hen-day basis (no mortality) 19 to 78 weeks of age

Figure 18-5. Feed Weighing System

18-L. FEEDING FOR EGG MASS VERSUS PHASE FEEDING

Phase feeding is a program where the nutrient content of the diet is lowered as the flock ages and production drops. The main purpose of phase feeding is to reduce feed costs as production declines. A key problem with phase feeding is the high probability of underfeeding hens that are laying at a higher rate than the flock average. Layers on a phase feeding program will have limits placed on body weight gain and egg size because of the reduction in nutrient density. Egg numbers can also be affected if the reduction is too severe for the high producing hens. Another problem with phase feeding is the arbitrary time selected to reduce nutrient density.

Some phase feeding programs are based on weeks from first egg and others on flock hen-day production.

Today, many layers are being fed based on empirical or factorial models. With these programs, the amount of egg mass produced by the layers has a big impact on the nutrient requirements. The models are primarily used for predicting the requirement of the flock for daily protein, certain amino acids, and ME. Instead of using an arbitrary time period or percentage production, performance criteria such as body weight, weight gain per day, and egg mass per day are used in these models. The Reading model actually allows for variation of individuals within a flock, thus providing compensation for the better producing hens. The Reading model also allows the nutritionist to add more or less of a specific nutrient depending upon the value of eggs and the costs of the nutrient.

18-M. EGG MASS AS A MEASURE OF EGG PRODUCTION

The use of egg mass rather than egg numbers will lead to better comparisons of flocks or strains of birds, along with feeding and management programs. To calculate egg mass it is first necessary to determine the average egg weight of eggs laid by the flock. It is necessary to weigh only a representative sample of the eggs laid. Total the weight of the entire sample, then divide this weight by the number of eggs in the sample. After the mean egg weight has been determined in grams, the following formula is used to compute egg mass on a daily basis:

$$P \times W = M$$

where

P = percentage hen-day egg production
W = average individual egg weight in grams per egg
M = average egg mass per hen per day in grams, e.g., 80% egg production $\times$ 60 g of egg weight = 48 g of egg mass / day.

Note: Average daily egg mass is a measurement not commonly used in the United States. The metric system (grams) is used to conform to the system used in other countries.

Egg Mass Equivalents

Inasmuch as eggs in the *large* category command a higher sales price than *medium* or *small* eggs, not only is it important that the flock produce a maximum number of eggs, but the eggs must be large. The combination of these two factors ensures the highest returns to the farmer. Some strains

Table 18-22. Average Daily Egg Mass for Various Egg Weights and Rates of Lay for Layers

Average Egg Wt.		Hen-Day Egg Production (%)							
		60	65	70	75	80	85	90	95
lb / case	g / egg	(Average Daily Egg Mass Per Hen in Grams)							
44.0	55.4	33.2	36.0	38.8	41.6	44.3	47.1	50.0	52.6
46.0	58.0	34.8	37.7	40.6	43.5	46.4	49.3	52.2	55.1
48.0	60.5	36.3	39.3	42.4	45.4	48.4	51.4	54.5	57.5
50.0	63.0	37.8	41.0	44.1	47.3	50.4	53.6	56.7	60.0
52.0	65.5	39.3	42.6	45.9	49.1	52.4	55.7	59.0	62.2
54.0	68.0	40.8	44.2	47.6	51.0	54.4	57.8	61.2	64.6

or flock of pullets produce large numbers of smaller eggs; others produce fewer eggs, but the eggs are larger.

Table 18-22 shows the comparative trade-offs of daily egg mass, when measured in grams per hen per day over a laying period of 365 days, using various average egg weights and percentages of hen-day egg production. When using the table, any combination of egg weight and percentage production producing the same average egg mass per hen per day would be comparable from the standpoint of nutrient requirements. However, the economic advantage of producing a higher number of eggs versus larger eggs will depend upon the prevailing market conditions within a particular season and / or marketing system.

Strains and Flocks Differ in Egg Mass

Generally, flocks are compared on the basis of hen-day or hen-housed egg production. These calculations are easy to make, but both disregard egg weight. A better procedure is to include egg weight as well as egg production and mortality. To bring the three measurements into focus as one index, the total egg mass produced on a hen-housed basis should be used to compare various strains or management programs. It will show differences where other comparisons fail. Comparisons must be made at a standard age to be meaningful.

Production, Egg Weight, Egg Mass, and Feed Consumption

These values have been projected in Table 18-23 for average white-shell egg layers by week, through 59 weeks of production. Average and total values from 19 to 78 weeks of age are as follows:

Total hen-housed egg production (number per hen)	333
Total hen-housed egg production (doz per hen)	27.75

Average hen-day egg production (%)	80.6
Average egg weight (oz/doz)	25.6
Average egg weight (g each)	60.4
Average feed consumed per 100 hens per day (lb)	21.55
Average feed consumed per 100 hens per day (kg)	9.80
Average feed consumed per dozen eggs (lb)	3.21
Average feed consumed per dozen eggs (kg)	1.46
Average feed consumed per unit of egg weight	2.01

Mortality from weeks 19 through 78 is assumed to be 6.9%.

Feed Consumption Needs Adjusting

As a general rule, except in extremely hot weather, layers will eat enough feed to meet their energy requirement, but depending on the calorie to protein ratio they may not be getting enough protein (amino acids). On an egg-mass basis (Table 18-23, col. 7) the period of probable inadequacy would be between 26 and 34 of age when the ratio of feed consumed to egg mass produced is the lowest. A protein deficiency at this time would, no doubt, reduce egg size and possibly even egg numbers. The relationship between feed consumed and egg mass produced would be a key indicator when developing an economical phase-feeding program. Therefore, an economical feeding program should incorporate a ration higher in protein through about 30 weeks of production, with a reduction in dietary protein for the remainder of the laying cycle that equates to increases in feed required/egg mass production.

Maintaining Body Weight During Lay

Caged layers must gain weight during the production year. As caging is conducive to heavier body weights than when on the floor, there is usually not a problem in maintaining the proper weight increase when the weather is cool or moderate. However, during periods of hot weather, birds do not consume as much feed, and there is always the possibility that body weight will suffer. Lowering the environmental temperature (if possible) and giving fresh feed early in the morning and later in the afternoon to increase feed consumption during the cool hours of the day, along with an ample supply of cool, fresh water, will help with consumption. (See *Consumption and Quality of Water,* Chapter 22.)

If there is a problem with excessive body weight, it may be best to initiate some form of nutrient reduction during the laying cycle. This may be achieved by reducing the nutrient density of the diet, increasing environmental temperature, or reducing the number of times that the feeder runs and stimulates feed intake.

Table 18-23. Hen-Day Egg Production, Egg Weight, Egg Mass, and Feed Consumption of White Leghorn Laying Hens by Week of Age

(1) Age (wk)	(2) Hen-Day Egg Production (%)	(3) Average Egg Weight (g)	(4) Ave. Egg Mass Per Hen Per Day (g)	(5) Feed Consumed Per 100 Hens/Day (lb)	(6) Feed Per Dozen Eggs (lb)	(7) Feed Per lb of Egg Mass (lb)
18	2.0	41.6	2.1	15.9	38.10	34.62
19	25.5	44.5	11.3	17.0	7.99	6.79
20	45.5	47.0	21.4	18.0	4.74	3.81
21	63.5	49.0	31.1	18.7	3.54	2.73
22	81.5	51.0	41.6	19.5	2.87	2.13
23	89.5	53.0	47.4	20.1	2.69	1.92
24	92.0	54.6	50.2	20.8	2.71	1.87
25	92.5	55.5	51.3	21.1	2.74	1.86
26	92.9	56.4	52.4	21.3	2.75	1.84
27	92.5	57.1	52.8	21.4	2.78	1.84
28	92.2	57.6	53.1	21.5	2.80	1.84
29	91.8	58.1	53.4	21.6	2.82	1.83
30	91.5	58.5	53.5	21.6	2.84	1.83
31	91.1	58.8	53.6	21.6	2.85	1.83
32	90.8	59.1	53.6	21.7	2.87	1.83
33	90.4	59.4	53.7	21.7	2.88	1.84
34	90.0	59.7	53.8	21.7	2.90	1.84
35	89.7	59.9	53.7	21.8	2.91	1.84
36	89.3	60.1	53.7	21.8	2.93	1.84
37	89.0	60.3	53.6	21.8	2.97	1.85
38	88.6	60.5	53.6	21.8	2.96	1.85
39	88.3	60.7	53.6	21.9	2.97	1.85
40	87.9	60.9	53.5	21.9	2.99	1.85
41	87.5	61.1	53.5	21.9	3.00	1.86
42	87.2	61.3	53.4	21.9	3.02	1.86
43	86.8	61.5	53.4	21.9	3.03	1.86
44	86.5	61.6	53.3	21.9	3.05	1.87
45	86.1	61.7	53.1	22.0	3.06	1.88

46	85.8	61.8	53.0	22.0	2.08	1.88
47	85.4	61.9	52.9	22.0	3.09	1.89
48	85.0	62.0	52.7	22.0	3.11	1.89
49	84.7	62.1	52.6	22.0	3.12	1.90
50	84.3	62.2	52.4	22.0	3.14	1.91
51	84.0	62.3	52.3	22.1	3.15	1.91
52	83.6	62.4	52.2	22.1	3.17	1.92
53	83.2	62.5	52.0	22.1	3.18	1.93
54	82.9	62.5	51.8	22.1	3.20	1.93
55	82.5	62.6	51.6	22.1	3.22	1.94
56	82.2	62.6	51.4	22.1	3.23	1.95
57	81.8	62.6	51.2	22.1	3.25	1.96
58	81.5	62.6	51.0	22.2	3.26	1.97
59	81.1	62.7	50.8	22.2	3.28	1.98
60	80.7	62.7	50.6	22.2	3.30	1.99
61	80.4	62.7	50.4	22.2	3.31	2.00
62	80.0	62.7	50.2	22.2	3.33	2.01
63	79.7	62.8	50.0	22.2	3.35	2.02
64	79.3	62.8	49.8	22.2	3.36	2.03
65	79.0	62.8	49.6	22.2	3.38	2.04
66	78.6	62.8	49.4	22.3	3.40	2.05
67	78.2	62.8	49.1	22.3	3.42	2.06
68	77.9	62.8	48.9	22.3	3.43	2.07
69	77.5	62.8	48.7	22.3	3.45	2.08
70	77.2	62.8	48.5	22.3	3.47	2.09
71	76.8	62.9	48.3	22.3	3.49	2.10
72	76.5	62.9	48.1	22.3	3.50	2.11
73	76.1	62.9	47.8	22.3	3.52	2.12
74	75.7	62.9	47.6	22.3	3.54	2.13
75	75.4	62.9	47.4	22.4	3.56	2.14
76	75.0	62.9	47.2	22.4	3.58	2.15
77	74.7	62.9	47.0	22.4	3.59	2.16
78	74.3	62.9	46.8	22.4	3.61	2.17

Source: Dekalb Poultry Research, Inc., 1999

Table 18-24. Commercial Egg-Layer Rations

	From Lighting to 50 Wks Lbs Consumed/100 Hens/Day (lb per 2,000-lb batch)				51 to 70 Wks Lbs Consumed/100 Hens/Day (lb per 2,000-lb batch)			
	19	20	21	22	21	22	23	24
Ingredient								
Ground yellow corn	1,257	1,326	1,349	1,387	1,379	1,388	1,423	1,450
Soybean meal (dehulled, 47.5%)	426	373	350	320	327	325	342	330
Meat and bone meal (50%)	100	100	100	100	100	78	26	6
Fat, animal-veg. blend or equivalent	30.6	12.7	9.6	3.1	4.3	7.9	0	0
Limestone-large particle	170	170	180	180	180	190	190	200
Salt	7.2	6.9	6.7	6.2	6.7	6.5	7.5	7.3
DL-Methionine or equivalent	3.61	3.44	3.15	2.89	2.25	2.19	1.96	1.52
Dicalcium phosphate (18.5%)	2	2	2	2	2	2	2	2
Vitamin premix	2	2	2	2	2	2	2	2
Choline chloride (60%)	1.39	1.31	1.14	0.99	0.71	0.56	0.37	0.24
Trace minerals	1	1	1	1	1	1	1	1
Vitamin and mineral supplements								
Vitamin A (Million IU)	6	6	6	6	6	6	6	6
Vitamin A (Million IU)	2.6	2.6	2.6	2.6	2.6	2.6	2.6	2.6
Vitamin E (IU)	3,760	3,760	3,760	3,760	3,760	3,760	3,760	3,760
Vitamin K (mg)	2,200	2,200	2,200	2,200	2,200	2,200	2,200	2,200
Vitamin B_{12} (mg)	5	5	5	5	5	5	5	5
Riboflavin (g)	3.76	3.76	3.76	3.76	3.76	3.76	3.76	3.76

Niacin (g)	20	20	20	20	20	20	20	20
Calcium pantothenate (g)	6	6	6	6	6	6	6	6
Zinc (g)	84	84	84	84	84	84	84	84
Manganese (g)	84.5	84.5	84.5	84.5	84.5	84.5	84.5	84.5
Selenium (mg)	136	136	136	136	136	136	136	136
Calculated nutrients								
Metabolizable energy (kcal/lb)	1,310	1,300	1,300	1,300	1,300	1,300	1,290	1,292
Protein (%)	17.5	16.5	16.0	15.5	15.6	15.0	14.2	13.5
Lysine (%)	0.91	0.84	0.81	0.77	0.78	0.75	0.71	0.67
Methionine (%)	0.46	0.44	0.42	0.4	0.37	0.36	0.34	0.31
TSAA (%)	0.75	0.72	0.69	0.67	0.64	0.63	0.61	0.57
Fat (%)	4.15	3.33	3.2	2.92	2.97	3.06	2.46	2.4
Fiber (%)	2.56	2.54	2.52	2.51	2.51	2.49	2.5	2.49
Calcium (%)	3.8	3.85	3.9	3.9	3.9	4.0	3.9	3.9
Total phosphorus (%)	0.54	0.54	0.53	0.53	0.53	0.48	0.36	0.31
Non-phytate phosphorus (%)	0.52	0.52	0.52	0.52	0.52	0.47	0.36	0.32
Vitamins and other								
Vitamin A activity (IU/lb)	3,685	3,723	3,736	3,756	3,751	3,757	3,776	3,791
Vitamin D_3 (IU/lb)	1,300	1,300	1,300	1,300	1,300	1,300	1,300	1,300
Vitamin E (IU/lb)	7.5	7.9	7.9	8.0	8.0	8.0	8.2	8.3
Vitamin K (mg/lb)	1.1	1.1	1.1	1.1	1.1	1.1	1.1	1.1
Riboflavin (mg/lb)	2.65	2.63	2.62	2.61	2.60	2.60	2.57	2.55
Niacin (mg/lb)	13.08	12.71	12.55	12.34	12.39	12.39	12.39	12.39
Pantothenic acid (mg/lb)	5.87	5.78	5.74	5.68	5.68	5.68	5.73	5.70
Choline (mg/lb)	620	590	560	530	475	475	450	425
Xanthophyll (mg/lb)	6.63	6.63	6.75	6.93	6.94	6.94	7.12	7.25

Source: Sandy Gretebeck, Bios Unlimited, 1999

More feed energy will not offset crowding. Feeds with higher energy values will not compensate for stress and lower body weights brought on when birds are overcrowded in cages.

18-N. LAYER RATIONS

Table 18-24 lists typical layer rations for two different ages of layers and also for different feed intakes. The formulation of diets must be based on daily intake to provide an adequate daily requirement of all nutrients. All diets include phytase enzyme in the vitamin premix to provide adequate available phosphorus. The rations show that as hens eat more per day the nutrient density can be decreased. Older hens represented by the 51- to 70-week-old columns are also being fed additional calcium to compensate for lower calcium retention of hens at these ages. The intake of protein and synthetic amino acids are also decreased in the later diets as the older hens will be laying less egg mass. Selection of the proper diet is first based upon the requirements associated with the age of the flock than upon the amount of feed intake expected for the upcoming feeding period (week) using current consumption data as a guide.

19

Feeding Broiler Breeders

by Craig N. Coon

The feeding and management of table-egg (white and brown shell) breeding birds was not included in this chapter because of the key management differences associated with feeding these breeders as opposed to the methods used for broiler breeders. The feeding of modern table-egg breeders does not require controlled feeding because, in general, the body weights are sufficiently small already to give efficient conversion of feed to hatching eggs. The specific nutrient requirements needed for good fertility and hatchability discussed for broiler breeders in this chapter also applies for shell-egg breeding stock. The nutrition and management specifications for white- and brown-shell egg breeding pullets and layers are listed in *Feeding Egg-Type Replacement Pullets,* Chapter 17, and *Feeding Commercial Egg-Type Layers,* Chapter 18, respectively.

19-A. FEEDING BROILER BREEDER PULLETS

Meat-type breeder females (broiler breeders), producing broiler offspring, possess the inherent ability to grow rapidly. When full-fed during the growing period they gain excessive weight and deposit too much internal fat for optimum fertility and maximum egg production. When full-fed the mortality is also increased for these breeders. Since the quantity of research on broiler breeder nutrition has been less than for feeding broilers or commercial layers, many people believe breeder nutrition is more of an art than a science. The author will also state that broiler breeders could probably be fed one diet during the entire rearing and breeding life cycle with only an adjustment of calcium being necessary. The *quantitative* amount of a diet to control feed during the rearing and breeding period is the main aspect of feeding broiler breeders.

When feeding broiler breeder females during the growing period, the nutritional object is to restrict the caloric intake to produce pullets that are leaner and older when they lay their first eggs, with less frame size thus requiring less nutrients for maintenance. The breeder should not be fat yet must be conditioned and fleshy enough to become sexually mature with increased lighting. The process of weight control must encompass the entire growing period; it cannot wait until just before egg production begins. Nutrients are used to produce different parts of the body during different stages of the grower period. Breeder pullets develop their skeletal system during the first twelve to fourteen weeks and then nutrients are partitioned later for the development of the reproductive tract. If the pullet is significantly underfed during the beginning period, she will not have adequate frame size when developing reproductive organs for egg production.

Feed Restriction During Growing

As early as 1937, it was found that restricting the feed intake of growing, meat-type birds would delay sexual maturity and increase the size of the first eggs laid. From this early beginning the methods of feed restriction have been improved; today, the results from the program show:

1. Restricting feed intake of growing birds will delay the onset of sexual maturity from a few days to 3 or 4 weeks, depending on the severity of restriction.
2. Feed restriction of a flock will reduce the body weight of the birds at sexual maturity, usually by reducing the amount of body fat.
3. Mortality during the growing period is reduced when feed is restricted.
4. Restriction of an ordinary growing diet may lead to nutritional deficiencies as certain nutrients may be overrestricted.
5. Restricting feed intake reduces the cost of growing pullets to sexual maturity, even though it may take three additional weeks before first eggs are laid.
6. Restricting the feed intake during the growing period produces better livability during egg production.
7. Egg production is not greatly affected during an equal number of months of lay, regardless of the feed regimens used during the growing period.
8. Egg weight is regulated by the age of the bird. Therefore,

birds reared using feed restriction will produce larger first eggs because they are older.

Comparison of Restricted and Full-Fed Broiler Breeder Growing Programs

It has been established that for broiler breeders to have top egg production the pullets need to have a mature body weight of around 4.85 lb (2.20 kg) at 20 weeks of age. This is the time when pullets are moved to the breeder facility and day length is increased. Many breeder flock managers don't transfer pullets to breeder facilities until 21 weeks of age in order to allow the smaller pullets more time to reach an optimum mature body weight. Other key body weights for breeders is an average body weight of 5.7 lb (2.60 kg) at 23 to 23 and ½ weeks of age when the first eggs are generally produced and approximately 6.2 lb (2.84 kg) at 5% hen-day egg production during the 24th to 25th week (168–175 days). Body weight at 20 weeks will be about 0.35 lb (180 g) heavier if the chicks are hatched during the warmer months and raised during the colder months (so-called out of season flocks).

If present-day, broiler breeder pullets are *full fed* a ration moderate in energy and protein, the average female flock weight at 24 weeks of age would be about 8.5 lb (3.89 kg). A high-calorie, high-protein diet will produce an average weight of up to 1 lb (454 g) heavier at the same age. Both of these are considered excessive weights for optimum performance.

The relationship between full feeding and restricted feeding is shown in Table 19-1 when both groups are fed the same ration. The last column in the table shows the percentage reduction in full feeding to get the recommended weekly average weights when controlled feeding is practiced. After 6 weeks of age these reductions in weight are between 41 and 57%.

Growing Feed Reduction Less Than Indicated

To evaluate this point, Table 19-1 shows that a flock of broiler breeder pullets where feed intake is restricted will average 4.3 lb (1.96 kg) in body weight and consume 20.3 lb (9.2 kg) of feed per 100 birds per day during the 20th week. This feed intake will be 49% lower than a full-fed flock of the same age.

The values may be better compared with a full-fed flock of the same average weight rather than the same age. For example, in Table 19-1 note that birds in a full-fed flock weigh 4.3 lb on the 10th week and consume 27.9 lb (12.7 kg) of feed per 100 pullets per day. On the basis of body weight, a restricted flock of the same average weight will eat only 11% less

Table 19-1. Comparison of Restricted Feeding vs Full Feeding of Growing Meat-Type Pullets

Week of Age	Restricted Feeding				Full Feeding				Restricted Feed Reduction (Weekly Basis)
	Feed Consumption per 100 Pullets per Day		Desired Body Weight		Feed Consumption per 100 Pullets per Day		Approximate Body Weight		
	(lb)	(kg)	(lb)	(kg)	(lb)	(kg)	(lb)	(kg)	(%)
4	9.5	4.3	1.1	0.50	11.1	5.0	1.3	0.59	14
6	11.0	5.0	1.5	0.64	15.9	7.2	2.2	1.00	31
8	12.1	5.5	1.9	0.86	21.2	9.6	3.3	1.50	41
10	13.3	6.0	2.3	1.05	27.9	12.7	4.3	1.95	49
12	14.7	6.7	2.7	1.23	34.3	15.6	5.2	2.36	57
14	16.1	7.3	3.1	1.41	36.6	16.6	6.0	2.72	56
16	17.5	8.0	3.5	1.59	37.9	17.2	6.7	3.04	54
18	18.9	8.6	3.9	1.77	38.7	17.6	7.3	3.31	51
20	20.3	9.2	4.3	1.96	39.6	18.0	7.8	3.54	49
22	21.7	9.8	4.8	2.18	40.4	18.3	8.2	3.72	46
24	23.1	10.5	5.5	2.50	41.2	18.7	8.5	3.86	44

Source: North and Bell, 1990

feed. This amount is the real criterion of feed restriction for it is doubtful if any flock could survive a restriction of 49% as calculated on an age basis.

Weekly Weight Gain During Growing Period

Table 19-2 shows the percentage gain in weight for the respective weeks when the pullets are on a restricted feeding program. Notice that to be effective, feed restriction must be started early in the flock's life.

Table 19-2. Weekly Percentage Weight Gain for Meat-Type Growing Pullets (Restricted Feeding Program)

Week of Age	Gain in Weight for Week (%)
4	22.2
6	15.4
8	11.8
10	9.5
12	8.0
14	6.9
16	6.1
18	5.4
20	4.9
22	4.4
24	4.2

Source: North & Bell, 1990

Meeting the Nutritional Requirements During Growing Period

Energy. The body weight of broiler breeder pullets must be controlled early in life, which calls for starter and grower diets moderately low in metabolizable energy (ME). To further reduce the growth rate, these rations must be restricted as early as 2 weeks of age, depending on the feeding program.

Leeson (1996) has suggested that the body weight of breeder pullets should be on the low side of body weight curves suggested by the Primary Breeder for the first 14 weeks, then when the pullets are approaching 15–16 weeks of age the pullet should have body weights equal to weights suggested by the Breeder (Figure 19-1). The pullet weight needs to be at target weight by this time in order to gain adequate weight prior to egg production. This provides that a pullet does not become overly framed or large during the growing period and also decreases the amount of energy needed for maintenance during this period. Leeson suggests flocks that are slightly heavier at the beginning of egg production tend to outperform lighter weight flocks. The additional body weight after 20 weeks would provide a pullet with more fleshing (breast muscle) and energy reserve that is necessary for reaching sexual maturity with increased lighting and then providing the pullet adequate energy to maintain egg production after sexual maturity without going into a negative energy balance.

Protein in the starter diet. A starter ration containing 18 to 20% is recommended. This percentage is somewhat higher than formerly suggested, but as the starter is fed for such a short period, the additional protein seems warranted.

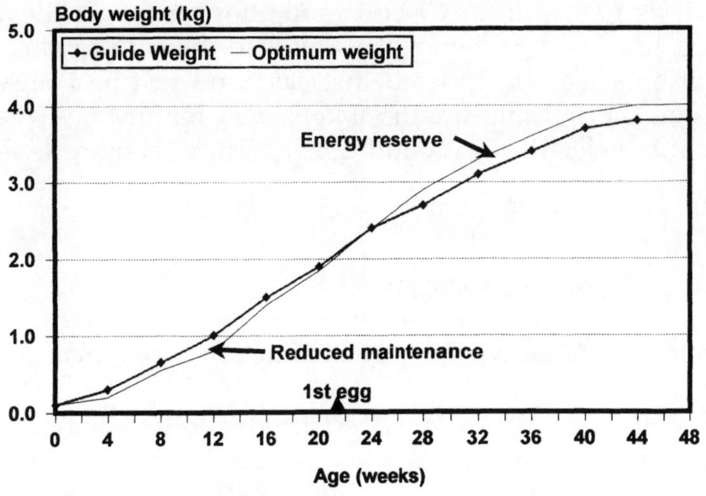

Figure 19-1. Growth Curve of Broiler Breeder Pullets

The sulfur amino acids, *methionine* and *cystine*, along with *lysine*, are as important as total protein. The daily requirements must be met.

The ability to control weight in the early period, as shown in Figure 19-1, with lower protein diets has been shown to be successful by Leeson, et al. (1984). The study suggested that higher protein starter programs must be followed by very severe quantitative restriction to slow growth rate. Research has shown that severe restriction during the growing period leads to increased mature body weight and a significant increase in the amount of time to reach sexual maturity. The pullets will have the same lean body mass at sexual maturity, but severely restricted pullets will have a significantly larger amount of body fat at sexual maturity. A concern of feeding low protein starter diets is a possible negative effect on the immune system and also on the increased probability of poor uniformity of the flock.

Recent research by Robinson, et al. (1999) indicates that breeder pullets fed higher protein diets during the starter period tend to have a better ability to partition nutrients to egg production than breeder pullets fed lower protein starter diets. The breeder pullets fed the lower protein starter diets had compensatory gain during the developer period and were equal in body weight at lighting but did not produce equal egg numbers during the production phase.

Protein in the grower diet. The pullet's daily protein need declines with age, starting with diets near 20% and lowering to diets as low as 14% near sexual maturity.

Although it is possible to reduce the protein in grower and developing rations as often as every 2 weeks, a more practical solution is to use a grower diet that contains between 14 and 15% protein, then feed this one diet from around 2 to 3 weeks of age up to 22 to 23 weeks of age.

Minerals in the grower diet. Calcium and phosphorus are the important minerals to consider when restricting feed during growing. Furthermore, it must be remembered that calcium must be increased after increasing the lighting to stimulate sexual maturity. A pre-breeder diet or a breeder diet with additional calcium should be fed at this time.

Programs for Feed Restriction

There are two general types of feed restriction programs:

1. Using skip-day feeding, a restricted (allotted) amount of feed is given on feed days and no feed is given on "skip" days. This is an effort to allow all birds to consume some feed by supplying enough feed that the more aggressive

birds will leave feed in the troughs for the less aggressive birds in the flock. To improve the uniformity of body weight is the most common reason for using such a program. (See Sections 19-B and 19-C.)

2. Feed the birds every day, but restrict the daily amount of feed given them. Under such a program it is essential that all the birds can consume feed at the same time to avoid excess competition at the feeder and poor flock uniformity. Using such a program will often reduce the amount of feed required to get pullets to their 20-week body weight by as much as 2.2 lb (1 kg). If the grower feed can be quickly delivered to the pullet flock with the use of a high speed feeding system, feeding every day could have advantages for the future. Some nutritionists feel that lower nutrient density diets may be more advantageous when feeding breeder pullets each day with a controlled amount of feed. The lower nutrient density diets would allow the non-aggressive pullets a greater opportunity to eat before all the feed is consumed. (See Section 19-D.)

19-B. SKIP-EVERY-OTHER-DAY PULLET GROWING FEED PROGRAM

The breeder starter should be fed the first 2 to 3 weeks. It is to be full-fed the first 2 weeks, then restricted during the third and fourth weeks, but fed every day. On the average, full-fed broiler breeder pullets should be eating about 8.8 lb (4 kg) of feed per 100 pullets per day at 14 days of age.

If the flock is in good health and on the targeted weight, beginning with the 5th week (29 days) feed a specified allotment on one day, no feed the next, feed the next, then no feed, and so on, so that feed is provided every other day. A guide for feed allotments on feed days is given in Tables 19-3 and 19-4 for two different breeder strains, along with the average desired live flock body weight of each. The feeding guides are slightly different for the two strains because of their different growth curves. One strain produces progeny that are high yielding and are extremely quick to put on their finish, whereas the other strain produces progeny that are high yielding and extremely lean and are known to be slower to reach a finish. Certain strains are known for producing progeny for whole bird and cut-up parts markets, whereas other strains of parent stocks are ideal for producing heavy males needed for further processing. Other variations on this classical skip-a-day program are also used where more feed days may be used in a week to reduce the quantity of feed fed on a feed day. In this manner the birds do not develop a large appetite which can be a concern once the birds are returned to everyday feeding at around 23 weeks of age.

It is considered important that growing pullets gain about 0.2 lb (91 g) per week from the 3rd week through week 11, then 0.25 lb (120 g) per week through the 24th week. However, the suggested daily feed allowances will be slightly greater for flocks hatched between April and September, and slightly less for those hatched between October and March in the Northern Hemisphere (reverse for Southern Hemisphere). The in-season flocks will be about 0.2 lb (91 g) lighter at sexual maturity than the standards given, and out-of-season flocks will be about 0.2 lb (91 g) heavier.

Exact Feed Allotments Depend on Body Weights

The feed allotments given in Tables 19-3 and 19-4 are only a guide; many things affect the exact feed amount: strain of birds, date of hatch, caloric and protein content of the feed, season of the year, ambient temperature, hours of light per day, physical condition of the flock, age of the pullets, etc.

Representative samples of the flock must be weighed weekly on the afternoon of non-feed days, beginning at the end of the 3rd week (21 days). For each 1% the flock average weight is below the standard each week, increase the daily feed allotment by 1%. If the flock average weight is above the standard by more than 1% in any one week, the feed allotment given on a day the birds are weighed should be maintained or increased only slightly until the correct body weight is reached. The feed allotment of a flock should never be reduced; for best results the weight of the pullets should increase some each week, and they should never be forced to lose weight.

Important measurements of the skip-every-other-day feeding program include daily consumption of protein and ME per bird as shown in Table 19-5. The accumulated amount of nutrients to produce a breeding pullet is also a good indicator for producing adequate fleshing and body composition of the pullet. The pullet needs to reach a *chronological age* as well as a *body physiological threshold* that is necessary for reaching sexual maturity with increased lighting.

19-C. IMPROVED SKIP-DAY BROILER BREEDER GROWING FEED PROGRAMS

Difficulties with Skip-Every-Other-Day Feeding Programs

Although skip-every-other-day feeding has been the most popular program in recent years, there have been some problems with it. For example, this program calls for feeding 2 days' supply of restricted feed allotment

Table 19-3. Recommended Female Body Weights & Feed Consumption (Cobb 500)

Light Controlled (In-Season)				Feed Allotments				
Age		Body Weight		Imperial (lb/100/d)		Metric (g/d)		
Days	Weeks	(lb)	(kg)	Daily (ED)	Skip (S)	Daily (ED)	Skip (S)	Key Points
	0–1			FULL	FULL	FULL	FULL	
7	1–2	0.25	0.12	FULL	FULL	FULL	FULL	
14	2–3	0.55	0.26	8.8 ED	8.8 ED	40 ED	40 ED	
21	3–4	0.85	0.40	10.0 ED	10.0 ED	45 ED	45 ED	
28	4–5	1.15	0.52	10.5	21.0 S	48	95 S	180 g cumulative protein at 28 days
35	5–6	1.40	0.62	11.0	22.0 S	50	98 S	
42	6–7	1.60	0.72	11.5	23.0 S	52	104 S	
49	7–8	1.80	0.82	12.0	24.0 S	54	109 S	
56	8–9	2.00	0.92	12.5	25.0 S	56	113 S	
63	9–10	2.25	1.02	12.7	25.4 S	57	115 S	
70	10–11	2.45	1.12	12.8	25.8 S	58	117 S	
77	11–12	2.65	1.22	13.4	26.8 S	61	122 S	
84	12–13	2.85	1.30	13.6	27.2 S	62	123 S	
91	13–14	3.05	1.38	13.8	27.6 S	63	125 S	
98	14–15	3.20	1.44	14.0	28.0 S	64	127 S	
105	15–16	3.35	1.52	15.0	30.0 S	68	136 S	
112	16–17	3.50	1.60	16.0	32.0 S	73	145 S	
119	17–18	3.70	1.70	17.5	35.0 S	79	159 S	
126	18–19	4.00	1.82	19.0	38.0 S	86	172 S	
133	19–20	4.30	1.96	20.5	41.0 S	93	186 S	
140	20–21	4.75	2.16	22.5	22.5 ED	102	102 ED	23,000 cumulative kilocalories at lighting
147	21–22	5.10	2.32	24.0	24.0 ED	109	109 ED	
154	22–23	5.50	2.50	25.0	25.0 ED	113	113 ED	
161	23–24	5.90	2.68	26.0	26.0 ED	118	118 ED	
168	24–25	6.25	2.84	28.0*	28.0 ED*	127*	127*	
175	25–26	6.50	2.95					
182	26–27	6.70	3.04					
189	27–28	6.90	3.13					
196	28–29	7.10	3.22					
203	29–30	7.20	3.26					
210	30–31	7.30	3.31					

Notes: *Feed increases should be in accordance with egg production
Weights 4 thru 20 weeks are off-feed weights. ED = Everyday feeding; S = Skip-a-day feeding
Source: Cobb 500 Broiler Breeder Management Guide

with no feed the next. Birds are very hungry following 1 day without anything to eat and are capable of consuming large quantities of feed in a short period of time, thus tending to gorge themselves when feed is dispersed. Because of this gorging, the crop and gizzard enlarge, and the birds not only can eat a larger amount of feed but can gorge themselves more, and the process is repeated over and over during the growing period. Problems with this can occur when the birds are changed to every day feeding just prior to the onset of egg production. Adding an additional feed day, once the quantity of feed to be given on a feed day reaches a certain level, can

Table 19-4. Female Body Weight, Feed Consumption, Lighting Schedule, and Type of Feed Relating to the Age of the Breeder Flock (Ross 508)

Week of Age	Body wt (lb)	Body wt (kg)	Feed lb 100 Every Day	Feed g/Bird Every Day	Calories/ Bird/Day	Light (hrs)	Diet
1	0.25	0.11	Full feed (2.5)	Full feed (11.3)	(33)	12	Starter
2	0.45	0.20	Full feed (5.5)	Full feed (24.9)	(72)	12	Starter
3	0.65	0.30	Full feed (7.8)	Full feed (35.4)	(91)	12	Starter
4	0.90	0.41	8.1	36.7	105	8	Grower
5	1.10	0.50	8.5	38.6	110	8	Grower
6	1.30	0.59	8.9	40.4	115	8	Grower
7	1.50	0.68	9.5	43.1	124	8	Grower
8	1.75	0.79	10.2	46.3	133	8	Grower
9	2.00	0.91	10.9	49.4	142	8	Grower
10	2.20	1.00	11.6	52.6	151	8	Grower
11	2.45	1.11	12.4	56.2	161	8	Grower
12	2.65	1.20	13.2	59.9	171	8	Grower
13	2.85	1.29	14.1	64.0	183	8	Grower
14	3.05	1.38	15.0	68.0	195	8	Grower
15	3.30	1.50	16.1	73.0	209	8	Grower
16	3.60	1.63	17.3	78.5	225	8	Grower
17	3.90	1.77	18.6	84.4	242	8	Pre-Breeder
18	4.25	1.93	19.9	90.3	259	8	Pre-Breeder
19	4.55	2.06	21.2	96.2	276	8	Pre-Breeder
20	4.90	2.22	22.5	102.1	293	8	Pre-Breeder
21	5.25	2.38	23.9	108.4	310	8	Pre-Breeder
22	5.60	2.54	25.0	113.4	325	14	Pre-Breeder
23	5.90	2.68	26.1	118.4	340	15	Breeder I
24	6.20	2.81	27.2	123.4	353	15	Breeder I
25	6.50	2.95	29.2	132.5	380	16	Breeder I
26	6.75	3.06	30.8	139.7	400	16	Breeder I
27	7.00	3.18	34.9	158.3	454	16	Breeder I
28	7.20	3.27	34.9	158.3	454	16	Breeder I
29	7.35	3.33	34.9	158.3	454	16	Breeder I
30	7.45	3.38	34.9	158.3	454	16	Breeder I
31			34.4	156.0	447	16	Breeder I
32	7.50	3.40	33.8	153.3	439	16	Breeder I
33			33.8	153.3	439	16	Breeder I
34			33.8	153.3	439	16	Breeder I
35			33.3	151.0	433	16	Breeder I
36	7.60	3.45	33.3	151.0	433	16	Breeder I
37			32.8	148.8	427	16	Breeder I
38			32.8	148.8	427	16	Breeder I
39			32.3	146.5	420	16	Breeder I
40	7.70	3.49	32.3	146.5	420	16	Breeder I
44	7.85	3.56	Per Weight/ Production		Per Weight/ Production	16	Breeder II
54	8.15	3.70	Per Weight/ Production		Per Weight/ Production	16	Breeder II
65	8.50	3.86	Per Weight/ Production		Per Weight/ Production	16	Breeder II

Source: Ross 508 Breeder Management Guide, 1999

Table 19-5. Estimated Protein and Metabolizable Energy Consumed by Broiler Breeder Pullets Housed in a Moderate Temperature (Skip-a-day Feeding)

Week of Age	Diet	Desired Body Weight End of Week (kg)	Feed Consumed per Bird per Day (g)	ME Consumed per Bird per Day (kcal)	Accumulated ME Consumed Per Bird End of Week (kcal)	Protein Consumed per Bird per Day (g)	Accumulated Protein Consumed Per Bird End of Week (g)
1	Starter 17.5% Protein 2,860 kcal ME per kg	0.25	Full feed (11.4)	(33)	231	2.0	13.9
2		0.45	Full feed (25.0)	(72)	735	4.4	44.5
3		0.65	Full feed (35.4)	(91)	1,372	6.2	87.9
4	Grower 15% Protein 2,860 kcal ME per kg	0.90	36.8	105	2,107	5.5	126.6
5		1.10	38.6	110	2,877	5.8	167.1
6		1.30	40.4	115	3,682	6.1	209.5
7		1.50	43.1	124	4,550	6.5	254.8
8		1.75	46.3	133	5,481	7.0	303.5
9		2.00	49.5	142	6,475	7.4	355.4
10		2.20	52.7	151	7,532	7.9	410.7
11		2.45	56.3	161	8,659	8.4	469.8
12		2.65	59.9	171	9,856	9.0	532.7
13		2.85	64.0	183	11,137	9.6	599.9
14		3.05	68.1	195	12,502	10.2	671.4
15		3.30	73.1	209	13,965	11.0	748.2
16		3.60	78.5	225	15,540	11.8	830.6
17	Pre-Breeder 15.5% Protein 2,860 kcal ME per kg	3.90	84.4	242	17,234	13.1	922.3
18		4.25	90.4	259	19,047	14.0	1020.3
19		4.55	96.3	276	20.979	14.9	1124.7
20		4.90	102.2	293	23,030	15.8	1235.5
21		5.25	108.5	310	25,200	16.8	1353.2
22		5.60	113.5	325	27,475	17.6	1476.4
23	Breeder I 16.0% Protein 2,860 kcal ME per kg	5.90	118.5	340	29,855	19.0	1609.1
24		6.20	123.5	353	32,326	19.8	1747.4
25		6.50	132.6	380	34,986	21.2	1895.9

Source: Calculated from Table 19-4 and Table 19-17

reduce this appetite. Ideally, the birds should not receive a feed allowance during rearing in excess of the maximum daily allowance that they will receive at the peak of production.

Once the feed is consumed on feed days, the birds spend their time drinking; they drink even more on the days no feed is given, in order to produce a degree of satiety to the digestive tract. This additional water results in loose droppings and wet litter. The only remedial measure is to restrict the *availability* of water on feed and non-feed days. To avoid birds choking when eating on a feed day it is wise to ensure that they have had access to water for at least 30 minutes before the feeders are filled.

Be sure to read *Consumption and Quality of Water,* Chapter 22, regarding water restriction.

Another criticism of the program is that the amount of feed allocated on feed days after the birds are about 18 weeks of age is more than they need or will eat. Consequently, the growing pullets leave some feed in the troughs or pans, to be eaten the next no-feed day. Evidence shows that when birds gorge themselves with feed, an increased amount passes through the digestive tract undigested, resulting in poor feed efficiency. Thus, there is more variation in body weights; there are more larger and more smaller birds.

Two Choices for Improving Skip-Day Feeding Programs

1. *Change to feed-every-day program.* When the birds begin to leave feed in the trough or pan at the end of a feeding day (about 18 weeks of age), change to a feed-every-day restricted feeding program. See Tables 19-3 and 19-4 for daily feed allocations starting at this age.
2. *Use decreased non-feed days per week program.* With this program the starter is full-fed for the first 2 to 3 weeks, then restricted for the next 3 weeks. A 15% protein grower/developer is fed beginning with the 7th week (43 days) through the 22nd week (154 days) using a different skip-day program. First feed every other day, then gradually increasing the feeding days to 5 days per week, and no feed on 2 days by the end of the 22nd week (Table 19-6). This program lowers feed gorging and reduces the daily feed intake so that birds clean up their *allotment* of feed each day.

19-D. DAILY RESTRICTED FEEDING GROWING PROGRAM

This program is fast gaining in popularity, but it calls for specialized flock management. Use the same starter, developer, and breeder rations

Table 19-6. Meat-Type Growing Pullet Feeding Program: Decreased No-Feed Days per Week

Weeks of Feeding	Type of Feed: Name	Protein (%)	ME per lb (kcal)	Feeding Program (Limited Feeding)
1–3	Starter	19	1,380	Self feed every day
4–6				Restrict, but feed daily
7–11	Grower	15	1,335	Skip every other day (e.g., 3½ days per wk)
12–19				Feed 2, skip 1 day (e.g., 4⅔ days per wk)
20				Feed 5, skip 2 days (e.g., skip Sun and Wed)
21–22	Breeder	16	1,320	
23–24				Self feed every day

Source: North & Bell, 1990

as utilized for skip-a-day feeding and the same overall daily intake of protein and ME shown in Table 19-5, but feed and restrict them *daily*.

During the 3rd week, sample weighing of the pullets must be started, because bird weight determines the daily feed allocation. If the pullets meet their target weight (see Tables 19-3 and 19-4) or similar figures furnished by the primary breeder, continue with the recommended daily feed allocations in the table. When the weekly average weight is below the target weight, slightly increase the daily feed allocations. The birds may not catch up to target weight for as long as three weeks. If the pullets' average body weights are above standard early in the growing period, there may be time to adjust feed quantities to bring the birds back to standard weights. It is suggested by primary breeders if a flock is overweight at 13–15 weeks of age do not attempt to reduce body weight to bring birds back to standard. Develop a new body weight curve and feed to achieve consistent weekly body weight gains. The pullets will be heavier than normal at the onset of lay.

Weekly Feed Consumption Does Not Change

Regardless of how many feed days are skipped, or which feeding program is used, the weekly feed consumption and standard weekly body weights must remain the same as those shown in Tables 19-3 and 19-4.

19-E. IMPORTANCE OF PULLET BODY WEIGHT UNIFORMITY DURING THE REARING PERIOD

It is necessary to begin sample weighing of the growing breeder flock during the 2nd or 3rd week and weigh every week thereafter. At this early age, weigh 1% of the birds or a minimum of 100 birds at each end of the house. Weigh the birds in the afternoon on the same day each week. When the skip-a-day feeding program is used, weigh on a non-feed day.

During the 2nd, 3rd, and 4th weeks, birds may be weighed in groups of 10, then on the 5th week individual pullet weights need to be determined. Individually weigh the sample of birds using a scale with no greater than 10 g graduations and determine the average weight, then compare this figure with the recommended weight. Not only should the flock average pullet weight meet the standard but flock uniformity must be high, as it may be a better measure of a quality pullet flock than average weight.

Uniformity of the pullet flock is best measured by determining the percentage of pullets within 10% (plus or minus) of the average weight of the birds in the sample. Degree of flock uniformity may be compared according to the following weight variance:

Terminology	*Percentage of Pullets Within 10% of Flock Average Weight*
Superior	81 and above
Excellent	77–80
Good	73–76
Average	69–72
Fair	65–68
Poor	61–64
Very Poor	60 and below

19-F. FEEDING GROWING BROILER BREEDER COCKERELS

Meat-type cockerels, to be used as broiler breeders, have standard weekly weights, and it is just as important that male weights be maintained as the weights of the growing pullets. To get these weights, the cockerel-growing feed must be restricted. In past years restriction was impossible when the cockerels were raised with the pullets that were on a restricted feeding program, since the robust males pushed the females away from the feeders. But with the advent of blackout housing, the males are raised in separate houses and the feed is controlled, so it is relatively easy to maintain the correct cockerel weight during the growing period.

Weigh a sample of the cockerels at the same time (age) the pullets are weighed, using the same system. Tables 19-7 and 19-24 give the target weights for growing males when the feed is restricted. At 24 weeks of age the males should average about 35% heavier than the females.

Growing Management Programs for Broiler Breeders

There are four broiler breeder rearing programs, the fourth involving blackout houses.

1. *Cockerels separate to 28 to 42 days.* Coming from small eggs, small meat-line cockerel chicks should be started

Table 19-7. Weights of Meat-Type Cockerels Fed on a Restricted Feeding Program (Moderate Temperature)

Week of Age	Guidelines for Approximate Male Body Weights			
	Aug-Jan Hatches		Feb-July Hatches	
	(lb)	(kg)	(lb)	(kg)
1	0.31	0.14	0.33	0.15
2	0.53	0.24	0.57	0.27
3	0.98	0.45	1.1	0.49
4	1.2	0.54	1.3	0.59
5	1.4	0.64	1.6	0.73
6	1.7	0.77	1.9	0.86
7	1.9	0.86	2.1	0.95
8	2.2	1.00	2.4	1.09
9	2.4	1.09	2.6	1.18
10	2.7	1.22	2.9	1.32
11	3.0	1.36	3.2	1.45
12	3.2	1.45	3.4	1.54
13	3.5	1.59	3.7	1.68
14	3.8	1.72	4.0	1.81
15	4.1	1.86	4.3	1.95
16	4.4	2.00	4.6	2.09
17	4.6	2.09	4.8	2.18
18	4.8	2.18	5.2	2.36
19	5.1	2.31	5.5	2.50
20	5.4	2.45	5.8	2.63
21	5.7	2.59	6.1	2.77
22	6.2	2.81	6.6	2.99
23	6.7	3.04	7.1	3.22
24	7.2	3.27	7.6	3.45

Note: Data are for the Northern Hemisphere. Reverse for Southern Hemisphere
Source: North & Bell, 1990

within guards under separate brooders using the same feed and feeding program for the cockerels and pullets. The earliest that cockerels should be mixed with pullets is 28 days and the cockerels should be at least 40% heavier than the pullets.

2. *Cockerels separate to 10 weeks.* During the first 7 days, keep the cockerel chicks under separate brooders confined to one part of the house by a high fence. Both cockerels and pullets should get a ration with the same formula as long as the starter feed contains at least 18% protein for the males. At 10 weeks of age, mix the cockerels with the pullets. Feed the pullets and cockerels the same feed in the same room.

Broiler breeder males should not be fed low-protein diets to reduce weight, particularly before 8 weeks of age, as such a practice reduces fertility later.

3. *Cockerels separate to 20 to 22 weeks.* Formerly, this program was the best of all growing programs; growth rate of each sex could be accurately controlled by feed allocation. Male aggression can be quite high using this program and so it is advisable to place some A-frame perches in each pen to allow the males to escape.

   At 20 to 22 weeks of age, move the cockerels with the pullets to the production house. Continue with the grower feed until the birds are 22 weeks of age or are up to standard weight; then feed a breeder feed.

4. *Feeding in blackout houses.* With this program the sexes are generally raised in *separate* houses that are environmentally controlled with forced-air ventilation, cooled, and capable of being fully blacked out.

   Full feed males an 18 to 20% protein starter for four weeks. The males should continue on a starter diet until they are 6 weeks of age but a skip-a-day program should be started during the 5th week. Males should be fed a pullet grower diet at 6 weeks of age. Females should be full fed an 18% protein starter for only two weeks then go to controlled feeding of the starter using daily feed restriction rather than a skip-a-day program. Initiate a skip-a-day program for the females on the 5th week similar to males. The pullets can be fed the grower on the 7th week. This procedure is easily accomplished because the males and females are in separate houses.

   This program will induce chicks hatched out of season to come into egg production earlier, resulting in smaller eggs at the start of production. To prevent this occurrence, care should be taken to see that pullets on this program do not become sexually mature (first eggs) too early. Cockerels should start mating when the first eggs are laid (during the 23rd or 24th week of age). Change to controlled feeding of a *breeder feed* when the first egg is laid.

19-G. FEEDING BROILER BREEDERS DURING THE CHANGEOVER PERIOD

From the end of the growing period until the flock is well into egg production is the *changeover period.* It is now recognized as an exceptionally important period as there are many changes in management, lighting, and

Table 19-8. Influence of Feed Allocation and Photoschedule from 20 to 25 Weeks of Age on the Onset of Sexual Maturity and Associated Carcass Characteristics

Main Effects		Sexual Maturity (days)	Body Weight (kg)	Fat pad Weight (g)	Ovary Weight (g)	Number of Large Follicles
Photoperiod	Fast*	169.6	2.64	87.6	51.0^b	8.0
	Slow	170.5	2.71	82.7	57.5^a	8.8
Feed Allocation	Fast	171.1	2.69	94.6	57.5^a	8.9^a
	Slow	169.0	2.66	75.6	51.1^b	7.9^b
SEM		1.2	0.03	10.7	2.2	0.3

^{a,b} Means within columns with different superscripts are significantly different ($P < 0.05$)
* Fast Photoperiod: 8L:16D to 15L:9D increased at 20 wks
Slow Photoperiod: 8L:16D to 15L:9D changed from 20 wks to 25 wks with weekly increases
Fast Feed Allocation: 125 g feed per pullet/day at 20 wks—increased to 130 g at 21–25 wks
Slow Feed Allocation: 100 g feed per pullet/day at 20 wks and increased 5–10 g weekly and reaching 130 g at week 25
Source: Robinson, 1997

feeding that are most critical to the flock and its future egg production. Scientists have made significant findings regarding the amount of feed as well as lighting programs that produce optimum results in producing hatching eggs and chicks. The period from 20 to 25 weeks is very important for a breeding pullet because it is a time in which the breeding pullet is stimulated to reach sexual maturity by increased lighting. The breeder hen appears to partition dietary energy differently than an egg-type hen and when additional energy is provided during ovarian development, extra ovarian follicles develop. The increase in ovarian follicles caused by increased feeding of dietary energy during the changeover period from 20 to 25 weeks has been shown to decrease the production of hatching eggs during the entire 25- to 64-week period. Robinson, et al. (1995) demonstrated that a slow photoperiod increase and a fast increase in dietary energy feeding both increased the ovary weight of the breeding pullet, and the fast feeding of dietary energy increased the number of large follicles (Table 19-8). These researchers also showed that breeding hens fed slow increases in dietary energy during the changeover period from 20 to 25 weeks produced a 10.9 egg advantage during the 25- to 64-week period compared to the fast-fed energy group (Table 19-9). The hatch of the fertile eggs and hatchability were also reduced in the fast-fed group compared to the slow-fed group. The conclusion from the research is that breeding hens are negatively influenced by overfeeding breeder pullets during the early laying period or by photostimulating flocks too early.

Body Weight Variability at Sexual Maturity

For simplicity, it has been stated that the female body weight should average 5.5 lb (2.5 kg) at 24 weeks of age. But on a seasonal basis this

Table 19-9. Influence of Feed Allocation and Photoschedule from 20 to 25 Weeks of Age on Various Performance Parameters[1]

Main Effects		Total Egg Output	Fertility (percent)	Hatch of Fertile (%)	Hatchability (percent)
Photoperiod	Fast	193.9	91.7[b]	86.7[b]	79.5[b]
	Slow	195.9	92.9[a]	89.6[a]	83.2[a]
Feed Allocation	Fast	189.4[b]	92.2	87.8	81.0[a]
	Slow	200.3[a]	92.4	88.5	81.8[b]
SEM		2.8	0.4	0.8	0.9

[1] See Table 19-8 for explanation of photoperiod and feed treatments
[a,b] Means within columns and main effects with different superscripts are significantly different ($P < 0.05$)
Source: Robinson, 1997

weight is variable, as shown in the following table (reverse for Southern Hemisphere).

Female Average Body Weight at Sexual Maturity (Northern Hemisphere)

Month of Hatch	*(lb)*	*(kg)*
Aug	5.5	2.50
Sept-Jan	5.4	2.45
Feb	5.3	2.41
Mar	5.4	2.45
Apr	5.6	2.55
May, June	5.7	2.59
July	5.6	2.55

Source: North & Bell, Commercial Chicken Production Manual, 1990

Although hatching date is used as the basis for the above variations in body weight at sexual maturity, it is the changing length of the light day and temperature during growing that are the cause. This variability, caused by differences in hatching date, necessitates changes in the feeding program during the changeover.

Normally, the largest birds in the flock are the first to produce eggs. About a week before a pullet starts to lay her body weight begins to increase rapidly. Between this time and 1 week after she lays her first egg she should gain about 0.5 lb (227 g) or 10%. During the next 8 to 10 weeks she should gain a similar amount.

At the time an individual pullet lays her first egg she should weigh between 5.5 and 6.0 lb (2.5 and 2.7 kg), depending on the month she was hatched. As the smaller birds reach sexual maturity, they too will attain a weight that approaches the weight of the first birds to lay. But there are still large, medium, and small birds in the flock, and always will be. The largest birds remain large; the smallest birds remain small.

First Week of Flock Egg Production

Even though today's breeder flocks are best brought into 5% hen-day egg production at 24 to 25 weeks of age, there will be variability because of hatching date, season, strain, temperature, ration, feeding program, etc., so flocks may vary 2 or 3 weeks from this age. Because of this variability, further feeding recommendations during the changeover period must be geared to the actual beginning of egg production rather than chronological age.

The common base for early egg production is the day when the flock first averages 5% hen-day egg production. About 8% of the flock will be in production at this time.

Pre-Breeder Feed

The utilization of a pre-breeder feed for breeders during the changeover period is practiced by many companies. The feed is commonly used between the periods of 19 to 23 weeks. The purpose of a pre-breeder feed is to increase the calcium levels in the diet above the levels fed in the grower/developer period to provide a calcium reserve for the initiation of eggshell formation. It is also used to stimulate the nutrient intake of protein and energy for pullets that are lower in body weight than standard body weights for the strain and season. At the beginning of this period, the calcium level of the feed is increased to approximately 2%. Leeson (1998) reviewed the different ways that pre-breeder rations are fed and stated that unless pullets are below target weight at the time of moving there is no advantage in using a separate pre-breeder diet. Leeson (1999) suggest that a nutrient dense pre-breeder diet may be useful to bring underweight pullets up to target body weights prior to maturity. The author suggests the diets may be useful in manipulating body weight but the late growth will not result in meaningful skeletal growth. The breeder pullets may be of the correct weight but of small stature. The reason a nutrient dense pre-breeder diet may be slightly better than simply increasing the daily quantity of the grower feed being fed in order to increase body weight is to ensure that breeders are not subjected to a step down in feed allocation at the time of first egg.

Timetable for Feed and Management Changes

The basic schedule is as follows, but remember it should be used only as a guide. Individual flocks will still require some adjustment in the schedule. The timetable is for pullets reared in season in dark-out housing and housed in light-controlled production facilities.

20/21 weeks: Pullets are moved to breeder facility and lights are increased from 8 hours to 13 hours. Breeder pullets are to be fed a restricted amount of feed every day for the remainder of the production period. Each pullet is to receive approximately 292 kcal ME per day at this time. Pullets should not be moved and photostimulated if the body weight is less than target weight. Proper fleshing as well as body weight of the breeder pullet is important in order for the pullets to reach sexual maturity after photostimulation. Breeder pullets should not be overfed during this time but gradual weekly increases of 1.0 to 1.5 pounds (0.45 to 0.68 kg) per 100 pullets/day should continue the flocks development.

7 days post housing: Increase the lighting one hour per day to a total of 14 hours for pullets reared in dark out houses. The pullets are each fed approximately 312 kcal ME per day.

Flocks lay first egg: Change from pre-breeder or grower diets to a breeder diet at this time. The breeder diet will provide the calcium needed to build a calcium reserve in the medullary bone.

23/24 weeks: Flock should be laying at a rate of about 1% production. The breeder pullets need to be fed approximately 337 kcal ME per day for the week.

5% production: Increase the length of the light day to 15 hours for pullets previously reared in dark-out housing. The flock should be between 24 to 25 weeks of age at this time. The breeder pullets need to be fed approximately 363 kcal ME per day.

Each 5% production increase: Increase the daily feed allotment for each breeder pullet 13 to 14 kcal ME per day for each 5% increase in the flock egg production. A good method of increasing the feed allotment to provide adequate nutrients is to make a decision when your particular strain should be fed peak feed levels, i.e., 35 to 65% production, and then divide the number of days normally required to reach this level of production into the amount of increased feed needed above feed fed at 5% production. An example would be a breeder pullet flock being fed 127-g feed per day at 5% production and peak feed for the flock will be 165 g per day. If the flock requires three weeks to reach 60% production from 5% production (pre-selected production point for flock to be fed the amount of feed for peak egg production), divide the 38 g increase by 21 days and increase the feed allocation approximately 0.4 lb/100 (1.8 g) per breeder per day during this period.

35 to 40% egg production: Each bird should be eating between 455 to 470 kcal of ME per day with all the nutrients needed for peak production. During this period birds will be consuming their maximum amount of between 34 and 36 lb (15.45 and 16.36 kg) of feed per 100 birds per day. This is called *lead feeding* because it takes 14 to 20 days to develop the ova needed for egg production. Feeding peak feed to pullets producing at 35 to 40% will work for uniform flocks, in-season flocks, and flocks reared in dark-out housing.

The amount of feed needed for peak egg production will depend on the dietary energy content, strain, rate of lay, egg size, body weight, and ambient temperature. Several primary breeders have suggested that some strains should not be fed peak feed until they reach 50 to 65% production because of their potential to gain weight rapidly at the onset of production and because some strains are slower to increase in egg production. A concern is overfeeding the pullets too early in the production cycle causing her to become overweight and consequently hurting overall performance. Also, if the temperature is colder, the peak feed may need to be fed earlier than 35 to 40% production compared to breeder pullets housed in warmer temperatures.

Peak feed may not need to be fed until the flock reaches 50 to 60% production when pullets are housed in warm to hot temperatures. Pullet flocks reared in dark-out housing may have a faster increase in egg production (4 to 5% increase in production/hen/day) compared to normal increases of 2 to 3% egg production per day. Flocks increasing in production faster should be fed peak feed earlier. Flocks that are less uniform because of disease or poor management during the rearing period should also be fed peak feed longer than uniform flocks.

60% egg production: Increase the length of the day to 16 hours about two weeks prior to the time the flock will peak in egg production, where it should remain throughout the laying period.

Note: If the birds are transferred to open-sided production houses in the summer the increases in hours of light will be determined by the date of transfer and the prevailing day length.

Important: If pullet flocks are overweight at the time first eggs are produced, do not make additional daily feeding reductions during the changeover period in an attempt to reduce body weight. Birds in such flocks should always remain heavier than normal, throughout the entire laying period. It is important to develop a new body weight curve for the overweight pullets compared to the standard weights as established in "Breeder Management Guides."

If pullets are underweight at start of lay, the feed allotment should be increased in order to bring the pullets up to their standard weight.

19-H. NUTRITIONAL REQUIREMENTS OF BREEDERS DURING EGG PRODUCTION

Energy Requirements During Egg Production

Broiler breeder strains tend to become too heavy if full fed during the laying period. Broiler breeder hens also will become overweight if fed the same energy levels used for peak production for an extended period following peak production. Broiler breeders produce fewer eggs than egg laying strains and tend to decrease in egg production more rapidly. Summers (1995) calculated the metabolizable energy requirements of breeder hens at various stages of egg production and body weight based on the hens being housed at 68°F (20°C) and showed that maintenance energy requirement is approximately 80% of requirement total (Table 19-10). The percentage utilized for maintenance decreases around peak egg production and continues lower through peak egg mass (% egg production × egg weight) production. The breeder hen only uses about 20% of her daily energy intake for egg mass production and if feed is not sufficient to meet the total energy requirement, the egg mass output (egg number and egg size) will be reduced. The breeder hen will utilize nutrients to meet her requirement for maintenance prior to partitioning nutrients for production.

Protein Requirement

An average protein level for a broiler breeder diet should be 16%, but is subject to some variation according to the environmental temperature (affects feed intake), caloric content of the diet, rate of egg production, size of birds, and so forth. Summers (1995) estimated the protein requirement for breeders with different body weights that were producing eggs of different sizes (Table 19-11). The protein required for egg production is based on the assumption that every hen was laying each day. The actual requirement for egg mass should be multiplied by the coefficient for % egg production of a flock (i.e., 87% egg production = 0.87) to be more realistic about the protein requirement of a real flock. An estimate for the daily protein requirement for a flock with an average weight of 3.5 kg, laying at a rate of 80% of eggs that weighed 60 g would be 18.6 g per day per breeder.

When breeders are fed diets containing 16% protein and provided 165 g feed/day, they are consuming approximately 26.4 g of protein. Researchers have reported that lower protein diets fed to breeders may improve the hatchability and increase the number of saleable chicks (Table 19-12). Summers suggests that many breeders are fed excessive levels of protein and this practice may be detrimental to performance and uneconomical. In Table 19-13 the effect of varying protein intake from 16.5 g to 27.0 g caused no response in breeders, whereas the effect of changing energy intake by 40% caused a marked reduction in performance.

Table 19-10. Predicted Energy Requirements of Broiler Breeder Hens from 20 to 68 Weeks with a Pen Temperature of Approximately 72°F (22°C)

	Age (weeks)												
	20	24	28	32	36	40	44	48	52	56	60	64	68
	Body wt												
(lb)	4.76	5.51	6.94	7.28	7.67	7.89	7.98	8.16	8.27	8.38	8.42	8.49	8.60
(kg)	2.16	2.50	3.15	3.30	3.48	3.58	3.62	3.70	3.75	3.80	3.82	3.85	3.90
	Egg production (%)												
	—	5	60	85	82	77	73	68	63	58	52	48	45
	Average egg wt (g)												
	—	47.2	54.4	58.6	61.6	63.3	65.2	67.1	68.4	69.5	70.3	71.1	71.5
	Predicted energy requirement (kcal/day)												
Total	300	350	400	450	450	450	450	445	445	440	440	435	435
Maintenance	250	285	323	335	343	350	350	352	352	353	353	354	354
Maintenance (% of Total)	83	81	80	74	76	78	78	79	79	80	80	81	81

Source: Summers, 1995

Table 19-11. Estimates of Dietary Protein Requirements at Various Breeder Body Weights When Producing Eggs of Different Size

Body wt (kg)	Protein Intake for Maintenance (g/b/d)	Average Egg wt. (g)	Protein Intake for Egg Production (g/b/d)
3.00	7.22	50	10.9
3.25	7.71	55	12.0
3.50	8.15	60	13.1
3.75	8.56	65	14.2
4.00	9.07	70	15.3

Source: Summers, 1995

Table 19-12. Influence of Dietary Protein Level on Performance of Broiler Breeders (26 to 60 Weeks of Age)

	Diet Protein Level (%)	
Trait	13.7	16.8
Hen day production (%)	60.3	57.8
Mean egg wt (g)	63.4	63.0
Fertility (%)	93.1	92.4
Hatch of fertile eggs (%)	88.6	85.5
Saleable chicks of fertile eggs (%)	84.5	80.5

Source: Whitehead, 1985

Table 19-13. Influence of Protein and Energy Intake on Production Parameters in Broiler Breeders

Energy Intake (kcal ME/d)	Body wt 60 wk (g)	Eggs/bird	Av. egg wt 21 to 60 wk (g)
449	3962	157.5	65.3
385	3587	156.6	64.0
315	2894	140.2	62.9
270	2688	100.7	61.6
Protein intake (g/d)			
27.0	3298	136.3	63.7
23.2	3284	137.4	63.6
19.5	3293	139.3	63.6
16.5	3258	142.1	63.0

Source: Pearson and Herron (1982)

Table 19-14. The Calculated Total Requirement of Amino Acids for a Broiler Breeder Hen at 29 and 64 Weeks of Age

Amino Acid	Total Requirement (mg/bird/day) 29 wk	64 wk
Arginine	1,005	870
Histidine	376	325
Isoleucine	767	661
Leucine	1,254	1,078
Lysine	1,121	973
Methionine	474	408
Methionine + cystine	794	687
Phenylalanine + tyrosine	1,316	1,130
Threonine	708	612
Tryptophan	239	205
Valine	882	763

Source: Fisher, 1998

Amino Acid Requirements

Fisher (1998) calculated the amino acid requirements of breeder hens in Table 19-14. The ideal amino acid profile of breeder hens in relation to lysine is also reported in Table 19-15. The factorial calculations include an additional requirement for the variation of egg mass output and body weight of the flock (the Reading model) thus providing for at least 97.5% of the birds in a flock. The calculated amino acid requirements are gener-

Table 19-15. The Ratio of the Calculated Amino Acid Requirements for a Broiler Breeder Hen at 29 and 64 Weeks of Age vs Lysine Which Is Taken as 100

Amino Acid	Calculated Requirement (vs Lysine = 100) 29 wk	64 wk
Arginine	90	89
Histidine	34	33
Isoleucine	68	68
Leucine	112	111
Lysine	100	100
Methionine	42	42
Methionine + cystine	71	71
Phenylalanine + tyrosine	117	116
Threonine	63	63
Tryptophan	21	21
Valine	79	78

Source: Fisher, 1998

Table 19-16. Nutrient Requirements of Meat-Type Hens for Breeding Purposes as Units per Hen per Day (90 percent dry matter)

Nutrient	Unit	Requirements
Protein and amino acids	g	19.5
Arginine	mg	1,110
Histidine	mg	205
Isoleucine	mg	850
Leucine	mg	1,250
Lysine	mg	765
Methionine	mg	450
Methionine + cystine	mg	700
Phenylalanine	mg	610
Phenylalanine + tyrosine	mg	1,112
Threonine	mg	720
Tryptophan	mg	190
Valine	mg	750
Minerals		
Calcium	g	4.0
Chloride	mg	185
Nonphytate phosphorus	mg	350
Sodium	mg	150
Vitamins		
Biotin	μg	16

Note: Margins for safety are not included in these guidelines. (For nutrients not listed, please see requirements for egg-type breeders as a guide)
Source: National Research Council, 1994

ally higher for breeding hens compared to the daily requirements suggested by NRC (1994) (Table 19-16).

Mineral Requirements

As with the commercial laying hen's need for large amounts of calcium for egg production, there is a similar requirement for calcium for the production of quality hatching eggs. This, along with other mineral requirements, are given in Table 19-17. The NRC requirements (1994) for calcium, non-phytate phosphorus, chloride, and sodium are also listed in Table 19-16 for broiler breeders.

Vitamin Requirements

Table 19-17 gives the vitamin requirements for hatching egg production suggested by a primary breeder. In order to secure good hatchability, the breeder feed requirement is greater than for other rations for poultry. Leeson (1997) reported a survey conducted by BASF, a major producer of pure

Table 19-17. Nutrient Specifications for Broiler Breeder Parent Stock

		Chick Starter 0 to 21 days	Grower 22 to 119 days	Pre-Breeder 120 to 154 days	Breeder 1, 155 to 314 days	Breeder 2, 315 days+
Crude protein	%	17.50	15.00	15.50	16.00	15.50
Metabolizable energy	kcal/lb	1,300	1,300	1,300	1,300	1,300
Fat	%	3–4	3–4	3–4	3–4	3–4
Fiber	%	3–4	3–4	3–4	3–4	3–4
Amino acids						
Arginine	%	1.10	0.83	0.78	0.78	0.78
Histidine	%	0.40	0.29	0.32	0.32	0.32
Isoleucine	%	0.80	0.50	0.81	0.81	0.81
Leucine	%	1.40	0.95	1.22	1.22	1.22
Lysine	%	0.90	0.75	0.80	0.80	0.75
Methionine	%	0.40	0.35	0.34	0.34	0.31
Methionine + cystine	%	0.72	0.60	0.58	0.58	0.55
Phenylalanine	%	0.70	0.48	0.74	0.74	0.74
Phenylalanine + tyrosine	%	1.27	0.82	1.05	1.05	1.05
Threonine	%	0.70	0.50	0.60	0.60	0.60
Tryptophan	%	0.22	0.17	0.18	0.18	0.18
Valine	%	0.95	0.57	0.70	0.70	0.70
Minerals						
Calcium	%	1.00	1.00	1.50	3.00	3.20
Total phosphorus	%	0.70	0.60	0.60	0.60	0.60
Available phosphorus	%	0.45	0.40	0.40	0.40	0.40
Sodium	%	0.16	0.15	0.15	0.15	0.15
Chloride	%	0.18	0.16	0.16	0.16	0.16
Potassium	%	0.40	0.40	0.60	0.60	0.60
Trace minerals						
Copper	ppm	4.00	4.00	16.00	16.00	16.00
Iodine	ppm	0.50	0.50	4.00	4.00	4.00
Iron	ppm	5.00	5.00	20.00	20.00	20.00
Manganese	ppm	70.00	60.00	100.00	100.00	100.00
Magnesium	ppm	250.00	250.00	250.00	250.00	250.00
Zinc	ppm	50.00	40.00	100.00	100.00	100.00
Selenium	ppm	0.15	0.15	0.20	0.20	0.20
Vitamins per lb. of final ration						
Vitamin A	IU	4550.00	4450.00	7300.00	7300.00	7300.00
Vitamin D_3	IU	1600.00	1600.00	1600.00	1600.00	1600.00
Vitamin E	IU	15.00	15.00	25.00	25.00	25.00
Vitamin K	mg	1.00	1.00	5.00	5.00	5.00
Thiamin (B_1)	mg	0.25	0.25	2.50	2.50	2.50
Riboflavin (B_2)	mg	2.25	2.25	7.00	7.00	7.00
Niacin	mg	9.00	9.00	23.00	23.00	23.00
Pantothenic acid	mg	3.70	3.70	9.00	9.00	9.00
Pyridoxine (B_6)	mg	0.50	0.50	1.75	1.75	1.75
Biotin	mg	0.12	0.03	0.20	0.20	0.20
Folic acid	mg	0.25	0.23	2.50	2.50	2.50
Vitamin B_{12}	mg	0.01	0.01	0.03	0.03	0.03
Choline	mg	45.00	45.00	140.00	140.00	140.00
Minimum specification						
Linoleic acid	%	1.00	1.00	1.25	1.25	1.00

Source: Ross 308 Breeder Management Guide, 1999

Table 19-18. Breeder Vitamin Levels and Costs per Ton of Feed

	Industry High Top (25%)		Industry Low Bottom (25%)		NRC 1994	
Vitamin	Level	$/ton feed	Level	$/ton feed	Level	$/ton feed
Vit A (IU/kg)	12,800	0.68	8,100	0.43	3,000	0.16
Vit D_3 (IU/kg)	3,500	0.08	2,100	0.05	300	0.01
Vit E (IU/kg)	36	1.08	14.3	0.43	10	0.30
Vit K_3 (IU/kg)	3.1	0.13	0.74	0.03	1.0	0.03
Thiamine (mg/kg)	3.2	0.11	1.0	0.04	0.7	0.03
Riboflavin (mg/kg)	9.9	0.53	5.6	0.29	3.6	0.29
Pantothenic acid (mg/kg)	17.3	0.38	9.3	0.20	7.0	0.20
Niacin (mg/kg)	43	0.27	23	0.14	10	0.14
Pyridoxine (mg/kg)	6.0	0.31	1.4	0.07	4.5	0.07
Folic acid (mg/kg)	1.3	0.13	0.63	0.06	0.35	0.06
Biotin (μg/kg)	220	0.77	88	0.31	100	0.31
Vit B_{12} (μg/kg)	17.5	0.09	10.0	0.05	8.0	0.05
Total		$4.56/ton		$2.10/ton		$1.65/ton
Cost/breeder hen 20–64 weeks		19.1¢		8.8¢		6.9¢

Source: Leeson, 1998

vitamins, that describes the levels of vitamins being fed to broiler breeders by the industry. The highest vitamin levels represent 25% of the industry and the lower levels represent the bottom 25% of the broiler industry (Table 19-18). The NRC (1994) values are also listed for comparison. There is approximately 100% difference in vitamin levels and costs per ton between the top 25% and the bottom 25%. Leeson suggest the 10 cents additional cost per breeder for feeding the highest levels of vitamins is equivalent in value to 0.5 chicks per breeder. The lowest levels of vitamins being fed by the industry are actually lower than NRC levels which may cause a negative effect on performance. Leeson believes that marginal vitamin deficiencies can easily result in the loss of 2 to 5 chicks per breeder, which is four to ten times the cost of the extra vitamins. Vitamin E, vitamin A, and biotin in the vitamin premix are the most expensive added vitamins and make up over 50% of the cost of the vitamin premix.

19-I. FEEDING BROILER BREEDERS BEFORE AND AFTER PEAK PRODUCTION

A program of continued feed restriction should be followed during the egg production period. One must be sure the flock is given ample amounts of feed necessary to produce the maximum number of eggs, but not amounts excessive to the extent that the birds gain too much weight. There are two segments to the program.

1. *Sustained Peak Production and Better Post-Peak Persistency May Require Challenge Feeding.* Leeson (1997) suggests that *challenge feeding* should be

initiated when the flock egg production is approximately 60 to 70% and should be discontinued after peaking when egg production falls below 80%. Challenge feeding allows a manager to establish a feeding program based on the needs of each flock and there is no standardized method. Leeson suggests that the amount of additional feed that should be provided above a base level should depend upon the uniformity of the flock, disease status, feed quality, nutrient density, and environmental temperature.

A challenge feeding program should lead hens into a sustained peak and the maximum amount of feed should coincide with peak egg production. Leeson suggests that the quantity of the challenge feed should not be more than 5% of the base feed amount and most often the quantity will be only 2 to 4%. The time taken for the feed to be completely consumed is a good indicator of how much feed the flock requires. If the time for feed consumption suddenly increases by 30 minutes or more without any change in the health or environmental temperature, that is a good indicator that the flock is receiving too much feed. When starting *challenge feeding* at 60% egg production, the least amount of added feed above the base allocation for a uniform flock fed a high nutrient density feed would be approximately 5 g/bird/day to be fed 3 times per week. Feeding the additional feed only 3 times per week will allow enough additional feed on the three days to increase the feeding time and help with uniformity. Leeson (1997) suggests a highly uniform flock in a warm environment would most likely require the minimum amount of challenge feed and be fed 8 g additional feed/bird/day 3 times a week at peak production and then be fed 5 g/bird/day 3 times a week when birds fall 2% below peak production.

The maximum amount of challenge feed and base feed would be needed for a situation with low nutrient density feed, poor ingredient quality, cooler temperatures, and average flock uniformity. A flock with these circumstances may be fed up to 38.5 pounds/100 birds of a base feed plus 8 g additional feed/bird/day 3 times per week when the flock reaches 60% production and then fed 14 g additional feed/bird/day 3 times per week when the flock is peaking. The flock could again be fed the 8 g additional feed/bird/day when the egg production drops 2% below peak production. In summary, Leeson suggests that birds subjected to stresses such as variable feed quality, mycotoxin challenge, fluctuating or extreme environmental temperatures may need a high base feed allowance plus an aggressive feed challenge. Lower feed inputs are possible when consistent high-quality energy feeds are used along with good environmental controls.

2. *Less feed.* When the flock has passed its peak and production has dropped to approximately 80%, challenge the flock to lower feed costs by reducing feed intake by 0.22 to 0.44 lb per 100 birds per day (1 to 2 g/bird/day). As a general rule, the amount of feed allocated is decreased weekly after feed reduction begins until approximately 10 to 13% less feed

is being fed compared to peak feed amounts. Broiler breeders will become obese, less fertile, and persistency of lay will suffer if continually fed the peak production allocations of feed. Small reductions of 0.22 lb per 100 birds per day compared to larger drops of 1.1 lbs/100 birds (5 g/bird/day) can be made weekly to reduce the risk of causing a drop in egg production. The feed reduction can be continued weekly but if production drops more than normal after a feed decrease, return the flock to the preexisting feed level. The feed reduction rate should be dependent upon the amount of feed fed during peak production, breeder's body weight, egg production, and environmental factors.

Continue this program throughout the remainder of the laying period. Do not make feed reductions during stress, disease outbreaks, sudden drops in temperature, or if body weight decreases.

Ambient Temperature and Feed Intake

Even though maximum and minimum guides have been presented for feed consumption throughout the laying period, there are times when these limits will not suffice, and the variations in ambient temperature are usually the cause. There is nothing that disrupts a feeding schedule more than temperature change. Extremes can cause variations in feed consumption of up to 40%. Variations are smaller for each degree of temperature change when the weather is cool than when it is hot.

A change in the daily feed clean-up time is a good indicator that hens may not be receiving the proper amount of feed. If the weather turns cold, the hens' demand for more energy means the feed supply will be consumed quicker, and to offset this, more feed must be given. When the weather turns warmer, the daily feed allotment will not be consumed as quickly or may not be completely eaten. It is then that less feed should be supplied. The ambient temperature primarily affects the maintenance energy requirements for breeders. The *Arbor Acre Breeder Management Guide* indicates that for every 1°C above 27°C, energy requirements decrease 5 kcal/bird/day. The Guide also indicates that for every 1°C below 20°C, energy requirements increase 5 kcal/bird/day. The environmental temperature should also be taken into consideration when implementing a feed reduction program following peak egg production. Feed quantities should be reduced more slowly in colder temperatures and more quickly in warmer temperatures.

19-J. ESTIMATED DAILY FEED ALLOWANCES FOR BROILER BREEDER FLOCKS

Estimated feed allowances for standard-size broiler breeder flocks are given in Table 19-19. A further guide for feed consumption of standard-

Table 19-19. Guide for Feed Consumption When Standard-Size Meat-Type Pullets Are Control-Fed During Egg Production

Week of Egg Production	Hen-Day Egg Production %	Feed Consumed per 100 Birds per Day		Female Body Weight	
		lb	kg	lb	kg
1	5	24–28	10.9–12.7	5.2–5.7	2.4–2.6
2	20	28–32	12.7–14.6	5.4–5.9	2.5–2.7
3	38	30–34	13.6–15.4	5.6–6.1	2.6–2.8
4	56	32–36	14.5–16.4	5.7–6.2	2.6–2.8
5	73	33–37	15.0–16.8	5.8–6.3	2.6–2.9
6	84	34–38	15.5–17.3	5.3–6.3	2.6–2.9
7	86	34–38	15.5–17.3	5.9–6.4	2.7–2.9
8	85	34–38	15.5–17.3	5.9–6.4	2.7–2.9
9	84	34–38	15.5–17.3	6.0–6.5	2.7–3.0
10	84	34–38	15.5–17.3	6.0–6.5	2.7–3.0
11	83	33–37	15.0–16.8	6.1–6.6	2.8–3.0
12	82	33–37	15.0–16.8	6.1–6.6	2.8–3.0
13	81	33–37	15.0–16.8	6.2–6.7	2.8–3.1
14	81	33–37	15.0–16.8	6.2–6.7	2.8–3.1
15	80	33–37	15.0–16.8	6.3–6.8	2.8–3.1
16	79	32–36	14.6–16.4	6.3–6.8	2.8–3.1
17	78	32–36	14.6–16.4	6.3–6.8	2.8–3.1
18	77	32–36	14.6–16.4	6.4–6.9	2.9–3.1
19	77	32–36	14.6–16.4	6.4–6.9	2.9–3.1
20	76	32–36	14.6–16.4	6.5–7.0	3.0–3.2
21	75	31–35	14.1–15.9	6.5–7.0	3.0–3.2
22	74	31–35	14.1–15.9	6.5–7.0	3.0–3.2
23	74	31–35	14.1–15.9	6.6–7.1	3.0–3.2
24	73	31–35	14.1–15.9	6.6–7.1	3.0–3.2
25	72	31–35	14.1–15.9	6.6–7.1	3.0–3.2
26	71	30–34	13.6–15.4	6.6–7.1	3.0–3.2
27	70	30–34	13.6–15.4	6.6–7.1	3.0–3.2
28	70	30–34	13.6–15.4	6.7–7.2	3.1–3.3
29	69	30–34	13.6–15.4	6.7–7.2	3.1–3.3
30	68	30–34	13.6–15.4	6.7–7.2	3.1–3.3
31	67	29–33	13.2–15.0	6.7–7.2	3.1–3.3
32	66	29–33	13.2–15.0	6.7–7.2	3.1–3.3
33	66	29–33	13.2–15.0	6.8–7.3	3.1–3.3
34	65	29–33	13.2–15.0	6.8–7.3	3.1–3.3
35	64	29–33	13.2–15.0	6.8–7.3	3.1–3.3
36	63	28–32	12.7–14.6	6.8–7.3	3.1–3.3
37	62	28–32	12.7–14.6	6.8–7.3	3.1–3.3
38	61	28–32	12.7–14.6	6.9–7.4	3.1–3.4
39	60	28–32	12.7–14.6	6.9–7.4	3.1–3.4
40	60	28–32	12.7–14.6	6.9–7.4	3.1–3.4
41	59	27–31	12.3–14.1	6.9–7.4	3.1–3.4
42	58	27–31	12.3–14.1	6.9–7.4	3.1–3.4
43	57	27–31	12.3–14.1	7.0–7.5	3.2–3.4
44	56	27–31	12.3–14.1	7.0–7.5	3.2–3.4

Source: North & Bell, 1990

Table 19-20. Metabolizable Energy and Protein Consumption of Meat-Type Breeders During Egg Production

Week of Egg Production	Hen-Day Egg Production (%)	ME Consumed per Bird per Day (kcal)	Protein Consumed per Bird per Day (g)
1	5	312–364	17.4–20.3
2	20	364–416	20.3–23.3
3	38	390–442	21.8–24.7
4	56	416–468	23.3–26.2
5	73	429–481	24.0–26.9
6	84	442–494	24.6–27.6
7	86	442–494	24.6–27.6
8	85	442–494	24.6–27.6
9	84	442–494	24.6–27.6
10	84	442–494	24.6–27.6
20	76	416–468	23.3–26.2
30	68	390–442	21.8–24.7
40	60	364–429	20.3–23.3

Source: North & Bell, 1990

size, meat-type breeder flocks showing the ME and protein consumed per day during egg production is given in Table 19-20.

Feed per Hatching Egg and Chick Produced

Table 19-21 shows the feed efficiency for breeder hens through 64 and 68 weeks of age. The best way to evaluate feed efficiency for a breeder flock is to measure the amount of feed per hatching egg or chick. The data in Table 19-21 is expressed for breeder hens alone and with 8% males. Leeson (1997) suggests that to compare feed efficiency for breeders on a worldwide basis, it is more meaningful to determine the amount of ME or protein per hatching egg or chick because of the variation in dietary energy levels. A rule of thumb is 1,000 kcal of ME per hatching egg or chick when including all of the rearing feed plus feed for the males. Leeson suggests feed efficiency can be improved in flocks by regulating the late increase in egg size. Egg size can be controlled with reductions in protein and methionine and minimize the overall need for nutrients required for egg mass. Another factor that will help feed efficiency is to minimize the energy needed for maintenance by controlling the environmental temperature of the breeders. Feed efficiency will be negatively affected with either high or low temperatures.

Table 19-21. Feed Efficiency for a Broiler Breeder Flock

		Age			
Females Only		0–64 wks	24–64 wks	0–68 wks	24–68 wks
Per hatching egg					
Feed	lb	0.70	0.57	0.71	0.58
	g	321	262	323	267
Energy	kcal	915	746	920	760
Protein	lb	0.11	0.09	0.11	0.09
	g	50	41	50	42
Per chick					
Feed	lb	0.81	0.66	0.82	0.68
	g	371	303	375	310
Energy	kcal	1,055	863	1,070	884
Protein	lb	0.12	0.10	0.12	0.10
	g	58	47	58	48
		Age			
Females + 8% M		0–64 wks	24–64 wks	0–68 wks	24–68 wks
Per hatching egg					
Feed	lb	0.76	0.61	0.76	0.63
	g	345	279	347	286
Energy	kcal	983	795	989	815
Protein	lb	0.12	0.09	0.12	0.10
	g	53	43	54	44
Per chick					
Feed	lb	0.87	0.71	0.88	0.73
	g	398	323	403	332
Energy	kcal	1,134	921	1,149	946
Protein	lb	0.13	0.11	0.13	0.11
	g	62	50	62	51

Assuming diets contain an average of 15.5% CP and 1292 kcal ME/lb (2,850 kcal ME/kg)
Birds maintained at about 72°F (22°C)
Source: Lesson, 1997

19-K. FEEDING THE BROILER BREEDER MALE DURING THE BREEDING PERIOD

In the past, recommendations for feeding and managing the meat-type breeding flock have been established for the female only; the male has been neglected. While the pullet's feed has been restricted, the cockerels have had all the feed they could eat. They became obese, and after a few weeks develop foot and leg problems that decreased their reproductive performance. Although overfeeding the male during the breeding period is a major concern, the initial period of the breeding period up to 30 weeks of age is critical for the male because they are still growing at a rapid rate. Leeson (1997) has reported the breeder male should gain almost as much weight (2.6 lb, 1.2 kg) from 20 to 30 weeks of age as from 10 to 20 weeks of age (3.1 lb, 1.4 kg).

Table 19-22. Pullet and Cockerel Breeder Rations for Dual Feeding (Major Feed Differences)

Item	Pullet Ration	Cockerel Ration
ME (kcal/lb)	1,300	1,275
ME (kcal/kg)	2,860	2,805
Protein (%)	16.0	12.0
Methionine (%)	0.31	0.22
Methionine + cystine (%)	0.56	0.41
Lysine (%)	0.78	0.50
Calcium (%)	3.00	0.90
Phosphorus, available (%)	0.46	0.40

Source: North & Bell, 1990

Dual Feeding System

McDaniel and Giesen (1991) reported that the remedy to the problem of the male being overweight is to feed two separate rations during the breeding season, known as *dual feeding*. The pullets are fed their regular ration, but the cockerels are fed a ration low in protein, calcium and slightly less energy. The main differences in the two rations are shown in Table 19-22. The nutrient requirements for feeding the male during the grower period and during the breeding period are listed in Table 19-23.

Table 19-23. Nutrient Requirements of Meat-Type Males for Breeding Purposes as Percentages or Units per Rooster per Day (90 percent dry matter)

		Age (weeks)		
	Unit	0 to 4	4 to 20	20 to 60
Metabolizable energy	kcal	—	—	350 to 400
Protein and amino acids				
Protein	%	15.00	12.00	—
Lysine	%	0.79	0.64	—
Methionine	%	0.36	0.31	—
Methionine + cystine	%	0.61	0.49	—
Minerals				
Calcium	%	0.90	0.90	—
Nonphytate phosphorus	%	0.45	0.45	—
Protein and amino acids				
Protein	g	—	—	12
Arginine	mg	—	—	680
Lysine	mg	—	—	475
Methionine	mg	—	—	340
Methionine + cystine	mg	—	—	490
Minerals				
Calcium	mg	—	—	200
Nonphytate phosphorus	mg	—	—	100

Source: National Research Council, 1994. National Academy of Sciences, Washington, DC

In order to be sure the pullets and cockerels get their respective rations, two automatic independent mechanical feeding systems must be installed in the same house. (See *Managing the Breeder Flock*, Chapter 34.)

Specifics for the Pullet-Feeding Program

Either trough-and-chain or auger-and-pan automatic feeders may be used. To prevent cockerels from eating from the pullet troughs or pans, male exclusion grills must be placed on the pullet feeders. See Figure 19-2 for an exclusion grill on a trough.

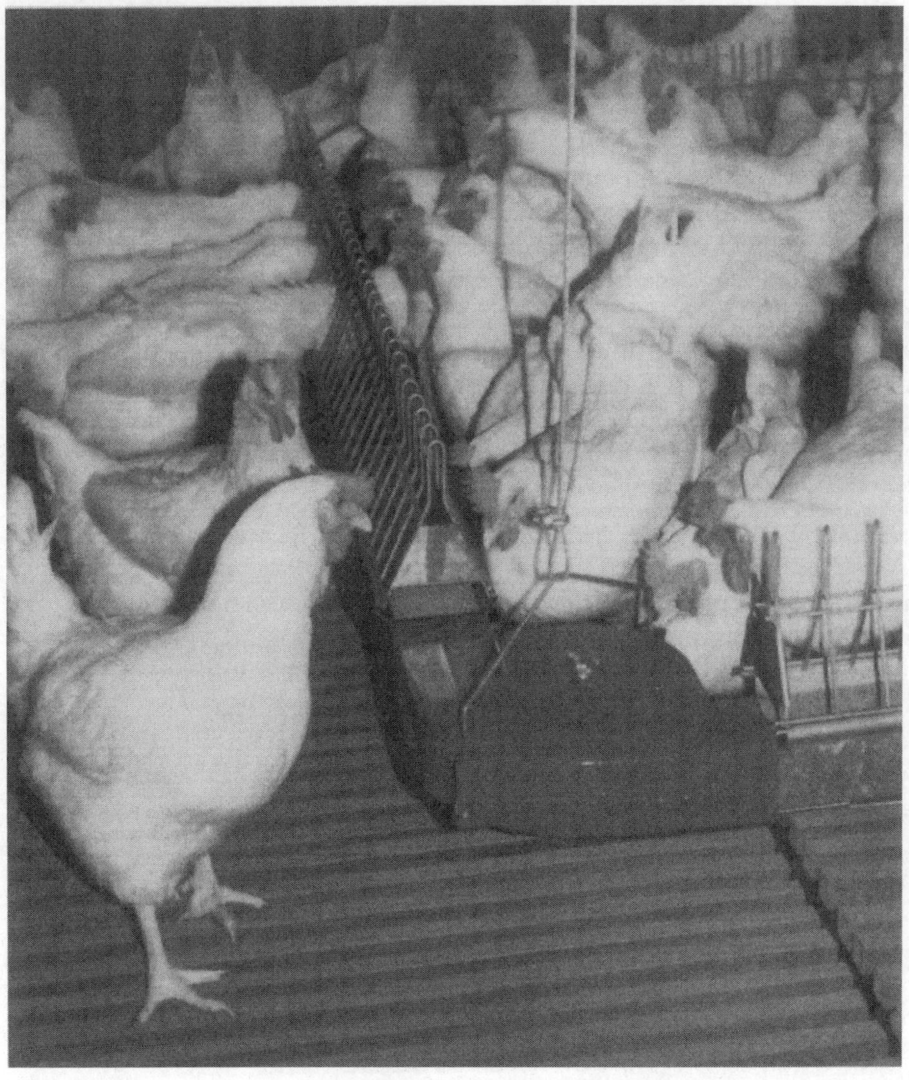

Figure 19-2. Dual Feeding System for Breeders—Female Feeders (courtesy of Big Dutchman)

If the house is a slat-and-floor type, place the pullet feeder on the slats with the bottom of the feeder trough or pan 1 inch (2.5 cm) above the slats. Pullets do not readily eat through an exclusion grill when the feeder is higher. The male exclusion grill should have an opening 1.7 inches (43 mm) wide through which the pullets can eat, but excluding the cockerels. Provide 6 inches (15 cm) of trough space per pullet or one 13-inch diameter (33 cm) pan for every 11 pullets. Each day, start the pullet feeders about 15 minutes before the cockerel feeders.

Specifics for the Cockerel-Feeding Program

A tube-and-pan automatic feeder must be used for the cockerel-feeding system, and it may be placed on the slats or on the litter floor. If the feeder is placed on the litter floor (the preferred location) then water should also be easily accessible for the males. It will take only 1 to 2 hours for the cockerels to consume their daily allotment of feed; therefore, it is very important that enough feeding space be available so all cockerels can eat at one time. Furthermore, when the feeder is started, feed must flow simultaneously to all pans to prevent cockerels from migrating to get feed.

When the cockerel feeders are first used, the lip of the feeders should be about 10 inches (25.4 cm) above the slats, then gradually increased to 18 inches (46 cm) by the time the pullets are added to the house. The male feeders on the litter should be raised so that the females are unable to eat out of them. Provide one 13-inch (33-cm) pan for every 10 cockerels. Some systems provide for raising the pan feeders by a winch between feedings.

Feeding and Management of the Dual System

The cockerels should be moved from the growing house to the dual system breeding house when about 19 to 20 weeks of age, and the inferior cockerels should be culled. Change the ration to a low-protein cockerel breeder feed, which is fed in the elevated male feeders.

Move the pullets to the breeder house 7 to 10 days later. Continue to feed the pullets a growing feed in the female feeders with the male exclusion grills. When the pullets lay their first eggs (about 23 to 24 weeks of age), change them to a regular breeder feed. Cull the cockerels again, retaining about 10 males per 100 females.

Body Weight Governs Feed Allocation

It is imperative that the body weights of both pullets and cockerels are on target when in the breeding house. Adhere to the weights suggested by the primary breeder or the example weights listed in Table 19-24.

Table 19-24. Recommended Male Body Weights and Feed Consumption (Cobb 500)

Age		Body Weight		Feed Allotment				
				Imperial (lbs/100/d)		(g/Metric bird/d)		
Days	Weeks	(lbs)	(kg)	Daily (ED)	Skip (S)	Daily (ED)	Skip (S)	Key Points
	0–1			FULL	FULL	FULL		Full feed 18% protein (minimum) starter for four weeks
7	1–2	0.50	0.22	FULL	FULL	FULL		
14	2–3	0.90	0.40	FULL	FULL	FULL		
21	3–4	1.10	0.50	FULL	FULL	FULL		
28	4–5	1.40	0.64	13.2	26.4 S	60	120 S	Four week target weight is MINIMUM
35	5–6	1.75	0.80	13.8	27.6 S	62	124 S	
42	6–7	2.10	0.95	14.4	28.8 S	65	130 S	
49	7–8	2.40	1.09	15.0	30.0 S	68	136 S	
56	8–9	2.70	1.22	15.5	31.0 S	70	140 S	
63	9–10	3.00	1.36	16.0	32.0 S	73	146 S	
70	10–11	3.30	1.50	16.4	32.8 S	74	148 S	
77	11–12	3.60	1.64	16.8	33.6 S	76	152 S	
84	12–13	3.90	1.77	17.2	34.4 S	78	156 S	
91	13–14	4.20	1.91	17.6	35.2 S	80	160 S	
98	14–15	4.50	2.04	18.0	36.0 S	82	164 S	
105	15–16	4.70	2.14	18.4	36.8 S	84	168 S	
112	16–17	5.00	2.27	18.8	37.6 S	85	170 S	
119	17–18	5.25	2.39	19.2	38.4 S	87	174 S	
126	18–19	5.50	2.50	19.6	39.2 S	89	178 S	
133	19–20	5.8	2.63	20.0	40.0 S	91	182 S	
140	20–21	6.10	2.77	21.5	21.5 ED	98	98 ED	
147	21–22	6.40	2.90	23.0	23.0 ED	104	104 ED	
154	22–23	6.75	3.07	24.5	24.5 ED	111	111 ED	
161	23–24	7.10	3.22	26.0	26.0 ED	118	118 ED	
168	24–25	7.50	3.40	27–35	27–35 ED	123–159	123–159 ED	footnote*
175	25–26	7.70	3.50					
182	26–27	7.90	3.59					
189	27–28	8.10	3.68					
196	28–29	8.30	3.77					
203	29–30	8.50	3.86					
210	30–31	8.70	3.95					

Notes: Weights 4 thru 20 weeks are Off-Feed Weights. ED = Everyday Feeding; S = Skip-a-Day Feeding
* Values from 27 to 35 lbs (123–159 g) will depend on how much feed the males are getting from the female feeders and also upon their body weights relative to the breeder guides. It will also depend upon the MEn content of the two diets
Source: Cobb 500 Broiler Breeder Management Guide, 1999

Both males and females should be on a program of daily feed restriction to maintain their target body weights. Do not use a skip-a-day program. Sample weigh the two sexes when they are moved to the breeding house, and each week thereafter. Weigh the birds in the afternoon after they have consumed their daily supply of feed.

Table 19-25. Examples of Feed and ME Intake for Male Breeders Consuming a Diet of Approximately 2,900 kcal ME/kg at Different Ages and Temperature

	Assuming males have access to hen feeders until approx. 28 wks of age				Assuming males totally excluded from hen feeders	
Temp	>95°F >35°C	68–82°F 20–28°C		<59°F <15°C	68–82°F 20–28°C	
Age (wk)	g per bird per day	g per bird per day	kcal ME/day	g per bird per day	g per bird per day	kcal ME/day
20	108	110	319	120	115	334
22	110	115	334	125	118	342
24	112	118	342	130	120	348
26	120	125	353	135	130	377
28	124	130	377	140	136	392
30	130	135	392	150	135	392
32	135	140	406	155	130	377
34	130	135	392	152	130	377
36	125	130	377	148	128	371
40	125	128	370	145	128	371
50	120	126	365	140	126	365
60	120	126	365	140	126	365

Source: Leeson, 1997; Hubbard Technical Report 10

Increase or decrease the daily feed allotment of both males and females so as to maintain their weekly standard body weights. Do not allow males or females to lose weight; their weight should increase regularly, but slowly. Leeson (1997) suggests that when males are fed with females the body weights and feed consumption will depend upon the type of exclusion system being utilized. During the early period in the breeder house (prior to 26 to 28 weeks of age) the young males may still have a small enough head to eat from the female feeders. It is difficult to determine the amount of feed that the males are consuming when they are consuming feed from the female feeders. The body weight gain for the male should be slowed down after 36 weeks of age. Leeson (1997) suggests that males can be almost totally excluded from the female feeders by using nose bars which are plastic rods inserted through the nostrils of the males. Table 19-25, developed by Leeson, lists the amount of dietary energy required by males from their own feed when totally restricted from female feeders by nose bars and also list the amount of energy required when the males are not totally excluded from female feeders. Table 19-25 also lists the energy needs of males housed at different environmental temperatures.

Improvement by Using the Dual Feeding System

According to laboratory and field tests, the dual system of feeding (or sex separate feeding) will accomplish the following:

1. From the first week on the program, the cockerel weights will be materially lower than when they are conventionally fed with the pullets. The differential will be more than 2 pounds (0.91 kilos) at 48 weeks of age and more than 3 pounds (1.36 kilos) at 60 weeks of age.
2. Compared with the old system, fertility will be better from the start of egg production, better increasing to 8% by the end of the breeding season.
3. Hatchability will increase from the start of egg production, the differential gradually increasing to 8 to 10% at the end of the breeding period.
4. About eight more broiler chicks will be produced by each female during the breeding season.
5. The cockerels will eat less of an expensive feed.
6. The cost of the additional equipment for the dual system will be paid back during one egg-production period.

19-L. BROILER BREEDER RATIONS FOR PULLETS AND HENS

Table 19-26 contains some practical industry type broiler breeder rations for starting pullets, feeding pullets during the grower/developer period, and two different breeder diets.

Table 19-26. Broiler Breeder Rations for Pullets and Hens

Ingredient		Starter (lb)	Developer (lb)	Breeder I (lb)	Breeder II (lb)
Ground yellow corn		1,326	1,424	1,399	1,372
Soybean meal (dehulled, 47.5%)		316	248	292	292
Rice mill feed		160	165	127	45
Meat and bone meal (50%)		100	100	100	100
Fish meal, Menhaden		50	25	40	25
Fat, animal-veg blend or equivalent		10			10
Limestone		20	24	127	142
Salt		6.5	7	7.5	8
DL-Methionine or equivalent		0.7	0.2	0.6	0.5
Defluorinated phosphate		7	4	2	2
Breeder vitamins		1.5	1	1	1
Liquid choline (70%)		2	1.2	1.9	1.9
Trace minerals		1	1	1	1
Vitamin and mineral supplements					
Vitamin A	(IU)	15,000,000	10,000,000	10,000,000	10,000,000
Vitamin D_3	(IU)	4,500,000	3,000,000	3,000,000	3,000,000
Vitamin E	(IU)	37,500	25,000	25,000	25,000
Vitamin K	(mg)	3,000	2,000	2,000	2,000
Vitamin B_{12}	(mg)	30	20	20	20
Thiamine	(mg)	3,000	2,000	2,000	2,000
Riboflavin	(mg)	10,500	7,000	7,000	7,000
Niacin	(mg)	37,500	25,000	25,000	25,000
Calcium pantothenate	(mg)	18,000	12,000	12,000	12,000
Pyridoxine	(g)	3,750	2,500	2,500	2,500
Folic acid	(g)	12,750	850	850	850

Biotin	(g)	225	150	150	150
Copper	(g)	1.5	1.5	1.5	1.5
Iodine	(g)	1	1	1	1
Iron	(g)	14	14	14	14
Zinc	(g)	45.4	45.4	45.4	45.4
Manganese	(g)	55	55	55	55
Selenium	(mg)	136	136.2	136.2	136.2
Calculated nutrients					
Metabolizable energy	(kcal/lb)	1,325	1,325	1,324	1,315
Protein	(%)	17.0	15.0	16.0	15.5
Lysine	(%)	0.93	0.78	0.86	0.82
Methionine	(%)	0.35	0.29	0.33	0.31
TSAA	(%)	0.65	0.56	0.62	0.59
Fat	(%)	4.07	3.66	3.35	3.75
Fiber	(%)	4.67	4.75	2.83	3.01
Calcium	(%)	1.15	1.101	3.09	3.35
Total phosphorus	(%)	0.67	0.605	0.584	0.56
Non-phytate phosphorus	(%)	0.47	0.403	0.399	0.38
Vitamins and others					
Vitamin A activity	(IU/lb)	8,494	6,067	6,048.2	6,028.8
Vitamin D_3	(IU/lb)	2,250	1,500	1,500	1,500
Vitamin E	(IU/lb)	22.25	16.2	16.16	16.06
Vitamin K	(mg/lb)	2.5	1.5	1.799	1.5
Riboflavin	(mg/lb)	5.89	4.09	4.1	4.08
Niacin, total	(mg/lb)	22.03	15.13	15.53	15.347
Pantothenic acid	(mg/lb)	11.48	8.42	8.28	8.24
Xanthophyll	(mg/lb)	5.46	5.82	5.74	5.63

Source: Dr. Richard Arnold, Southwest Nutrition Service, Dallas, TX

20

Vitamins, Minerals, and Trace Ingredients

by Craig N. Coon

Besides carbohydrates, proteins, and fats, there are many nutrients in the chicken's diet that are necessary in smaller quantities to enable the bird to live, produce meat and eggs economically, and to reproduce efficiently. The list includes the vitamins, macro- and micro-minerals, and certain other additives.

20-A. VITAMINS

Generally speaking, vitamins are organic chemical compounds that are usually not synthesized by the body cells, but are necessary for maintenance, growth, and egg production. They are required in small amounts, and when they are deficient or absent from the diet, characteristic manifestations result. Many of them are enzyme-associated.

There are 13 vitamins usually listed as necessary for the chicken; they occur in feedstuffs in varying quantities and in different combinations. Some vitamins are produced by microorganisms of the intestinal tract, one by irradiation at the area of the bird's skin, while others are manufactured synthetically. As vitamins are definite chemical compounds, commercially produced vitamins are interchangeable with those found in natural feedstuffs.

Vitamins are segregated into two groups: fat-soluble and water-soluble.

Fat-soluble vitamins. The fat-soluble vitamins are A, D, E, and K.

The fat-soluble vitamins contain only carbon, hydrogen, and oxygen, and require some body fat for their metabolism. They are often present in plants as *provitamins* which are quickly converted to the true vitamins in the body of the chicken. They are easily stored in the fat cells of the bird and excesses are excreted through the feces. Only one, vitamin K, is synthesized in the intestinal tract.

Water-soluble vitamins. The important vitamins in the water-soluble group are:

ascorbic acid (C)	pyridoxine (B_6)
thiamin (B_1)	choline
riboflavin (B_2)	biotin
pantothenic acid	folic acid (folacin)
niacin	cobalamin (B_{12})

Water-soluble vitamins contain carbon, hydrogen, and oxygen plus sulfur, cobalt, and nitrogen. They are primarily needed for the transfer of body energy. Chickens require all the known water-soluble vitamins in their diet except vitamin C. Although vitamin C is produced by chickens, there are situations such as heat stress that may produce an increased requirement above the bird's ability to produce the vitamin.

When feed contains more water-soluble vitamins than the bird needs, excesses of all but one are excreted in the urine. Vitamin B_{12} has the capacity of being stored. Those not stored must be included in the daily diet; the bird has no reservoir on which to draw.

Vitamin A

True vitamin A exists only in the animal kingdom. Its precursor, carotene, is found in the plant kingdom, and it too is fat-soluble. Carotenes are consumed as plant sources, where they undergo conversion to provitamin A, then to vitamin A, which can be stored in the body, mainly in the liver. This process is poor in young chicks. Vitamin A is essential for normal vision, growth, egg production, and reproduction.

Deficiency Symptoms

1. *Retarded growth*
2. *Weakness, ruffled feathers*
3. *Absence of liquid from the tear glands.* Xerophthalmia and blindness may result. There are cheesy exudates from the eyes in adult birds.
4. *Impaired egg production and hatchability*
5. *Increased incidence of blood spots in eggs*
6. *Lowered resistance of the bird to some diseases*

Sources of Vitamin A

Precursors of vitamin A are found in green leafy plants, alfalfa meal, yellow corn (maize), and corn gluten meal. A good, dehydrated alfalfa meal (17% protein) should contain at least 100,000 IU of vitamin A activity per lb (454 g). Broiler rations are best prepared by using an alfalfa meal containing 20% protein and a minimum of 150,000 IU of vitamin A. Yellow corn contains about 2,200 IU per lb (454 g); corn gluten meal, about 12,000 IU per lb. Vitamin A is found in many fish and animal liver meals. It is produced commercially as a synthetic product of high and variable concentrations.

Oxidation of Vitamin A

Carotenes and vitamin A are easily oxidized, reducing their potency in the feed ingredient in which they are present. To reduce this oxidation, alfalfa meals are pelleted, antioxidants are sometimes added to certain feedstuffs, and synthetic vitamin A products are manufactured as small particles, then coated with fat, oil, or wax to produce stabilized forms.

Vitamin D

This vitamin has several forms, but D_2 and D_3 are the most important. Vitamin D_3 (cholecalciferol) is utilized by birds, humans, and four-footed animals, while vitamin D_2 is of value to humans and four-footed animals. Thus, D_3 becomes essential for poultry. Vitamin D aids in the absorption of calcium and phosphorus from the intestinal tract, thus increasing the amounts of these two minerals available for bone development, and calcium for eggshell deposition.

Under natural conditions, the ultraviolet rays of sunshine or fluorescent light act on 7-dehydrocholesterol, synthesized by the bird, to produce cholecalciferol, which in turn is absorbed to become the birds only source of vitamin D. With the advent of commercial poultry production, chickens are closely confined to houses with no irradiation from the sun. Even glass does not allow the ultraviolet rays to pass. Therefore, vitamin D_3 supplements must be added to the feed.

Deficiency Symptoms

1. *Rickets:* Calcium and phosphorus are not deposited in the bones in normal amounts. The hock joints are enlarged, the ribs are "beaded," and the beak and shanks in young chicks are soft and pliable.
2. *General unthriftiness*
3. *Soft-shelled eggs*
4. *Calcium crystals on eggshells*
5. *Lowered egg production*

6. *Reduced hatchability*

Sources of Vitamin D_3

Good natural sources of vitamin D_3 are the fish liver oils. However, 7-dehydrocholesterol can be irradiated with ultraviolet light to produce cholecalciferol. This process is the basis for manufacturing commercial vitamin D products, some of which have potencies of 400,000 IU per gram.

Vitamin E

Vitamin E (tocopherol) is necessary for maintaining the integrity of cellular membranes in the body and for blood formation. When the diet is deficient in vitamin E, there are several manifestations, which can vary as other dietary components can affect the requirement for this vitamin. The measure of alpha-tocopherol is in IU (International Units).

Deficiency Symptoms

1. *Nutritional encephalomalacia,* evidenced by a twisted neck, prostration, curled toes, and "crazy chick" disease.
2. *Exudative diathesis:* There is some indication that selenium is involved, as additions of this mineral have been shown to reduce this difficulty when it results from a lack of adequate vitamin E.
3. *Male sterility:* With prolonged deficiencies, the sterility may be permanent as a result of a degeneration of the testes.
4. *Nonproduction in the female:* Birds stop laying on diets low in tocopherol, but cessation is not permanent; additions of vitamin E to the diet restore the bird to normal egg production.
5. *Embryonic mortality:* There is a circulatory failure at about the fourth day of incubation.

Sources of Vitamin E

Whole grains and alfalfa meal are the best natural sources of vitamin E. Synthetic tocopherols are available and these are usually added to chick starter and breeder rations.

Destruction of Vitamin E

This vitamin is easily oxidized and the process is increased when minerals and unsaturated fatty acids are present in the feed. An antioxidant should be used in the ration when vitamin E is added.

Vitamin K

Vitamin K, the antihemorrhagic vitamin, is necessary for the synthesis of prothrombin, a chemical necessary for blood clotting. When vitamin K is low or lacking, the blood vessels rupture, causing excessive bleeding. There are several types of vitamin K, all of which are active but at varying potency; three of these are:

1. Vitamin K_1 (phylloquinone): Present in plant tissues.
2. Vitamin K_2 (menaquinone): Synthesized in small amounts in the intestinal tract, and found in fish meal.
3. Vitamin K_3 (menadione).

Deficiency Symptoms

There is a hemorrhagic condition. Hemorrhages that are first pinpoint in size, and then become enlarged later occur on the flesh. If the skin is removed, bleeding may be seen on the breast, thighs, and ribs.

Sources of Vitamin K

A good natural source of vitamin K is alfalfa meal. Meat scrap and fish meal are fair sources. A synthetic water-soluble form of vitamin K_3 is available.

Thiamin (B_1)

Thiamin is necessary to stimulate appetite, to form certain enzymes necessary for digestion, and to prevent nervous disorders that culminate in *polyneuritis*. However, as many feed ingredients carry abundant supplies of this vitamin, the symptoms are seldom seen.

Sources of Thiamin

Thiamin is relatively abundant in cereal grains, mill by-products, vegetable oil meals, and alfalfa meals. It is also available as synthetic thiamin.

Riboflavin (B_2)

This vitamin is of major importance, not only because of its effect on body processes, but because it generally is inadequate in rations composed of ordinary feedstuffs. Riboflavin is a part of over a dozen enzymes needed by all living cells. Commercially it is synthetically manufactured by a process of requiring fermentation. Most rations contain added riboflavin.

Deficiency Symptoms

1. *Curled-toe paralysis:* The toes curl and sometimes the legs are affected producing paralysis.
2. *Poor hatchability:* Non-hatched embryos are dwarfed, having abnormal down called "clubbed down."

Sources of Riboflavin

Although fish meals, fish solubles, alfalfa meal, and milk products are relatively high in riboflavin, their limited use in most poultry feed formulas means that practically all poultry rations must include supplemental (synthetic) riboflavin.

Pantothenic Acid

Pantothenic acid in combination with coenzyme A is important for the metabolism of protein, carbohydrates and fats required by all cells. It is relatively unstable. The requirements of young and growing chicks for pantothenic acid are high.

Deficiency Symptoms

1. *Retarded growth in young chicks*
2. *Ruffled feathers*
3. *Granulated and stuck eyelids in young chicks*
4. *Scabs at the corner of the mouth*
5. *Dermatitis of the feet*
6. *Lowered egg production*
7. *Lowered hatchability*

Under field conditions, diets are seldom so low in pantothenic acid as to cause major exemplification of many of the symptoms.

Sources of Pantothenic Acid

Several feedstuffs are good sources of pantothenic acid, including liver meal, peanut meal, milk products, mill by-products, and alfalfa meal. Calcium pantothenate is manufactured commercially to serve as the supplementary source of this vitamin. It is added to most rations.

Niacin (Nicotinic Acid)

This vitamin is important for the metabolism of carbohydrates, proteins, and lipids. Nicotinic acid is an important part of two enzymes, and is found as nicotinic acid in the plant kingdom and nicotinamide in the animal kingdom. The amino acid, tryptophan, is a precursor of niacin, but the conversion is very poor (40 to 1), yet must be considered when adding niacin to feed.

Cereal grains contain little niacin, which is in a bound form, unavailable to the chicken. High-energy diets, young chickens, and chickens under stress have a higher niacin requirement. In the presence of adequate tryptophan (0.215%), and diets of medium to high energy when less feed is consumed, the niacin requirement is about 30 mg/kg of feed for chicks and broilers, 10 mg/kg for layers, and 15 mg/kg for broiler breeders.

Deficiency Symptoms

1. *Swollen hocks, similar to perosis, but the tendon seldom slips from the condyle*
2. *Reduced growth*
3. *Inflamed tongue and mouth (black tongue)*
4. *Scaly skin and feet, ruffled feathers*
5. *Hysteria*

Sources of Niacin

Yeast, fish meal, fish solubles, distiller solubles, and a synthetically produced product.

Pyridoxine (B_6)

This vitamin plays a part in protein, carbohydrate, and lipid metabolism. It forms a part of several enzymes and is a muscle conditioner.

Deficiency Symptoms

Diets low in pyridoxine show reduced chick growth.

Sources of Pyridoxine

Most natural feed ingredients contain some pyridoxine so dietary deficiencies are very uncommon.

Choline

The chick's requirement for choline is high. Choline is needed for the synthesis of phospholipids in the liver. Many times choline is not considered a true vitamin. Choline may be synthesized by the chick, but the amounts are small and usually inadequate. Lesser amounts of dietary choline are needed as the chicken ages.

The vitamin has a great many functions in the body: It helps in fat movement in the bloodstream; it has a sparing action on methionine; it aids in growth; it prevents a type of slipped tendon; and it helps to reduce excessive fat deposits in the liver.

Deficiency Symptoms

1. *Perosis*
2. *Fatty liver (syndrome)*
3. *Retarded growth*

Sources of Choline

Fish meal, fish solubles, yeast, liver meal, soybean oil meal, and distiller solubles are good sources, but a choline chloride product is the usual feed supplement (contains 25 to 50% choline).

Biotin

Biotin may seem adequate in the diet when composed of normal feedstuffs, however, only about half is available to the chicken. The availability of biotin in wheat diets is very low. Thus, it is possible to have deficiencies when some diets are fed. At times, biotin may be synthesized in the intestinal tract, but this process can be highly variable.

Deficiency Symptoms

1. *Scaly dermatitis*
2. *Mild perosis*
3. *Retarded growth*
4. *Reduced hatchability*

Sources of Biotin

Good sources are alfalfa meal, yeast, liver meal, and soybean oil meal.

Folic Acid

Folic acid, an antianemia factor, is a complicated chemical compound necessary for many physiological functions: growth, muscle formation, blood formation, and feather growth. Diets are seldom low in this vitamin.

Deficiency Symptoms

1. *Depressed growth*
2. *Poor feathering, with feathers lacking pigment*
3. *Anemia*
4. *Necrosis*
5. *Dermatitis*
6. *Perosis*
7. *Increased embryonic mortality*

Sources of Folic Acid

Good sources are alfalfa meal, yeast, liver meal, and soybean oil meal.

Vitamin B_{12} (Cobalamin)

This vitamin is associated almost entirely with ingredients of animal and fish origin. Plant products contain little or no vitamin B_{12}. It is synthesized by microorganisms of the intestinal tract as a cobalt-containing compound. However, the process is inadequate to supply the amount needed by most chickens. Lack of a normal amount causes anemia (pernicious anemia in humans). Most rations contain a supplemental supply, particularly chick-starting and breeder diets. The bird's own droppings are a source of vitamin B_{12}. Therefore, birds raised on wire are more likely to show a deficiency than those kept on a litter floor.

Deficiency Symptoms

1. *Anemia*
2. *Reduced chicken growth*
3. *Poor hatchability*
4. *Fatty liver*

Sources of Vitamin B_{12} (Cobalamin)

Meat scrap, fish meal, fish solubles, and poultry manure.

Ascorbic Acid (Vitamin C)

In all probability vitamin C is not required in the feed of chickens, for birds synthesize an adequate amount. Ascorbic acid helps embryo growth, aids bone development in young chicks, and stabilizes body fat. Ascorbic acid is sometimes added to poultry feed in heat stress situations to reduce mortality.

Vitamin Fortification of Diets

Vitamins are found in small amounts in many feedstuffs but the supplementation of diets with different feedstuffs to supply the needed vitamin requirements would be impractical. Current dietary formulations emphasize both quality and quantity of metabolizable energy and digestible amino acids for high-producing birds so formulation space cannot be wasted. Therefore, nutritionists will need to add a vitamin supplement or premix to all poultry diets. The amount of vitamin fortification used by the poultry industry is different for each country because of natural ingredient variations, government policies on maximum levels, and different concerns about welfare. Table 20-1 shows the trend of the industry for increasing vitamin fortification for broilers since 1977. Table 20-2 shows the potential fortification levels used in the industry when using either wheat- or corn-based diets. The concern nutritionists have is the vitamin requirements suggested by the National Research Council, 1994 (NRC), haven't changed much over the past 40 years yet growth rate of broilers and average egg mass for layers have been steadily increasing. The suggested requirements by NRC do not take into consideration (1) the effect of environment on vitamin stability and retention, (2) the effect of stress on requirements, and (3) the margins of safety that nutritionists often use to equalize feedstuff variability in vitamin concentration. There is also a con-

Table 20-1. Average Commercial Broiler Starter Vitamin Fortification Relative to NRC Recommendations[1]

	Year								
	1977			1984			1994		
Vitamin	NRC Total	NRC–Feed Ingr. Suppl.[2]	Ind. Avg. Suppl.[3]	NRC Total	NRC–Feed Ingr. Suppl.	Ind. Avg. Suppl.	NRC Total	NRC–Feed Ingr. Suppl.	Ind. Avg. Suppl.
Vit. A, IU	1,500	—	5,511	1,500	—	7,165	1,500	331	8,840
Vit. D_3, IU	200	200	1,543	200	200	2,205	200	200	2,811
Vit. E, IU	10.00	—	7.16	10.00	—	9.92	10.00	1.21	17.92
Niacin, mg	27.00	14.70	30.86	27.00	16.64	34.17	35.00	24.53	45.76
Pantothenic acid, mg	10.00	6.06	8.82	10.00	6.94	9.92	10.00	7.05	12.04
Riboflavin, mg	3.60	1.10	5.51	3.60	1.21	6.06	3.60	1.32	7.10
Menadione, mg	0.50	0.50	1.10	0.50	0.50	1.32	0.50	0.50	1.85
Thiamine, mg	1.80	—	1.10	1.80	—	1.32	1.80	—	1.77
Pyridoxine, mg	3.00	0.19	1.18	3.00	0.22	1.43	3.50	0.85	2.62
Folic, mg	0.55	—	0.33	0.55	—	0.50	0.55	—	0.86
Biotin, mg	0.15	0.03	0.02	0.15	0.03	0.04	0.15	0.03	0.08
Vit. B_{12}, mg	0.009	0.009	0.010	0.009	0.009	0.011	0.010	0.010	0.013

[1] Units/kg
[2] Supplements required in a CORN/SBOM/meat-meal diet to meet NRC recommendations
[3] Typical industry supplementation of similar diets
Source: Adapted from McKnight, 1994, NRC, 1977, 1984 and 1994

Table 20-2. Possible Vitamin Fortification Levels for Poultry Diets Based on Wheat or Corn (per kg of diet)

Bird Type	Broiler				Broiler		Hen	
Diet Type	Starter		Finisher		Breeding		Laying	
Cereal Base	Wheat	Corn	Wheat	Corn	Wheat	Corn	Wheat	Corn
Vitamin A (IU)	10,000	9,000	10,000	9,000	12,000	1,100	7,000	6,000
Vitamin D_3 (IU)	5,000	5,000	5,000	5,000	3,000	3,000	3,000	3,000
Vitamin E (mg)	30	30	30	30	30	30	5	5
Vitamin K (mg MSB)	3	3	3	3	3	3	2	2
Thiamin (mg)	1	1	1	1	1	1	—	—
Riboflavin (mg)	6	6	5	5	12	12	3	3
Niacin (mg)	60	65	30	35	12	17	5	10
Pantothenic acid (mg)	12	14	10	12	12	15	5	7
Pyridoxine (mg)	3	1	2	1	6	3	1	—
Biotin (mg)	180	60	100	20	200	150	—	—
Cyanocobalamin (mg)	12	12	10	10	15	15	10	10
Folic acid (mg)	2.5	2.5	2.5	2.5	2	2	—	—
Choline (mg)	250	450	—	100	—	200	—	—

Source: Whitehead, 1994

cern with some of the new commercial expander and pelleting systems that vitamin retention of mixed feeds are decreased with increasing heat levels. Also, during the past decade there has been increased research indicating the levels of vitamins and other nutrients that are adequate for growth, egg production, and feed conversion are not adequate for normal immunity and optimum disease resistance.

20-B. MINERALS

Besides proteins, carbohydrates, fats, and certain vitamins, many other elements form a part of the bird's nutritional requirements. In most cases, the necessary quantity of each is small, often infinitesimal. Many have interrelationships with other nutrients. In some instances, trace amounts are necessary, but excesses are toxic. Although most of these elements must be added to the diet in their inorganic form, sometimes organic forms are important sources. High percentages of some are absorbed through the intestinal wall; in other instances, the amount is small.

Minerals are divided into macrominerals and micro or trace minerals. The macrominerals are *calcium, phosphorus, sodium, potassium, chloride,* and *magnesium.* Microminerals that are needed for poultry are *zinc, manganese, iron, iodine, selenium, copper,* and *molybdenum.* Trace elements chromium and fluoride have been shown to be essential in other species but the nutritional significance for poultry remains unclear. There are a few research reports suggesting that *nickel, vanadium, tin,* and *silicon* are essential for poultry, but the needed amounts of the elements are so small that a practical concern for these nutrients seems unlikely.

Feed ingredients, such as cereal grains and oil seed protein sources, usually provide adequate amounts of magnesium, potassium, and molybdenum for poultry. Magnesium levels in some limestone sources (Dolomite) need to be evaluated to ensure that an excess isn't fed. Molybdenum is also an element where the concern is primarily focused on not feeding excessive amounts. Molybdenum has been shown to interfere with copper bioavailability in the tissue. The remaining elements that are considered essential for poultry are found in natural feed ingredients but the quantities may be low or the bioavailability of the element may be of concern.

A number of factors have been shown to affect the requirement of different minerals. Phytate levels in feeds not only keep phosphorus from being available but also form complexes in the digestive tract with calcium, zinc, and iron. Research has indicated that phytate in feeds may also interfere with the availability of copper and manganese. Dietary fiber, such as cellulose and lignins in feedstuffs, have also been shown to bind to copper and zinc and decrease their retention. Dietary fat has been shown to form insoluble soaps with minerals such as calcium, decreasing the absorption of the mineral. The disease coccidiosis may also decrease the absorption of minerals by damaging the villi and microvilli in the intestine. The requirement of minerals are greatly affected by the growth rate and feed intake of poultry, which is why mineral nutrition like other nutrients is often linked to dietary energy levels.

Calcium

Calcium is primarily necessary for bone and eggshell formation but also has certain other functions. The mineral is deposited in bone mainly as calcium phosphate, but there is some calcium carbonate. Eggshells are almost entirely calcium carbonate.

Calcium and phosphorus are important for the development of the skeletal system as well as for many biochemical functions. Vitamin 1,25 dihydroxy cholecalciferol, which is a metabolite of vitamin D_3, is needed for calcium absorption in the gastrointestinal tract and also for resorption of calcium from bones. The potential exists for a laying hen weighing only 1,500 g to produce approximately 460 g of eggshell calcium (260 eggs $\times$ 1.77 g), or one-third of her average body weight, for a laying period from 18 to 65 weeks of age. Laying hens need at least 50% of their calcium as a less soluble large particle (limestone or oyster shells) to provide calcium during the dark hours when layers are forming eggshells. The larger particles of calcium will stay in the gizzard until the acid from the proventriculus solubilizes the calcium particles. The calcium requirements of growing birds are provided by feeding limestone along with dicalcium phosphate or defluorinated phosphate for both calcium and phosphorus requirements. Rendered animal by-products and fish meal also provide a good source of these two important elements.

Eggshell Formation

At the approach of egg production (sexual maturity) estrogens are released from the ova in greater abundance, which in turn increase the level of blood calcium. The parathyroid secretes hormones to keep the blood level of calcium constant. Calcium is then deposited in the medullary bone to be released later for eggshell formation. The amount deposited is not related to the amount of calcium fed during the growing period. In fact, too much calcium during the growing period is detrimental to maximum egg production, probably because of injury to the developing parathyroid gland. An increase in dietary calcium is needed only about 1 to 2 weeks before the first egg is laid. Feeding more oyster shell or other forms of calcium during this period is the accepted field practice. Once egg production begins, the source of calcium for eggshell formation comes from both the dietary calcium and the medullary bone (see *Feeding Commercial Egg-Type Layers,* Chapter 18).

Phosphorus

Although a major constituent of the blood, phosphorus plays an important part in metabolic processes and is found in the cells, enzymes, and other body compounds. Not all the phosphorus in the feed is available to the chicken. Normally, the phosphorus content of the ration is represented by two measurements:

1. Total phosphorus
2. Non-phytate phosphorus

Phosphorus derived from plant sources is approximately 2/3 phytate phosphorus which is less available than other sources. The amount of phytate phosphorus that is available to chickens depends on the age, calcium:phosphorus ratio, and type of bird. Research suggests that the availability of phytate phosphorus ranges from 0 to 30%. The retention of non-phytate phosphorus in plant sources may be around 60%.

Availability of Phosphorus

Biological values are determined with 3-week-old chick feeding studies with diets containing several levels of the test phosphorus source and also with diets containing different levels of a standard phosphorus source. The weight gain and bone ash response of chicks fed the test phosphorus feeds are compared to the same responses in chicks fed the standard phosphorus source. In general, the feed grade monocalcium phosphates have a 5% higher biological value than dicalcium phosphates, and the latter have a 5% higher biological value than defluorinated phosphates. Biological values can be quite variable and when possible, the biological value should be determined for individual sources of phosphorus. An example of biological values for several commonly used phosphates with broilers (Huyghebaert, et al., 1980) are as follows:

Phosphate Source	*Biological Value*
Hydrated dicalcium phosphate	97
Monocalcium phosphate	96
Anhydrous dicalcium phosphate	86
Defluorinated phosphate	96
Meat and bone meal	90
Disodium phosphate	100 (reference standard)

Calcium: Phosphorus Ratio

Not only must the diet contain minimal amounts of calcium and phosphorus but there must also be a correct ratio between the two minerals, particularly for starting and growing rations. A ratio of 1.5 to 2.0 parts of calcium to one part of *total phosphorus* is optimum for starting and growing rations, with 2.5 to 3.5 being too great and generally producing rickets.

A better meaning is given when it is the *dietary* ratio of calcium to *non-phytate phosphorus*. Examples of the latter are given below:

Calcium: Available Phosphorus	
Type of Ration	*Ratio*
Starting	2.2:1
Growing	2.5:1
Laying	9.0:1

Calcium, Phosphorus, and Vitamin D

Not only are calcium and phosphorus essential dietary minerals for the production of bone, and calcium for the deposition of eggshell material, but vitamin D also plays an important part in the processes. Evidently, vitamin D helps to form a protein in the intestinal tract to keep the calcium in solution so that it can pass the intestinal wall and reach the cells. Vitamin

D also aids in other ways to get the calcium to those areas of the body that need it.

Amount of Calcium and Phosphorus in the Diet

The dietary amounts of these two minerals must be maintained within close limits according to the age and type of bird involved. The following are typical examples:

		% Phosphorus	
Type of Ration	*% Calcium*	*Total*	*Non-Phytate*
Starting	0.9	0.60	0.40
Growing	0.9	0.60	0.35
Laying, egg-type	3.5–4.0	0.50	0.42
Laying, meat-type	2.9–3.1	0.50	0.42

Phosphorus and Phytase

There is some natural phytase in feedstuffs but because of the short digestive tract in poultry the phytase breakdown of phytin phosphorus is minimal. Commercially prepared phytase is currently becoming more economically feasible to add to poultry diets. The ability to utilize both phytin and non-phytin phosphorus will allow nutritionists to feed needed phosphorus levels while minimizing phosphorus excretion, which is an increasing environmental concern. Phosphorus requirements are normally expressed as non-phytate phosphorus, which is considered by many nutritionists to be the same as available phosphorus. Retention of minerals is a more accurate description when describing mineral digestibility or availability because a significant percentage of minerals in the excreta are from endogenous losses from the intestine and kidney (see *Feeding Commercial Egg-Type Layers,* Chapter 18).

Trace Minerals

Trace minerals are important as cofactors in many metabolic enzymes or as structural components of proteins like hemoglobin. They are primarily provided to poultry in the form of a premix. Adequate trace minerals are added to supply at least 100% of suggested NRC requirements, and nutritionists concerned with stress will often increase the fortification of trace minerals from 2 to 3 times NRC levels. Most mineral premixes contain zinc, manganese, copper, iron, and iodine. Since selenium toxicity is a major concern for poultry, a separate premix containing 0.06% selenium can be added at a level of 1 pound per ton to provide 0.3 ppm in the diet.

Trace elements are just as physiologically important as the macroelements but the daily amounts needed are much less. It is important to remember that in some cases adverse interactions can occur between various minerals. For instance, some elements interfere with the absorption and utilization of others while in other situations increasing the tissue level of a trace element may actually aid in the absorption or utilization of others. *Nutritionists should keep the vitamin premix separate from the trace mineral premix because the minerals will affect the stability of certain vitamins.*

Sodium, Chlorine, and Potassium

These three elements are involved with acid-base equilibria in the body. Natural feedstuffs usually require supplemental feeding of salt (NACL) to satisfy the bird's requirement for sodium and chlorine. The amount of salt added to the ration should seldom be over 0.25 to 0.35%. Too much salt produces a laxative effect. Additions over 8% are lethal.

Potassium is a necessity, but ordinary poultry rations are seldom deficient in this element. No supplementation is given.

Sulfur

Sulfur is a part of two amino acids, cystine and methionine, which are often in short supply in natural feedstuff protein. The naturally occurring form, cysteine, is readily oxidized in the body to form cystine. Sulfur is important to certain enzymes and hormones. Cystine and methionine are often grouped together as total sulfur-containing amino acids (TSAA) because of their complementary action.

Note: Chickens have specific TSAA and methionine requirements, but extra cystine will not substitute for a deficiency of methionine.

Iodine

Iodine has a relationship to the thyroid and its hormone, thyroxine. When the ration is low in iodine, the thyroid increases in size, producing a goiter. Besides being a part of many metabolic functions in growing and adult birds, iodine is needed by the developing embryo. When the iodine content of hatching eggs is low, hatchability is reduced. Iodine usually is added to the diet as potassium iodide in iodized salt.

Fluoride

Fluorine, a dietary necessity of the chicken, is associated with proper bone development. Small amounts are required. In many areas the requirement is supplied naturally by the fluorine in the water the chicken drinks, since many soils contain an abundance of fluorine which finds its way into the water supply. But in other areas the water supply is almost void of fluorides. Fluorine is usually not added to poultry rations because it occurs with many minerals such as limestones and phosphates. Before being fed to chickens, these materials must be processed to lower their fluorine content. The products are sold as *defluorinated rock phosphate* or *high calcium limestone.* Most of these have a fluorine content of less than 0.5% and can be safely fed to chickens.

Iron and Copper

Nutritional anemia occurs when there are deficiencies of copper and iron. The red blood cells contain iron. The mineral also is needed to pigment the feathers of certain breeds of chickens. Copper is necessary for iron utilization when hemoglobin is formed; therefore, if absent from the diet, anemia results. The amount of iron and copper needed in the diet of the chicken is quite specific; excesses may be toxic. About 5 to 10 times as much iron as copper is required. Usually, only small amounts, if any, are ever added to commercial feeds.

Manganese

The chief function of manganese is to prevent perosis (slipped tendon), a condition where the hock joint becomes enlarged and the gastrocnemius tendon at this location slips from the condyle, twisting the shank to one side. But manganese is also needed for normal growth, eggshell deposition, egg production, hatchability, and to prevent ataxia. Since all rations composed of normal feedstuffs are deficient, manganese is added to the feed as manganese sulfate or manganous oxide. A 70% feeding grade is commonly used. From 30 to 50 g of manganese are added to a ton (2,000 lb) of feed to prevent perosis and 50 to 75 g to increase eggshell strength.

Magnesium

Magnesium is one of the essential trace elements of nutrition. When it is absent from the diet, chicks grow slowly, exhibit convulsions, and may eventually die. Deficiencies in laying rations produce a rapid drop in egg production. Calcium is poorly utilized in the absence of magnesium. An excess of magnesium in the diet is about as detrimental as too little. One pronounced effect of an excess is wet droppings. Some limestones (the dolomites) contain a high percentage of magnesium and are to be avoided.

Selenium

This element is required in small amounts by the chicken. Not only is it essential in itself, but it reduces some of the symptoms of vitamin E deficiency. *Exudative diathesis* is one evidence of the lack of vitamin E, yet selenium has been shown to be capable of eliminating the condition. It does not, however, affect *encephalomalacia,* another condition often present when vitamin E is lacking. More vitamin E is needed only when selenium is deficient.

The selenium content of plants is closely related to the amount of selenium in the soil on which the plants grow. Where the soil is deficient, the plants and their seeds are deficient. Excesses of selenium are toxic, reducing growth, hatchability, and increasing the incidence of embryonic malformations. The optimal dietary level of selenium is 0.1 ppm for chickens up to 16 weeks of age. Diets low in selenium decrease egg production and hatchability and produce anemia. Sodium selenite is a compound that can supply selenium; 1 lb (454 g) to 2,250 lb (1,023 kg) of ration will supply 0.1 ppm.

Caution. In some countries it is illegal to add selenium to poultry feed because of its residual effect in poultry meat and eggs. In others, the quantity fed is rigidly controlled. In some it cannot be fed to birds producing eggs for human consumption. Check with authorities before making additions. It will take 4 weeks after the removal of a selenium supplement in the ration for selenium to disappear from the bird's body tissues and eggs.

Vanadium

Too little vanadium is of no consequence to chickens, but too much is harmful. Some types of dicalcium phosphate and defluorinated rock phosphate have reduced the quality of egg albumen, with the difficulty being traced to high amounts of vanadium in these minerals. Either should not contribute more than 4 ppm of vanadium to diets of laying hens.

Zinc

A small amount of zinc is needed by the chicken to foster good egg production, hatchability, proper feathering, and growth. As feedstuffs generally are low in zinc, this mineral is usually added to the ration as zinc carbonate (about 57% zinc) or zinc oxide (about 80.5% zinc). Normally, from 15 to 30 g of zinc are added to 1 ton (2,000 lb) of feed.

20-C. AMINO ACIDS

Of the 22 amino acids, six are more likely to be deficient in most poultry rations than the others. The other amino acids are usually adequate from combinations of feedstuffs found in most poultry rations, or by internal synthesis. The six likely deficient amino acids are *methionine, cystine, lysine, tryptophan, threonine* and *arginine.*

When a poultry ration is low in one or more of these six amino acids, supplements of feed grade pure amino acids must be added to the feed formula to make up the deficiencies. Since methionine is most often lacking, most formulas call for supplementation in pure form as DL-methionine or methionine hydroxyanalogue. One cause of methionine deficiency is the fact that large amounts of vegetable protein supplements are now used in feeds, plus low levels of the animal and fish proteins. Lysine and cystine often are inadequate when normal feedstuffs are used in poultry feeds.

Proper amino acid formulation requires minimum levels of all, with little excess of any. This is practically impossible; there is some waste. Usually the value of the protein portion of the ration is determined by the limiting amino acid. Large amounts of others are usually of no value to the limiting one.

Some amino acid characteristics of the major feed ingredients are as follows:

Barley	low in threonine and lysine
Corn (maize)	low in lysine and tryptophan
Milo (sorghum)	low in lysine
Soybean oil meal	low in methionine; high in lysine
Corn gluten meal	low in lysine; high in leucine

Amino Acid Relationships

As dietary protein is expensive, most laying rations have but a minimum of protein in them. Anything that reduces the daily consumption of feed per bird reduces the daily intake of protein, and egg size may suffer. If continued for some time, egg production will also be affected.

But more important than total protein is the daily intake of amino acids necessary to produce the protein of the egg yolk and albumen. If only one such essential amino acid is deficient, the production of egg protein will be reduced, as will egg size. The amino acids most often involved and lacking for laying hens are lysine, cystine, methionine, isoleucine, and valine.

Normally, 100 laying Leghorn hens will consume about 22 to 24 lbs (10 to 11 kg) of feed per day. At peak egg production of 90% or over for Leghorns, the bird average intake of the critical amino acids should be about 720 mg of lysine, 360 mg of methionine, and 650 mg of sulfur amino acids per day. Normally, a bird eating 17 g of quality protein per day would consume these levels of amino acids, yet 16 g or less are often shown to be adequate. But how?

On an individual bird basis, when the flock peaks in egg production some birds are just beginning to lay, and consequently are eating less feed and amino acids. The precocious individuals have a high demand for feed because of higher egg production, and thereby eat more, resulting in the consumption of more protein and amino acids. On the average, however, it appears that all birds are consuming the average amount of protein.

20-D. OTHER FEED CONSTITUENTS

There are other items added to feed in relatively small amounts. Some are directly associated with metabolism; others are not.

Antioxidants

Fat rancidity in a feed tends to destroy the fat-soluble vitamins A, D, and E. Most of this oxidation of the fats may be prevented by adding an antioxidant to the mixture. The two usually added at 0.0125% level are ethoxyquin and butylated hydroxytoluene (BHT).

Arsenicals

Supplements of 3-nitro-4-hydroxyphenylarsonic acid, arsanilic acid, or sodium arsanilate are added to broiler diets to promote growth and to improve the yellow color of the skin and shanks. Although the metal is deposited in the tissues, the amount is very small, and once the arsenical is withdrawn from the feed the tissues are depleted. Any dietary source should be withdrawn from the feed at least 5 days before broilers are slaughtered. Arsenicals in combination with certain other drugs should not be used in laying rations.

Caution. Be sure to check with authorities before using any arsenical. Its use is illegal in many countries.

Coccidiostats

Coccidiostats, both ionophore and non-ionophore compounds, are added to feed for growing poultry reared on the floor to prevent a build-up of intestinal coccida causing coccidiosis (see *Diseases of the Chicken,* Chapter 27). Some coccidiostats must be removed from the feed 5 to 7 days prior to marketing meat birds. The time period of feeding diets without coccidiostats at the end of the growing period is called the *withdrawal period.* Nutritionists should be aware of government regulations in different countries concerning the use of antimicrobials as growth promotants and also regarding the regulations on feeding different types of coccidiostats. Levels of these coccidiostats should be those designated by the manufacturer.

Electrolytes

Body water has substances called electrolytes dissolved in it. They may be divided into two groups:

1. Extracellular (sodium, chloride, and bicarbonate)
2. Intracellular (potassium and phosphate)

The electrolytes regulate enzyme activity, regulate the osmolarity of some body fluids, and help control the body pH.

In mammals, the loss of water from sweating will upset the electrolyte balance, but this is not a factor with chickens because they have no sweat glands. However, during certain diseases and other stresses, there will be excessive water and mineral loss from the body. Under these circumstances some remedial physiological improvement may be shown through the addition of electrolytes to the diet to maintain a cation and anion balance.

Enzymes

Commercial enzymes are often added to poultry diets to enhance the digestion of carbohydrates and proteins. The enzymes used are primarily enzymes not found in the gastrointestinal tract of poultry. The enzymes help digest specific pentosan structures called *xylans* in wheat and also hydrolyze β-glucans in barley. Several commercial enzymes are "cocktails" or combinations of enzymes that may also contain proteases to help with the digestion of proteins. These enzymes can be added for specific types of cereal grains that are in the rations. Phytase is also a very important enzyme that is being added to poultry feeds to help break down phytate phosphorus in order for the phosphorus to be utilized.

Growth Promotants

Low quantities of antibiotics are used as a growth promotant as they potentially suppress disease, decrease microbes that are utilizing essential nutrients and producing excess ammonia in the intestine, and increase feed and water intake. In general, the use of antibiotics added to feed on a sub-therapeutic level as growth promotants has been reduced because of growing concern that this practice may cause bacteria resistant to antibiotics for the human population. Their action is indirect; they alter the microbial environment of the intestines, thereby increasing the availability of certain other feed constituents. A common growth promotant used in the industry that has no opportunity of causing bacteria resistance is zinc bacitracin.

Mold Inhibitors and Mycotoxin Binders

Organic acids such as propionic acid are often added to feeds to control mold growth. Some of the organic acids may cause corrosion of metal feeders and feed lines. Mycotoxin binders such as aluminosilicates or zeolites are often added to the feed as an absorbant of the mycotoxin. The aluminosilicates have no nutritional value.

Pellet Binders

Pellets processed from some mash mixtures tend to crumble. To increase their hardness certain binders are commonly added.

Xanthophylls

Xanthophylls compose a group of chemicals within a larger group of plant pigments known as *carotenoids.* The xanthophylls impart yellow color to the fat deposits and skin of the birds and to egg yolks. Lutein is a natural xanthophyll in alfalfa meal and in marigold petals.

The amount of carotenoids added to mixed feed depends upon the consumer demand in each area. Many parts of the United States are accustomed to chickens being fed grain sorghum, barley, or wheat and are more conditioned to seeing less yellow pigmentation in broiler skin and egg yolks. Corn, corn gluten meal, and dehydrated alfalfa meal are good natural sources of xanthophylls. Once dehydrated alfalfa meal was added to feeds to provide xanthophylls and vitamin K, there was no space for low energy ingredients such as dehydrated alfalfa meal. Marigold meal or extracts of marigold meal are important sources of xanthophylls when large concentrations are required. Synthetic pigments are also being used in poultry feeds *outside* the United States.

21

Feed Formulation and the Computer

by Gene M. Pesti

One of the first applications of computers to agricultural problems was in least-cost feed formulation. *Linear Programs* were developed that could choose the combination of ingredients that would satisfy the animal's nutrient needs at the least cost. Least-cost feed formulation programs are management tools that make fast responses to changing prices possible. Feed ingredient prices change daily. Feed formulation programs allow managers to quickly calculate the effects of potential formulation changes on the cost of the finished feed. Changes that may be evaluated include substituting one ingredient for another and feeding different nutrient levels.

21-A. THE BASIC FEED FORMULATION PROBLEM

The essential elements of a typical least-cost linear feed formulation problem are shown in Table 21-1. It is necessary to give the computer:

- a list of possible ingredients to use and their nutrient compositions.
- the cost of each ingredient in the same units, in this case $/Imperial ton, the first row of the matrix.
- any limits for each ingredient, the minimum and maximum.
- the nutrient limits, in this case all nutrient limits are minimums although it is possible to include maximums as well.

Figure 21-1. Nutritionist Formulating Feed

The solution. The weight of each ingredient is included with a value of 1.00 and the "Weight" row is fixed at 1.00 to keep the nutrients in proper proportion to each other. This row is hidden in most commercial programs. When a formulation problem is solved, the least cost combination of ingredients that meet the nutrient limits is reported. In Table 21-1 the least-cost solution to the problem is in the "Quantity" row: the feed will contain 0.6744 units of corn for each 1.00 units of weight or 67.44% corn, etc. The total cost of one ton of feed meeting these ingredient and nutrient restrictions is presented; in this case the feed costs $163.28 per ton. The "Final Values" are the actual nutrient levels in the least-cost solution, not necessarily the nutrient minimum levels specified.

The underlying assumption. The least-cost feed formulation is that the minimum nutrient "requirements" for the birds to be fed are known. It is usually further assumed that supplying minimum nutrient requirements will result in maximum profits. Neither of these assumptions is necessarily true: Minimum requirements are usually the lowest amounts of nutrients that result in maximum growth or performance. However, maximum performance may not always result in maximum profits. The point of economic efficiency is dependent on actual performance and the costs of inputs and value of outputs. The chicken's responses to different nutritional intakes are dependent on the genetic

Table 21-1. Example of a Least-Cost Feed Formulation Matrix with Limits and Solutions for a Broiler Finisher

	Linear Feed Mix Problem											
	Corn	SBM-48	Poultry Meal	Poultry Fat	Limestone	Dical. Phos.	Salt	V & M Premixes	DL-Met	Nutrient Min's	Nutrient Max's	Final Values
Cost ($/ton)	128	220	280	286	34	220	55	2,136	2,200			
Weight	1.000	1.000	1.000	1.000	1.000	1.000	1.000	1.000	1.000	1.0000	1.0000	1.0000
M.E. (kcal/g)	3.430	2.440	2.670	8.200					3.606	3.1500		3.1500
Protein (%)	8.800	48.500	58.000						57.520	19.8000		19.8000
Calcium (%)	0.020	0.270	3.000		38.000	21.300	0.300			0.8500		0.8500
Avail. phos. (%)	0.084	0.240	1.700			21.300				0.4250		0.4250
Sodium (%)	0.010	0.300	0.400		0.050	0.060	39.000			0.1835		0.2505
Arginine (%)	0.500	3.680	4.000							1.2395		1.3702
Lysine (%)	0.240	3.180	2.700							1.0336		1.0336
Met + cys (%)	0.350	1.450	1.690						98.000	0.7999		0.7999
Threonine (%)	0.390	1.910	2.000							0.6455		0.7969
Tryptophan (%)	0.090	0.670	0.530							0.1980		0.2428
Limits—Min			0.0300				0.0040	0.0010				
Limits—Max			0.0300	0.0700			0.0040	0.0010				
Quantity	0.6744	0.2481	0.0300	0.0178	0.0111	0.0121	0.0040	0.0010	0.0016			
Least Cost ($/ton) = 163.28												

strain, age, and environmental conditions. Also, costs and prices are almost constantly changing; therefore, the feeding levels that result in maximum profits need to be re-evaluated often. Incremental changes in profits must be weighed against the higher feed cost associated with that change. Modem microcomputers make it fast and easy to update information and make new calculations of maximum profit formulations as the need arises.

21-B. FEATURES OF FEED FORMULATION SOFTWARE

Commercial least-cost feed formulation programs have many functions in addition to solving for the least cost diet or feed. For instance, they can calculate *shadow prices* for ingredients and nutrients. The shadow price for an ingredient is the maximum price at which the ingredient would be included in a least-cost formula. If an ingredient is more expensive than its shadow price, it isn't used in the formulation. A shadow price for wheat in the problem illustrated in Table 21-1 was $118/ton. Wheat would have replaced corn if its price were to fall below $118/ton. This is only true for the exact situation specified. If the amount of protein or any other nutrient in the corn or wheat (or the price of any ingredient) were to change, the shadow price of wheat would also change.

The shadow price for a nutrient indicates the amount the formula cost would increase if any particular nutrient would increase by 1 unit. If protein were included in a matrix in %, the shadow price would be for an increase from 19 to 20% protein, however, if the units of protein were g/kg the shadow price would be for an increase from 190 to 191 g/kg. In this case, if the protein level of the feed were to increase by 1.0 unit, from 19.8% to 20.8%, the cost of the formula would increase by $2.80/ton. These shadow prices allow the nutritionist to evaluate how changing feed specifications impact the economics of poultry production.

Parametric analyses are used in the same way as shadow prices, but they give a more complete picture of the levels of use of an ingredient, or nutrient, in relation to cost. Parametric ingredient analyses show how much of an ingredient would be used at different prices; while parametric nutrient analyses show what formulas would cost with different levels of nutrients.

Multi-blend problems can also be solved by linear programming. Multi-blend software was designed to solve a practical problem. Sometimes there is a limited amount of an ingredient and it isn't clear which diet it should be used in (starter, grower, finisher, layer, etc.). Multi-blend software can solve many formulations simultaneously, putting any limited ingredient in the feed where it lowers the cost of production for the entire mill.

21-C. MAXIMUM PROFIT FEED FORMULATION

As stated earlier, the assumption that maximum performance results in maximum profits isn't always correct. There are situations when it may be more economical to not achieve the maximum performance and feed efficiency possible. The example below illustrates how different diets may each result in maximum economic performance under different cost situations. It is important to recognize that the producer still must strive for the maximum performance for birds fed each diet.

Understanding the response to different diets. The grower-withdrawal diets shown at the top of Table 21-2 were fed to a high yield strain of male broilers from 18 to 39 days of age. At 39 days of age, the birds had different body weights, feed intakes, and feed conversion ratios. The birds fed the highest protein levels were the largest and ate the most feed. Therefore, they would be expected to have the highest costs and highest returns. The producer's goal is normally to produce a certain size broiler for a particular market. For this example, the data were standardized to what would be expected if birds fed each diet were processed when they were 5.0 pounds instead of 39 days old. Broilers fed the highest protein level are expected to reach market weight at 37.0 days of age, 5.5 days before broilers fed the lowest protein level.

Higher feed cost vs more output. Multiplying the pounds of feed consumed by each bird times the feed cost per pound gives the feed cost to produce 5.0 pound birds fed each protein level. With the set of prices at the top of Table 21-2 it appears that the 16% protein diet is the most economical to feed. However, in practice, integrators may want to make additional considerations: The cost of extra time in the house may need to be considered.

For the integrator, who pays the grower for each pound produced, there is a cost for extra mortalities during the extra time birds spend in the house. For the grower there is a cost for lost capacity. With 5 days per batch and 7 batches per year (35 days per year), nearly 10% could be saved by the grower to increase output and decrease the average costs of their buildings and equipment by using the shorter grow-out period.

The bottom of Table 21-2 illustrates how changing ingredient prices changes the most economical feed formulation to use. If the cost of corn were to increase from $95 to $105/ton, there is practically no difference in the cost of feeding the three protein levels; but if the cost of corn were to increase to $115/ton, then there would be a $0.016 per bird advantage to feeding the 24% protein diet.

Nutrient levels affect growth rates. Because of all the indigestible carbohydrates in soybean meal, higher levels of oils or fats need to be added to the diet when higher protein diets are fed. Thus the costs of oils

Table 21-2. The Influences of Different Diets and Several Prices on the Most Economical Broiler Diet to Feed

Illustration of Maximum Profit Feed Formulation					
Protein Level %			16	20	24
Energy Level (kcal/g)			3.2	3.2	3.2
	Cost ($/ton)				
Ground Corn	95	%	72.12	59.97	47.83
Soybean Meal	201	%	20.32	30.66	40.97
Poultry Oil	300	%	3.50	5.38	7.25
Salt	55	%	0.40	0.40	0.40
DL Methionine	2,200	%	0.03	0.07	0.11
Defluorinated Phosphate	240	%	1.78	1.72	1.66
Limestone	34	%	1.45	1.41	1.38
Premixes	534	%	0.40	0.40	0.40
Formula Cost		$/lb	0.0638	0.0716	0.0794
Body wt. @ 39 days		Pounds	4.466	4.928	5.316
Feed intake @ 39 days		Pounds	8.184	8.492	8.624
Feed conv. @ 39 days		lb/lb	1.833	1.723	1.613
Age @ 5.0 pounds		Days	42.5	39.4	37.0
Feed intake @ 5.0 pounds		Pounds	9.558	8.661	7.881
Feed conv. @ 5.0 pounds		lb/lb	1.912	1.732	1.576
Feed cost @ 5.0 pounds		$/Bird	0.610	0.620	0.626
With Corn Costing $105/ton					
Formula Cost		$/lb	0.067	0.075	0.082
Feed Cost @ 5.0 Pounds		$/Bird	0.644	0.646	0.644
With Corn Costing $115/ton					
Formula Cost		$/lb	0.071	0.078	0.084
Feed Cost @ 5.0 Pounds		$/Bird	0.679	0.672	0.663
With Corn Costing $115/ton and Poultry Oil Costing $450/ton					
Formula Cost		$/lb	0.075	0.083	0.091
Feed Cost @ 5.0 Pounds		$/Bird	0.712	0.719	0.720

and fats can affect the use of soybean meal and therefore protein levels in broiler diets. This is illustrated at the bottom of Table 21-2, where increasing the price of poultry oil again favors the 16% protein feed. This change shows how important it is to use proper costs in feed formulation. If a broiler producer is rendering his own poultry oil, should he use the cost of production or the cost of purchase oil on the open market? The formulator has to decide based on maximizing profits to the firm.

Potential savings. The analysis in Table 21-2 is only an example of the complexity of the feed formulation problem. The technical responses, in particular, will be different for each genetic strain of bird. Also, overall growth rates will vary by climate, location, and housing type. Carcass fat and yields are affected by protein level and must be included in comprehensive models. All these factors need to be carefully considered. The potential savings (over $0.015 per bird or over $1,000,000 in many complexes per year) are great and justify the re-

Figure 21-2. Test House for Testing Feeds

search necessary to determine nutritional responses precisely under each integrator's conditions to truly maximize profits.

Protein and energy levels are important. The example in Table 21-2 deals with changing dietary protein levels. Changing dietary energy levels have some similar effects to protein level. Increasing either protein or energy levels increases growth and improves the feed conversion ratio. However, increasing the dietary protein level decreases carcass fat, exactly the opposite of increasing dietary energy level, which increases carcass fat!

Excessive carcass fat may lead to adverse conditions such as oily bird syndrome as well as decreased carcass yields. Figure 21-3 graphically depicts an equation developed by Gonzalez, et al. (1994), illustrating the trade-offs possible. The graph shows that birds of the same body weight can be produced with different combinations of protein and metabolizable energy (ME) at different ages. For instance, it is possible to produce a 2.0-kg broiler at 45.1 days of age with a diet containing 3.4 kcal ME/g with 177 g protein/kg or 2.8 kcal ME/g and 188 g protein/kg. However, if higher protein levels are fed, the birds will reach market age sooner: at 42.2 days with 2.8 kcal ME/g and 210 g protein/kg, or with 3.2 kcal ME/g and 220 g/protein/kg (and still weigh 2.0 kg). Broiler profit maximizing models should make the choice of the best combinations of protein, ME and time

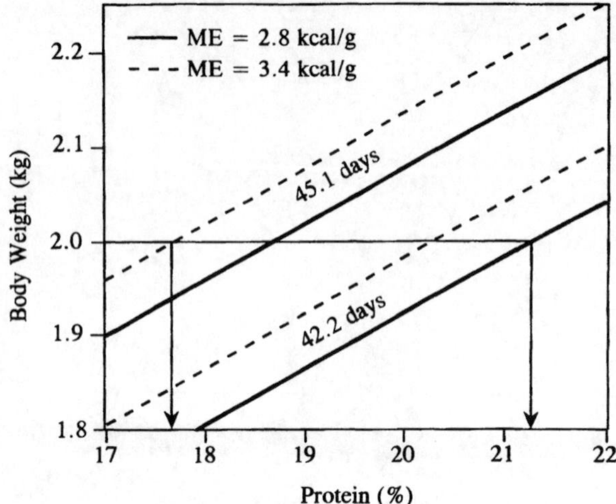

Figure 21-3. Graphical depiction of the broiler growth model of Gonzalez, et al. (1994). A 2-kg bird could be produced by feeding many combinations of protein and energy: for instance, 17.7% protein and 2.8 kcal ME/g for 45.1 days, or by feeding 21.3% protein and 3.4 kcal ME/g for 42.2 days.

based on costs of inputs and values of outputs. Many other factors need to be added to profit maximizing models for commercial broilers. The genetic stock being used and the environmental factors that impact growth have to be considered when predicting how to maximize profits.

Computers and computer models are necessary to help managers deal with such complex problems quickly. If prices of the major inputs and outputs were stable, companies would eventually find the best conditions to produce under, but they are not. Costs and returns change daily, so managers need computers to react quickly to these changes. The better the computer model can predict how changes in input levels affects changes in outputs, the better their reaction will be.

21-D. RISK MANAGEMENT AND STOCHASTIC FEED FORMULATION

Because of variability in ingredient composition in feedstuffs, there is variability in the composition of mixed feeds. Feed is normally formulated to contain some minimum limit or level of each nutrient, using *average ingredient composition values.* Because half the loads of each ingredient will contain less than the average level of any nutrient, half the resulting batches of feeds will normally contain less than the specified minimum limit. If the composition of the nutrients in each ingredient is normally

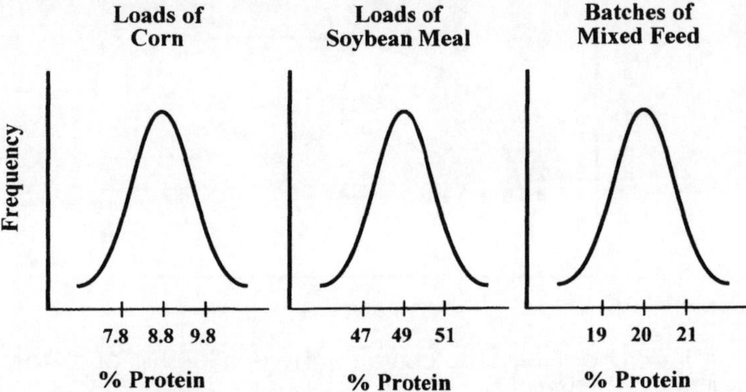

Figure 21-4. Normal frequency distribution of loads of corn and soybean meal result in batches of feed being normally distributed.

distributed, the nutrients in the batches of the finished feed will also be normally distributed. For example, if the protein contents of corn and soybean meal are normally distributed, their mix into a 20% protein feed will also be normally distributed as in Figure 21-4.

Stochastic programming is a method of non-linear programming that can be used to specify the frequency with which a feed will contain more (or less) than a specified amount of a nutrient. Stochastic programming models include a measure of the variability of one or more nutrient in each ingredient, in addition to the average composition values.

Stochastic programming models may be set to meet the minimum specification any proportion of the time (Figure 21-5). In this example, the 50% stochastic restriction is identical to the simple linear programming solution. As the restriction is increased to assure that individual batches of a feed meet the 20% protein restriction 80 or 95% of the time, the average protein concentration of the feed is increased, thus increasing the cost of producing the feed while reducing the risk of producing feeds low in protein or any other nutrient.

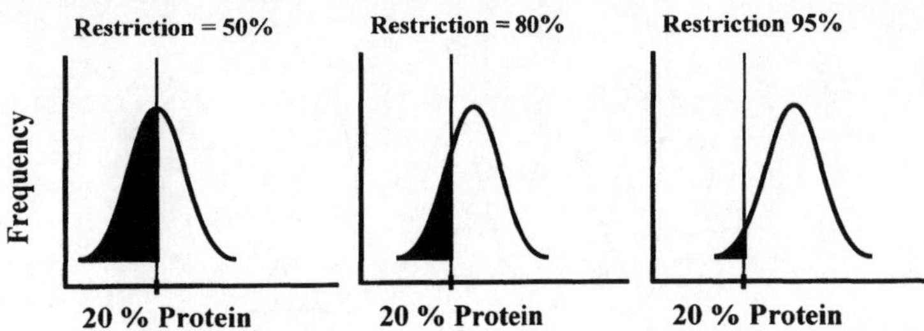

Figure 21-5. Stochastic programming solutions result in the specified restriction (20% protein) being achieved at different frequencies; higher restrictions result in higher average values.

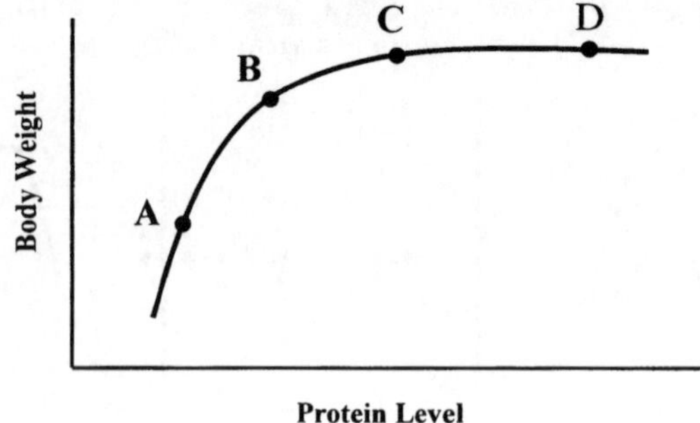

Figure 21-6. A theoretical response curve for the relationship of dietary protein level and resultant body weights.

The resulting response of animals should be evaluated when interpreting the utility of stochastic programming. For instance, consider the following response to dietary protein for broiler chickens (Figure 21-6).

The distribution from feeding protein levels A, B, C, and D on a specified day might be as shown in Figure 21-7. The live weight responses from birds fed protein levels C and D are identical since there is no difference in response to protein in this range. The frequency distribution is due only to genetic variability within the flocks and environmental differences among the flocks.

Bird uniformity depends on diet uniformity. The frequency distributions of birds produced at points A and B, however, would be expected to have more variability (higher coefficients of variation) because variability due to protein level of the diet has been added to genetic and environmental (housing) influences (Figure 21-6). Further, the change in growth due to protein level near point A is greater than near point B, so the coefficient of variation [standard deviation divided by the mean, times 100] of birds is expected to be larger near point A than B. In other words, there is less uniformity from feeding at point A: more extra small and extra large birds and fewer near the mean. Feed formulation should always be in the context of the animals being pro-

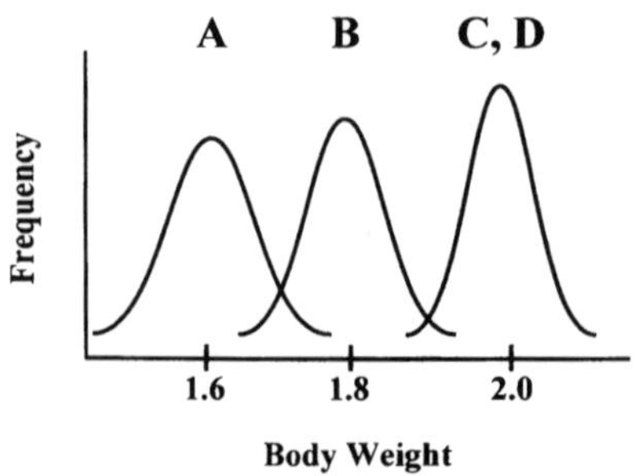

Figure 21-7. A distribution of live body weights that would result from feeding various protein levels in Figure 21-6.

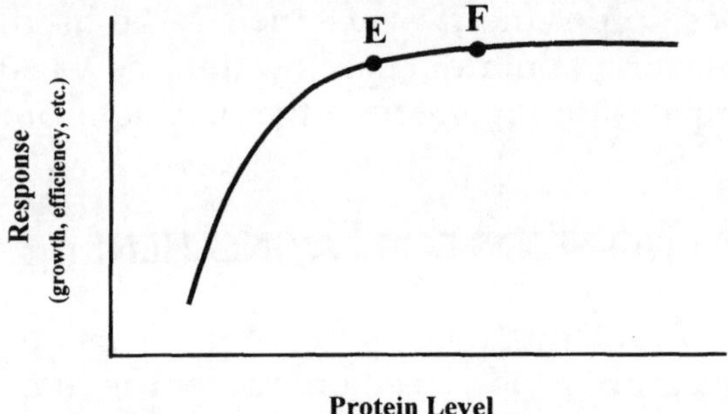

Figure 21-8. A theoretical response curve for broiler body weight as a function of dietary protein level with the nutritional requirement (Point E) and margin of safety (Point F) levels identified.

duced and these examples illustrate the importance of good feed ingredient quality control to minimize the variability in the products (broilers). Maximizing uniformity may or may not be an issue depending on the marketing strategy that a company is following. The difference in cost between providing protein levels A, B, C, or D needs to be interpreted in comparison to the returns expected from selling the broilers represented by distributions A, B, C, and D. This can most easily be done with comprehensive bioeconomic models of response to dietary nutrient levels.

Setting requirements. Another issue to be considered with stochastic programming models is how to determine the minimum specification or "nutrient requirement." Consider again the theoretical response to dietary protein (Figure 21-8).

If the requirement is specified at point E, where the ascending portion of the response meets the plateau, some decrease in response may be expected from batches of feed that have protein levels below the average. How much the response is reduced will depend on the shape of the curve at that point.

However, nutritionists often include a "margin of error" when setting nutrient requirements. If point F in the figure is chosen as the "requirement" to be sure that each bird gets enough protein, no benefit of stochastic formulation may be realized by the producer. Margins of error are usually small for stable nutrients like trace minerals—perhaps 10%—but may be 50% or more for unstable vitamins.

Guaranteeing minimums and maximums. Stochastic formulation models can be very useful to firms selling feeds that have to guarantee minimum nutrient compositions to the consumer. The stochastic method is preferable to less formal means such as choosing an arbitrary margin of error in the requirement or specifying that the ingredients contain less of each nutrient than they actually do [many nutritionists use 0.5 standard deviations less than actual]. To effectively use stochastic programming, the formulator must have good, current information on the composition of the raw materials to be used, and appreciate the costs of meeting minimum specifications 50 vs 80, 95, and 99% of the time, compared to the value of the outputs from each feed.

21-E. FORMULATING FEEDS FOR LAYING HENS

Least cost feed formulation models for laying hens are similar to those for broilers. More emphasis is placed on nutrient intakes for layers since feed intake and egg output on a daily basis are very predictable. Feed formulation for laying hens is based on knowing how much of each nutrient is needed for the number of eggs and the size of eggs being produced, and then providing that quantity of nutrients in the amount of feed the hen is expected to eat.

Formulation affects egg size. Dietary protein and amino acid levels may be adjusted to influence egg size. Egg size can be increased by feeding higher protein levels at the beginning of lay, and decreased by lowering protein levels at the end of lay. Similar responses can be obtained by reducing house temperature for some flocks and increasing temperatures in other flocks.

Prediction equations. *Equations* derived from an experiment conducted by Pesti (1991) were used to construct Table 21-3. This table illustrates the effects of different protein levels fed in combination with two me-

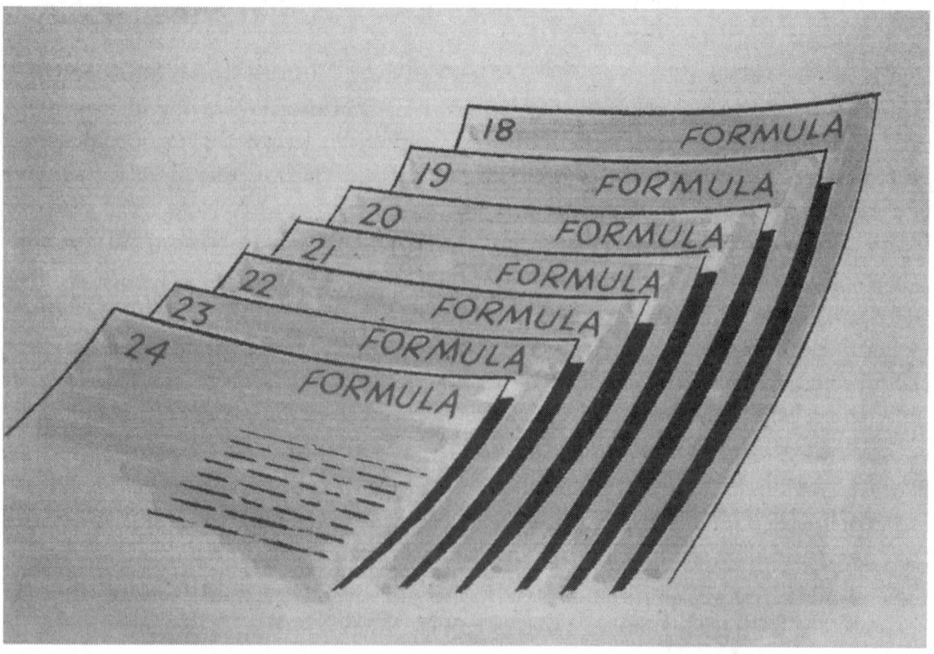

Figure 21-9. Multiple Aged Flocks Require Many Different Feed Formulas

Table 21-3. Predicted Commercial Laying Hen Performance for Hens Fed Different Protein and Energy Levels (based on an experiment reported by Pesti, 1991)

	1273 kcal/pound 2.8 kcal/g diet			1364 kcal/pound 3.0 kcal/g diet		
	Weeks			Weeks		
% Protein	20–28	28–36	36–44	20–28	28–36	36–44
			% Hen-Day Egg Production			
14	63.0	60.5	28.5	57.5	65.8	47.4
16	62.4	75.5	58.6	58.4	78.3	69.6
18	61.8	86.4	80.8	59.3	86.8	83.8
20	61.2	93.3	95.2	60.2	91.2	90.1
22	60.6	96.2	96.2	61.1	91.5	88.6
			Egg Weight (g/egg)			
14	50.3	51.2	57.5	50.0	52.2	56.2
16	51.8	53.8	57.4	51.8	54.8	56.7
18	52.4	55.4	57.4	52.9	56.4	57.1
20	52.3	55.9	57.1	53.2	56.9	57.6
22	51.3	55.4	57.0	52.6	56.4	58.0

tabolizable energy levels. In the first eight weeks of lay, dietary protein level has an effect on egg size, especially with higher energy levels. The increase in egg size is especially important in the beginning of the laying cycle since there may be big differences in the prices of medium compared to large eggs. Neither feed consumption nor egg production are greatly affected by dietary protein level at the beginning of lay. Non-statistically significant effects on feed consumption at the beginning and end of the laying cycle lead to lack of differences in the predictions in Table 21-3. This indicates that the differences were small and producers need to develop data that show consumption trends over shorter intervals.

For the second eight weeks of lay, just after peak production has been reached, protein and energy levels have big effects on egg production rates and feed consumption, and even bigger effects on egg size. However, in the third eight weeks of lay, it is very important to feed higher levels of protein to maintain egg production. Protein level does not appear to influence feed consumption, but greatly influences feed utilization.

Formula costs and output levels. As is the case with broilers, the layer diet that should be fed is the one that results in the biggest difference between returns and costs. High protein rich ingredient costs tend to favor lower protein diets, while high premiums for large eggs tend to favor high protein diets. Managers need to react to changing prices to be sure maximum profits are realized. Seasonal price changes and the price differences for smaller egg sizes compared to larger egg sizes differs considerably from one country to another.

Temperature also plays a large role in determining the feed consumption of laying hens. Raising the temperature in laying houses is a very effective tool for increasing feed efficiency, as long as care is taken to not decrease egg production or egg size to a level that more than offsets feed savings. Several models of feed intake in response to changing temperatures have been developed to help nutritionists make dietary adjustments when temperatures change. If nutritionists predict temperatures will increase, they can calculate how much feed intake will decrease. Nutrient levels in the feed can then be increased to ensure the hens continue to consume enough nutrients to maintain their level of output.

Models of egg production relating feed intake to egg output were pioneered in the 1940's (Byerly, 1941). The influence of temperature on feed utilization and egg production is especially important, so factors including temperature have been added to egg production equations. Equations using historical data can be used to predict feed consumption as it relates to temperature (Pesti, et al., 1994) (Table 21-4). If the

Table 21-4A. Predicted Feed Consumptions (g/hen/day) of Laying Hens with Various Body Weights and Kept at Different Temperatures (from the equation of Pesti, et al., 1992)

Temperature		Body Weight		
°C	°F	3.7 1.7	4.0 1.8	4.4 lbs 2.0 kg
			g/hen/day feed consumption	
20	68.0	103.4	105.7	109.7
23	73.4	98.9	101.2	105.2
26	78.8	93.9	96.2	100.2

Egg Output = 50.7 g/day; Dietary ME = 2.85 kcal/g or 1,295 kcal/lb

Table 21-4B. Predicted Feed Consumption (g/hen/day) of Laying Hens Fed Different Metabolizable Energy Levels and Kept at Different Temperatures (from the equation of Pesti, et al., 1992)

Temperature		Dietary Metabolizable Energy		
°C	°F	11.5 2.75 1,250	12.0 2.87 1,300	12.5 MJ/g 2.99 kcal/g 1,300 kcal/lb
			g/hen/day	
20	68.0	104.7	103.2	101.2
23	73.4	100.1	98.6	96.7
26	78.8	95.1	93.7	91.8

Egg Output = 50.7 g/day; Body Weight = 3.74 lbs/bird, 1.7 kg/bird

costs of keeping hens at different temperatures is determined, these equations can then be used to determine the combination of diet and temperature that result in the most profitable egg production.

Table 21-4 shows the results of predictions from an equation derived by Pesti, et al., 1992. This equation predicts feed intake from known values for temperature, body weight, body weight gain, daily egg output, and dietary metabolizable energy. If producers know the cost of maintaining different temperatures in the house, a computer can evaluate and choose the ideal temperature that is economical in terms of feed savings. Care must be taken that the house isn't too warm to decrease egg production or egg size to unacceptable levels and that air quality isn't sacrificed during times of minimum ventilation.

Equations describing hen performance as a function of temperature, diet, and many other factors have been added to so-called *Expert Systems* such as the one described by Schmisseur, et al. (1989). These systems are quite complex and are designed to identify production problems and to make management recommendations based on strategies to maximize profits. In the expert systems, feed formulation is considered as it relates to the overall profitability of the firm, not just a tool to minimize feed costs.

22

Consumption and Quality of Water

by Donald D. Bell

Water is considered as the most critical nutrient and yet its availability to all birds and its quality are commonly ignored. This chapter describes the needs for water, factors affecting its use, water quality, and the effects of water deprivation.

22-A. NEEDS FOR WATER

In general, the poultry farm must be concerned with the amount of water available, the reliability of the source, and the quality of the water as determined by contaminants.

Water requirements for a commercial poultry farm are large and must be carefully considered when new facilities are being planned. Water is needed on the poultry farm for the following purposes:

1. Daily requirements of the flock
2. Cooling of the flock (evaporative pads or foggers)
3. Sanitation programs
4. On-site egg processing
5. Fire protection

The general needs of the flock can be fairly well estimated by the fact that chickens of all ages generally consume about twice as much water by weight as they eat feed. Under normal conditions, the ratio of water to feed intake will range from about 1.5–2.5 to 1, and at extremely high temperatures, an even higher ratio may be seen. Interestingly, both meat- and egg-type chickens have similar water to feed ratios.

For planning purposes, a layer facility needs to allow at least 0.5 pounds (0.23 kilos) of water per day per hen. This is equivalent to 60 gallons per 1,000 hens per day. A broiler farm would require about the same for six-week-old birds and more if birds were kept to older ages. These requirements do not allow for abnormal wastage of water.

In hot regions, a large reliable supply of water is also required to cool the flock with sprinklers, misters (foggers) or pad and fan cooling. A typical large layer house (100,000 hens) with foggers may use more than 1,000 gallons of water per hour during the hot period of the day for fogging alone. If cooling is needed for six hours a day, this would total 6,000 gallons per day—an amount equal to that required for drinking purposes.

Water requirements for cleaning and disinfection of houses are difficult to estimate and vary with the individual cleaning program. In many cases, mobile washer-sprayers are used requiring relatively small amounts of water. Egg processing plants use water for washing eggs and for plant cleanup. Because a recycling system is used for wash water, the total amount of water usage is minimal (2,500 to 3,000 gallons per day) for a plant processing eggs from one million layers—an amount equal to about 0.02 pounds per hen per day. This would be equivalent to 2.5–3.0 gallons per 1,000 layers per day.

On-site water storage may be recommended in some regions to insure a continuous supply in case of water delivery problems. Most large farms try to have at least a two-day supply on hand at all times. This would also provide adequate reserves for fire protection. To provide the needs listed above, a one million hen-laying farm should have storage for at least 250,000 gallons of water. Standby generation for emergency power and spare pump parts are also necessities for any poultry farm.

22-B. FACTORS AFFECTING WATER CONSUMPTION

The daily water intake of a flock is useful information for the flock manager in helping to diagnose flock performance and wet litter problems. Meters are excellent tools to detect leaking waterers and / or breaks in the water lines. In addition, water consumption data is necessary to predict per bird intake of medications when added to the drinking water.

How Much Water Will a Broiler Flock Consume?

Water consumption is highly correlated to the amount of feed a flock consumes which, in turn, is associated with the age of the flock, body weight, environmental temperature, and the energy content of the feed. Water consumption also varies with the type of watering system.

Table 22-1. Water Consumption Relative to Age in Broilers

Estimated water consumption based upon: Age (days) × 5.28 grams*

	Age (days)					
	7	14	21	28	35	42
Gallons/100 birds/day	1.0	2.0	3.1	4.1	5.1	6.1
Liters/100 birds/day	3.9	7.7	11.6	15.4	19.3	23.1
Gallons to date/100 birds	4.1	15.3	33.6	59.1	91.7	131.5
Liters to date/100 birds	15.4	57.9	127.3	223.8	347.2	497.7

* *Note*: 5.28 represents the daily increase in water consumption based upon an annual average. It is suggested that a factor of 5.1 be used for flocks raised in the cooler months and 5.7 for flocks raised during the summer
Source: Pesti, et al. (1985)

Studies by Pesti, et al. (1985), with commercial broilers raised to 42 days, show a water to feed ratio during different seasons ranging from 1.61 to 1.91 using trough-type drinkers (average 1.77). This research also showed a significant positive linear age to water consumption relationship. In flocks between 1981 and 1983, daily water consumption per bird increased by 5.28 grams (ml) with each additional day. This is equivalent to an increase of approximately one gallon (3.8 liters) per 100 birds each week (see Table 22-1). These age-related curves have probably increased slightly as body weights have changed during subsequent years.

Caution: Estimation of water consumption is best when it is based upon data from the individual farm in question. Differences in wastage, feed composition, location of waterers, and other factors all contribute to variation in water consumption from one site to another. If company data is not available, use breeder estimates adjusted for the season.

Temperature Affects Feed and Water Consumption

Traditional water consumption research is done in controlled environment chambers with constant temperatures. One of the early studies of this type over a range of temperatures from 26° to 94°F (−3.3° to 34.4°C) showed very high water intake and water to feed ratios in 180-day-old White Leghorn pullets at 94°F (34.4°C) and very low water intake and water to feed ratios at 26°F (−3.3°C). Such extremes are rarely seen in commercial housing, and if they do occur they are not constant over a multi-week period as in this research (see Table 22-2).

Recent research (University of California, 1997) showed water consumption and water to feed ratios in commercial layer flocks considerably lower than the earlier studies done in constant temperature chambers. Whereas the chamber work showed water to feed ratios of about 2.20, the current research in commercial housing indicated a ratio of only 1.77—a difference of about 20% (see Table 22-3).

Table 22-2. Water and Feed Intake Relative to Environmental Temperature in White Leghom Hens

Degrees F C	94 34.4	84 28.9	74 23.3	64 17.8	56 13.3	47 8.3	34 1.1	26 −3.3
Trait								
Age (days)	180	200	225	245	268	289	311	335
Feed (lbs/hen/day)	0.14	0.19	0.20	0.23	0.23	0.25	0.25	0.25
(grams)	64	86	92	102	103	115	115	114
Water (gallons/100 hens/day)	7.9	5.8	5.4	5.5	6.0	6.1	5.1	4.0
(liters/100 hens/day)	30	22	20	21	23	23	19	15
Water to feed ratio	4.67	2.60	2.24	2.03	2.19	1.99	1.69	1.32

Source: USDA, ARS 42–43, 1961

Table 22-3. House Temperature As It Affects Feed and Water Consumption in White Leghorn Hens

Degrees F C	<67.5 <19.7	70 21.1	75 23.9	80 26.7	82.5+ 28+	Average
Trait						
Daily Feed Consumption (lb/hen)	0.234	0.234	0.231	0.227	0.218	0.228
(g/hen)	106.1	106.1	104.8	103.0	98.9	103.4
Water Intake (gallons/100 hens/day)	4.93	5.03	4.92	5.16	5.65	5.03
(liters/100 hens/day)	18.7	19.0	18.6	19.5	21.4	19.0
Water to Feed Ratio	1.70	1.74	1.72	1.83	2.08	1.77
Water to Egg Mass Ratio	3.49	3.59	3.49	3.65	4.05	3.57
Water (% of body weight)	11.51	11.29	11.09	11.52	12.71	11.29

Source: University of California, 1997 (167 flocks)

Table 22-4. Water Consumption as Affected by Age of Flock and Temperature in White Leghorn Hens

	Temperature					
Degrees F C	<67.5 <19.7	70 21.1	75 23.9	80 26.7	85+ 28+	Average
Age (wks)	Gallons/100 hens/day					
25–29	4.81	4.90	4.94	5.26	5.67	5.01
30–34	4.96	4.89	4.94	5.34	5.85	5.06
35–39	5.09	5.03	4.92	5.21	5.76	5.06
40–44	5.08	5.00	4.95	5.15	5.65	5.05
45–49	4.61	5.18	4.88	5.11	5.70	5.03
50–54	4.58	5.09	4.92	5.06	5.38	5.01
55–60	5.39	5.12	4.91	5.01	5.51	5.01
Av	4.93	5.03	4.92	5.16	5.65	5.03

Source: University of California, 1997 (167 flocks)

It is interesting to note that the exact same water to feed ratio (1.77) was observed in Pestis's research with broilers and the University of California research with laying hens.

The 1994 Nutrient Requirement of Poultry (NRC) states that water consumption in broilers increases about 7% for each 1°C increase in ambient temperature above 21°C (1.8° and 70°F, respectively).

Water Consumption as Affected by Temperature and Age in White Leghorn Hens

Table 22-4 illustrates the relatively constant water consumption levels observed in US layer houses with flocks of differing ages and at different house temperatures. Flocks between 25 and 60 weeks of age showed no significant differences in daily water consumption when temperature was not considered. All ages averaged 5.03 gallons/100 hens per day (19.0 liters). Within normal house temperature ranges, though, daily water consumption varied from 4.93 gallons (18.70 liters) at house temperature below 67.5°F (19.7°C) to 5.65 gallons (21.4 liters) at temperatures above 85°F (28°C).

22-C. WATER TEMPERATURE

It is important for drinking water to be cool during the warmer months. Chickens will not drink hot water, and consequently feed consumption will decrease and performance will suffer. Closed-pipe systems using cups or nipples can have water temperatures as high as the house temperature. Water pipes must be protected from the sun and should not be placed in hot attics or in areas where heat buildup is possible.

The effect of water temperature on the performance of chickens has been studied by numerous researchers. Different water temperatures in constant ambient temperature conditions and different water/ambient temperature combinations have been studied. Harris, et al. (1975) found that body weight gain and feed consumption of broilers during an eight-week experiment were significantly higher at water temperatures of 75°F (23.9°C) compared to 95°F (35°C); feed conversion, though, was unaffected. In addition, livability was poorer during the first three weeks of brooding at the higher water temperature. The authors concluded with the recommendation that the broiler chick should receive water at a temperature similar to the initial house air temperature. House temperatures significantly above the cooler water temperature depressed gain and feed consumption and house temperatures below the hotter water temperature also suppressed gain and feed consumption.

A more recent report by Hulet, et al. (1998), using broilers, showed that body weights between 21 and 35 days increased approximately 1.7% when

Table 22-5. Effect of Drinking Water Temperature on Feed Intake, Egg Production, and Egg Weight

Trait	Water Temperature– 95°F (33°C)	36°F (2°C)
Feed/hen/day (g)	63.8	75.8
H.D. egg production (%)	81.0	93.0
Av. Egg weight (g/egg)	49.0	48.5

Source: Leeson and Summers (1975)

water was cooled 5° to 14°F (3° to 8°C) below normal room and water temperatures. A similar (1.5%) improvement in body weight was also seen between 35 and 49 days. Feed consumption, though, was only marginally affected.

Leeson and Summers (1975) conducted a 38-day experiment with 30-week-old White Leghorn pullets comparing 95°F (33°C) with 36°F (2°C) drinking water in an environmental chamber at 95°F (33°C). Results favored the cold water treatments (see Table 22-5).

Kuney and Bell conducted three brief experiments in the early 1980's which demonstrated improvements in performance of White Leghorn laying hens as a result of cooling the drinking water between 5° and 10°F (3° to 6°C). Experiment #1 used young pullets just starting into production; experiment #2 used the same birds 12 months later. In the first experiment, water was cooled by circulating it through a home refrigerator. Birds were watered by commercial cup drinkers. In the second experiment, a bleed-line was attached to a temperature release valve (set at 68°F, 20°C) at the end of the test line. Whenever the water at the end of the line reached the set point, the valve was released allowing a small runoff of water from the system. Results are shown in Table 22-6.

Glatz (1996) studied the effects of four water temperatures on 59-week-old brown-egg layers in an 86°F (30°C) environment over an eight-week period. Daily feed intake, egg weight, shell weight, and shell thickness progressively decreased as water temperatures increased from 41° to 86°F (5° to 30°C). The rate of egg production was depressed only in birds on the highest water temperature treatment.

22-D. WATER QUALITY

Other than by distillation there is no such thing as pure water. However, a supply of good quality water is essential for confined animal production. All water contains foreign substances in solution or suspension, which may affect its palatability, functionality, or the well-being of the animals consuming it. The micro-element composition of water varies with the geologic nature of the region from which it originates, and/or travels to its final destination. Contamination may occur as a result of "normal" annual events within the water table or along the water course or periodically

Table 22-6. The Effect of Cooling Drinking Water on Performance of White Leghorn Hens

Experiment Age (weeks)	Average Ambient Temperature (90°F, 32°C) Egg production (%) Water (90°F, 32°C)	Water (81°F, 27°C)
25	64	74
26	74	79
27	77	86
28	76	84
29	88	93
Mean	76	83
Feed intake (g/bird/day)	83	90
Experiment #1 (22 to 38 weeks of age)	Cold water	Ambient water
Feed/hen/day (g)	96.6	92.1
H.D. egg production (%)	67.3	64.3
Av. Egg weight (g/egg)	55.0	54.6
Experiment #2 (72 to 88 weeks of age)	Cold water	Ambient water
Feed/hen/day (g)	110.2	104.3
H.D. egg production (%)	79.3	74.5
Av. Egg weight (g/egg)	66.7	66.0

(Kuney and Bell, 1984)
* All experiments were conducted in open housing in the summer months.

when improper drainage of surface waters occurs. Further contamination of local water from poultry farm runoff must be avoided.

Much of what is known about water quality for poultry and livestock was developed from drinking water standards for humans. Relatively little of the threshold standards for poultry were established in research utilizing growing and laying chickens. Even less is known about the additive or interactive effects of two or more water components. Quantitative models which illustrate progressive effects are not available.

Standards for water quality include elements associated with taste and appearance of the water and health and demonstrated toxic effects in animals or humans consuming the water. Toxic effects can be at specific concentrations in the water for different classes of animals or humans and may be cumulative, i.e., building up to a critical level over time. Effects may be difficult to measure, observable only under laboratory conditions, and seen in commercial settings as abnormal behavior or performance, or in severe cases, death.

Water Sampling

In order to know a farm's baseline for water quality, samples of water should be submitted to a laboratory to determine the level of various chemicals and the presence of harmful microbes. The accuracy of results will depend upon the taking of representative samples, care of the samples, and the choice of a suitable laboratory. Consult with your laboratory about procedures and the type of containers to be used. When samples are collected, allow the water to run for several minutes before sampling. The outlet should be first sterilized by flaming with a portable torch and the water collected in a sterilized (not sanitized) container. Samples should arrive at the testing laboratory within 24 hours of collection.

Some of the determinations made to evaluate water quality are as follows:

Color. Pure water is colorless and therefore any color may be considered a sign of contaminants. Such substances include various synthetic chemicals, natural minerals, or organic components of the soils in the surrounding area. During the process of water movement, specific contaminants may be concentrated and create harmful levels.

Turbidity. Particles of silt, clay, algae, or organic material in suspension rather than in solution cause the water to be turbid (as in muddy water). A suspension results in a relatively short-time dispersal of microscopic particles, whereas in a solution, the particles are invisible and permanently dispersed in the carrier (water). Turbid water may be unpalatable and can cause clogging of the water system. Such water should be filtered prior to use. If the particles are in solution, filtration would be ineffective.

Hardness. Salts of calcium and magnesium form scale and sludge and cause water to be "hard." Hardness affects the taste of water and the action of soaps and detergents. Excesses of calcium and magnesium in the watering system may cause scale buildups within pipes and drinkers and eventual malfunctions leading to stoppages and leakage. Hard water should be treated using systems that do not substitute sodium for calcium and magnesium. High sodium levels can result in wet droppings.

Iron. Although iron in water seldom affects chickens, it stains almost everything with which it comes in contact. Iron levels in excess of 2 ppm are conducive to the multiplication of certain types of bacteria. Iron compounds may also form within the drinking system which may contribute to water leakage problems.

pH. The pH of a solution is a measure of its acidity or alkalinity. On a scale of 0 to 14, below 7 is acid; above 7 is alkaline. Water is normally about 7.0 to 7.2. A water pH of 6.5 to 8.0 is considered acceptable for poultry. Low pH (acid) makes water less palatable, corrodes metal parts of the watering system, diminishes the effectiveness of vaccines, and may affect performance of the flock. High pH (alkaline) water has been reported to adversely affect feed consumption, but animals can build a tolerance over a period of use. Nevertheless, excessively high or low pH water should be avoided.

Total solids. Total solids represent the total amount of solid material in a suspension or solution. In itself, it only measures the total number of ions (both cations and anions) in the water and is not associated with specific problems apart from the concentration of individual components. Total solids are indirectly measured by electrical conductivity.

Nitrogen. Nitrogen determinations are indicative of decaying organic material, and is a measure of contamination. High nitrate levels are associated with poor oxygen utilization in the animal consuming the water (see Table 22-8).

Poisonous metals. Excesses of as little as 0.1 ppm of certain metals in the drinking water may accumulate in the bird to produce pronounced difficulties and cause illness. These include elements such as lead, selenium, and arsenic.

Bacteria. Type of bacteria, rather than number, is important in a water analysis. Some bacteria may be detrimental to humans and chickens and others are not. Wells and other water sources must be protected from the possibility of bacterial contamination, especially from animal or human wastes.

In recent research with broilers, there appears to be an association between total aerobic plate bacteria numbers and poor feed conversion, lower body weights and post-mortem condemnations (Zimmerman, 1997).

Dissolved oxygen. This measurement can be a key test for water pollution. At levels below 3 ppm, fish may die. Normal levels are considered to be 7 to 14 ppm. Levels in excess of 14 ppm may indicate algae growth and/or pollution.

Typical Poultry Farm Water Analyses

Table 22-7 summarizes the average and range of results for water composition in three different regions of the US. Samples were analyzed in three different laboratories over a ten-year period. The purpose of including the range of results is to demonstrate the wide differences in water quality within and between regions and farms.

Maximum Levels of Contaminants in Poultry Farm Drinking Water

Table 22-8 lists typical limits for each measurement. These limits represent a consensus of opinions from different researchers and are, by no means, precise. In addition, they do not represent the same degree of risk to the flock. A comparison of Tables 22-7 and 22-8 illustrates numerous samples that exceed the recommended limits.

Table 22-7. Comparison of Poultry Farm Drinking Water Quality in Different Regions of the US

Variable Measured	Arkansas Broiler Farms	Delmarva Broiler Farms	Washington Broiler Farms
No. of samples	300	71	83
pH	6.5	6.1	7.0
Hardness (ppm)	107	68	75
Conductivity	441	348	235
Iron (ppm)	0.35	1.35	0.12
Manganese (ppm)	0.27	0.10	0.20
Dissolved oxygen (ppm)	6.2	4.1	7.7
Bicarbonate (ppm)	129	117	74
Calcium (ppm)	19.3	15.8	12.5
Magnesium (ppm)	5.1	7.1	6.9
Sodium (ppm)	18.3	45.8	14.7
Potassium (ppm)	1.7	8.0	2.7
Sulfate (ppm)	19.7	18.0	5.5
Chloride (ppm)	32.2	26.5	15.4
Copper (ppm)	0.14	<0.07	0.05
Phosphate (ppm)	0.19	<1.40	0.58
Nitrate-N (ppm)	1.2	7.1	0.9
Total aerobic bacteria*	n/a	191	1511
Coliform bacteria*	n/a	1.0	60
Result Ranges Between Farms Within States			
pH	n/a	3.7–8.2	5.7–8.4
Hardness (ppm)	n/a	0.25–335	0–342
Conductivity	n/a	22–2,273	30–1,500
Iron (ppm)	n/a	0–22.5	0–1.26
Manganese (ppm)	n/a	0–0.8	0–0.6
Dissolved oxygen (ppm)	n/a	1.0–9.0	2.0–10.4
Bicarbonate (ppm)	n/a	0–620	2.6–762
Calcium (ppm)	n/a	1.8–75.0	0–87.3
Magnesium (ppm)	n/a	0.24–68.8	0.01–33
Sodium (ppm)	n/a	5.0–500	1.82–120
Potassium (ppm)	n/a	0.1–39	0.23–38.1
Sulfate (ppm)	n/a	0–337	0–58.9
Chloride (ppm)	n/a	2.8–237	1.0–247
Copper (ppm)	n/a	n/a	0–0.47
Phosphate (ppm)	n/a	0–9.9	0–1.82
Nitrate-N (ppm)	n/a	0–82.6	0–30.3
Total aerobic bacteria*	n/a	0–2,100	0–39,600
Coliform bacteria*	n/a	0–40.0	0–3,680

* cfu per ml
n/a = not available

Possible Beneficial Effects of Water Micro-elements

Several studies addressing the possible relationships of water quality on flock performance have been conducted in recent years with variable results. Barton, et al. (1986) examined water quality and broiler flock performance and found numerous correlations between water components

Table 22-8. Suggested Maximum Limits of Water Components for Chickens

Factor	Maximum Level	Comments (at levels in excess of listed)
pH	<6.5 or >8.0	Possible performance problems
Hardness (ppm)	100 ppm	Reduces effectiveness of soaps, disinfectants and some administered medications
Conductivity		None established
Iron (ppm)	0.3 ppm	Bacteria growth is enhanced as iron increases
Manganese (ppm)	0.3 ppm	Deposits in drinking systems—leaky valves
Dissolved oxygen (ppm)		A measure of possible water pollution
Bicarbonate (ppm)	500 ppm	No known effect
Calcium (ppm)	500 ppm	Decreased livability
Magnesium (ppm)	125 ppm	Laxative effect
Sodium (ppm)	20 ppm	Loose droppings—consider level when formulating feeds
Potassium (ppm)	500 ppm	None known
Salt	2,000 ppm	Water palatability problems; defective egg shells
Sulfate (ppm)	250 ppm	Laxative effect
Chloride (ppm)	250 ppm	None known
Copper (ppm)	0.5 ppm	Liver damage
Phosphate (ppm)		No recommended level
Nitrate (ppm)	20 ppm	Possible performance problems
Total aerobic bacteria (colony forming units per ml) (cfu/ml)	0	Low levels may be present without problems
Coliform bacteria (cfu/ml)	0	Possible fecal contamination and disease exposure

and various performance parameters (see Table 22-7 for average results). Zimmerman (1997) studied drinking water quality and performance in broiler flocks in the Delmarva region of the eastern US and found an improvement in feed conversion associated with increases in potassium, water hardness, and water conductivity levels. In earlier studies with table egg layer flocks, he showed that higher egg production was associated with lower water conductivity and low levels of sodium, chloride, sulfate, and nitrates in the drinking water. Inconsistent results were noted when various studies were compared.

22-E. MANAGEMENT OF WATERING SYSTEMS

The management of watering systems is based upon the need to thoroughly understand the system and all of its weaknesses, and to recognize the absolute necessity of providing a continuous supply of good quality water for all birds (accept for certain periods with broiler breeders). Disruption of water supplies for only a few hours may affect the performance of a flock. Routine monitoring of the watering system must occur daily. Monitoring should include:

1. a check of water pressures on individual lines,
2. replacement of clogged filters and drinkers/nipples,
3. chlorine levels at the drinker,
4. random sampling of drinkers in every line, and
5. inspection of pumping, water treatment, and storage sites.

Water system failure alarms should be based upon an expected flow rate. Excessively low water flow rates indicate lack of water; excessively high flow rates are an indication of leakages. Alarms must be associated with individual houses. Spare parts must be available to correct any problem with the system within minutes of its discovery. Each cage unit should be supplied with a minimum of two watering devices. Floor-housed flocks require an evenly spaced water system in close proximity to the feeders. Ideally, a farm should have two sources of water in case one system fails.

Water Delivery Systems

Water systems require careful installation and maintenance to provide the flock with an adequate water supply without leakage. Nipple drinker systems must be kept level to prevent air locks that could deprive birds of water. In general, nipple systems are designed to operate at low pressures. As birds grow, drinkers are raised and water pressure is increased.

Filters should be used to exclude foreign particles from waterers and water pressure should be carefully managed to assure proper operation of the system. Leaking valves should be replaced as soon as discovered to avoid wet manure problems. Any wet manure resulting from leaking waterers should be removed and replaced with fresh bedding or nearby manure to prevent fly breeding (see *External Parasites, Insects, and Rodents,* Chapter 12).

Water Space

Regardless of the type of watering system used, the system must be convenient to the birds, and sufficient in number or length based on the number of birds it is to serve. In general, provide at least one inch (2.5 cm) of watering space for each growing chicken in floor operations or 10 nipples or cups for each 100 birds.

One cup located at the partition of every other cage is adequate in small cages (3 to 4 birds) but when larger colony sizes are used, additional cups may be necessary. The number of cup drinkers provided should be similar

Figure 22-1. Nipple Watering System
(courtesy of Chore-Time)

Figure 22-2. Water Filtering and Medication System

in number to nipples systems, but remember that one cup will accommodate more birds at one time than one nipple. Some producers prefer and some animal welfare specifications require a minimum of two drinkers per cage to provide a backup in case of single drinker malfunctions.

Running water systems are still very common in various regions of the world. In the US, very few running water systems are seen. Systems which use running water should have a slope of 3 inches per 100 ft (2.5 cm per 10.0 m) of trough. In cage systems, it is imperative that the trough be rigid so that a uniform depth of water may be maintained for all cages. Poorly beak trimmed birds may find it difficult to consume adequate amounts of water if water levels are not adequate.

Note: The water consumption amounts listed earlier in this chapter do not apply to *continuous flow* systems.

Trough space is usually more than adequate in cages because any trough runs across the entire width of the cage, but the minimum is 1.5 inches (3.8 cm) of waterer space per pullet, which will be ample in hot weather.

More detailed recommendations are available from the systems manufacturers and from the major chicken breeders. (see Chapters 34, 43, 51, and 52 for specifics about watering for breeders broilers, replacement pullets and laying hens).

The Use of Water Meters

Practically all new poultry installations use water meters to determine daily water consumption. Some meters are installed and measure water

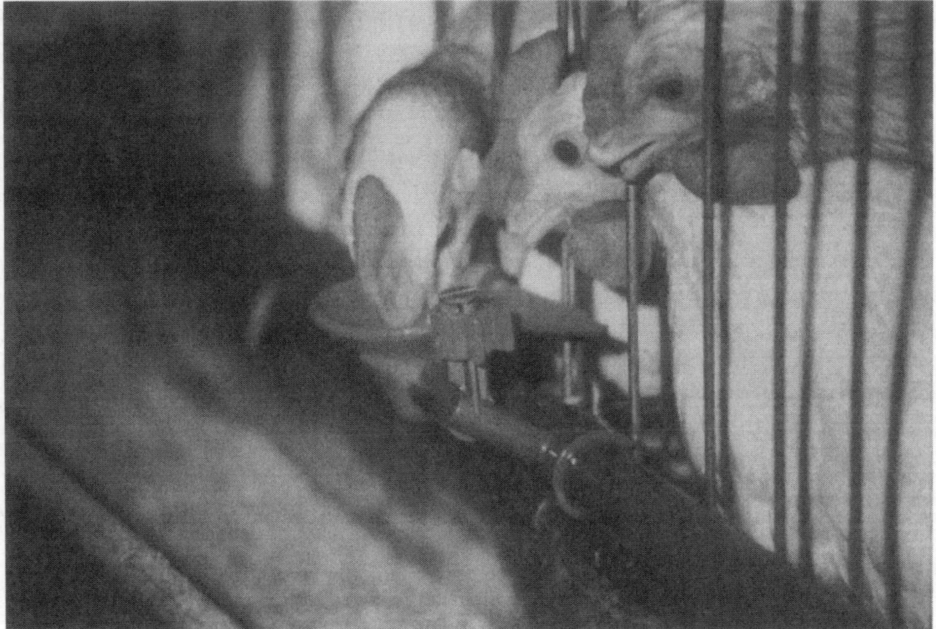

Figure 22-3. Cup Watering System

consumption for the entire house; others may be installed in each cage bank. In computer monitored houses, recording is continuous with periodic summaries provided at set intervals. Abnormal patterns may even trigger an alarm system. A sudden drop in water consumption is often the first indication of trouble in a flock. A large increase in water consumption is usually associated with leakage.

22-F. WATER TREATMENT

Water may be treated with sanitizers for disinfection or with cleaners or other processes to remove specific impurities. Treatment may consist of filtering incoming water or the addition of chemicals to kill bacteria or algae or to counteract undesirable chemical components found in the water supply.

Filtering Water

Water supplies in some regions may contain a variety of filterable materials including sand, plant debris, insects and their parts, and algae. For this reason, many producers choose to filter all incoming water with a large farm filtering system in addition to filtering the water into each house or lines within the house with in-line filters.

Small valve drinking systems and fogger systems are particularly susceptible to malfunctioning when foreign materials get into the watering system. Normally, this means that leakage or stoppage will occur, both are harmful to the operation.

Chlorinating Water

When water is microbiologically contaminated it should be chlorinated. There are several suitable chlorinators on the market, most of which operate by super-chlorinating the water at the farm source. In-line systems are also available. These systems provide a satisfactory level of chlorine in the water in the poultry waterers. The addition of chlorine to water also reduces oxidation of any iron, thus eliminating some of the rust which develops in pipes and valves.

In general, a level of 2 to 3 ppm of chlorine should be maintained at the farthest drinker from the chlorinator. The chlorine level should be periodically tested to assure that chlorine is being added as planned.

The selection of a water-quality maintenance program is a very important decision for management and must be based upon the best advice available. If water vaccination is to be used, include your veterinarian in the discussion.

Cleaning the Watering System

Cleaners are used to remove lime and scale deposits, rust, algae, and to bind calcium or magnesium in the water. The use of acid cleaners helps to remove lime, scale, and algae deposits on the drinkers and inside the lines. Sanitizers will work more effectively when the interiors of water distribution systems are routinely cleaned. Most poultry consultants recommend that lines be thoroughly cleaned following flock removal and before and after each medication or vaccination. Each cleaning should be followed by a thorough flushing of the lines to remove any dislodged deposits and to prevent cleaner residue from inactivating vaccines.

Water Proportioners

Many vaccines and medications are added to the drinking water. At other times disinfectants are added to sanitize the water. When automatic waterers are used it is often difficult to add these treatments directly to the drinkers. To facilitate the addition of these compounds, a water proportioner may be used. Usually this is a pump that operates when water from the water supply is forced through a cylinder. When the proportioner is installed in the incoming water line, the flow of water to the fountains in the poultry house operates the pump. The pump has the ability to draw liquid from a container and inject it into the water line at a rate proportionate to the flow of water to the drinkers. Thus, the proportion of the medicated or treated liquid is always constant. The proportion may be adjusted by altering controls on the proportioner. Be sure that your stock solution is properly prepared and that the use of the stock solution is in proportion to the consumption of the birds for a predetermined period of time.

Water Vaccination

Many vaccines are administered in the drinking water (see Chapter 25). The presence of any sanitizer in the water will affect the viability of the vaccine, often making it worthless.

Warning. Do not add vaccine to any drinking water containing a sanitizer. First flush the water system several times until it is free of any sanitizer, then provide clean water to which the vaccine has been added.

Skim milk prolongs the effectiveness of water vaccines. Where there is no supply of unsanitized water, the sanitizing agent negative effect on the vaccine may be partially offset by adding dried skim milk to the drinking water. Add 3.2 oz (90.7 g) of dried skim milk to each 10 US gal (37.9 liters) of water. This is approximately 1 part of dried skim milk to 400 parts of water. Add the vaccine to the milk / water solution and mix thoroughly.

Consult your veterinarian. Vaccines differ in their concentration and ability to withstand different environmental factors, therefore when uncertain, check with a veterinarian or similar professional.

22-G. WATER DEPRIVATION AND RESTRICTION

Water deprivation and restriction may be *unintentional* as a result of inadequate water availability or failure of the watering system or *intentional* as part of a program to produce drier droppings or to initiate a molt. In either case, water availability is limited by removing it completely for a specified time period or by allowing the flock to access it only intermittently.

Unintentional deprivation occurs when an insufficient number of drinkers or amount of watering space is provided or when breakdowns in the system occur. In such cases, performance tends to diminish in direct proportion to the amount of water consumed by the entire flock. For this reason, it is recommended that daily water consumption be compared with previously gathered information from the same flock and that these data

Table 22-9. Effect of Water Restriction on Various Traits in 8 Week-Old Broilers

	Level of Restriction					
Trait	0	10%	20%	30%	40%	50%
Water consumed (lb)	10.6	9.5	8.4	7.3	6.4	5.4
(kg)	4.8	4.3	3.8	3.3	2.9	2.5
Feed consumed (lb)	7.7	7.0	6.7	6.2	6.0	5.7
(kg)	3.5	3.2	3.0	2.8	2.7	2.6
Final body weight (lb)	3.3	2.9	2.7	2.5	2.4	2.1
(kg)	1,500	1,318	1,227	1,136	1,091	955
Overall feed conversion	2.3	2.4	2.5	2.5	2.5	2.7
Water:Feed ratio	1.4	1.4	1.2	1.2	1.1	1.0

Source: Kellerup, et al., 1965

be compared to standards. Oftentimes, performance may only be affected to a small degree and managers are totally unaware of its occurrence.

Kellerup, et al. (1965) demonstrated a depression in performance in broilers with water restriction. Table 22-9 shows results of 0, 10%, 20%, 30%, 40%, and 50% restriction. Even with levels of only 10% restriction, feed consumption was decreased by 9% and body weight was reduced by 12%.

Researchers (Bierer, et al., 1965) have also studied the effects of total water or feed deprivation on light and heavy breeds of chickens at different ages. In general, chickens of both breeds withstood feed removal much better than they did water removal.

Bell (1976) studied the effects of short-term feed and/or water removal on various egg traits and found that 24 hours without feed reduced egg weight by 2.8% compared to 9.2% for hens without water for a similar length of time. Most of this appeared to be due to smaller yolks. On the other hand, 24 hours without feed reduced eggshell weight by 24.9%, compared to 7.9% for 24 hours without water.

Methods to molt chickens have been studied for many years and water removal has been one of the methods commonly used to induce flocks to cease egg production. Today, many countries forbid this practice and in the US, most producers no longer use this practice (see *Flock Replacement Programs and Flock Recycling,* Chapter 54).

Water Restriction of Laying Hens

As a method of reducing the moisture content of the feces, it may be practical to restrict drinking water intake. Usually this is accomplished by allowing birds to drink for 15 to 30 minutes, followed by a period of from 2 to 4 hours without water. The procedure is then repeated during the light hours. The moisture content of the feces can be reduced significantly when proper water-restriction programs are used.

Reducing water consumption. Pullets in cages quite often consume and waste more water than they actually need, causing the droppings to become wet. This is especially true in deep water systems. Intermittent watering can eliminate some of the problem, but do not restrict water prior to peak protection (before 35 weeks of age) or during hot weather. Water restriction is rarely practiced in cages with cup or nipple systems because of leakage problems.

Be careful. Water restriction during hot weather may be disastrous. Birds may drink two times as much water during hot summer months compared to other periods of the year, and restricting water consumption may result in mortality or loss of production. Water restriction may be impractical in nipple or cup watering systems due to the problem of leakage associated with turning the system on and off. High cage densities also make this practice less practical.

Many experiments have compared restricted watering with full watering programs with commercial layers. In most of these experiments, egg production was not significantly affected. If restriction is practiced, pay close attention to possible effects on egg size.

Water consumption highest late in the day. Caged layers drink 2 to 3 times as much water per hour during the 3 hours immediately preceding the time lights go off than during other periods of the day. Consumption during the remainder of the daylight hours shows a fairly uniform hourly intake pattern (Mongin and Sauveur, 1974).

Results with a typical water restriction program is shown in Table 22-10. In this experiment, comparisons were made between 4, 8, and 15 fifteen-minute-watering periods per day. Four such periods produced the best results. Many different water-restriction programs

Table 22-10. Average Egg Production, Feed Efficiency, Feed Consumption, Livability, and Manure Moisture as Influenced by Restricted Watering Time (White Leghorns)

Waterings per Day (15 min ea)	Hen-day Egg Production (%)	Feed per Dozen Eggs (lb)	Feed per Dozen Eggs (kg)	Feed per 100 Birds per Day (lb)	Feed per 100 Birds per Day (kg)	Livability (%)	Manure Moisture (%)
15	61.5	4.99	2.27	25.6	11.6	92.1	78.2
8	61.6	5.04	2.29	25.8	11.7	92.1	78.5
4	62.2	4.90	2.22	25.4	11.5	91.3	75.6

Source: Maine's Timely Topics, March 1976

have been tested and results generally show dryer droppings and rarely any harmful effects on egg production or egg size.

Water Restriction of Broiler Breeder Pullets and Cockerels

With broiler breeder pullets and cockerels on restricted feeding programs, water should be turned on and available for one hour prior to feeding and turned off or removed one hour after all feed is consumed. If a skip-a-day feeding program is used, water should also be available for approximately two hours on the non-feed day. During hot weather, watering times on both feed and non-feed days should be increased accordingly.

Section III. Poultry Health

23

Microorganisms and Disease

by Gregg J. Cutler

Microorganisms constitute a large group of living organisms that complete their life cycle as single cells or as small groups of attached single cells. They do not form organized tissues as are found in the higher forms of life. Most such cells remain in a single form. Some may align themselves with other similar cells to form long filaments or chains. Being living organisms, they function similarly as higher organisms in that they digest and assimilate food and excrete waste products.

As do higher forms of life, microorganisms must also fight for survival. Being small, they have been provided by nature with a survival potential through their ability to reproduce in extremely large numbers. These large numbers mean that dissemination throughout nature is great. Most microorganisms are abundant everywhere. Some microorganisms are necessary to perform various functions required to sustain life. Others are pathogenic in that they produce certain reactions that are, or can be, detrimental to the higher forms, among which is the chicken. Many of the pathogens can be excluded from flocks of chickens thus preventing disease (see *Biosecurity on Chicken Farms,* Chapter 28).

23-A. IMPORTANT TERMS

Active immunity. Immunity produced by the bird against an antigen introduced either by natural exposure or by vaccination.

Acute disease. A disease having a short duration and possibly severe consequences.

Agglutination test. A test for the presence of antibodies, and performed by mixing blood or serum with an antigen.

Air sacculitis. Inflammation of the air sacs.

Anamnestic response. A boost in immunity seen when an individual is exposed to an antigen after it has previously produced a primary immune response to that antigen.

Anthelmintic. A compound capable of expelling or destroying parasitic worms, especially in the intestines.

Antibiotic. A substance that has the power to kill or inhibit the growth of microorganisms.

Antibody. A protective substance formed in the body as the result of infection or administration of suitable antigens (vaccines).

Antigen. A substance that produces antibodies when introduced into the bird.

Antiserum. A serum containing antibodies specific to a certain antigen.

Aseptic. Free from pathogenic organisms.

Ataxia. Uncoordinated voluntary muscular movements.

Attenuated. A disease organism that has been modified to reduce its virulence.

Autogenous vaccine. A vaccine prepared from cultures derived from infected birds on a specific premise and used to immunize subsequent flocks.

Avirulent. An organism that is not capable of producing disease.

B-cells. Cells of the immune system that are transformed and matured in the bursa of Fabricius.

Bacteria. Microscopic organisms that are composed of a single cell that contain no membrane bound organelles.

Bactericide. A substance that kills bacteria, but not necessarily their spores.

Bacterin. A suspension of killed bacteria (antigen) that brings about immunity when injected into the chicken.

Bacteriostat. A substance that inhibits the growth of bacteria without killing them.

Broad-spectrum antibiotic. An antibiotic that inhibits the growth of many kinds of microorganisms (the more numerous the kinds, the wider the spectrum).

Bursa of Fabricius. A sac like organ adjacent to the upper part of the cloaca involved in the processing and maturation of cells of the immune system.

Carrier. A chicken that shows no evidence of a disease yet harbors the organism and is capable of transmitting the disease to others.

Caseous. Cheesy in appearance.

Catarrhal. Capable of producing an inflammation of the mucous membranes usually of the respiratory tract.

Cellulitis. Inflammation of cellular tissue.

Chronic disease. One that has a long duration, usually evidenced by morbidity rather than mortality.

Coccidiostat. A chemical compound added to the feed or drinking water to control coccidiosis by restricting coccidial growth.

COFAL test. A complement fixation test for the detection of avian leukosis viruses.

Contagious disease. An infectious disease that is readily transmitted to other birds.

Culture (noun). Microorganisms grown on artificial media in a laboratory.

Culture (verb). A procedure whereby organisms from the bird are grown on and isolated from artificial media.

Cyanosis. Bluish color of an organ as a result of lack of oxygen.

Disease. An impairment of the normal function of any body organ or part of the bird.

Disinfectant. A compound used to kill pathogenic organisms and is usually applied to inanimate objects.

Edema. An excess of fluid in body tissues.

Endemic. A disease which occurs with predictable regularity in a population.

Enteritis. An intestinal inflammation.

Erythrocyte. A red blood cell used for transporting oxygen.

Flagella. Whip-like, filamentous appendages from the surface of bacteria or protozoa that serve as organs of locomotion.

Fomites. Inanimate objects such as clothing, equipment, etc., that mechanically convey disease-producing organisms.

Gram negative/positive. Referring to the staining techniques used as the basic beginning of the identification process of bacteria.

Hemorrhage. A condition occurring when blood escapes from the circulatory system.

Hepatitis. An inflammation of the liver.

Histology. Study of the microscopic structure of tissues.

Host. An animal that supports a parasite or a pathogenic organism.

Host-specific. An organism confined to a single host species.

Immune. Protected against a particular disease (antigen).

Immunity. The ability to resist or overcome an infection to a specific disease (antigen).

Inclusion body. An accumulation of virus particles found in the cell contents when the bird is infected with certain diseases.

Induced immunity. Immunity resulting from vaccination.

Infection. The introduction of a pathogen into susceptible tissue.

Infectious disease. A disease produced by the introduction of a pathogenic organism.

Latent. An infection that remains hidden or dormant but may become apparent at a later time.

Lesion. A variation in the normal appearance of tissue as the result of a pathogen or injury.

Lymphocyte. White blood cells produced during an infection that are involved in disease control.

Lyophilized. Stabilization of biologic materials by rapid freeze drying.

Memory cells. T-cells that "remember" previous immune responses and help create an anamnestic response to antigen exposure by activating B cells.

Microorganism. Living organisms that are too small to be seen without the aid of the microscope.

Morbidity. Diseased birds in a flock.

Mortality. Birds that have died, commonly expressed as a percentage.

Mycosis. Any disease caused by a fungus.

Necropsy. Pathologic examination of an animal after death.

Necrotic. Those tissues or cells that are dead.

Parasite. An organism that lives in or on another organism, from which it derives its nourishment.

Passive immunity. Parental immunity passed from the female parent to offspring through the egg (by antibodies).

Pathogen. An organism capable of causing disease.

Pathogenicity. The capability of an organism to produce a disease.

Pericarditis. Inflammation of the sac surrounding the heart.

Peritonitis. Inflammation of the abdominal cavity.

Perosis. Deformity of the leg bones.

Plasma. The clear liquid remaining after the cells have been removed from the blood.

Polyvalent. An antigen or bacterin containing several strains of an organism or organisms.

Protozoa. Minute protoplasmic accellular or unicellular animals with varied morphology and physiology.

Renal. Relating to the kidneys.

Sanitizer. A preparation capable of reducing the number of bacteria present.

Saprophytic. Something that grows on decomposing organic matter. Usually not a pathogen.

Septicemia. Invasion of the bloodstream by pathogenic microorganisms.

Serological test. A test performed on blood serum to determine the presence or absence of specific antibodies.

Serotype. A particular strain of a microorganism.

Serum. The clear fluid remaining after the cells and clotting properties have been removed from the blood.

Stress. Anything that negatively affects the bird's well-being.

Syndrome. A group of symptoms common to a specific disease.

T-cells. Cells of the immune system that are transformed and matured in the thymus.

Thymus. An organ located in the neck involved in the processing and maturation of cells of the immune system.

Titer. A value placed on the potency of a biological agent as in a vaccine. When applied to a serological test, it is the relative concentration of the antigen being measured.

Tumor. A mass of abnormal tissue that grows independently of the tissues around it.

Vaccine. A preparation of microorganisms (killed or living) that when introduced to the bird produces or increases immunity to a specific disease.

Variant. In microorganisms, one that is serologically or physiologically different from the original form, often the result of a mutation.

Vector. A living entity that carries and transmits a disease or parasite to poultry, such as the earthworm, which carries the chicken tapeworm eggs.

Viremia. The presence of viruses in the blood.

Virulence. The relative ability of a microorganism to invade the host or produce disease.

Virus. Infectious agents that are ultramicroscopic in size, some of which are capable of causing disease, that can multiply only in living host cells.

23-B. CAUSES OF DISEASE IN CHICKENS

The organisms producing diseases in chickens are too small to be seen with the naked eye and, therefore, require a microscope. All such organisms are not similar in structure, size, chemical composition, mode of nutrition, or in the manner in which they attack their host. These organisms can be divided into four classes: (1) bacteria, (2) viruses, (3) protozoa, and (4) fungi. Although the organisms within each of these groups are similar in many respects, most may be identified only by using intricate laboratory techniques.

1. *Bacteria*

Disease-producing bacteria are abundant in the poultry world. These organisms are capable of invading the chicken, where they multiply rapidly through cellular division and produce physiological changes in the host bird that it cannot withstand. Sickness follows, and if the reaction is severe enough mortality may result.

How Bacteria Grow

Commonly, the bacteria divide by fission with each cell separating into two equal halves. Once the disease-producing bacteria gain entrance into the chicken, multiplication of the cells is rapid in susceptible birds. One cell divides into 2, 2 into 4, 4 into 8, 8 into 16, and so forth. But geometric multiplication (commonly called exponential growth) does not continue forever; soon there is competition for a food supply and oxygen forcing many cells to become incapable of division. But the numbers are soon so large that their ability to produce disease in the bird is far beyond the quantity necessary.

Organisms grow similarly in the laboratory. Bacteria may be made to grow and reproduce in the laboratory by using artificial means. The bacteria are removed from the host, or from other laboratory cultures, and placed on or in a medium known to furnish the proper food material and moisture. The medium is then warmed to a certain temperature to create optimum growth. When the bacteria are placed on a medium such as agar, cell division is rapid and colonies of the growing organisms may easily be observed with the naked eye. Each organism produces its own pattern of colonies with different shapes, colors, and makeup, providing one means of identifying the type of bacteria.

Variant forms of bacteria. Some bacteria, particularly those in certain *Pasteurella* species (e.g., *P. multocida*), have through the years developed variations in their makeup, giving rise to several types of the same organism. Some of these variations may be due to mutations, and others to differences in the environment in which the bacteria live. These variations have caused many misinterpretations in the reading of tests and in the production of antibodies. These differences are very important in the selection of the individual strains of *P. multocida* that are used in vaccines. If strains other than the ones found on a premise are used, the vaccine may not protect the birds against the resident strain.

How Bacteria Produce a Contagious Disease

Poultry diseases may be classified as *contagious* or *noncontagious.* When contagious, they are capable of being transmitted from one bird to another. Some bacteria are primary pathogens and produce a disease. Other bacteria are secondary pathogens and only produce disease when birds are stressed by other microorganisms or by a disruption in their environment or nutrition.

In most instances a disease is the result of the entrance of the disease-producing bacteria into the bird and the ensuing multiplication. The organisms produce toxins that are in turn antagonistic to the host. Being quantitative, the greater the amount of the toxin, either by increased cellular production or because the quantity of bacteria is overwhelming, the greater the effect of the disease.

The *virulence* (pathogenicity) of an organism is a measure of its ability to produce disease. Each organism varies according to its power to invade the tissues and produce the necessary toxins. The rate of multiplication of the disease-causing organisms within the host is also a contributing factor. Virulence within a species of bacteria can vary. Many disease-producing organisms remain in a quiescent state within many birds, but when conditions are right, can multiply and produce toxins in abundance. On farms where successive groups of chicks are started, the virulence of the organism may be increased in each successive flock once an outbreak occurs.

2. *Viruses*

A second group of disease-producing organisms is represented by the viruses. They are exceptionally small in size. They are capable of passing through specific filters that bacteria are too large to pass through. It is impossible to see viruses using the light microscope. With proper preparation the electron microscope is capable of photographing them.

Parts of a Virus

A mature virus is known as a *virion* and is made up of a central core of genetic material (nucleic acid) plus a protein coat surrounding the nucleic acid for protective purposes and to aid some viruses in penetrating the host cell. Other viruses have a coat of lipoprotein called an *envelope* that develops when the virus buds from the host cell. All viruses contain RNA (ribonucleic acid) or DNA (deoxyribonucleic acid) but not both. Thus, viruses may be classified as an RNA or a DNA virus.

Properties of the Viruses

Although infinitely small, there is a great variation in the size of viruses, from about 20 nanometers (Nm) to 300 Nm. They invade the host cells, and it is this invasion that is thought to precipitate the respective disease, rather than any toxin produced.

Secondary invaders. When a virus attacks the cells of the lining of the respiratory tract and the air sacs, the cell walls are ruptured and become a point for invasion for other viruses and bacteria. In many cases the disease developed by the primary invader virus may produce little damage to the bird, but the secondary invader may cause serious disease.

How Viruses Multiply

Surprisingly, little is known about the exact life cycle of most viruses. Their small size has complicated experimental work. A virus must rely on the cell it invades for its reproduction, using the enzymes from the cell to complete the process. Viruses are specific in their need for certain enzymes and seek out a location in the body where they are produced. Thus, some viruses may invade liver, some the bursa of Fabricius, others the respiratory tract, and so forth.

Viruses multiply only in cells. Bacteria can multiply almost anywhere in the body, whereas a virus can live and reproduce only within a host cell.

Treatment of a viral disease is difficult. Because viruses use the machinery of the host cell for replication it would be necessary to stop that machinery to prevent multiplication of the virus. This would in turn damage the cells that are being targeted thus causing damage to the chicken. There are antiviral drugs but they are aimed at enzyme systems specific to certain viruses, but these are very expensive. At this time, the use of antiviral drugs in chickens is not economically feasible.

3. *Protozoa*

The third group of disease-producing organisms are the protozoa. They are animals and are similar to higher forms of life except for the fact that all the functions are carried out within a single cell. In poultry they are parasitic in nature. They live on the contents of the cell, eventually destroying it.

Protozoa may have a complex life cycle. An example would be coccidiosis in chickens. In this case oocysts are expelled in the fecal material. Heat, oxygen, and moisture cause them to sporulate. They are then consumed and enter the intestinal tract. The cell wall of the sporulated oocyst ruptures; eight infectious organisms exude and soon penetrate the cells of the intestinal lining. In turn, these grow until the intestinal cells rupture and hemorrhage. This is the *asexual* portion of the life cycle, and may be repeated several times; but eventually a *sexual* phase occurs, giving rise to both male and female organisms. After other cell invasions, more oocysts are produced and leave the body by way of the fecal material, and the life cycle is complete.

Number of oocysts extremely large. A teaspoonful of droppings from a bird with severe coccidiosis will contain several million unsporulated oocysts. Consumption of as few as 10,000 sporulated oocysts will produce evidence of coccidiosis.

Coccidiosis is treated with a number of different drugs. There are also ways to prevent coccidiosis by the use of coccidial vaccines usually given at a very young age. Coccidiosis is still one of the largest causes of economic losses in chickens.

4. *Fungi*

Fungi represent a group of organisms including *molds* and *yeasts.* Generally, they grow and produce toxins outside the chicken. When consumed they may continue to grow, with the toxin causing a distinct setback in the bird's well being. In other cases, continued ingestion of the mold and toxin (mycotoxins) produces a disease-like effect, as well as feed efficiency losses.

Fungi can also cause infectious disease in chickens. These diseases can affect the respiratory tract, skin, and intestinal tract. In some cases, these infections can become generalized and cause serious problems.

23-C. HOST MUST BE SUSCEPTIBLE FOR A DISEASE TO DEVELOP

Conditions must be proper, and the host susceptible, for the disease causing organisms to invade, establish themselves, and produce the disease. Nature has endowed the bird with protective immune systems to prevent invasion, and with certain other factors to help reduce the incidence of disease.

1. *Those that prevent (or attempt to prevent) organisms from entering the body.* The three means available to the chicken to help prevent organisms from invading are secretions, skin, and mucous membranes. However, these protections can have flaws with invasion occurring through skin abrasions, via the respiratory tract, and by other means (or methods).
2. *Those that fight the organisms that have entered the body.* Each type of organism causes the production of an antibody that acts to destroy the organism. This self-destruction is, in most instances, complete to the extent that all the disease-producing organism in the body are destroyed and the disease in the flock subsides.
3. *Species resistance.* Certain organisms may invade one species but not others. These organisms may grow and multiply in the resistant species but will produce no discomfort or other evidence of the disease. Some others may invade and cause disease symptoms in several host spe-

cies. Some diseases common in other poultry will produce no disease in chickens.

4. *Age susceptibility.* Some diseases attack chickens of a certain age and produce disastrous results, yet show few, if any, symptoms in birds of other ages.
5. *Climate and season.* Some poultry diseases affect birds more during cold weather than during warm weather; others are more prevalent during periods of warm or hot weather.
6. *Freedom from stress.* The physiological well-being of the individual bird has an effect on the incidence and severity of disease outbreaks. Proper nutrition and adequate housing that provides for the proper control of temperature, relative humidity, and noxious gases improves conditions for the birds thus making disease outbreaks less likely.

24
Immunity
by Gregg J. Cutler

Once a bird undergoes a natural infection from a disease-producing organism or from a vaccination, chemical compounds (antibodies) are produced in the body that attempt to kill that organism. The process of building protection against pathogens is known as *immunity*. The acquired protection also allows the bird some level of immunity from future invasions of similar organisms. The intricacies of the development of immunity are complex and varied, as seemingly each organism has its own program.

24-A. THE IMMUNE SYSTEMS

The ability to develop immunity to an antigen (foreign proteins that cause the production of antibodies), is present in the very young chick. It is a highly specialized defense system, called the immune system, and is nature's way of resisting disease caused by the early invasion of many infectious organisms such as bacteria, viruses, fungi, and protozoa.

The immune system owes its origin to specialized cells of which lymphocytes, and other cells derived from lymphocytes, are the most important. Actually, two immune systems develop in the body of the chick, with each having the responsibility of producing two types of lymphocytes. These two systems are:

B-system (Bursal System)

Thymus (T) cells and other lymphocytes in the young chick pass through the *bursa of Fabricius*, a small gland located above the cloaca, where maturation takes place. The T-cells locate in the region near the bursal duct opening. The bursa may also have a secondary function similar to a lymph node.

Plasma cells develop in the B-system, including the bursa, spleen, and cecal tonsils. These cells are responsible for the production of most of the antibodies in the young chick. In fact, the B-system produces more than 700 times the number of antibodies produced by the T-system. These B-cells live for less than a week and, therefore, must be continually replenished to maintain the bird's defenses.

T-system (Thymus System)

In the very young chick, certain immature lymphocytes originating in the yolk sac and bone marrow pass through the *thymus* (in the neck) and are known as T-lymphocytes. Here they mature, then grow and accumulate in the lymphoid organs such as the spleen, cecal tonsils, and Harderian gland, for a few weeks after hatching. The T-lymphocytes do not produce antibodies, but they do have the ability to develop *lymphokines,* chemicals that can destroy foreign cells by direct contact without the presence of an antibody. This is called *cell-mediated immunity* or *cellular immunity.*

The B-lymphocytes passing the bursa of Fabricius soon locate throughout the body, including the blood system where they no longer require the bursa for maturation. A similar procedure takes place with the T-lymphocytes. Any disease that affects the thymus or bursa in the very young chick disrupts the early development of both the B- and T-systems.

Anamnestic Response

The B-system is also responsible for developing lymphocytes that act as *memory cells.* This is referred to as the *anamnestic response.* These cells have great longevity and cause the B-cells to "remember" former immune responses and accelerate repeated responses. An example is the greater response of the second vaccination against a specific disease as opposed to the lower level of response to the first. The body defenses are produced more quickly after the second response than after the first. Evidently the cells that produce the antibodies do not forget how, and when restimulated start production more rapidly.

The Immune Systems and Vaccination

Although the immune systems are nature's way of protecting the birds against early disease, they are far from perfect. Massive numbers of infectious organisms can produce an overwhelming effect. The infectious organism must first be present in the bird in order to initiate the immune systems. There is also a lag time after the infection is introduced before the systems produce adequate antibodies to stop the infectious organism. It is during this time that some morbidity or mortality may occur. Early vaccination against the infectious organism is the most successful method of reducing this lag time (see *Vaccines and Vaccination,* Chapter 25).

Suppression of the Immune Systems

Certain diseases and other conditions are known to affect the developing thymus and bursa of Fabricius in the young chick. These conditions may cause varying amounts of gland destruction followed by a lessening of the immune systems, known as *immunosuppression.* Among those having been shown to have such an effect are the following:

Suppression of the B-system

1. Infectious bursal disease
2. Environmental temperature extremes
3. Lymphoid leukosis
4. Nutritional deficiencies
5. Toxins
6. Inclusion body hepatitis
7. Aflatoxins

Suppression of the T-system

1. Marek's disease
2. Environmental extremes
3. Heredity
4. Incomplete vaccine reaction
5. Aflatoxins

24-B. PROGRESSION OF A DISEASE OUTBREAK

The progression of most poultry diseases is the same. There are three phases:

Infection. Disease-producing microorganisms invade the chicken, and if no or inadequate immunity is present these organisms attack various parts of the body and produce a sickness in the birds. The type of sickness is specific for the particular disease involved. Morbidity usually occurs first. Mortality may follow, depending on the characteristics of the microorganism and other factors within the chicken.

Development of immunity (resistance). Once the microorganisms establish themselves in the host chicken, the production of a protective compound (antibody or lymphokine) begins with the compound being specific for the organism involved.

Disease subsides. Protection develops and the protective agent produced destroys the causative organisms. In the case of most diseases the outbreak subsides or is reduced to a very low level. The bird recovers except for any permanent damage developed during the course of the disease.

24-C. ANTIBODIES AND IMMUNITY

Physiology of Antibodies

Once a foreign substance enters the chicken, the body acts to eliminate it. Some such substances never are assimilated, as the body eliminates them through the feces. Others, such as some of those generated by the bird, are eliminated through the urinary tract.

Bacteria are composed of proteins, and are formed from one or more protein molecules that are foreign to the bird. These foreign protein particles produce a toxic reaction, resulting in what we call *disease*. In trying to eliminate them, the body system generates a chemical compound that reacts with the organisms or inactivates them. This compound is known as an *antibody*. Each antibody is specific for the bacterium or virus that initiated its production. As the antibody is especially coded for the protein of the invading organism, the fact that there may be one or more protein molecules involved with a single microorganism means that there may be one or more antibodies formed.

Antibody Behavior

Independent antibody for each disease. The antibody is specific for each type and strain of microorganism that caused its production. The antibody produced as the result of one organism will not protect the bird from other organisms, e.g., antibodies against infectious bronchitis will have no effect on the Newcastle disease virus.

Length of time for antibody production. Once the disease-producing organism establishes itself within the bird, antibody production varies according to:

1. Number of organisms involved in the invasion
2. Virulence of the organism

3. Condition of the bird (freedom from stress) at time of invasion
4. Type of organism

For the most part, the time required for immunity to develop (through the production of antibodies) is dependent on the organism involved. Some produce a response in a short time. Others may take a longer time.

Length of immunity varies. The value of antibodies in protecting the bird against future invasions of the same disease organisms is great and the immunity lasts for a long period. In some instances, it continues for a lifetime. In others, it lasts for several weeks or months. Immunity is relative, and absolute immunity is probably nonexistent.

Antibodies may not destroy all organisms. In the case of certain diseases, pullorum disease for example, the *Salmonella pullorum* bacteria in the bird may not be completely destroyed. In this instance, the atrophied ova have no blood supply, which normally would bring in antibodies produced elsewhere. Consequently, any pullorum bacteria harbored in the degenerated tissue of the ova would continue to live on and multiply. This allows the possibility of the development of individuals that may become chronic carriers of a disease.

Variability in antibody production. In the course of a normal outbreak of a disease, antibody production reaches a maximum, after which it decreases and eventually, in the absence of reexposure to the disease organism, may be reduced to zero. The rate of this decrease, and the longevity of adequate immunity, is a function of the type of antibody involved. It will also vary according to the disease.

Quantitative measure of antibodies present. It is possible to determine the amount of antibodies present in the bird. This determination involves computing the *antibody titer*. Titer is a quantitative determination of the presence of antibodies. Titers are usually expressed as a reciprocal of the highest dilution of the solution showing specific activity. Thus, readings of 1:16, 1:64, and 1:256 would be examples. The higher the titer, the greater the number of antibodies (see *Diagnostic Testing,* Chapter 30).

Titer determines resistance. Following certain disease outbreaks, when the antibodies slowly decrease, the bird gradually loses its ability to withstand future invasions of the bacterium or virus. If the antibodies are reduced to a very low level, reinfection may occur and the bird again may contract the disease. The severity of the second invasion will be determined by the number of antibodies and memory cells in the bird at that time. If it is in the medium range, the bird may show only slight effects from the disease, but if the quantity of antibodies is low, the disease may be severe. The titer of antibody as measured in the laboratory is generally, but not always, proportional to the ability to resist infection.

Age of the bird and antibodies. In respect to many diseases, young chicks seem more susceptible than older birds. At a young age, the immune system may not be sufficiently mature to produce a response great enough to overcome the invading bacteria or viruses. In these cases, the birds are said to be *immunologically incompetent.* As the bird grows older, the immune system matures and the ability to mount a thorough and fast immune response reaches a maximum and then begins to diminish.

24-D. MATERNAL IMMUNITY

Protection for the prevention of many diseases in newly hatched chicks can be attributed to *maternal immunity,* sometimes called *passive immunity.* If the breeding hen has had exposure to a specific disease, either by natural infection or vaccination, she has a large number of antibodies circulating in the blood and present in other organs. Many of these localize in the yolk and albumen, and consequently are passed to the chick through the hatching egg. These antibodies aid in protecting the chick from invasions of the specific disease-producing bacterium or virus when it is quite young. Only actual antibodies are passed, not the cells that produce the antibodies. These antibodies have a definite half life. Consequently the level of antibody decreases over the first weeks of the chick's life to near zero by about 3 weeks of age.

Antibody number in chick correlated with number in mother. When the amount of antibodies in the hen producing the hatching eggs is high (high titer), the number in the day-old chick will also be high. Usually, the concentration of antibodies in the egg yolk is identical with that in the hen. However, there is a reduction by the time the chick is hatched. The number of antibodies found in the yolk then is only about half that in the yolk of the fresh-laid egg and remains at this concentration for about 3 days, or until the yolk is absorbed. Thus, to provide ample immunity in the day-old chick, the titer must be high in the chick's mother.

Chicks with no maternal immunity are easily invaded. When the hen producing the hatching egg has not had a specific disease or been vaccinated for that disease, she has produced no antibodies against the infection. A chick hatched from an egg laid by such a mother would contain no antibodies and would have no protection against an invasion of the specific bacterium or virus. Therefore, the chick would be completely susceptible to the disease.

Passing live organisms through the egg. Not only are antibodies passed from the dam to the chick through the egg but live disease-producing organisms are similarly transferred. One example is avian encephalomyelitis (AE) in which viruses are passed from the dam to the chick, along with some antibodies. However, the production of antibodies is not great enough in the dam to kill all the viruses so some viruses may be found in the chicks, inducing an outbreak of AE as early as the first day after hatch.

What may be passed through the egg? There are several possibilities, such as:

- *No bacteria, viruses, or antibodies* where the female parent has neither been vaccinated nor had the disease.
- *Live viruses or bacteria,* where the dam is in the early stages of a disease or where a live-virus vaccine has been used and antibody production has not yet had a chance to destroy the organisms.
- *Antibodies,* where, from an outbreak of a disease or a vaccination against it, antibodies have been produced in the parent and are transmitted to the chick.

Length of Parental Immunity

Any maternal immunity begins to wane after the chick hatches. Half is lost in about 3 days. By the end of the second week of the chick's life it is minor. It is generally ineffective by the end of the third week and completely disappears by the end of the fourth. Thus, at best, maternal immunity can be considered an effective means in preventing disease outbreaks before the chicks are 2 weeks of age, but even during this period it must be recognized as a highly variable protective system.

24-E. THE IMMUNE SYSTEMS OF PROTOZOA

After an outbreak of a protozoan disease, most birds develop limited immunity. One challenge to the organism will not produce complete immunity, but repeated challenges generally cause it to occur. Although the immunity from one challenge is short-lived, most protozoan disease outbreaks reinfect the bird, and seemingly the immunity lasts a lifetime.

Although antibodies are produced after the bird has been exposed to protozoa, these do not represent the sole immune system operating to alleviate the effect of the disease. Evidently, localized tissue immunity continues in the intestine because the host will show immunity long after the antibodies have disappeared.

25

Vaccines and Vaccination

by Gregg J. Cutler

Immunity is the result of antibody production and the effects of certain cells directly or indirectly on disease agents. Immunity occurs after a natural infection with a disease but because of the time lag between natural infection and immunity, some morbidity or mortality may occur before immunity is sufficient to stop disease. Planned exposure to a disease agent at the correct time with the correct agent is a method of producing immunity. The process is known as *vaccination.* The agent used is called a *vaccine.*

25-A. HOW VACCINES WORK

In most instances vaccines are used to produce a mild infection, with few or no signs of the disease, by using a specific disease agent. Normally, the production of immunity by the use of a vaccine will duplicate that from a natural outbreak. Usually, for effective vaccination protection from a disease, the vaccination must be given well in advance of any natural exposure or infection. There are a few exceptions. These are usually in very slow spreading diseases or ones in which immunity develops very quickly.

The virulence of the vaccine and the number of vaccine or virus particles per individual dose will determine the effectiveness of a vaccination. The immune competence of the chicken and host factors such as stress, genetic background, and other concurrent diseases also play an important role in the outcome of a particular vaccination.

25-B. TYPES OF VACCINES

Vaccines may be classified as:

- live vaccines
- attenuated live vaccines
- inactivated (killed) vaccines
- genetically engineered vaccines

All vaccines are produced from live agents specific for the disease in question. Each vaccine is the result of harvesting bacteria or viruses produced from specific disease agents and grown in the laboratory. They may be treated in a manner to eliminate or reduce the severity of disease (or reaction) produced by the agent, they may be killed by chemical or other means, or portions of the agent may be genetically placed in another agent.

1. *Live vaccine.* The organisms in the vaccine are alive and completely capable of producing the disease in birds not infected or previously vaccinated. Because the vaccine contains a live agent, the vaccine is also capable of transmitting the disease to any susceptible bird that comes in contact with it. Live vaccines (including attenuated live vaccines) replicate after application giving high numbers of the vaccine agent over a fairly long period of time. This helps produce a strong immunologic response.
2. *Attenuated vaccine.* The active organisms used to prepare a vaccine may be weakened (attenuated) by various methods so that when administered to a bird, a mild form of the disease will be produced. In many cases there is no evidence of the disease. Vaccines of this type are tested carefully in the laboratory for their ability to revert to a virulent form of the disease. Only those that have little or no ability to revert to virulence are used as vaccines.
3. *Inactivated (killed) vaccine.* The organisms used to produce these vaccines have been grown in culture in the laboratory and then inactivated by chemical or physical means. Inactivated vaccines are not capable of spreading the agent to other birds. They do, however, have the capacity to produce immunity when used through vaccination. In some instances, however, their ability to do this is impaired, and immunity will not reach as high a level as with live or attenuated vaccines. In such cases, immune stimulators called *adjuvants* are added to killed vaccines to help overcome the limited amount of the infectious

agent present. Killed vaccines are frequently given after a primary vaccination with a live vaccine. With proper adjuvants, high levels of long lasting immunity are produced.

4. *Vectored (genetically engineered) vaccines.* These represent a new type of vaccine that places immune stimulating components (antigens) from one agent onto another agent known as a vector. The vector is usually a very large virus such as a pox- or herpesvirus. While this method of vaccination holds promise, it has been difficult to get the vector to express the antigens added from the second disease agent. It is hoped that in the future, antigenic components of many agents can be placed on a single vector, greatly simplifying vaccination. Since only specific, nondisease producing components are added to the vector, disease reactions from the vectored vaccines cannot be produced.

25-C. HOW VACCINES ARE ADMINISTERED

Vaccines may be classified according to methods of administration:

1. *Intramuscular:* Vaccine is injected with a syringe and needle into the muscle. This route is usually reserved for killed vaccines.
2. *Subcutaneous:* Vaccine is injected under loose skin via a syringe and needle. Typical sites are the loose skin on the back of the neck and the loose skin of the leg fold between the thigh and abdomen (Figure 25-1).
3. *Ocular:* Used for live vaccines, a liquid is placed by eye drop such that the solution flows through the lacrimal duct to the respiratory tract.
4. *Nasal:* A drop of live vaccine is placed in the nostril.
5. *Water:* Vaccine is dispensed into the watering system either from a tank or via a proportioner. Vaccine goes into the respiratory and digestive tract by way of the throat. This method allows mass application of vaccine without handling each chicken.
6. *Wing web:* A needle, usually double, is dipped into the vaccine. There is a small depression in the side of the needle that holds vaccine by surface tension. The vaccine is applied by puncturing the skin in the web of the wing. This method is used for fowl pox and some *Pasteurella multocida* (cholera) vaccines.

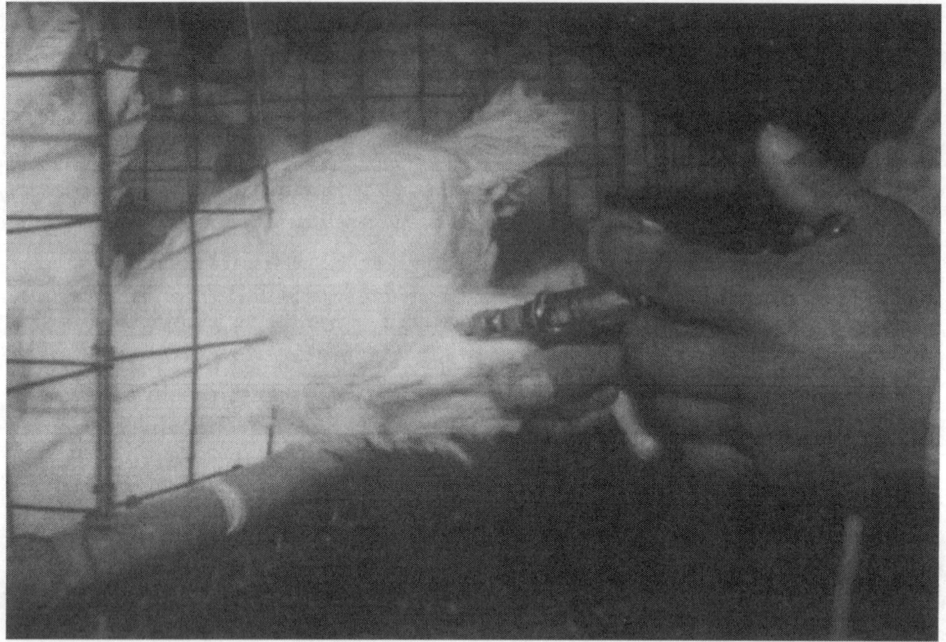

Figure 25-1. Vaccine Injection

7. *Aerosol:* Vaccine is sprayed in the air over birds on the floor or aimed at the face of chickens in cages. Chickens do not need to be individually handled. A number of machines for this purpose are available. Particle size is very important. Initial vaccination should be done with a large particle size. Subsequent vaccinations can be done with finer particle size to get deeper into the respiratory tract.

   Aerosol vaccination is usually practiced for the respiratory disease vaccines such as Newcastle disease, infectious bronchitis, infectious laryngotracheitis and *Mycoplasma gallisepticum.*
8. *Spray cabinet.* Vaccine is sprayed on the birds by a special machine at the hatchery. This is a very good means of administering respiratory disease vaccine such as Newcastle and infectious bronchitis to birds without being individually handled.
9. *In-Ovo*: This is a method of vaccination done at the hatchery. Eggs are transferred at 18 days from the incubator to the hatcher. During transfer the eggs are placed in a machine that aseptically punches a hole in the shell and injects vaccine into the 18-day embryonating egg. This is the usual method of vaccinating broilers for Marek's dis-

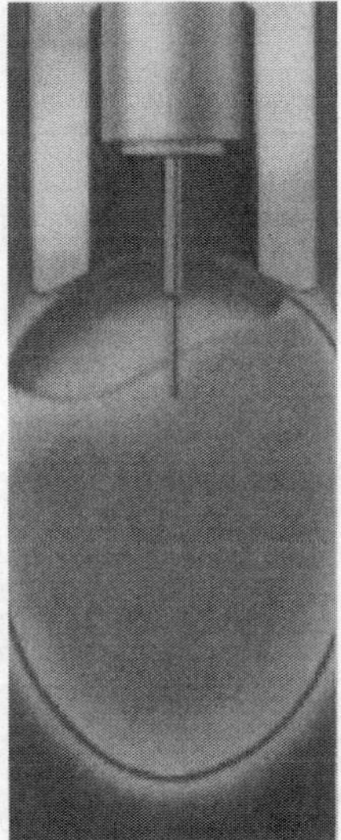

Figure 25-2. In-Ovo Vaccination (Diagram of Injection Site) (courtesy of Embrex, Inc.)

ease (Figure 25-2 and 25-3). This method avoids the need to handle every chick at hatch for Marek's vaccination.

When choosing the route of application for a vaccine there are several factors that must be considered. The natural route of infection is important in determining how well vaccination will stimulate immunity. Poxvirus, for instance, must be given by a route that causes the vaccine to penetrate the skin. No other route has proven effective in inducing immunity. A respiratory vaccine such as Newcastle disease, may be much more effective if applied as an aerosol to the respiratory tract. An enterovirus such as avian encephalomyelitis virus might be best applied via the drinking water (although it is commonly applied effectively in combination with fowl pox by wing web).

A secondary consideration may be one of ease of application and the reduction of stress on the chickens. One of the mass application techniques such as aerosol or water may be easier, cheaper, and less stressful than handling every bird to give it vaccine by the eye drop or oral route.

Location of the vaccination is also important. Some killed vaccines are very irritating and can cause scarring. Birds that are to be used as meat are not good candidates for intramuscular vaccination in the breast. There is the possibility of abscess formation and consequent condemnation of some muscle at slaughter. Therefore, subcutaneous vaccination in the loose skin of the neck or the leg fold might be better for some killed vaccines. The recommendations of the manufacturer should always be followed. If there are questions about the route of application of a specific vaccine, contact the manufacturer or a poultry veterinarian.

Figure 25-3. In-Ovo Vaccination Machine (courtesy of Embrex, Inc.)

25-D. MATERNAL IMMUNITY AND INTERFERENCE WITH VACCINATION

The function of maternal immunity is to prevent pathogenic (and other) organisms from producing the effects of disease in young chicks when the incidence of disease might overwhelm their ability to survive. Because the level of antibodies declines after vaccinating the mother, parental immunity in the chicks produced from a vaccinated flock can be variable. The longer the time since the vaccination of the mother, the lower the maternal immunity. If hatching eggs are coming from several flocks of breeders, each in a different period of egg production, some chicks would have a higher while others would have a lower maternal immunity.

It has been shown that high levels of maternal antibodies can interfere with vaccination in many diseases. It is therefore important to know the level as well as the uniformity of maternal antibodies when planning the vaccination program (see *Diagnostic Testing,* Chapter 30). Vaccination before maternal antibodies are gone may be done to protect individuals with low maternal antibodies. Strains of some vaccines are available that are strong enough to overcome maternal antibodies. Maternal antibody interference is not considered as critical as the necessity for uniform early protection.

25-E. REVACCINATION

The first exposure to a vaccine produces a moderate immune response (*active immunity*) of relatively short duration in the chicken. Since it may have been given at a time when maternal antibodies were still present, all birds may not have developed an adequate response. In order to get a uniform response in all birds and to encourage the creation of memory cells for long-term immunity, it is necessary to revaccinate the chicken one or more times. It is important when planning to use a killed vaccine that several live vaccinations be given at intervals of at least three weeks apart, to prime the chickens to produce a strong response to the killed vaccine.

When chickens are molted, it is important to check antibody levels and revaccinate if levels are low. If vaccination is required, it should be timed so that the immune response occurs approximately three weeks before the molt begins or after the molt has been completed and when birds are back on full feed. The lowered plane of nutrition during the molting period is not conducive to a good immune response.

25-F. STRESS AND VACCINATION

Most present-day vaccines produce a mild response in a normal healthy bird. However, stress can accentuate the effect, producing a greater physiological change in the bird, sometimes with disastrous results.

What are the stresses? There are many periods of stress during which birds should not be vaccinated, such as:

- When the birds are "off feed."
- During periods of extremely hot weather.
- When birds have some other disease, such as coccidiosis.
- When birds will be moved before they would recover from a vaccination.
- When birds are in a stage of recovery from another vaccination or have recently been moved.
- When birds are being medicated or are diseased.
- Following beak trimming.
- During the first few weeks of an induced molt.
- When young birds are chilled.
- When parental titers are high.

Contaminated Vaccines

When vaccines first became available some of the vaccines on the market were contaminated with impurities, particularly other organisms. Many vaccines are prepared from material harvested from growing embryos. There was always the chance that other organisms, particularly those that are egg transmitted, would find their way into the commercial vaccines. With modern techniques, however, this possibility is rare.

Purchase vaccines from a reliable company that will stand behind its products and supply the service necessary to accomplish the best vaccination. In some countries all vaccines must be COFAL-negative and *Mycoplasma*-negative. Eggs for vaccine manufacture usually come from specific pathogen free (SPF) flocks.

25-G. LACK OF IMMUNE RESPONSE

Some flocks respond poorly to a vaccine. There are three main reasons for failure of a flock to respond to a vaccination. They are:

- Lack of immune competence of the chicken.
- Poor potency or lack of antigenicity of the vaccine.
- Improper application of the vaccine.

The ability of the bird to respond may have been reduced because of an infection with infectious bursal disease. If the bursa of Fabricius has been damaged at an early age, inadequate numbers of B-cells may not be present in the various tissues to provide the immune response (see *Immunity*, Chapter 24).

One must assume that when dealing with a reputable vaccine manufacturer, the potency of the vaccine when it is shipped is adequate to stimulate an immune response. In reality there is usually much more antigen than necessary to accomplish this.

After the vaccine leaves the manufacturer it can suffer loss of potency from mishandling. Vaccines should be kept refrigerated until used. Do not freeze vaccines unless the label indicates to do so. Frozen vaccines should be kept frozen. Do not open vials of vaccine until you are ready to use them.

Vaccines must be applied correctly to ensure the best immunity. Always keep a record of the vaccine manufacturer and the serial number of the vaccine. Follow these recommendations for application:

- Mix vaccines thoroughly.
- Vaccinate no more birds from a vial than the directions recommend.
- Follow the manufacturer's procedures for vaccination.
- Do not rush the vaccination job.
- When using water-type vaccines be sure there are no

sanitizers in the water. (Neutralize the water with non-fat milk powder)

- When spraying, use the particle size recommended by the manufacturer.
- Certain vaccines may be mixed (e.g., bronchitis and Newcastle). With certain others, the vaccinations must not be mixed, but can be given at the same time.
- Handle birds carefully for individually applied vaccines.

25-H. TITER OF VACCINES

Just as titers can be established for blood serum, they can also be determined for a vaccine, where the titer indicates the concentration of the virus in the product when it is prepared for use.

Definition. Vaccine titer means the number of virus particles, either living or dead, in 1 milliliter (ml) of the product. The number of times the original virus can be diluted and still infect 50% of the chicken embryos when injected into fertile eggs under incubation is the measure of the vaccine titer. The virus titer is generally given in logarithms to the base 10. For example, a titer of 6 or (10^6) indicates six zeroes after the number one, or 1,000,000.

25-I. VACCINATION PROGRAMS

Not only are there many vaccines, but there are numerous vaccination programs. These programs involve:

- Type of bird (breeders, layers, broilers) involved.
- Management style used (all in, all out versus multiple ages).
- Diseases known to exist in the area.
- Two or more vaccines given simultaneously so as to reduce the number of handlings of the birds.
- Age of the birds when vaccinations are given.
- Route of vaccination
- Strains of vaccine to use
- Regulatory considerations: Which vaccines are authorized for use in a specific area.

Important. There are many types of vaccines and vaccination programs, with examples given in Tables 25-1, 25-2, and 25-3. It must be understood that they may not be practical for all areas of the world or under all conditions. Diseases in the area, availability of vaccines, periods of

Table 25-1. Example of Broiler Vaccination Program

Age	Vaccine	Strain	Route
18-day embryo	Marek's disease	HVT and SB-1	In-ovo
1 day in hatchery	Newcastle disease	B1, B1	Spray cabinet
	Infection bronchitis	Massachusetts–Connecticut	Spray cabinet
14 days	Newcastle disease	B1, B1	Water
	Infectious bronchitis	Massachusetts–Connecticut	Water

stress, climatic conditions, and many other factors are involved when developing a vaccination program for a specific area. Consult your veterinarian and vaccine supplier before initiating any vaccination program.

1. *Broiler Vaccination Program*

The fact that broilers are marketed at about 7 weeks of age and are not subject to many of the poultry diseases that affect older birds allows broilers to be vaccinated differently from birds used for laying and breeding (see Table 25-1). The program should be kept simple as extra vaccinations add stress that can reduce production efficiency.

2. *Layer Vaccination Program*

Growing birds used later for the production of commercial eggs require a long-lasting vaccination program (see Table 25-2) as these birds will reach an age of 80 to 120 weeks before being sold. Since egg production is of prime importance, the program will most likely contain a vaccine for avian encephalomyelitis. Vaccinations for infectious laryngotracheitis and *Mycoplasma gallisepticum* should only be used where there is risk of expo-

Table 25-2. Example of Commercial Egg-Type Grower Vaccination Program

Age	Vaccine	Strain	Route
1 day in hatchery	Marek's disease	HVT and SB-1	Subcutaneous
18 days	Newcastle disease	B1, B1	Water
	Infection bronchitis	Massachusetts–Connecticut	Water
	Infectious bursal disease	Intermediate	Water
28 days	Infectious bursal disease	Intermediate	Water
6 weeks	Newcastle disease	LaSota	Coarse spray
	Infectious bronchitis	Massachusetts–Connecticut	Coarse spray
9 weeks	Pox	Fowl	Wing-web
	Avian encephalomyelitis	Calnek	Wing-web
12 weeks	Newcastle disease	LaSota	Fine spray
	Infections bronchitis	Massachusetts–Connecticut	Fine spray

Table 25-3. Example of Breeder Replacement Vaccination Program

Age	Vaccine	Strain	Route
1 day	Marek's disease	HVT	Subcutaneous
14 days	Newcastle disease	B1, B1	Water
	Infectious bronchitis	Massachusetts–Connecticut	Water
	Infectious bursal disease	Intermediate	Water
28 days	Newcastle disease	LaSota	Water
	Infectious bronchitis	Massachusetts–Connecticut	Water
	Infectious bursal disease	Intermediate	Water
7 weeks	Newcastle disease	LaSota	Coarse spray
	Infectious bronchitis	Massachusetts–Connecticut	Coarse spray
	Infectious bursal disease	Intermediate	Coarse spray
10 weeks	Pox	Fowl	Wing-web
	Avian encephalomyelitis	Calnek	Wing-web
12 weeks	Newcastle disease	LaSota	Fine spray
	Infectious bronchitis	Massachusetts–Connecticut	Fine spray
	Infectious bursal disease	Intermediate	Fine spray
16 weeks	Newcastle disease	Killed oil emulsion	Subcutaneous
	Infectious bronchitis	Killed oil emulsion	Subcutaneous
	Infectious bursal disease	Killed oil emulsion	Subcutaneous

Note: Breeders are sometimes vaccinated during lay to keep maternal antibodies in the chick at a high level. Antibody titers in the hens should be periodically checked to determine if revaccination for one or more of the diseases listed above is necessary.

sure. In some areas killed vaccines may be added if necessary after the last live Newcastle and bronchitis vaccination.

3. *Breeder Replacement Vaccination Program*

This program is similar to the vaccination program for commercial laying strains. Because of the desire for high levels of maternal antibodies in the chicks, killed vaccines for Newcastle disease, infectious bronchitis, and infectious bursal disease are usually given. Other killed vaccines that are often given to broiler breeder replacement candidates are viral arthritis and fowl cholera. Breeders may be revaccinated during lay to keep maternal antibody levels high (see Table 25-3).

26

Medication for the Prevention and Treatment of Diseases

by Carol J. Cardona and Gregg J. Cutler

The judicious use of medications is a part of any balanced management program. Even so, disease outbreaks should be considered failures of nutrition, management, biosecurity, cleaning and disinfection, or vaccination programs. Preventative measures most effectively limit the adverse effects of disease but when they fail, medication can, and should be used to end the clinical signs of disease and return the flock to its normal balance. Medications should never be applied to flocks without also addressing the management failures that led to the outbreak (see *Diseases of the Chicken*, Chapter 27).

The proper use of medications can end the mortality and lesions associated with a disease outbreak, thus limiting financial losses. However, the improper use of medications can be costly, ineffective, and sometimes harmful. Proper use requires the consideration of:

- Class of bird (breeder, layer, broiler)
- Age of the bird
- Possible residues
- Disease diagnosis
- Sensitivity of the disease agent to medications
- Other ongoing treatments

26-A. HOW DRUGS ARE GIVEN

In Water

Medications used to treat birds have properties that make them acceptable for administration in the drinking water, in the feed, or as injections.

Administration in the drinking water is the preferred method of delivery in most cases. During a disease outbreak some birds will not eat but often they will still drink. Chickens can be encouraged to drink by removing water for one to a few hours prior to the administration of medications. However, water removal must be carefully monitored to avoid dehydration of the chickens.

The administration of medications in the water requires that the factors affecting water consumption be considered (see *Consumption and Quality of Water,* Chapter 22). Weather conditions or other factors that may increase the consumption of water must be taken into consideration when calculating how much medication to put into the water.

Medications given in the drinking water either form solutions or suspensions when added to water. Solutions are especially well suited for administration in the drinking water because they completely dissolve in water. Some drugs go into suspension when added to water, that is, they do not dissolve but rather float in the water. Suspensions require extra labor to keep the drugs suspended evenly in the water throughout the period of administration.

In Feed

For the most part, drugs administered in the feed are insoluble in water and their use is confined to the feed. Soluble drugs can alternatively be given in the feed, but water administration is usually preferred.

The amount of medication put into feed is calculated based on the drug's optimal dose and an assumed average feed consumption for the type of bird being medicated. As with water delivered drugs, factors that may increase or decrease the consumption of feed, such as environmental temperature and production rate, should be considered when calculating doses.

Injection

Medications are rarely given to commercial poultry in injections because, although it is a highly effective method of drug delivery, it is very labor intensive. In addition, training workers to give injections properly can be time-consuming and expensive. Improperly trained crews risk injury with self-injection and harm to the chickens. *Self-injection is very dangerous and when it occurs, should be immediately followed by a doctor's examination and proper medical care.*

When administered, injections are usually given subcutaneously or intramuscularly. Subcutaneous (SQ) injections are given under the skin of the chicken, not in the muscle or body cavity. Subcutaneous injections are commonly given under the skin of the neck or in the skin fold of the leg. Intramuscular (IM) injections are given into the largest part of a muscle. Intramuscular injections can be given into any muscle but economic considerations play an important part in determining the best site of injection. The thigh and leg are commonly used for IM injections.

The site of injection is very important for several reasons. First, care should be taken not to inject medications in sites that are the most valuable to the consumer, like the breast. The carriers and adjuvants associated with injections often leave residues that have to be trimmed at slaughter. Second, injectable drugs should always be administered according to the manufacturer's directions. If the directions say that the medication should be injected intramuscularly, then care should be taken to ensure that the drug is delivered into a major muscle, like the thigh, rather than as a subcutaneous injection. Subcutaneous injection of a drug that should be given intramuscularly can result in injury to the bird and even death. The chemical formulation of a drug determines its route of delivery.

26-B. THERAPEUTIC LEVELS

In order to be effective, a drug must reach the site of infection at therapeutic levels and must remain at those levels for several days. The final distribution of the drug will depend on its chemical nature and how it is eliminated from the body. Some drugs remain in the intestine and are directly eliminated in the feces while others are absorbed into the bloodstream, processed and inactivated in the liver, or eliminated through the kidneys in the urine. Still others are processed and eliminated very rapidly while others are eliminated more slowly. Recommended doses and dosing schedules have been determined by the manufacturer based on these factors and should be followed carefully.

26-C. DOSING

Once a drug is administered, the concentration reaches a maximum level in the blood in 3 to 4 hours, after which elimination of the drug begins and its concentration in the blood diminishes rapidly. In many instances the drug is completely eliminated in 24 hours. If the drug is effective only at maximum concentrations, then the period during which it will kill microorganisms is short. To administer less of the drug than the recommended amount will result in concentrations less than those required for treatment and overdosing is costly and potentially dangerous.

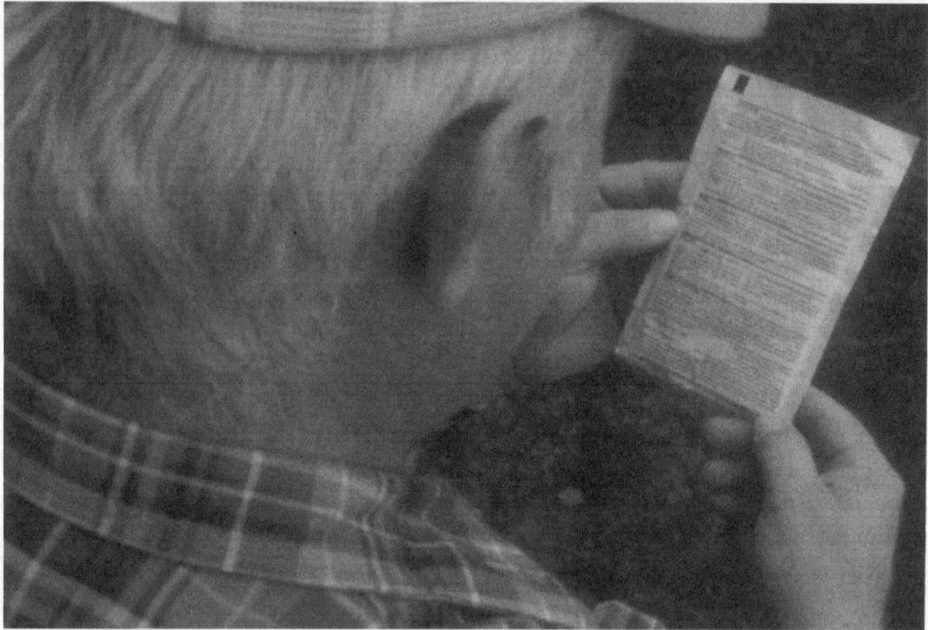

Figure 26-1. Read Instructions Carefully When Medicating Flocks

Improper Use of Drugs

Overdosing with drugs can be detrimental. First, they may cause a toxic reaction or affect various physiological functions of the bird. Second, overdosing may result in some of the drug not being broken down during the process of digestion and metabolism, and therefore, is not eliminated. If not eliminated, drugs can accumulate in tissues, gradually becoming toxic to the bird, or remaining as residue in the tissues.

26-D. WITHDRAWAL PERIODS FOR MEDICATIONS

Drugs that are not completely eliminated from the chicken can accumulate in muscles and eggs. Drugs and their by-products in food are called residues and can be harmful to humans if they are consumed. In many countries, governments have established tolerance levels for drug residues in meat and eggs. Withdrawal periods for selected antibiotics, anticoccidials, and anthelmintics are given in Tables 26-1, 26-2, and 26-3, respectively.

The withdrawal period given on a drug label is the recommended number of days to wait before consuming meat or eggs from treated birds. All medications should be withdrawn from birds several days prior to slaughter or before collecting eggs for human consumption. The recommended withdrawal periods have been carefully determined and should always be followed. If medications with withdrawal periods indicated are given to chickens in production, eggs must be discarded during the period of treatment and during the withdrawal period. The cost of administering drugs with withdrawal periods to egg layers must include the cost of the lost production. This should be considered when choosing what drug to use for treatment of a disease outbreak.

26-E. ANTIBIOTICS

An antibiotic is a substance that can inhibit growth of or kill a microorganism. Bacteria produce antibiotics as a mechanism of defense against other bacteria. There are hundreds of naturally occurring antibiotics, but only a relatively few have been found to be of value in treating diseases or conditions in chickens.

Antibiotics are used to control *bacterial* diseases in chickens. They prevent bacterial multiplication, provided they are used at therapeutic levels and the organism is not resistant. Antibiotics are of no value against infections caused by viruses. However, antibiotics are sometimes given in association with viral disease outbreaks in order to prevent secondary infections with disease agents like *Escherichia coli.*

Classes of Antibiotics and Their Activity

Sulfonamides

Sulfonamides are a group of synthetic antibiotics that inhibit the use of paraaminobenzoic acid (PABA), a chemical necessary for the synthesis of folic acid. Bacteria require folic acid to divide, thus, the inhibition of PABA prevents their multiplication. Sulfonamides, though, can be quite toxic for chickens, occasionally causing high mortality even when used at prescribed levels. Sulfonamides should be used with extreme caution in chickens and only with the guidance of a poultry veterinarian.

Fluoroquinolones

Fluoroquinolones attack a specific bacterial enzyme important in shaping bacterial DNA into its proper form, *DNA gyrase.* Precisely how the inhibition of DNA gyrase leads to bacterial death is not known.

Tetracyclines

This class of antibiotics inhibits protein synthesis required by growing or multiplying bacteria. Tetracyclines inhibit bacterial growth but do not

Table 26-1. Selected Antibiotics and Their Characteristics

Antibiotic	Administration	Notes
Bacitracin	Feed and water	No withdrawal period.
Chlortetracycline	Feed and water	Withdrawal period varies by manufacturer.
Erythromycin	Feed, water, injection	Not for use in layers; 2-day withdrawal period.
Lincomycin	Feed and water	No withdrawal period.
Oxytetracycline	Feed, water, injection	24-hour withdrawal period when used at therapeutic levels.
Enrofloxacin	Water	Not for use in layers; 2-day withdrawal period.

directly kill bacteria. They have a broad spectrum of activity against many microorganisms. Tetracyclines are readily absorbed from the intestine and once absorbed, they are widely distributed in the body.

Examples of Antibiotics

Table 26-1 lists several examples of antibiotics currently approved for use in chickens in the United States. This list may differ in other countries and may vary from year to year. It is always advisable to check the list of approved drugs for use in poultry before undertaking a treatment regimen.

Resistance to Antibiotics

When antibiotics are administered to a flock over a long period of time, particularly at a low level, bacteria can become resistant, making the antibiotic ineffective. For this reason, antibiotics should only be used medicinally after a specific disease diagnosis has been made. Use of antibiotics for growth promotion will be discussed later. A sensitivity test using the bacteria causing the disease will determine which antibiotic should be used for maximum effectiveness. Using the proper dose of an effective antibiotic to eliminate a specific bacterial disease agent will maximize the antibiotic's effectiveness and minimize the development of resistant bacteria.

In recent years, an increasing number of bacteria have become resistant to an increasing number of antibiotics. The management of diseases caused by resistant bacteria can be problematic if there is not an alternative treatment available. This has become a major concern in the treatment of human bacterial infections. Some resistant bacteria have developed from the use and misuse of antibiotics in humans. However, resistant bacteria can also be transferred from food animals to humans. To minimize the potential for transfer to human, antibiotics should always be used properly when treating the diseases of chickens.

Use as Growth Promotants

Some antibiotics are added to feed continuously at a low level to improve growth and feed conversion. This supplementation is not to be confused with therapeutic uses of antibiotics in which high levels are administered to treat a specific disease problem. The development of resistance by bacteria has been associated with this practice and, therefore, is not permitted in many countries.

Caution. Check with authorities before adding low-level antibiotics to a feed.

Increasing Activity of an Antibiotic

Calcium from the feed forms an insoluble salt when combined with oxytetracycline and chlortetracycline in the intestinal tract. The salt that is formed is insoluble, and therefore cannot be absorbed. If calcium in the feed is reduced, absorption of the antibiotic is increased, since a smaller portion forms an insoluble salt. Removing added calcium from the feed can double the amount of antibiotic absorbed.

Caution. Low-calcium feeds should not be fed to laying hens.

26-F. ANTICOCCIDIALS

The anticoccidial drugs are those that are used in the treatment of coccidiosis (see *Diseases of the Chicken,* Chapter 27). Therefore, coccidiostats fall into this group of therapeutic agents. Drugs that are not specifically listed in this category may also have some effect on coccidial infections.

Anticoccidials are generally used in floor-raised, and free-range chickens. Caged chickens rarely have clinical disease caused by coccidia. Coccidia are transmitted from bird to bird by the ingestion of infected fecal material (see *Diseases of the Chicken,* Chapter 27). Caged birds do not generally have enough contact with the feces of other birds to get coccidiosis and, hence, are not usually medicated with anticoccidial medications. However, floor-raised pullets that have not been on an effective parasite control program can have clinical coccidiosis even after being caged.

Classes of Anticoccidials and Their Activity

Amprolium

Amprolium has a structure similar to thiamine, an essential vitamin. This drug competitively inhibits the transport of thiamine into the coccidium and into chicken cells. Coccidia are fifty times more sensitive to thiamine deficiency than are chickens. Hence, at therapeutic levels coccidia are inhibited but the chicken is unaffected.

The Polyether Ionophores

The polyether ionophores include *monensin, salinomycin, lasalocid, narasin, maduramicin,* and *semiduramicin.* The polyether ionophores act by disrupting the internal balance of electrolytes in the coccidium. These ionophores are sometimes used in combination with other drugs. However, combinations should be used with care since the polyether ionophores can be toxic if used with certain antibiotics and some other coccidiostats.

Other Anticoccidials

Roxarsone is primarily used as a growth promotant but also has anticoccidial effects. It is commonly used in conjunction with other anticoccidial drugs to increase their activity. *Nicarbazin* and *halofuginone* have unknown mechanisms of action against coccidia.

Examples of Anticoccidials

The list of anticoccidial agents for poultry is quite long. A few selected agents currently approved for use in chickens in the United States and their characteristics are listed in Table 26-2. The list of approved coccidiostats changes every year and is different for every country. Before beginning any medication program, the current regulations should be checked.

Resistance to Anticoccidials

Resistance to coccidiostats has been a problem in the poultry industry. Resistance to some anticoccidial drugs appears very quickly, as early as

Table 26-2. Selected Anticoccidials and Their Characteristics

Anticoccidial	Administration	Notes
Monensin	Feed	Not for use in layers. No withdrawal period.
Lasalocid	Feed	No withdrawal period. Wet litter may be a problem if fed at higher levels.
Salinomycin	Feed	Not for use in layers. No withdrawal period.
Amprolium	Feed and water	No withdrawal period.
Nicarbazin	Feed	4-day withdrawal period. May effect egg quality, therefore do not use in layers or breeders.
Halofuginone	Feed	3-day withdrawal period. May interfere with skin repair and lead to skin tears.
Roxarsone	Feed and water	5-day withdrawal period. Not for use in layers.

weeks or months. Resistance to the polyether ionophores has taken years to develop. Shuttle or rotational programs can be used to prevent the development of resistance. *Shuttle programs* use the concept of changing anti-coccidial agents within the grow-out period of a flock. *Rotational programs* change coccidiostats *between* flocks. Shuttle and rotational programs should alternate *between* anti-coccidial agents with different modes of activity to prevent the development of resistance.

26-G. ANTHELMINTICS

Anthelmintics are drugs that are used for the treatment of nematode or cestode infections. Nematodes are roundworms such as *Ascaridia galli* (small intestinal roundworm) and *Heterakis gallinarum* (cecal worm). Cestodes are tapeworms such as *Raillietina cesticillus* and *Choanotaenia infundibulum.*

Tapeworms are transmitted through the ingestion of an insect intermediate host that has fed on the feces of infected chickens. Roundworms are directly transmitted by the ingestion of infective eggs in the feces of other chickens (see *Diseases of the Chicken,* Chapter 27). Transmission of these parasites occurs readily in birds housed on the floor but only rarely in caged birds. Hence, control of these parasites and the use of anthelmintics are limited to floor-raised and free-range chickens.

Anthelmintics and Their Activity

Piperazine. Piperazine blocks the transmission of signals from nerves to muscles in nematode worms. The worms are paralyzed by piperazine and are eliminated in the feces because they can no longer actively position themselves in the intestine of the bird. Mature worms are the most sensitive stage in the life cycle, so, repeated treatments are recommended to be fully effective.

Hygromycin B. When fed continuously, hygromycin B is an antibiotic that is effective in treating roundworms, cecal worms, and tracheal worms, if fed continuously. Its mechanism of action against nematodes is not known.

Anthelmintic Characteristics

There are only two anthelmintics currently approved for use in chickens in the United States. They are listed in Table 26-3 along with some of their characteristics. Like other drugs, the list of approved anthelmintics is subject to change and, therefore, it should be verified prior to treatment.

Table 26-3. Anthelmintics and Their Characteristics

Anthelmintic	Administration	Notes
Piperazine	Water	Effective against round worms but not cecal worms. No withdrawal period.
Hygromycin B	Feed	3-day withdrawal period.

Summary

New drugs are constantly being developed and some will become available for use in poultry. The emergence of a new drug always has a certain amount of optimism associated with it. That is, there is a sense that a new drug will be the answer to all disease problems. The truth is that there is no drug that can cure lapses in biosecurity, management problems, or vaccine failures. These problems must be addressed in conjunction with the use of any therapeutic agent.

27

Diseases of the Chicken

by Gregg J. Cutler

While there are many diseases affecting chickens, only the most important ones will be discussed in this chapter. Since this text is on management, the disease section will emphasize recommendations and directions for the prevention and treatment of each disease discussed. The material involving the cause, symptoms, transmission, and diagnosis has been reduced to include only information necessary to the important management programs. Diseases are listed alphabetically.

27-A. ASCITES OR PULMONARY HYPERTENSION SYNDROME (PHS)

Ascites syndrome (PHS) was first seen in rapidly growing young broiler chickens being grown at high elevation. It is an important cause of mortality in broilers. First identified in 1968, it has become increasingly prevalent since then and is most likely due to genetic improvement in growth rate. *Ascites* is an accumulation of fluid in the peritoneal cavities of the bird's interior. It is a noninflammatory fluid. In PHS, fluid most commonly accumulates in the ventral hepatic, peritoneal, or pericardial spaces. Mortality from PHS is usually less than 2% but in some flocks of roasters it can be as high as 15 to 20%.

Cause

PHS is caused by increased hydrostatic vascular pressure (blood pressure). Rapid growth and improved diets cause the bird to have a high metabolic rate. This creates a high oxygen demand and increases the work load on the heart. The muscle mass of broilers has increased much more than their lung capacity. High oxygen demand, combined with the relatively poor lung capacity, causes an increase in pulmonary blood pressure which causes the right ventricle of the heart to enlarge and thicken.

The enlargement of the right ventricle leads to its dilation, followed by right ventricular failure, congestion, and then ascites. Because of the anatomy of the right atrioventricular valve, thickening of the right ventricular wall causes the valve to leak. This causes more congestion and ascites.

Anything that reduces the oxygen availability in the broiler house such as high altitude, poor ventilation, or respiratory disease should be addressed. Cold-rearing temperature increases metabolic rate and thus oxygen demand leading to more cardiac strain and ascites.

Liver damage resulting in reduced venous return to the liver can also cause ascites. The damage to the liver can be caused by hepatic toxins. Some mycotoxins have been suspected, but their role is unclear.

Symptoms

Symptoms from PHS usually do not occur until right ventricular failure and ascites have occurred. Birds with ascites stop growing and are smaller than the normal birds. Affected birds will be depressed, have ruffled feathers and are reluctant to move. They may have reddened skin with peripheral blood vessel congestion. Distended abdomens and respiratory distress are common. Birds frequently die on their backs. Not all that die will have ascites. Some die acutely and will only be found to have an enlarged right heart.

Diagnosis

Clinical symptoms in rapidly growing broilers along with enlargement and dilation of the right ventricle and fluid in the heart sac with or without ascites in the abdomen lead to a diagnosis of PHS. More chronic cases may have a fibrotic liver.

Prevention

Reducing the metabolic rate and thus the demand for oxygen will help prevent PHS. This is accomplished with a combination of feed restriction and lighting programs to control feed intake. It should be remembered that this puts the grower in a difficult position. The goal is to have a program that allows good feed efficiency and low mortality without increasing the number of days to market. Broiler houses should not be too cold. Feed free of mycotoxins and other toxins will prevent PHS due to hepatic toxicity.

Treatment

There is no treatment for PHS.

27-B. ASPERGILLOSIS

Also known as *brooder pneumonia,* aspergillosis is most commonly seen as a disease of very young chicks. It is seen less frequently as a systemic or localized disease of older breeder flocks. The disease usually occurs when there are poor sanitary practices in place in the hatchery and on the farm. High mortality can occur in young chicks.

Cause

The disease is caused by a fungus, *Aspergillus fumigatus.* Normally this fungus grows on decaying organic material in the chicken house and in the hatchery, but also has the ability to reproduce itself in certain tissues of the bird.

Symptoms

The lungs are the major area of internal infection. A close examination of the lung tissue will show small nodules that are hard and yellow. There may only be a few in some cases; in others there may be hundreds. Fungal nodules will frequently be visible in the trachea, particularly at the syrinx. In some instances the fungus gains entrance into the air sacs, and labored, open mouth breathing is observed. Chicks have few external symptoms, except for gasping for air. As the fungus continues, older birds are able to withstand the fungal growth, and few birds are affected.

Transmission

Spores from the fungus dry and are transported easily from chick to chick in the air. This may occur in the hatchery or in the brooding house. The incubator may become a source of contamination, as can contaminated hatching eggs.

Diagnosis

In many instances a visual examination of the trachea or a sliced lung will show the nodules, and form a basis for diagnosis. In others, it will be necessary to submit chicks to the laboratory. Here, the suspected tissue is cultured and examined microscopically, or the fungus is isolated.

Treatment

Poultry house. Other than providing adequate ventilation, there is no treatment for birds that are infected. A thorough cleaning of the brooding premises will eliminate the source of infection for future flocks. Any moldy feed should be removed, bulk feed bins cleaned, old litter removed from the house and replenished with new, and waterers and feeders cleaned and disinfected.

Hatchery. A general cleanup of the hatchery and incubators will remove this source of infection. Be certain to use a disinfectant that is effective against fungi.

27-C. AVIAN ENCEPHALOMYELITIS (EPIDEMIC TREMOR) (AE)

This disease is found in virtually every area of the world. Avian encephalomyelitis causes neurologic signs including tremors in young chicks. Mature birds may show only a decrease in production which can be significant, but of relatively short duration. It is a disease of the intestinal tract and central nervous system that is caused by an *enterovirus.*

Symptoms in Young Birds

The disease is one that affects young chicks between the first and third week of age. They show nervous symptoms of varying proportions, including paralysis. The body quivers. Quivering is especially noticeable when the chick is held in the palm of the hand. Many chicks lie on their sides and cannot move. Mortality is high; death results not from the disease itself but because the chicks cannot get to feed and water. Birds over 4 weeks of age seldom show evidence of AE.

Symptoms in Adult Birds

With most outbreaks in adult flocks there will be no noticeable symptoms in the birds. However, egg production may drop, sometimes precipitously. Production usually returns to preinfection level in two weeks or less.

Transmission

The disease is transmitted by two methods:

1. *Fecal transfer of the virus.* As the virus multiplies in the intestinal tract, it is shed in the feces. Since the virus is very resistant to environmental factors, complete eradication is impractical. Contaminated water and feed are sources of transmission from bird to bird and house to house. Similarly, transfer of fecal material to uninfected houses or premises will cause an outbreak. The disease can be transmitted by people and equipment contaminated with virus-containing feces.
2. *Transmission through the hatching egg.* The virus is shed by the infected breeder hen through the hatching egg to the newly hatched chick. Unfortunately, because infection in the breeder flock may go unnoticed, the first indication of trouble can come from the growers of the chicks. As the infected eggs were laid at least 3 weeks prior to hatching, the disease has almost always run its course in the breeders by the time chick infection is noticed in the field.

Eggs in incubators must be destroyed. It takes just a few days for the infected breeder hen to recover, but because all birds in the flock do not have AE at the same time, most flocks will lay eggs containing the virus for a period of about 3 weeks. Therefore, under most conditions all eggs in the incubators from the infected flock must be destroyed.

Hatching eggs from a recovered flock. If a flock has had a natural outbreak of AE, and has recovered, hatching eggs will contain antibodies and may be used to produce chicks.

Diagnosis

A presumptive diagnosis of AE in young chicks can usually be made in the poultry house. However, other diseases that cause neurologic signs, such as Newcastle disease, deficiency of vitamins E, A, or riboflavin, Marek's disease, and toxicities such as with salt and pesticides, must be ruled out. Suspect chicks should be submitted to a laboratory. The following tests are used for the identification of AE:

1. *Histopathology.* Brain tissues are fixed, sliced thin, stained, and examined under a microscope.

2. Fluorescent antibody test.
3. Enzyme-linked immunosorbent assay (ELISA) test.
4. Virus isolation, although expensive and time-consuming.

Treatment

There is no effective treatment for this disease.

Prevention

Prevention of the disease is by using hatching eggs from breeder flocks that are immune to AE. A natural outbreak in young flocks produces this, but vaccination is the practical method.

When to vaccinate. Growing birds to be used for breeders should be vaccinated after 8 weeks of age and at least 4 weeks before the first eggs are laid. It is also advisable to vaccinate commercial layer pullets to prevent later drops in egg production. Vaccination can be done via drinking water. It is also often given by wing-web stab. Immunity to AE is long lasting. Breeder males should be vaccinated the same as the pullets.

Type of vaccine. Avian encephalomyelitis vaccines have been used for some time, and there are two types:

1. *Live virus vaccine.* Freeze-dried (lyophilized) vaccines are used.
2. *Killed-virus vaccine.* The live virus will usually cause a drop in egg production if it is administered to laying birds. A killed-virus vaccine may be used for flocks in production. There is less effect on egg production.

Determine status of vaccinated birds. About 4 weeks after vaccination, serum should be submitted to the laboratory for an evaluation of the presence of antibodies. The enzyme-linked immunosorbent assay (ELISA) test is most frequently used. Recheck breeders about four weeks before setting hatching eggs. Flocks that are positive on the ELISA test should produce chicks that are protected from clinical signs of AE.

27-D. AVIAN INFLUENZA (AI)

Avian influenza is a viral disease infecting many different species of poultry, affecting the respiratory, enteric, and nervous systems. In its most highly pathogenic form, it results in extremely high mortality. Low pathogenicity forms which result in almost no clinical signs also occur.

Cause

The AI virus is classified as an orthomyxovirus-type A. There are also numerous sub-types that are identified by hemagglutinin (H) and neuraminidase (N) typing. Each sub-type differs in its pathogenicity, ability to infect different species, and transmissibility.

Symptoms

Outbreaks of low pathogenicity AI may cause almost no symptoms. Some flocks that are infected with low pathogenicity AI will experience mostly respiratory symptoms including coughing, sneezing, and sinusitis. Highly pathogenic AI is a short-duration, generalized disease that may cause 90% or higher mortality rates. Egg production may suffer and there may be diarrhea, edema of the face and head, and / or nervous symptoms. Eggshell pigmentation and egg quality may also suffer.

Transmission

Waterfowl are the natural hosts of AI. The virus has been isolated from both wild and domestic species and from the water in lakes and ponds used by waterfowl. The disease is easily transported from farm to farm on contaminated clothing, crates, filler flats, and equipment.

Diagnosis

Serologic evidence may be useful in determining the exposure to AI viruses, but positive confirmation requires virus isolation and identification.

Treatment

There is no known treatment for this disease. Treatment with broad-spectrum antibiotics may reduce complications from secondary infections.

Control

Exclusion of waterfowl is essential. Because the disease is so easily transmitted, strict biosecurity is critical (see *Biosecurity on Chicken Farms,* Chapter 28). Highly pathogenic AI is a disease reportable to international disease control organizations. Outbreaks of highly pathogenic AI may cause trade embargoes and economic disruption. Strong control measures for highly pathogenic AI are required. Local authorities may require depopulation of infected premises. Vaccination is practiced in certain regions. Professional advice and diagnosis should be sought immediately if AI is suspected.

27-E. CAGE LAYER FATIGUE

It is known that the breaking strength of bones from layers held in cages is less than that of bones of birds kept on a litter floor. This difference is great enough at times to cause difficulties with caged layers. Furthermore, the brittleness of the bones of caged birds at the end of their laying year may be so great as to make the spent hens unacceptable for poultry processing. In the deboning process, their bones may disintegrate leaving fine bone splinters in the meat removed.

Cause

There is some indication that the difficulty may be due to an inadequate amount of inorganic phosphorus in the ration. However, this is not shared by all scientists. Feeding experimental diets with little or no inorganic phosphorus will not produce the symptoms; nor will adding inorganic phosphorus to the ration increase the strength of the bones. There is some indication that inadequate calcium during the period of peak egg production may be a precursor to the difficulty.

Symptoms

After long periods of egg production, layers may experience skeletal problems. They may lose control of their legs and lie on their sides. Birds showing this behavior usually have no loss of egg production, shell quality, or interior egg quality. Some of the bones may be fractured; some will break when the bird is handled. The ribs may be beaded at their cartilaginous junctures.

Treatment and Control

There is no generally recognized method of treatment. In some instances, adding extra vitamin D and calcium to the ration may be of benefit. One experiment showed that feeding a ration containing 6% calcium just before the birds were slaughtered increased bone strength. Other tests have indicated some alleviation when the phosphorus content of the ration is increased.

27-F. CAMPYLOBACTERIOSIS

Campylobacteriosis is a chronic, contagious bacterial disease of chickens and other animals including man. Also known as *avian vibrionic hepatitis,* campylobacteriosis causes liver disease, unevenness and increased culls in young growing flocks, and reduced egg production in older birds. Clinical disease in chickens is not now commonly seen but subclinical infections may be common. Special attention should be paid to this disease because of the association of serious enteric disease in humans and consumption of poultry products.

Cause

Campylobacteriosis is caused by *Campylobacter fetus spp. jejuni.* This is a gram-negative bacteria that falls into the vibrio group. There may be more than one strain. When the disease is seen it is usually in floor-reared flocks.

C. fetus spp. jejuni is shed in the feces. Spread is by ingestion of the organism, directly, or through contaminated feed or water. While subclinical infections may be common, stress, due to poor management, or other diseases may cause the appearance of clinical disease. Infected birds shed the organism for weeks to months.

Symptoms

Usually only a few birds appear affected. Culls increase in number. Affected birds appear unthrifty, are anemic, and have dry, crusty combs. Mortality increases slightly. In growing broilers, reduced weight gain and poor feed efficiency are noted. Layers show decreased egg production.

Internally there may be enteritis and swelling of the liver. Bone marrow may be pale and watery. A few birds may show distinctive necrotic lesions in the liver. The liver may also be brown or show hemorrhages under the capsule. Ascites and fluid around the heart may also be present.

Diagnosis

Clinical disease can be diagnosed by typical symptoms, flock history, and identification of the organism by culture of bile in the laboratory. The disease must be differentiated from other liver lesions producing diseases such as lymphoid leukosis, fowl cholera, inclusion body hepatitis, fowl typhoid, and pullorum.

Prevention

There are no vaccines available at this time for protection from Campylobacteriosis. High-quality sanitation programs and biosecurity should be practiced to prevent introduction of the organism (see *Biosecurity on Chicken Farms*, Chapter 28, and *Cleaning and Disinfection of Poultry Facilities*, Chapter 29). Care should be taken to prevent concurrent disease or management stress. Depopulation of infected flocks followed by thorough cleaning and disinfection will help prevent recurrence.

Treatment

Treatment is not very effective. Antibiotics such as tetracycline and dihydrostreptomycin may be helpful (see *Medication for the Prevention and Treatment of Diseases*, Chapter 26). Prevention through sanitation, biosecurity, and stress reduction, both management and disease, are the best methods of controlling campylobacteriosis.

27-G. CHICKEN INFECTIOUS ANEMIA

Chicken infectious anemia is a disease primarily of chickens less than 3 weeks of age causing anemia. It may also cause hemorrhages beneath the skin and in the muscles. The disease is immunosuppressive.

Cause

The disease is caused by a virus. The virus has not been classified.

Symptoms

Anemia is the main symptom. Infected birds can be depressed, pale, experience slow weight gain, and are very susceptible to infections from other agents due to immunosuppression. Mortality can be as high as 60% but usually is in the 5 to 10% range. Infections with other diseases that cause immunosuppression such as infectious bursal disease and Marek's disease cause high mortality. The bone marrow is typically pale and yellow. The bursa of Fabricius and thymus will be small. The liver is often swollen and hemorrhages are seen under the skin and in the muscles.

Transmission

The disease is transmitted vertically from the breeder flock through the hatching egg to the chick. If the breeder flock is undergoing viremia, infected eggs are produced. Chicks that hatch without maternal antibodies are most susceptible. These chicks then spread the virus rapidly through the flock. There are no known hosts other than the chicken.

Diagnosis

Diagnosis is based on observation of clinical signs. The virus is very difficult to isolate. An ELISA test is available to check breeder flocks and chicks for presence of antibodies.

Treatment

No effective treatment is known. Treatment of secondary infections with appropriate antibiotics is helpful. The addition of vitamins to the drinking water is a general supportive therapy.

Control

The use of seropositive breeder flocks for hatching eggs is the best way to prevent flocks from contracting chicken infectious anemia. Most flocks turn seropositive in the grow period. Vaccination is recommended to immunize the flock before they begin producing eggs.

27-H. COCCIDIOSIS

Coccidiosis is a term used to identify the disease in chickens produced by a group of protozoan organisms in the genus *Eimeria.* There are hundreds of types, but only nine are important to the raiser of chickens. It is one of the most devastating of poultry diseases. As larger groups of birds were raised in confinement, the significance of coccidiosis increased. By the 1930s and 1940s much information about the disease was understood. It is now recognized that concurrent infection with one of the immunosuppressive diseases such as infectious bursal disease or Marek's disease increases the severity of coccidiosis.

Cause

Coccidia have specific hosts, and each species produces its own type of coccidiosis. The coccidia that inhabit the chicken are all in the genus *Eimeria.* Coccidiosis is spread by unicellular bodies known as *oocysts.* These are shed in the droppings, but are not infectious. They first must be *sporulated,* a process that takes place when conditions of air, temperature, and moisture are opportune. This sporulation, therefore, occurs outside the body, requiring 2 to 4 days. The sporulated oocyst is eaten by the chicken and finds its way to the intestinal tract, where a series of divisions and multiplications take place. Eventually more oocysts are produced, most of which again are expelled from the body in the fecal material, and the life cycle is complete. The time from ingestion to expulsion varies between 4 and 7 days, depending on the species.

Coccidia, being parasitic, live in the epithelial tissues of the intestinal tract, and there they inflict their damage. Usually one oocyst in the intestinal tract can destroy only a few cells. Therefore, the extent of destruction of the intestinal wall is closely related to the number of oocysts present. But when the disease is at its height, there will be millions of oocysts, although not all will be sporulated. Continued ingestion of the sporulated oocysts means that more cell tissue is destroyed.

External Symptoms

Symptoms depend on the species of *Eimeria* involved. They can be mild to severe. They include bloody droppings, ruffled feathers, paleness, loss of appetite, lowered growth, poor feed conversion, drop in egg production, diarrhea, and many other manifestations. Most of the external symptoms are the result of general disability due to the destruction of the lining of the intestinal tract. This is turn prevents the absorption of food material from the intestines to the bloodstream. In many cases hemorrhaging of the intestinal lining occurs, and blood in many forms is deposited with the droppings.

Internal Symptoms

Internal symptoms usually are confined to the intestinal tract, including the ceca. Inflammation, hemorrhages, lesions, mucus, and exudates are all

Table 27-1. Characteristics of *Coccidia* Species

Species (Common Name)	External Symptoms	Intestinal Area Most Affected	Mortality	Morbidity
E. necatrix (Intestinal coccidiosis)	Diarrhea Bloody droppings Ruffled feathers Weight loss	Whitish lesions on upper third of small intestine	Heavy	Heavy
E. tenella	Bloody droppings Drop in feed Droopy Fewer eggs	Hemorrhagic ceca	Heavy	Heavy
E. acervulina (Layer coccidiosis)	Diarrhea Fewer eggs Drop in feed Weight loss	Upper half of small intestine	Light	Medium
E. brunetti (Intestinal coccidiosis)	Diarrhea Emaciation	Lower half of small intestine, ceca, and cloaca	Light	Light
E. maxima (Intestinal coccidiosis)	Diarrhea Bloody droppings Drop in feed Pale color	Middle and lower sections of small intestines	Light	Medium
E. mivati (Intestinal coccidiosis)	Fewer eggs Ruffled feathers	Lower half of small intestine	Light	Heavy
E. hagani	Diarrhea Drop in feed	Upper half of small intestine	Light	Light
E. praecox	Diarrhea Weight loss	Upper third of small intestine	Light	Light
E. mitis	Diarrhea	Entire small intestine	Light	Light

Source: North and Bell, 1990

indicative of coccidiosis, but most species produce specific internal symptoms. These are outlined in Table 27-1.

Transmission

The only method of transmission is for the bird to consume sporulated oocysts. Within the pen this goes on constantly because birds have access to the droppings. The transfer of oocysts from house to house and farm to farm is by mechanical means. In the active stage of the disease, millions of oocysts are contained in each teaspoonful of fecal material. These are easily transferred to a new location by shoes, feed trucks, crates, pets, rodents, and moving equipment. Once sporulated, the oocysts soon cause an outbreak of coccidiosis in the new premises.

Diagnosis

Appearance of the birds, along with intestinal lesions, may be sufficient for a diagnosis of most outbreaks. However, many symptoms are similar to those produced by other diseases. A laboratory diagnosis therefore is necessary. Scrapings are made of the infected area of the digestive tract, and a microscopic examination is made for the presence of coccidia.

Control

Coccidiosis is far more easily prevented than treated. Certain chemicals, known as anticoccidials, suppress the life cycle of the protozoan or actually kill it at a specific stage of the life cycle. Those chemicals that suppress the life cycle are known as coccidiostatic. Those that actually kill the protozoa are known as coccidiocidal. Anticoccidials usually are added to the feed at a designated percentage. However, all such chemicals do not have equal ability to suppress or kill all of the *Eimeria* spp. Such drugs reduce or eliminate the shedding of oocysts in the droppings, thus reducing or preventing oocyst contamination of the poultry house floor. Some species of *Eimeria* have become resistant to certain anticoccidial drugs, mainly those that have been used consistently for treating several generations of birds. It is best to rotate anticoccidial drugs. When withdrawing coccidiostatic drugs, the disease may return since the protozoa were only suppressed and not killed (see *Medication for the Prevention and Treatment of Diseases,* Chapter 26).

Most outbreaks of coccidiosis are produced by three *Eimeria:*

1. *E. tenella*
2. *E. necatrix*
3. *E. acervulina*

The others account for less than 15% of the cases, and usually involve only slight economic loss. Any good anticoccidial should at least be specific for the above three *Eimeria,* plus perhaps *E. brunetti* and *E. maxima.*

Properties of Good Anticoccidials

1. Prevent infection from as many species of *Eimeria* as possible.
2. Do not interfere with reproduction (egg production and fertility).
3. Should be nontoxic, palatable, and stable.
4. Should be economically acceptable.

Coccidiosis Control Program for Broilers

Broilers should be fed an anticoccidial that will completely suppress coccidiosis. This is difficult inasmuch as coccidia eventually develop some degree of resistance to specific drugs. When this happens it becomes necessary to change the anticoccidial. Some broiler producers make this change every 8 to 12 months; others change only when they see a failure with the drug being fed; others resort to a "shuttle" program in which one anticoccidial is used the first 10 to 15 days of the growing period, then another for the remainder. Some use three anticoccidials.

Caution. Get professional advice before mixing anticoccidials. Remember that in the United States the use of all medications, including anticoccidials, are regulated by the Food and Drug Administration (FDA).

Side effects. Some anticoccidials reduce feed consumption and feed conversion. Others reduce the absorption of specific nutrients from the intestine. Skin pigmentation may be reduced by others. Consult a specialist in this field before selecting a drug for your coccidiosis-control program.

Withdrawal period. Most anticoccidials must be withdrawn from broiler feed for a specified period prior to slaughter to allow time for the drug to be eliminated from the tissues.

Immunity is not to be produced in broilers. There is not enough time prior to marketing to develop immunity to coccidiosis and allow the birds to recover. All drugs should be fed at full strength (necessary for full suppression of coccidiosis) from 1 day of age to near broiler marketing age.

Coccidiosis Control Program for Replacement Birds

Under natural outbreaks, chickens with coccidiosis develop immunity to the *Eimeria* species that caused the disease. This immunity does not last a lifetime, but as the bird is continually consuming more sporulated oocysts, immunity is continually stimulated throughout the life of the bird.

Immunity may be developed artificially in the bird without the stress of its enduring an attack of acute coccidiosis. This process is made possible by the fact that if the number of oocysts in the intestinal tract is kept at a low level there will be no danger of serious coccidiosis, yet the number will be adequate to enable the bird to build an immunity.

Most coccidiocidal drugs suppress oocyst reproduction completely when they are fed at a designated level in the feed. No immunity can build up during such feeding. Before the bird can acquire immunity, a coccidiostat must be used at a low level in order that some sporulated oocysts, consumed from the litter, may be allowed to complete their life cycle. A small reduction builds a little immunity and the coccidiostat may then be further reduced, producing more immunity. Gradually, all the coccidiostat is removed, and immunity is complete. The trick is to reduce the coccidiostat fast enough to produce just a very little coccidiosis. There must be a little, or no immunity will result.

How much moisture in the litter? At 20 to 30% moisture, optimum oocyst sporulation occurs. This moisture level should be maintained. Most types of litter should have approximately the same consistency and feeling as shavings cut from green lumber. Litter moisture should be carefully monitored.

Coccidiosis Control Program with No Anticoccidial

Producers of egg-type replacement pullets have largely gone to cage rearing of replacements. Coccidiosis is seldom a problem in this type of housing. It can be a problem if manure builds up, particularly in the corners of the cages. If there is a problem with coccidiosis, a treatment regimen is used.

Coccidiosis Inoculation

Inoculation of birds for coccidiosis is a practical procedure. Products composed of live oocysts are given in the water or fed to chicks. Eventually, oocysts are deposited in the litter, they sporulate, the bird consumes relatively small numbers, and immunity begins. The program gives assurance that inoculation can start at an early age, under controlled conditions. Immunity may develop in 5 or 6 weeks.

Treatment

In spite of many excellent programs and chemicals for the control of coccidiosis, there are many outbreaks of the disease. Treatment, therefore, becomes a necessary part of any control program. Several drugs are used; some are to be given in the feed and some in the water (see *Medication for the Prevention and Treatment of Diseases,* Chapter 26).

Cause of Coccidiosis Outbreaks

Why is there still so much difficulty from coccidiosis? Why are there so many outbreaks? Some of the reasons are:

Broilers

1. Anticoccidial too weak for the *Eimeria* spp. involved.
2. Incorrect percentages of the anticoccidial in the feed.
3. *Eimeria* spp. have become resistant to the drug.
4. Litter too damp, causing increased sporulation of the oocysts.

Breeder replacements (or for laying purposes)

1. Anticoccidial too weak for control of specific *Eimeria.*
2. *Eimeria* have become resistant to the drug.
3. Litter too dry or too wet.
4. Coccidiostat fed at too high a level, thus making it impossible for the bird to develop proper immunity.
5. Coccidiostat removed too fast.
6. Other drugs are being fed, potentiating the effects of the coccidiostat.

How to handle outbreaks. Outbreaks of coccidiosis, regardless of the cause, should be handled quickly. Time is most important. As feed consumption is reduced during coccidiosis outbreaks, medications should be given in the water. A chicken will drink even though it will not eat. Most applications are made through the water. After this, the flock is returned to the normal coccidiosis-control program. If there are reasons for failures in the first place, make corrections at this time.

27-I. COLIBACILLOSIS (*E. COLI* INFECTION)

The coliform organisms are responsible for a variety of poultry diseases with a variation in manifestation. The *Escherichia coli* (*E. coli*) are bacteria that represent one of many of the coliform group of organisms inhabiting the lower intestinal tract. Most are harmless and are called *saprophytic.* These aid in the process of digestion. Others are *pathogenic* and produce certain poultry diseases. Colibacillosis in chickens is usually a secondary disease occurring when host defenses have been compromised or overwhelmed. *E. coli* occurs in a variety of serotypes. Most that are pathogenic to chickens are not important in other animals, including humans. However, chickens can become infected with *E. coli* O157:H7 which is an important pathogen in humans and is important from a food safety standpoint.

Specific Diseases

E. coli are responsible for several types of diseases, namely:

1. airsacculitis	4. salpingitis
2. omphalitis	5. cellulitis
3. synovitis	6. coli septicemia

Only the most important diseases produced by *E. coli* organisms will be discussed.

1. *Airsacculitis*

E. coli organisms may find their way into the air sacs by a direct path. They may enter the upper respiratory tract through the process of breathing, and soon settle in the thoracic air sacs, then eventually find their way to the abdominal air sacs. When infection reaches its height, the air sacs become filled with a yellowish cheesy material. A similar material may also surround the heart and lungs. Sometimes when *E. coli* infects the air sacs, morbidity, rather than mortality, becomes the economic problem. In broilers, birds with infected air sacs will be condemned in the processing plant as unfit for human consumption. Other organisms such as the mycoplasmas may be involved as primary or secondary pathogens in airsacculitis.

2. *Omphalitis*

This condition is also known as *navel infections.* Newly hatched chicks may be depressed, have diarrhea and increased mortality. The navel is reddened and swollen. The yolk is abnormal looking and may have a foul odor. There are often other bacteria contributing to this condition. The cause is usually poor hatchery sanitation or dirty hatching eggs.

3. *Synovitis*

Birds will have infected, swollen joints and be lame. Birds usually recover quickly but some will become chronic and emaciated. There are usually other infectious agents involved such as *Staphylococcus, Mycoplasma,* or viruses.

4. *Salpingitis*

The oviduct becomes enlarged and filled with foul smelling cheesy material. Birds may die without other symptoms. The abdomen feels hard and full. This may occur after an air sac infection or as an ascending infection from the cloaca.

5. *Cellulitis*

This is also known as *infectious process.* It is usually found at slaughter. There is yellowish discoloration of the skin with some dimpling noted. When the skin is cut, the subcutaneous area has a grayish yellow exudate present. Bacterial colonies can be seen in the exudate upon histopathology. This condition requires trimming, downgrading and condemnation of some birds. It can occur at a rate of 5% or more. There appears to be a correlation with flocks that have exhibited *E. coli,* airsacculitis, scratches, and navel infections.

6. *Coli Septicemia*

This is an acute, generalized infection causes by *E. coli.* It can resemble other similar generalized infections such as fowl typhoid or fowl cholera. There is variable mortality. Livers are green and swollen. Other internal organs may be swollen. There can be congestion of the breast muscles and small hemorrhages on the heart sac and peritoneum.

Transmission

There are several means of *E. coli* transmission:

1. *Fecal.* Organisms in the intestinal tract are continually being shed through fecal material, and in turn these bacteria dry and float in the air and gain entrance to uninfected individuals by way of the respiratory tract.
2. *Eggshell contamination.* As the shelled egg lies in the cloaca prior to being laid, it becomes contaminated with the excrement of the intestinal tract, including *E. coli.* More bacteria are added to the shell when the egg remains in the nest. Subsequently, some organisms may contaminate the egg through the pores in the shell and reach the developing embryo with resultant loss in hatchability and chick quality.
3. *Ovarian.* Transmission through the ovary is possible when birds are shedding the *E. coli* organisms through uterine infection. Infected breeder hens thus transmit the disease to the newly hatched chick.

4. *Feed.* Although not a primary route of infection, coliforms may gain entrance into the body through contaminated feed.

Diagnosis

A laboratory test is the only satisfactory method of accurate diagnosis. Coliforms are isolated and identified on artificial media in the bacteriology laboratory.

Treatment

Any treatment must begin with a complete sanitation program as most *E. coli* infections start with dirty surroundings and various environmental insults. Broad-spectrum antibiotics may be helpful in treating the conditions. Many *E. coli*, though, are resistant to the commonly used poultry antibiotics. An antibiotic sensitivity test is essential to help identify the best drug to use in a treatment program.

27-J. EGG-DROP SYNDROME (EDS 76)

Egg-drop syndrome is a disease of laying hens that causes sudden losses of egg production, depigmentation of brown-shelled eggs, and increased incidence of thin-shelled and shell-less eggs.

Cause

Egg-drop syndrome is caused by an *adenovirus,* first described in 1976 in Holland. It has been seen in numerous countries, but at the present time has not occurred in the United States. The organism is commonly recovered from waterfowl but the disease does not cause reproductive problems in these species.

Symptoms

The symptoms are almost exclusively associated with the reproductive performance of the flock. Egg production may drop by 40 to 50% for periods of up to 10 weeks. Brown-shelled eggs are laid without pigment. Many soft-shelled and shell-less eggs are laid.

Transmission

While waterfowl are the natural hosts of the disease, their involvement in the transmission of EDS 76 is extremely rare. Contaminated vaccines and needles were responsible for earlier cases of transmission. Vertical transmission through breeders has been a source in the spread of this disease. Infected chicks excrete no virus and often show no serologic evidence of the disease until egg production reaches 50%. Horizontal transmission is slow in cages.

Diagnosis

Along with the specific clinical signs observed, the hemagglutination inhibition (HI) and serum neutralization (SN) tests are the procedures most commonly used. Immunofluorescence tests can be used on tissues in the laboratory.

Treatment

There are no known treatment procedures available to combat the disease.

Control

An inactivated vaccine is available and is used in replacement pullet flocks at 14 to 16 weeks of age in areas where the disease is present. Strong biosecurity should be practiced. Visitors should not be permitted to enter poultry houses. Waterfowl should be excluded from all poultry premises.

27-K. FATTY LIVER SYNDROME (FLS)

Fatty liver syndrome is a metabolic disorder in laying hens, causing excess fat in the liver. Factors that will cause an increased deposition of fat in the cells of the liver are toxins, high egg production, nutritional imbalances, excessive consumption of high energy diets, deficiency of nutrients that mobilize fat from the liver (lipotrophic agents), and endocrine imbalances. There may also be a genetic component.

Cause

The exact cause of increased fat deposits in the liver of the chicken affected with FLS is not known. It has been impossible to produce the conditions experimentally. However, there is now adequate evidence to show that FLS is of nutritional origin, and that it is not pathogenic. There appears to be a breakdown of the metabolic processes involved in the synthesis and mobilization of lipids, particularly during the stress created by heavy egg production.

Symptoms

Fatty liver syndrome appears only in good-laying flocks. Most birds appear in good physical condition. For this reason there is little indication of the disease until egg production drops between 10 and 40% or the flock fails to have a high peak of egg production. Body weight increases from 20 to 25% are noted. The problem is more acute with birds housed in cages than with those housed on the floor.

Postmortem examination will reveal an enlarged, fatty, and friable liver, tan in color. Obesity, as indicated by internal fatty deposits, will be evident. The normal liver will contain about 36% fat, while one with FLS will have about 55% fat. Death usually occurs from rupture of the liver along with gross hemorrhaging.

Treatment

Various nutritional supplements have been tried with mixed results. The addition of the following to the feed has been shown to alleviate the condition in some laying flocks, but not all:

Add to a ton (2,000 lb) of regular feed

10,000 IU vitamin E	12 mg vitamin B_{12}
1,000 g choline chloride	900 g inositol

The success and failure of different treatments is an indication that FLS is not a single-source problem, but rather one that involves complex nutritional and possibly environmental interactions that are still unknown.

Control

Reducing energy intake, either by feed restriction or by lowering the ME content of the diet, has merit. The addition of alfalfa meal and wheat bran will help control the problem if it has not been recurrent.

27-L. FOWL CHOLERA

Fowl cholera is a bacterial disease that affects chickens, turkeys, waterfowl and other birds. The disease usually occurs as an acute septicemia, but a chronic form also exists. Although better sanitation programs, along with less range-rearing of growing birds, have brought about a reduction in the incidence of the disease, there are still many outbreaks, and they still are difficult to control. The disease is very important with turkeys, probably because many are grown on the ground, and turkeys seem to be more susceptible than chickens.

Cause

The disease is due to a gram-negative bacterium, *Pasteurella multocida*. There are many immunologically different serotypes associated with different avian species.

External Symptoms

Acute form. Usually, the first observation of the disease is a high incidence of birds dead on the floor or roosts, or in the nests—many times without any obvious external or previous symptoms. Mortality is rapid; 50% or more of the birds may die. Birds between 12 and 18 weeks of age seem very susceptible. There may be a greenish colored diarrhea.

Chronic form. Death losses are relatively low from the chronic form. The most obvious symptom is a swelling of the wattles, particularly in male birds. One or both wattles may be swollen containing a cheesy, hard deposit, often resembling a marble. The disease may affect the inner ear, causing a twisting of the head, and an unsteady gait. Some birds may appear lame. Adult birds show more of this symptom than do young birds.

Internal Symptoms

In the acute form there may be few if any lesions. There may be very small hemorrhages on the liver, and sometimes in the heart and intestines. The liver may be enlarged. The disease takes on the form similar to that encountered with chronic respiratory disease (CRD). The air sacs may contain lesions, and the heart sac is enclosed in a yellowish-appearing film. Other organs can also be affected.

Transmission

The organism responsible for fowl cholera may be passed easily from bird to bird. It will enter the body through either the respiratory or digestive tract. Transmission can be by people, clothing, or footwear and through contaminated feed and water. Some mammalian pests, such as rodents, may also transmit fowl cholera. A healthy bird pecking at a contaminated bird may be a means of spreading the disease through blood or nasal exudate. Artificial insemination is a means of spread through contaminated equipment and people. Ulcerated wattles are a source of infection. The disease is not known to be egg-transmitted. Recovered birds may become chronic carriers.

Diagnosis

Fowl cholera can be diagnosed accurately only in the laboratory. The method is bacteriological, where the organism is isolated and identified. The test should take no longer than a few days.

Treatment

Treatment involves the use of sulfonamides or antibiotics. Sulfadimethoxine, sulfaquinoxaline, tetracyclines, erythromycin, and penicillin have been found useful. Although the treatment may bring the disease under control and reduce mortality, relapses are frequent once the drug is withdrawn. Extreme care should be used when treating with sulfonamides as birds may be forced out of egg production, and under certain conditions mortality may occur from kidney damage.

Control

Prevention of the introduction of fowl cholera is the best method of control. Good sanitation and the reduction of stress are important to controlling infection. Dead birds should be picked up frequently to reduce spread. Although bacterins have not always been effective, they have helped with control. The bacterin must be produced from the correct serotype to be effective. Live vaccines are now available. They are relatively effective but proper administration is important to prevent adverse vaccine reactions.

Complete Control

The only satisfactory method for the control of cholera is by eradication. Treatment of growing pullets or laying hens either with drugs or bacterins should be considered a temporary measure. The organism must be eliminated from the farm.

27-M. FOWL POX

Fowl pox may be found in most poultry regions, but due to the use of vaccines it is not often seen in commercial flocks. However, in certain areas of the world it still remains a disease of major importance and is commonly seen in numerous small unvaccinated flocks. It is caused by the fowl pox virus, a member of the genus *Avipoxvirus*. Strain variations do exist but there are varying degrees of cross protection. It now appears that some variant strains do exist that are not protected by vaccination.

Symptoms

Pox may occur at any age; parental immunity has no effect. There are two forms of the disease, and the symptoms of each are different:

1. *Cutaneous type (dry pox).* Wartlike scabs are found on the comb, wattles, eyelids, and earlobes. There may be a loss of appetite and general unthriftiness. Egg production drops and fertility is impaired. In most instances little mortality occurs from the skin type of fowl pox.
2. *Diphtheritic type (wet pox).* Yellowish, cankerous, and cheesy lesions appear on the internal wet surfaces of the mouth, tongue, esophagus, trachea, nasal passages, and sometimes the crop. When such lesions are removed, profuse bleeding occurs. Breathing is hampered by the diphtheritic membranes and exudates, and the birds may suffocate. Egg production is retarded in laying birds and fertility is lowered. Mortality is much higher with the diphtheritic type than with the skin type.

Transmission

The skin acts as a barrier to the entrance of the pox virus. The skin must be broken for the organism to gain entrance. Transfer of the virus from bird to bird is relatively easy. Spread may be categorized:

1. *Bird to bird.* Both forms spread slowly from one bird to another in the pen. Most of the transmission is the result of birds picking, fighting, or scratching one another.
2. *Mosquito.* The mosquito is definitely a means of spread.

It punctures the lesions of affected birds, then transfers the virus to other birds when it "bites" them. Consequently, fowl pox is more prevalent during the mosquito season. In some instances, the virus may enter the body of the mosquito and it will remain a carrier for several months.

Diagnosis

Pox lesions on the facial tissues of the bird are a definite indication of dry fowl pox. No other poultry disease produces such lesions. But, the diphtheritic type of the disease is difficult to diagnose by visual means. Several types of laboratory techniques are used to make a diagnosis:

1. *Bird transfer.* A small amount of scab material from an infected bird is scratched onto the surface of the comb of an uninfected bird. About 5 days later, typical pox scabs will arise when dry pox is involved.
2. *Virus isolation.*
3. *Inclusion bodies.* Bollinger bodies are easily demonstrated inside the cytoplasm of affected cells in material collected from pox lesions during the early period of the disease.

Treatment

Treatment generally is ineffective, although some drugs may be used to reduce morbidity, particularly from secondary pathogens. Although technically not a treatment, vaccinating a flock at the earliest signs of a fowl pox outbreak with a wing-web fowl pox vaccine will stop most outbreaks in a short period of time.

Control

Control is readily accomplished by vaccination. There are several types of vaccines:

1. *Fowl pox vaccine.* This is a virulent type of vaccine. Being a live virus, it is capable of spreading the disease and therefore must be used carefully. The entire flock should be vaccinated in a few days to prevent the vaccine from spreading to other areas in the house.
   a. Because of its virulence, birds must be 5 weeks of age or older when vaccinated.

b. Birds must not be under stress when vaccine is administered.
c. Fowl pox vaccine is available commercially in combination with avian encephalomyelitis vaccine. It also is frequently given at the same time as the laryngotracheitis vaccination.

2. *Attenuated fowl pox vaccine.* This is a mild form of fowl pox vaccine and gives good immunity without many of the side effects of the fowl pox vaccine. However, it may not stand up to severe challenge.
3. *Pigeon pox vaccine.* This vaccine is milder than fowl pox vaccine. Consequently, it may be used in cases where fowl pox vaccine produces a severe reaction, as in:
   a. Day-old or very young chicks.
   b. Birds undergoing stress.
   c. Birds in egg production.

In areas that have problems with the diphtheritic form it has been found that pigeon pox vaccine may be helpful as a means of control.

Fowl pox vaccine is applied by the wing-web puncture method. An application tool comes with the vaccine. It is a double needle with a depression to which the proper amount of vaccine adheres by surface tension. Use both needles. Do not remove one to save vaccine.

Examination for a "take." About 10 days after vaccination, the birds should be examined. If the vaccination has "taken" a definite pox scab will be obvious on the wing-web where the puncture was made. If there is no scab the birds should be revaccinated.

Vaccinating broilers. If it becomes necessary to follow a vaccination program, pigeon pox vaccine should be used. Chicks can be vaccinated at 1 day of age. A mosquito-control program should also be made a part of the preventive procedure.

27-N. GANGRENOUS DERMATITIS

This disease is also known as *necrotic dermatitis.* It causes death of skin over various areas of the chicken. There is severe infection under these areas. There is evidence immunosuppression occurs in affected chickens. Occurrence is more common in warm, humid facilities and it tends to affect birds 4 to 16 weeks of age.

Cause

Bacteria such as *Clostridia* spp., *E. coli,* and *Staphylococcus* spp. are usually involved in this disease. Birds may receive small injuries to the skin which can then become infected by the bacteria. When there is immunosuppression due to infection with infectious bursal disease or chicken infectious anemia virus, the bird cannot overcome these infections and severe necrosis of the skin occurs. Mycotoxins, poor sanitation, and nutritional deficiencies may also contribute to the disease.

Symptoms

Depression and mortality are usually the first symptoms seen. Birds may be down and lameness may be seen. Birds with dark patches of skin with feather loss is common. These areas of skin often have underlying gas bubbles, especially in clostridial infections. There is also blood tinged watery fluid under the skin. Mortality is variable, but it can sometimes be very high. There may also be necrotic areas in the liver. Often there is a marked reduction in size of the bursa of Fabricius.

Diagnosis

Observation of clinical symptoms and bacterial culture are usually used for diagnosis. Histologic examination of tissues will confirm presence of bacteria and necrosis.

Treatment

Broad-spectrum antibiotics are used to treat gangrenous dermatitis. They should be added to the water for fast action. Follow up can be accomplished by the addition of antibiotics to the feed. Prevention is much more effective for controlling this disease.

Control

Efforts should be taken to prevent injuries to the skin. House sanitation should be carefully controlled. Feeder, water, and litter management are very important. Vaccination programs for infectious bursal disease will help eliminate the immunosuppression factor in this disease. Good nutrition and prevention of stress are also helpful.

27-O. INCLUSION BODY HEPATITIS (IBH)

Inclusion body hepatitis is a disease of young chickens characterized by the sudden onset of increased mortality, anemia, and hepatitis. There are often intranuclear inclusion bodies in the cells of the liver. The anemia sometimes seen with this disease may actually be caused by interaction with the chicken infectious anemia virus (see section 27-G).

Cause

Inclusion body hepatitis is caused by a group 1 adenovirus. Adenoviruses are very common in chickens. Many produce no or little disease and therefore, are often found in healthy chickens. Inclusion body hepatitis can be caused by a number of different serotypes of adenovirus. The disease is much more severe when there is underlying immunosuppression caused by infections such as infectious bursal disease or chicken infectious anemia virus.

Symptoms

The disease is commonly seen in chickens 3 to 18 weeks of age but most commonly in those aged 4 to 8 weeks. There is usually a sharp increase in mortality followed by a return to expected levels of mortality over a 2-week period. Total mortality may reach 10%. General depression and listlessness is observed although only a few birds may appear sick. Some die in just a few hours after the onset of signs. The skin is pale and may contain hemorrhages over the legs and breast. The liver and kidneys are frequently swollen, bone marrow is yellow, blood is thin and watery, and hemorrhages are often seen in skeletal muscles.

Transmission

Adenoviruses are quite widespread and transmission from farm to farm has been observed. The virus is excreted in the feces and can be mechanically spread by humans. It is suspected that the organism is also transmitted through the egg. It can then be spread horizontally through the flock.

Diagnosis

The agar-gel-precipitin test is commonly used to demonstrate antibodies to IBH, which in conjunction with mortality patterns and gross lesions are used to identify the disease.

Treatment

No treatment for the disease is available.

Control

Normal sanitation recommendations should be followed. Breeding flocks with IBH should not be used for producing hatching eggs. Wild birds should be excluded from the poultry house. If infectious bursal disease (IBD) predisposes the flock to IBH, a vaccination program for IBD should be considered. The role of chicken infectious anemia virus should also be investigated.

27-P. INFECTIOUS BRONCHITIS (IB)

Infectious bronchitis is an acute respiratory disease affecting chickens in every part of the world. It is a serious disease of young chicks, causing high mortality. In laying birds it causes great economic loss through reduced egg production, watery albumen, and poor eggshell quality. Some strains cause kidney lesions which lead to higher than expected flock mortality. The chicken is the only bird known to be susceptible.

Cause

The disease is caused by a *coronavirus.* There are 20 or more recognized serotypes. Some of the serotypes produce cross-immunity while others do not. Coronaviruses have a propensity toward genetic mutation. This can lead to new field strains which established vaccines may not protect against. It is best not to add new vaccine strains to geographical regions when the specific serotype has not been isolated as doing so may introduce new pathogenic strains or induce mutation of field strains present. The best known serotypes are:

Massachusetts
Connecticut
Holland
Arkansas-99
JMK
Florida

Properties of Important Types

1. The Massachusetts strain produces the severest type of the disease.
2. The Massachusetts strain produces cross-immunity with the Connecticut strain.

3. The Connecticut strain produces a poor cross-immunity with the Massachusetts strain.
4. The Holland strain produces cross-immunity with Massachusetts, Connecticut, JMK, Florida, and SE-17 strains.

Symptoms

The disease produces symptoms varying with age. Chicks and adult chickens are affected differently.

Infectious bronchitis in chicks. In young chicks there is a noticeable wheezing and sneezing, particularly at night. There may be a nasal discharge, watery eyes, and swollen sinuses. The birds gasp for breath. The disease has extremely rapid dissemination following almost instantaneous onset. Mortality may run as high as 50%. Morbidity affects practically all the remaining birds. The disease affects the immature reproductive system of the young pullet chick, leading to reduced egg quality and egg numbers in the laying house.

Secondary invaders follow. The severity of the disease is associated with the damage done by secondary diseases, particularly those produced by the coliforms or from other viruses. The incubation period of infectious bronchitis is from 18 to 36 hours. Usually the disease will run its course in 5 to 20 days, but the effects of secondary invaders may linger for long periods of time.

Infectious bronchitis in adult birds. As in chicks, the disease starts fast, without notice. Infection from bird to bird is rapid. However, there are few noticeable external symptoms; the disease is manifested by a severe drop in egg production. After the disease subsides, the return to normal egg production may take several weeks. Egg quality undergoes a drastic change. Eggs are softshelled, misshapen, and wrinkled and eggshells are porous and chalky. Brown eggs are light in color. Albumen quality is poor. Even though egg production eventually may return to normal, egg quality seldom does. The disease lasts from 4 to 10 days.

Some serotypes of infectious bronchitis cause kidney disease. These serotypes are known as *nephrotropic.* Damage to the kidney is done at the time of infection and may not be noticeable for weeks to months following infection. Mortality in layers may increase to four to ten times expected levels.

Internal symptoms. Necropsy reveals mucus in the trachea, nasal passages, and sinuses. The air sacs of young chicks may contain cheesy deposits. In older birds there may be a low prevalence of lesions, and in many instances there are none. Nephrotropic serotypes may cause the kidneys to first become enlarged and later become plugged with stones (urolithiasis) and then atrophy. A white granular material, *urates*, may coat the internal organs. This condition is called *visceral gout*.

Transmission

Birds contract the disease through contact with the virus. The disease is transmitted:

1. *By the air.* It takes just a few of the virus organisms to infect a bird. As the virus is easily airborne, inhaling infected air is the most important means of spread. The virus is known to spread over long distances by this route.
2. *By people and equipment.* This represents a major means of spread from house to house and farm to farm.
3. *By carrier birds.* Birds may shed the virus for as long as 4 weeks after recovery.

Infectious bronchitis has not been shown to be egg-transmitted.

Diagnosis

Diagnosis of infectious bronchitis can be difficult. It is sometimes made by eliminating other similar diseases as causative agents. Newcastle disease and laryngotracheitis are two such diseases. Virus isolation or rise in titer of serum samples taken at the time of the disease and again several weeks later will show evidence of the disease. Serum samples can be tested by either the enzyme-linked immunosorbent assay (ELISA) or the hemagglutination inhibition test (HI) (see *Diagnostic Testing*, Chapter 30).

Determination of which serotype of infectious bronchitis caused a disease break can only be done in a sophisticated diagnostic laboratory. If following apparent successful vaccination for infectious bronchitis a disease break occurs, it would be important to identify the serotype involved. This information could be used to construct a more appropriate vaccination program utilizing a more specific serotype or procedure.

Treatment

There is no known treatment for IB. However, when secondary infections are in evidence, treatment for these may alleviate some damage to the bird. Laboratories making postmortem and laboratory examinations usually will analyze the situation thoroughly for the presence of accompanying diseases.

Control

Control of IB is by vaccination with a vaccine of the appropriate strain(s). Live attenuated vaccines are recommended. The vaccine used will depend on the age and type of bird, as well as the serotypes known to exist in the area. As so many variant strains of IB virus have been found, the vaccination program must be one that will provide the greatest protection in a particular location.

General recommendations for type of vaccine. Since the Massachusetts strain will produce cross-immunization against many strains, it should be considered as the main component of most bronchitis vaccination programs. It must be remembered that some serotypes will not be affected by either the Massachusetts or Connecticut antibodies, including the Arkansas-99, Florida, and JMK viruses. However, many mixed vaccines, containing both the Massachusetts and Connecticut strains, are on the market, and are the vaccine strains predominantly used.

Other strains of IB vaccine are available. They include Holland and Arkansas-99. These strains can help provide broader protection from a number of serotypes. There are two types of Holland strain vaccine: (1) mild and (2) virulent. Do not use the virulent strain as the first vaccination.

Method of vaccinating broilers. Broilers should be vaccinated only where the disease has taken on acute proportions. The reasoning behind this is that the stress and side diseases produced by bronchitis vaccination may cause greater flock morbidity than the disease itself. When needed, most integrators vaccinate broilers at day of age in the hatchery using a spray cabinet. Many broilers are also vaccinated in the water or by spray from 10 to 14 days of age, after most maternal immunity has dissipated.

Method of vaccinating breeders and egg-type replacement pullets. Replacement pullet growers usually vaccinate three or four times for IB between 2 and 14 weeks of age. The vaccine is usually given in the drinking water or as a spray in combination with Newcastle disease vaccine (see *Vaccines and Vaccination,* Chapter 25).

Combination vaccinations. Bronchitis vaccine is sometimes combined with Newcastle disease vaccine and birds are vaccinated for both diseases at the same time.

Comments concerning the combining of IB and Newcastle vaccines:

1. Bronchitis virus vaccine multiplies more rapidly than Newcastle vaccine and the bronchitis growth may interfere with the growth of the Newcastle virus.
2. Newcastle virus vaccine is more stable than bronchitis vaccine therefore, deteriorated or older vaccine is more

likely to produce immunity to Newcastle disease and not to bronchitis.

3. There is more variability in the potency of bronchitis vaccine than Newcastle virus vaccine.

Killed autogenous vaccines. When it is determined that live commercial strains of infectious bronchitis vaccine do no provide protection against a strain found on a particular premise, killed vaccines can be prepared from the virus isolated. These are usually prepared with an oil emulsion adjuvant. They are used only in breeder and egg-type replacement pullets. Be certain to use a biologics manufacturer of high reputation for the production of these products.

Establish the titer. There may be failures of bronchitis disease vaccinations due to improper vaccinating methods, deteriorated vaccine, interference from maternal immunity, and other causes. In order to ascertain the level of the antibodies, titers should be established by a competent laboratory. If the titers are low, the birds should be revaccinated.

Seek professional help with IB vaccination programs. Because of the complexities of bronchitis variants, and the available vaccines, along with area-differentiating types of the disease, a competent veterinarian should be contacted to help design the best IB vaccination program for a particular location. Some states require permits to use vaccines made from other than the three common strains of the virus.

27-Q. INFECTIOUS BURSAL DISEASE (IBD)

This is an important disease of the immune system of chickens. It was formerly called *Gumboro disease* because it was first found near the town of Gumboro, Delaware, US. Infectious bursal disease is highly contagious. It develops rapidly but has a short duration. Clinical disease is not seen in chicks under 3 weeks of age although subclinical infections before three weeks can have a devastating effect on the immune system. The disease is caused by a *birnavirus.*

The bursa of Fabricius plays an important role in the disease. The gland is the location where B-cells develop (see *Immunity,* Chapter 24). Infectious bursal disease localizes in the bursa and causes severe damage to the organ. If chicks are infected before 3 weeks of age there can be permanent damage to their immune systems. The humoral antibody response can be reduced or lacking altogether. Later infections at least cause a transitory reduction in the immune system's ability to respond to antigens. Often the bursa of IBD-infected chicks triples in size and becomes edematous. It may return to normal size once the disease subsides. However, in the course of infection the tissues of the bursa may be partially or permanently destroyed.

Because the bursa of Fabricius is the origin of cells that produce circulating antibodies to many diseases, IBD may reduce the flock's immune response to vaccination. The morbidity and mortality resulting from these other diseases in many instances is greater than that from IBD itself. A poorly timed outbreak of IBD can disrupt an entire disease-control program.

Symptoms

The disease is limited to growing birds. Subclinical or inapparent infections from this disease during the first three weeks are frequently the most important. Onset of the clinical disease is rapid and occurs between 20 and 60 days of age, and recovery will take from 1 to 3 weeks. Birds are listless, nervous, sleepy, dehydrated, and have a whitish diarrhea. The vents seem irritated, and birds continue to pick at their own vents.

Mortality is variable. Although most flocks of chicks harbor the virus or its antibody, most cases of the disease are so slight as to go unnoticed. Only on occasion are there severe outbreaks, with mortality running as high as 30%. Mortality increases with age up to 10 weeks.

Transmission

The virus is very stable, and is capable of remaining viable outside the host for several months. Unclean poultry houses and poultry equipment are definite sources of infection. The disease may be spread to other areas by people, equipment and some insects. It is not egg-transmitted.

Diagnosis

Visible symptoms may be minimal and do not justify a positive diagnosis by observation alone. Birds must be sent to the laboratory. Several tests are used to arrive at a decision:

1. *Virus isolation (and identification).* The virus is isolated in embryonating eggs with final identification in the laboratory.
2. *Serological.* The ELISA test is often used to help with diagnosis.
3. *Chick inoculation test.* The bursa of noninfected birds will enlarge appreciatively a few days after inoculation with infective material.

4. *Histopathology.* Microscopic lesions of the bursa of Fabricius specific to IBD are seen early in the disease. There are also fluorescent antibody techniques that may be used.

Treatment

Treatment with drugs and antibiotics has been of little value except for some possible reduction of morbidity from secondary infections.

Control

Control is by vaccination.

1. *Vaccination of growing and adult breeders.* This is done to produce a high level of maternal immunity in the chicks. Breeder flocks receive live vaccines during the growing period followed by killed oil emulsion vaccine prior to lay. The breeders may receive live booster vaccinations during lay to keep maternal antibody titers high.
2. *Vaccination of chicks to prevent IBD.* Chicks can be vaccinated as early as 1 day of age. There is a concern relative to timing of vaccination and maternal antibody titer in the chicks. Some vaccines will not overcome maternal antibodies. Other intermediate IBD vaccine strains may be able to overcome some maternal antibodies. Since this disease is so important to the lifelong functioning of the immune system, a competent veterinarian should be consulted to determine need for and type of vaccine program on a particular premise.

27-R. INFECTIOUS CORYZA

Infectious coryza is an acute respiratory disease primarily of chickens causing swelling of the infraorbital sinus, discharges from the eyes and nostrils, and edema and bruising of the face. Chickens in egg production can have production drops of up to 70%. The production drop may last three weeks or more if complicated by other respiratory diseases. With the practice of more rigid sanitation on poultry farms, the incidence has reduced. However, in highly concentrated poultry areas, particularly those involving caged layers housed on multiple-age farms, it is still a major problem.

Cause

The disease is due to a gram-negative, nonmotile bacterium, *Haemophilus paragallinarum*. Three serotypes have been isolated. It is a relatively fragile type of organism, and although easily spread from bird to bird, it cannot live outside the body of the chicken longer than 5 to 6 hours.

Symptoms

External. The disease may affect birds of all ages. Usually the first sign is sneezing. This is followed by a watery condition of the eyes, then a discharge of the nasal and sinus passages. Gray mucus may be squeezed from the nostrils. In cases complicated with other organisms such as *Mycoplasma gallisepticum*, these areas may become filled with cheesy exudates, particularly the sinuses. Swelling occurs and lumps of material appear in the sinuses below the eyes.

Internal. Although the symptoms are usually external, there may be some inflammation of the air sacs.

Mortality. Death loss from coryza usually is low, but continued infection in the laying flock creates a loss of appetite, and drops in egg production. The disease may persist for months.

Transmission

Transmission can occur via aerosol droplets from one bird to another. Exudates can contaminate feed and water, but the causative organism only survives outside the chicken for a short time.

Diagnosis

Although visible evidence of the symptoms in birds is usually suggestive of a diagnosis of infectious coryza, laboratory techniques should be employed for definitive diagnosis. Culture and identification of *Haemophilus paragallinarum* is diagnostic.

Treatment

Drugs such as oxytetracycline, erythromycin, and streptomycin have been used to treat infectious coryza, however, treatment is usually not very effective. After treatment the disease sometimes returns. Chronic carriers perpetuate the disease. Because of low feed consumption in chickens with infectious coryza, it is recommended that rations very dense in nutrients be fed.

Control

Preventing outbreaks of infectious coryza is most difficult because of the nature and ease of transmission of the organism responsible. Several items are involved in such a program.

1. *Keep only birds of the same age on the farm.* Practice *all-in, all-out management*. Depopulation of the premises under such a program prevents older carrier birds from infecting younger birds.
2. *Use a bacterin.* An effective, killed, polyvalent bacterin, specific for several isolates of *H. gallinarum* is available. It is used mainly with growing birds. Two vaccinations are usually necessary. The first is given at about 10 weeks and the second 4 weeks later. It is important that the bacterin have an appropriate adjuvant.

27-S. INTERNAL PARASITES

Internal parasites cost the poultry industry many millions of dollars annually. Losses are from pathology to various organs, mortality of chickens, and from reduced feed efficiency as a consequence of the chicken supporting the particular parasite.

Large Roundworms

The large roundworm, *Ascaridia galli* is the major roundworm parasite of the chicken's small intestine. A few worms cause little damage. Larger numbers, though, can cause the birds to look unthrifty, reduce growth rate and cause poor feed conversion. Large roundworms may cause enteritis in young chickens.

Transmission is by direct life cycle. Chickens ingest the infective eggs which hatch in the proventriculus or duodenum. After 9 days the larvae penetrate the lining of the duodenum. Young worms enter the lumen of the duodenum after the 17th or 18th day. They remain there until maturity at 28 to 30 days. The females then produce eggs which are shed with the feces and the cycle begins again.

Control and treatment. Roundworms, in large part, can be prevented by raising birds in cages. Floor-raised birds can be treated with anthelminthic drugs (see *Medication for the Prevention and Treatment of Diseases,* Chapter 26).

Capillaria Worms

This infection is causes by a number of *Capillaria* spp. They are very small roundworms that live in the small intestine. The life cycle can be direct or indirect through the earthworm. They imbed themselves in the mucosa where they spend their entire life. The wall of the intestine shows hemorrhages and becomes thickened. The greater the number of worms, the more pronounced the lesions.

Growth and feed conversion are affected in growing birds. Egg production may suffer in layers. Utilization of vitamin D is impaired which causes poor growth and feed efficiency. Control and treatment are the same as for large round worms.

Cecal Worms

The small roundworm *Heterakis gallinarum* is known as the cecal worm. Its life cycle is similar to the large roundworm except that it ends up in the ceca instead of the small intestine. In the chicken, this disease is of little economic importance except that its eggs can carry the disease blackhead, which is important in the turkey. Little attention is paid to treatment.

Tapeworms

There are many species of tapeworms that affect the chicken. Most are quite large, up to 4 inches (100 mm) in length and several millimeters in width. They are easily seen with the naked eye. The head imbeds in the intestinal wall and new segments grow behind it. Segments at the end contain eggs. These segments are excreted with the feces. There may be some lesions at the point of attachment of the head. Tapeworms use nutrients that would otherwise go to the chicken. The seriousness of their pathogenicity is in dispute. It is probably less than most believe.

Control and treatment. Most tapeworms are transmitted through an intermediate host such as an insect, earthworm, or snail. Tapeworms can be controlled by controlling the intermediate host. Tapeworms are often seen where fly populations are high. Controlling flies usually controls the tapeworms. Some countries, including the United States, have no approved drugs for the treatment of tapeworms.

Trichomoniasis

Trichomoniasis is caused by the flagellated protozoa, *Trichomonas gallinae.* The organism is a pear-shaped protozoa that typically has four flagella. The disease affects the upper digestive tract as far down as the proventriculus. The liver is frequently invaded. The lesions appear as round, well-defined caseous areas on the mucosal surfaces. The caseous material may build up to the point that the digestive tract is blocked. In the liver, lesions appear as solid, circular, white-to-yellowish masses.

Control and prevention. The disease is spread through feed and water by oral secretions. Good water and feed sanitation should be practiced. There are few drugs to treat trichomoniasis but in some countries, including the US, there are no approved drugs for their treatment. Antibiotics may be effective against secondary bacterial infections.

27-T. LARYNGOTRACHEITIS (LT)

Laryngotracheitis is an acute respiratory disease caused by a *herpesvirus.* Although the disease is found in every area of the world, it is commonly only sporadic in nature. Some areas do not seem to be affected for long periods, then the disease will appear. It is found in chickens, pheasants, and peafowl.

This respiratory disease is one that spreads rapidly. The virus produces only limited disease in chicks under 1 month of age. It may vary in virulence, thus producing severe attacks in some outbreaks and mild in others. The virus has an incubating period of from 6 to 10 days. Usually the disease runs its course in about 14 days, but in some cases it may linger on for a month. The virus may be present in the chicken for many months without exhibiting diseases (*latency*). The LT virus does not enter the blood of the infected chicken, but forms lesions only where it contacts the tissue.

External Symptoms

Laryngotracheitis produces a very severe respiratory disease. The birds cough and have difficulty breathing. The trachea becomes clogged and the birds gasp trying to deliver air to their lungs. This gasping is so severe and common that it becomes a major criterion of the presence of the disease. Affected birds often are seen with their head and neck extended upward with the mouth open. A loud wheezing sound is heard on inspiration. In young chickens there is an infection of the eye, causing pain. Tears flow freely and the eyes are watery.

Internal Symptoms

The trachea is filled with an exudate loosely attached to the tracheal lining. In most cases, it may be removed easily. This differentiates LT from diphtheritic pox in which lesions are usually firmly attached to the respiratory tract. With LT there is severe hemorrhaging in the trachea and blood is coughed up as the birds try to breathe. With milder strains of LT there may be only a slight respiratory disease. This may occur without hemorrhage in the trachea, but exhibiting only a mild tracheitis. Egg production may be reduced.

Mortality

As the disease is of varying severity, death losses fluctuate from flock to flock and from year to year. In some cases as many as 50% to 70% of the birds will succumb. In other cases the mortality may be light. Birds under 1 month of age seem to be somewhat resistant.

Transmission

The disease spreads rapidly from bird to bird. Modes of spread are:

1. *By air.* Although airborne for short distances, as in a pen of birds, seemingly the organism cannot be carried in the air over long distances. Thus, this is not a means of transferal from farm to farm. It is doubtful if the virus will be carried by the air to adjacent poultry buildings on the same farm.
2. *By people, trucks, birds, rodents, and so forth.* Mechanical means of dissemination of the virus is, no doubt, the major means of spread.
3. *Latent infection.* Since the virus may persist for many months in otherwise healthy birds, this may be a source of infection. Stress may cause the virus to shed again thus infecting other birds.

Diagnosis

Clinical signs often lead to a rapid diagnosis. Few other diseases are capable of creating the bloody tracheitis seen in LT. To confirm the diagnosis a laboratory may use the following tests:

- *Virus isolation (and identification).* The virus is grown in embryonating eggs. Only birds in the early stages of the disease should be sent to the laboratory for this test.

- *Histopathology.* During the early stages of the disease, intranuclear inclusions can be seen in the cells of the trachea.
- *Enzyme linked immunosorbent assay (ELISA).* A rise in titer on paired serum samples several weeks apart helps to establish a diagnosis.
- *Challenge of susceptible chickens.* Virus can be inoculated into the trachea. If typical lesions are observed this will help with diagnosis.

Control

Control is by vaccination. There are two main types of vaccines:

1. *Modified chick embryo origin LT vaccine.* This vaccine is the one most commonly used in areas where LT is endemic. It produces very good immunity. It can be administered to young birds and to birds in egg production with only slightly noticeable effects. It is usually given by eyedrop or in the water but some vaccines are labeled for spray application. Occasionally, one to four weeks after vaccination, lesions very similar to clinical disease will be seen. These are vaccine-induced breaks. Morbidity and mortality are usually low.
2. *Modified tissue culture origin LT vaccine.* This vaccine is much less likely to cause clinical symptoms of the disease. It is also less effective in protecting from very strong challenges.

   In some locations where LT is not endemic, vaccination may be prohibited. Check with local authorities before vaccinating with LT vaccine.

Management and Eradication

Although it is probably possible to eliminate LT from the premises by the use of the attenuated vaccine, management should become a part of any eradication program. Isolation of affected flocks after vaccination, or even complete isolation of the farm, should be stressed as important. Since the disease can exist in wild pheasants, it may not be possible to eradicate the disease from endemic areas.

Some Birds Remain as Carriers

After a natural outbreak, some birds remain as carriers for life. These few make up a reservoir of infection and are capable of spreading the disease to uninfected flocks.

Vaccinating in the Face of an Outbreak

If LT has been diagnosed early in an outbreak, the disease can be stopped with vaccination. Start the vaccination with those birds farthest from the infected house and work back toward the original site of infection. Immunity develops rapidly while LT has a fairly long incubation period.

27-U. LYMPHOID LEUKOSIS (LL)

Lymphoid leukosis is a tumor-producing viral disease, usually of chickens 16 weeks of age and older. It is commonly a problem of breeders and laying hens, although some new strains are known to affect young broiler chicks. Most commonly, transmission is through infected breeders shedding virus into the egg.

Cause

Lymphoid leukosis is caused by a family of *retroviruses*. The subgroup A virus is most common in the United States. A new subgroup labeled J has recently been identified. It has been shown to cause myeloid tumors in breeders and young broilers, and so far has not been a significant problem in egg-type layers.

Symptoms

Birds suffering from LL are commonly pale and emaciated. The abdomen may be very large because of the bird's massive, tumorous liver. The bursa of Fabricius is commonly enlarged and lymphoid tumors are seen in the liver, kidney, ovary, and bursa. Flocks have progressive, higher than expected mortality which reduces productivity.

Transmission

Egg transmission is considered to be the primary means of spreading the virus.

Diagnosis

Lymphoid leukosis is usually diagnosed by clinical signs associated with the age of the affected flock. Diagnosis is made more difficult because of the similarity of lesions to Marek's disease. Tumors of the bursa of Fabricius are usually indicative of LL. Tumors of the skin and nerves are usually indicative of Marek's disease. Tumors involving the liver, kidney, spleen, and other viscera are difficult to differentiate without histopathology and other sophisticated means. Serological and virological methods are not considered to be useful.

Treatment

There is no known treatment for lymphoid leukosis.

Control

Eradication of individual infected breeding hens is considered the best means of eliminating or greatly reducing the incidence of LL. Since egg transmission is the number one method of spreading the disease, elimination of carrier birds from the breeding flock has been highly successful in reducing the infection in their progeny. In such cases, mortality due to tumors has been reduced and egg production has increased.

27-V. MAREK'S DISEASE (MD)

Marek's disease is a tumor-causing disease of chickens. It is a transmissible disease caused by a herpesvirus. It causes tumors of the nervous system and sometimes the visceral organs. It usually occurs in birds 2 to 8 months of age but occasionally occurs after the birds have come into egg production. This form is referred to as *late Marek's*.

Transient paralysis is another manifestation of MD. Young pullets appear to be paralyzed but if watched seemingly recover in a few days.
Skin leukosis in broilers is another form of MD which can be responsible for large economic losses due to condemnation.

Cause

Marek's disease is caused by a DNA-cell–associated herpesvirus. Several serotypes have been identified:

- *Serotype 1* is present in virtually all chickens. It is the serotype responsible for the clinical disease but there are many causes for the variations in the severity of the disease.

- *Serotype 2* is common in chickens but does not cause tumors.
- *Serotype 3* is a nonpathogenic herpesvirus found in the turkey. It is called the herpesvirus of turkeys (HVT). This virus has been found to protect against tumor formation in chickens.

Birds may become infected early in life, and undoubtedly remain infected until death. While all infected birds show the presence of the disease, the lack of visible evidence does not mean the bird does not harbor the virus. Why some birds develop tumors and others do not is the subject of important on-going research.

Transmission

Marek's disease virus replicates readily in the cells lining the feather follicles. The feather dander is a rich source of virus. Bird-to-bird transmission is via inhalation of the infective dander. The respiratory route is the major mode of transmission. Secretion can contain virus but this is a less important method. Some established means of transmission are:

- *Airborne* Marek's disease is highly contagious. Infected dander can transmit the disease over great distances. Material from the feather follicles is sloughed off as fine particles which float in the air and are inhaled by the birds.
- *Mechanical.* When feather dust and dander settle on clothing, equipment, etc., the virus is easily transmitted to other poultry houses and to poultry-producing areas.
- *Other means of spread.* Undoubtedly there are means of spread other than sloughed material from the feather follicles. Certain beetles have been identified as carriers.
- *MD virus not embryo-transmitted.* From a practical standpoint it may be assumed that this method is of very little importance in the transmission of the disease.

Marek's disease affects the bursa of Fabricius, a gland necessary in the production of antibodies, and therefore may reduce the bird's ability to resist other infectious diseases.

Diagnosis

Marek's disease can often be diagnosed by the presence of enlarged peripheral nerves that may be yellowish and have lost their typical cross striations. Lameness and paralysis are typically seen. These lesions may or may not be accompanied by tumors of the visceral organs. Tumors of the proventriculus are also usually an indication of MD. The presence of tumors, particularly in the liver, spleen, gonads, kidneys, and heart may easily be observed. Those in birds under 18 weeks of age are usually caused by MD but *late Marek's* may continue for another ten weeks. Some cases of tumors are difficult to differentiate from lymphoid leukosis. Laboratory confirmation is necessary to make a definitive diagnosis.

Control

There are now several vaccines used in the control of Marek's:

- Serotype 3. *Turkey herpesvirus (HVT) vaccine.* The turkey has a herpesvirus but this virus is not the same one that causes MD in chickens, however, they are similar. Vaccines have been developed using the turkey virus as the origin. When injected into young chicks they have the ability to stimulate the bird's resistance to the production of tumors or other lesions of MD. The HVT virus will produce a viremia in the chicken, and such birds will not develop the symptoms of Marek's even if the bird is later infected with the MD virus.
- Serotype 2. *SB-1 or 301 B.* These naturally occurring nonpathogenic MD viruses are frequently used in combination with HVT.
- Serotype 1. *Rispens and other similar viruses.* These are viruses whose origin is the pathogenic MD virus. They have been treated in the laboratory to make them nearly nonpathogenic.

Serotype 1 and or 2 vaccines are now usually used in combination with HVT. These combinations provide better protection against the virulent forms of MD. Although vaccination is essential in the control of MD today, it is important not to overlook other management practices that help in the control of the disease. These include proper cleaning of brooder houses to remove virus containing dander and disinfection. Keeping the chicks under strict biosecurity for the first week also allows vaccines to stimulate immunity in the chicks before the MD virus infects them.

Quantitative vaccine measure. The amount of the MD vaccine administered is very important, as it determines the potency of the vaccine. The quantitative measure is in plaque-forming units (PFUs). Proper and adequate immunization in the layer-type chick requires at least 5,000 PFUs. When the dosage is 0.2 ml, each dose should contain 5,000

PFUs, or more. In broilers the vaccine is commonly diluted to one-half or one-third strength. Too much dilution increases the risk of MD disease breaks.

Method of Vaccine Administration

There are two methods of vaccination:

1. *Subcutaneous.* The vaccine is injected subcutaneously in the nape of the neck. This is done at 1 day of age while the chicks are being processed in the hatchery.
2. *In-ovo.* This is done at 18 days of incubation when the eggs are being transferred from the incubator to the hatcher. A machine aseptically punctures the shell and delivers the vaccine inside the egg. This method saves labor and may give the vaccine more time to act before the chick is infected with the MD virus.

Combination Vaccines

Marek's disease vaccine is sometimes combined with infectious bursal disease and or viral arthritis vaccines. Genetically engineered vaccines are being developed using HVT as a vector.

Failures with HVT Vaccines

For several years following its introduction, the HVT vaccine gave almost complete suppression of tumors in the vaccinated chicks, but lately there have been some failures. The following causes have been suggested:

1. There is failure of the vaccine to cause a viremia before exposure to the MD virus.
2. Variant strains of MD virus are capable of overcoming immunity from the vaccine used.
3. Vaccine has been stored improperly.
4. Vaccine is not used in a timely manner after being mixed with diluent
5. Improper diluents or combination vaccines are used with the MD vaccine.
6. There is a deficiency of immune response in vaccinated chicks.

If vaccine failures are encountered, professional help should be sought.

27-W. *MYCOPLASMA GALLISEPTICUM (MG)*

This disease is found everywhere and is extremely important to both the broiler grower and the table-egg producer. While not a catastrophic disease it is one of significant economic importance. It is an important egg-transmitted disease. *Mycoplasma gallisepticum* is often a co-infection agent with other agents that makes the clinical signs of the other disease much worse. Infection of the air sacs in broilers is cause for condemning the dressed birds as unsuitable for human consumption. Laying flocks positive for *MG* have been shown to produce as many as 20 fewer eggs per year than *MG*-negative flocks. Although much progress has been made in the control of *MG* it remains a disease of considerable importance.

Cause

The *Mycoplasma gallisepticum* organism is very small and delicate and has no cell wall. It has been found in the chicken, turkey, duck, and other birds. Carrier birds are responsible for transmitting the disease. *Mycoplasma gallisepticum* may remain dormant and cause no disease until the chicken undergoes some stress. At this point the disease becomes active. Other viral or bacterial diseases may trigger *MG* infection. Sometimes live virus vaccines cause a flare-up of *MG* causing a severe reaction to the vaccination.

Symptoms in Young Birds

Mycoplasma gallisepticum is a respiratory disease, affecting the entire respiratory tract, particularly the air sacs, where it localizes. All the air sacs may be involved, become cloudy in appearance, and filled with mucus. In the latter stages of the disease, this mucus develops a yellow color and a cheesy consistency. Similar exudates may encircle the heart and heart sac.

Mycoplasma gallisepticum in itself is not a killer, in fact, even morbidity is not great. However, an outbreak may be followed quickly by many secondary infections, and it is these that do the damage. Coliform bacteria and respiratory viruses are particularly involved. Thus, visible identification of MG is often confused by symptoms of secondary invaders.

In young chicks there is a rattling, sneezing, and sniffling, all indicative of a respiratory difficulty. If complicated by other similar respiratory diseases these symptoms are accentuated. In severe cases, mortality may run as high as 30%.

Symptoms in Mature Birds

Visible evidence of the disease in adult birds may go unnoticed. Occasionally, the birds will appear depressed and inactive. There may be a definite diarrhea during the intestinal phase of the disease. Feed consumption drops. Egg production is reduced. Production losses may continue for many months. Unthrifty birds are seen in the flock. Mortality in adult birds is low.

Transmission

There are several methods whereby *MG* may be transmitted:

1. *Through the hatching egg.* This is the major means of spread. The organism passes from one generation to the next. Transmission rate is not uniform during the life of the bird. It is greater during the period of active infection in the adult birds, then subsides. Only a few organisms in a few infected chicks are necessary to cause the disease to spread to practically all young chicks in the flock.
2. *Through the air.* The organisms may move short distances through the air. This is enough to cause infection of birds within a pen, but probably has little consequence in the transfer from house to house.
3. *On clothing, feed, egg filler flats, poultry equipment, and trucks.* Personnel are significant carriers. Probably 60% of cross-contamination between houses is the result of organisms being transported by people and the equipment they use.
4. *Through infected chickens.* Complete depopulation of poultry houses and farms is essential to prevent exposing the new flock to the disease. Where possible, practice *all in, all out management*.

Diagnosis

Mycoplasma gallisepticum may be fairly accurately diagnosed by clinical signs coupled with the following laboratory tests:

1. *Rapid serum plate or tube agglutination test.* While very sensitive at detecting true positive flocks, it is not very specific and will sometimes call negative flocks positive. If

positive, this test should always be confirmed with one or more of the other tests below.

2. *Hemagglutination inhibition test.* This is somewhat difficult to run. It is now generally replaced by other methods.
3. *ELISA test.* This test is fairly rapid and can be run in high numbers with an automated system.
4. *Mycoplasma culture.* While mycoplasmas are difficult and slow to grow in artificial media, the isolation of *MG* in culture confirms the diagnosis.
5. *Embryonic examination.* Embryos infected with *MG* have lesions in their air sacs. A few cull chicks or pipped embryos should be examined at hatching time. Positive evidence of infection is an indication that the parents have an *MG* infection.

Treatment

Tylosin is an antibiotic specific for the treatment of birds infected with *MG*. Aureomycin, erythromycin, spectinomycin, or streptomycin may also be used as drugs for the treatment of the disease in growing and adult birds. Be certain to observe national and local regulations on antibiotic use. Egg-dipping using tylosin or gentamycin solutions is used for treating hatching eggs (see *Maintaining Hatching Egg Quality,* Chapter 38).

Control

Medication can only be considered a temporary solution and is usually quite expensive. Control can be by vaccination, controlled exposure, or eradication. Eradication is built around complete curtailment of embryo transmission. Thus, infected female parents must be removed as a source of infection. Unlike pullorum disease, *MG* eradication is difficult. First, it is so contagious that one or two infected birds in the pen will infect practically all others in a very short time. Second, an infected bird continues to be a carrier, and passes the organism through the egg. Thus, if any infected birds are found in the breeding pen, all the birds in the pen must be eliminated as a source of hatching eggs. There is no such thing as testing and retesting, as with pullorum disease. Furthermore, breeder flocks may remain clean (no infected birds) for long periods, but suddenly break with infection. Eradication at the breeder level is the best method of handling the disease.

In the United States, the National Poultry Improvement Plan (NPIP) is a voluntary program with poultry breeders striving to eradicate *MG* and other diseases from breeder flocks. Publications of this organization have detailed plans and regulations on eradicating diseases.

Vaccine to Control *MG* Infection

Primary breeders of broiler and egg-laying stocks have gone the route of eliminating *MG* from their flocks. Most multiplier flocks are also now negative for *MG*, but it is still a common disease in many parts of the world including table-egg flocks in the United States because of the multiple-age flock practices in common use.

In order to protect table–egg-laying chickens from becoming infected with *MG*, several vaccines have been developed. The F strain is a low virulence vaccine strain. It does cause some egg production loss but results are much better than from non-protected flocks in high risk areas. There are at least two other newer live vaccines that have virtually no virulence. Vaccines should be used only on premises that are known to be subject to *MG* infection.

There is also a killed bacterin vaccine for the prevention of *MG*. It prevents the clinical signs of the disease without the threat of spreading the organism in the flock. Administration of the vaccine is by intramuscular or subcutaneous injections. Breast injections are not recommended because of vaccine residue problems.

27-X. *MYCOPLASMA SYNOVIAE (MS)*

Also known as *infectious synovitis*, *MS* usually causes inflamed joints with exudate in the chicken. It can be either acute or chronic. The disease is also often seen as a mild upper respiratory disease. It is most common in broilers. *Mycoplasma synoviae* is an important egg-transmitted disease.

Cause

The disease is caused by *Mycoplasma synoviae.* It is small, delicate, and has no rigid cell wall.

Symptoms

Actually, *MS* is a respiratory disease but seldom is the respiratory tract involved with visible signs or mortality. Air sac infection is noticed when broilers are processed causing condemnation and economic losses. However, the organisms soon locate in the footpad and the synovial fluids of the hock and joints of the foot. These three areas become swollen and inflamed. In some severe cases, the wing joints may also become affected.

In most instances, this is a disease of young and growing chickens between 6 and 14 weeks of age, but it can infect older birds. There is a loss of appetite and weight. Birds are lame when the joints become inflamed, causing them to sit on their hocks. A persistent tenosynovitis may be evident in layers. Morbidity, rather than mortality, is a problem; seldom is there a very high death rate in older birds. Experimentally, hens exposed to aerosol MS have shown a drop in egg production. This is not a common occurrence in commercial layers.

Transmission

There are several known means of transmission and probably others that are not clearly understood:

1. *Through the hatching egg.* The disease is definitely egg-transmitted, and although the percentage of infected eggs is very low, these will give rise to enough infected chicks so that eventually a high percentage of the flock becomes infected. After infection, breeders will shed MS organisms for as long as 14 to 40 days.
2. *Through the air.* The organisms are easily transmitted from bird to bird within the pen by air, but probably not from house to house.
3. *On clothing, trucks, equipment, etc.* Mechanical means of transfer are of major importance in transporting the organisms over long distances.

Diagnosis

Swelling of the joints of the hock and footpads is a symptom of *MS*, but not necessarily a diagnosis; there are too many diseases that produce similar conditions. Birds should be submitted to the laboratory where tests may be conducted:

1. *Rapid serum plate or tube agglutination test.* While very sensitive at detecting true positive flocks, it is not very specific and will sometimes call negative flocks positive. If positive, this test should always be confirmed with one or more of the other tests below.
2. *Hemagglutination inhibition test.* This is somewhat difficult to run. It is now generally replaced by other methods.

3. *ELISA test.* This test is fairly rapid and can be run in high numbers with an automated system.
4. *Mycoplasma culture.* While mycoplasmas are difficult and slow to grow in artificial media, the isolation of *MS* in culture confirms the diagnosis.

Treatment

Some broad-spectrum antibiotics are of value. Relatively high levels of chlortetracycline and oxytetracycline administered in the feed or water have been found somewhat effective in treating *MS* infection. Birds with established synovitis do not respond well to treatment.

Control

Complete eradication is the solution for controlling *MS,* but to completely eliminate the disease from a farm is subject to some difficulty. The problem lies in the fact that some birds do not produce antibodies against the organism promptly and therefore the presence of *MS* cannot always be detected by the blood test. The US National Poultry Improvement Plan has detailed procedures for the eradication of *MS.*

27-Y. MYCOTOXICOSIS

Mycotoxicosis represents a broad range of diseases caused by toxins which, in turn, are produced by various molds. Mycotoxicosis is a major problem in the poultry industry as it may complicate the recovery from other diseases, such as coccidiosis and viral infections. It can also produce specific syndromes and can be immunosuppressive. As a result of these multiple effects, it is responsible for fairly large economic losses.

Cause

Mycotoxicosis, in its various forms, is caused by the toxins produced by various molds. The molds grow on grains, and under certain conditions during storage, toxins may be produced. Certain geographic areas of grain production favor one mold over another and in this age of ready transportation, new toxins are easily introduced into local poultry areas with transported feedstuffs. There are many types of toxins, each producing its own group of lesions.

Symptoms

Although morbidity and mortality are usual, the symptoms of the four most toxic mycotoxins in the chicken are as follows:

Aflatoxin

- Decreased growth and feed efficiency
- Impaired immunity
- Increased bruising
- Increase in blood clotting time
- Altered protein and fat metabolism
- Lowered resistance to other diseases
- Reduced pigmentation

Ochratoxin

- Reduced feed consumption
- Dehydration and emaciation
- Deposits of urates throughout the body cavity
- Impaired kidney function

Fusariotoxin

- Sores in the mouth
- Scabby lesions on the feet and shanks
- Reduced egg production and shell thickness
- Reduced feed consumption
- Reduced weight
- Poor feathering in growing birds

Citrinin

- Decreased feed consumption
- Reduced growth
- Increased water consumption
- Diarrhea
- Pale and swollen kidneys

Treatment

The best treatment is to find and remove the source of the mycotoxin. Replace the feed or use a new source of feed or feedstuffs if feed is the suspect. Treat with fat-soluble vitamins via the drinking water.

Control

There are several suggestions, all being a way of preventing mycotoxin consumption by the bird:

1. Monitor feed ingredients for levels of mycotoxins
2. Use corn of low moisture content.
3. When the feed is not of high quality, add a mold inhibitor.
4. Do not allow feed ingredients or mixed feed to have a temperature buildup.
5. Disinfect all feed milling and handling equipment.
6. Do not allow moldy feed to build up in feed troughs or storage bins.
7. Use mature cereal grains, high in quality, in the feed.
8. Blend batches/loads of ingredients to reduce overall levels.

27-Z. NECROTIC ENTERITIS

Necrotic enteritis is a bacterial disease of the intestinal tract of the chicken. There are sudden deaths that are caused by severe necrosis of the inner intestinal lining. Mortality is high.

Cause

The cause of necrotic enteritis is a bacteria called *Clostridium perfringens*. There are two types involved in the disease. They are type A and C, identified by the type of toxin produced by each. Although *C. perfringens* is present in virtually all chicken intestines, the disease is produced when there is damage to the inner intestinal lining. This allows the bacteria to grow and produce toxins which damage the intestines further and quickly cause death.

Coccidiosis is a frequent precursor to necrotic enteritis. Other diseases that make the disease possible are ascarid larval migration and *Salmonella* infections. Immunosuppression from infectious bursal disease may also precede necrotic enteritis.

Changes in the intestinal microflora may also lead to the disease. These changes can be brought about by rapid changes in feed components. Sudden changes to high levels of fish meal or wheat may also be responsible for enough upset of the intestinal microflora to allow the clostridia to grow.

Symptoms

Flocks experiencing necrotic enteritis show a rapid onset of ruffled, depressed birds followed quickly by an increase in mortality. Internally, the middle part of the small intestine is usually affected. The intestine is distended and filled with a foul-smelling brown fluid. The inner intestinal lining is covered by a brownish membrane. The liver may be swollen. There is often dehydration.

Diagnosis

In the field, typical symptoms with a rapid increase in mortality can lead to a presumptive diagnosis. In the laboratory, histopathology will demonstrate the high number of clostridia attacking and causing necrosis of the intestinal inner lining.

Prevention

The best prevention method for necrotic enteritis is to identify the predisposing factor and be certain that it is controlled. Particular attention should be given to coccidiosis and infectious bursal disease. Good management and sanitation practices should be employed.

Treatment

Although prevention is the best protection from necrotic enteritis, treatment with antibiotics can be helpful. Clostridia respond well to bacitracin, penicillin, and lincomycin (see *Medication for the Prevention and Treatment of Diseases,* Chapter 26). Predisposing conditions must be controlled.

27-AA. NEWCASTLE DISEASE (ND)

Newcastle disease is a highly infectious viral disease of many species of birds, including poultry. It was named for the town in which it was first diagnosed—Newcastle, England, in 1926. The disease is extremely variable in its severity. It can affect many organ systems including respiratory, neurologic, gastrointestinal, and reproductive.

Cause

The disease is caused by a *paramyxovirus,* and although there is only one serotype, there are four general forms classified according to their pathogenicity. The virus may be associated with different systems of the body: neurotropic (nervous), pneumotropic (respiratory), or viscerotropic (internal organs).

Death in the subacute form is usually the result of paralysis. In some cases the organisms invade the respiratory tract; in others, they may locate in the intestines and proventriculus.

Symptoms

Symptoms vary according to the age of the bird and the form of Newcastle disease virus involved. Three symptoms are common:

1. Respiratory difficulty
2. Nervous disorders
3. Egg production and eggshell quality reduced

The virus localizes in the respiratory tract, and all affected birds show evidence of the respiratory type. If nervous symptoms are involved, they arise later. Older birds seldom show any manifestations of the nervous disorders. Egg production and eggshell quality are affected quickly in laying birds.

There are three forms of the disease. They are identified as:

Velogenic, high pathogenicity
Mesogenic, intermediate pathogenicity
Lentogenic mild pathogenicity.

Examples of their symptoms are

1. *Velogenic. Viscerotropic velogenic* (VVND) (Exotic ND) is very highly pathogenic. It is sometimes known as the Asiatic type. It is highly virulent with high mortality. Hemorrhagic and necrotic lesions are seen in the intestinal tract. Respiratory and nervous signs are less evident than in other forms. Young chicks may exhibit spasms and twisted necks.

   Velogenic neurotropic is highly pathogenic. There is sudden onset with high morbidity and mortality, evidence of neurologic signs (twisted neck), and respiratory disease. Lesions are usually found in the respiratory tract only. Much of the ND outside the United States is of this type, therefore, vaccination programs for the United States may not be effective elsewhere.
2. *Mesogenic.* This form causes intermediate pathogenicity. There is acute disease in young chicks with respiratory

and nervous symptoms. These are usually not seen in older birds. This is the most common type of ND in the United States.

3. *Lentogenic.* This form causes only mild pathogenicity. Birds of all ages may have unnoticed infections or mild respiratory difficulty. Egg production declines. Eggshell quality deteriorates rapidly.

Vaccines and Vaccination Procedure

Vaccines against Newcastle disease are usually made from the lentogenic or mesogenic form of the virus, and to some degree give protection to all forms of the disease. However, each vaccine is highly variable in the response it produces. A summary of these effects follows:

Lentogenic strains (B 1-type). In the United States the lentogenic vaccines will prevent drops in egg production and respiratory signs, but this is not always true outside the United States.

1. *F strain.* The vaccines from the F strain have the lowest virulence of the common lentogenic strains. They are most effective when birds in a flock are vaccinated individually.
2. *B 1-strain (Hitchner).* This is one of the two common strains used in the United States. It is slightly more effective than the F strain. It is usually given in the drinking water or by the spray method. There is very little reaction seen when this strain is given to a flock. It is generally safe to give to broilers and can be used at a very young age. It may be given at day-old but then must be followed by a La Sota-type vaccine to give stronger immunity.
3. *La Sota strain.* Of the lentogenic vaccines, this is the one most often used. It is usually not used for the first vaccination but usually as a booster after one or more B1 type, B1 strain vaccines. After La Sota-based vaccines are given, there is mild bird-to-bird spread. It will not prevent the drop in egg production when adult birds are challenged with a velogenic strain.

Mesogenic strains

1. *Mukteswar strain.* This strain is particularly pathogenic and its use for vaccines should be confined to those birds previously vaccinated with one of the lentogenic vac-

cines. It has had wide acceptance in tropical climates and in southeast Asia.

2. *Hartfordshire* (H) and *Komarov* (K) strains. Vaccines prepared from these strains are less pathogenic than those from the Mukteswar strain. The H strain may be administered by subcutaneous or intramuscular methods.
3. *Roakin strain.* Vaccines prepared from this strain of the virus are often called *wing-web vaccines.* They are isolates that are attenuated, but still very virulent. They are administered by dipping a needle in the vaccine, then forcing the needle through the web of the wing. Roakin-strain vaccine cannot be given to young chicks. If given, Roakin-type vaccine should be administered after one or two vaccinations with a lentogenic-type vaccine. It then has merit when there is a field infection of the Newcastle virus as it will help prevent mortality and will help prevent a drop in egg production and in eggshell quality. The Roakin strain is no longer used in the United States.

Transmission

Newcastle disease virus particles are easily spread. The disease is highly contagious. Methods of transmission are:

1. *Through the air.* Coughing dislodges the virus from the respiratory tract. It easily becomes airborne. It travels quickly from bird to bird and from house to house over short distances.
2. *On clothing, unsanitized filler flats, feed trucks, and equipment.* This category probably represents the major means of transfer of the virus to uninfected flocks and farms.
3. *No cleanup period on farm.* Poultry operations that start chicks on a regular basis, resulting in several ages of birds on the farm at one time, have continuous problems. The older birds infect the younger. Farms with an *all-in, all-out* program of management break any cycle of infection when the premises are depopulated.
4. *Feed*
5. *Wild birds, neighboring poultry*
6. *Exotic birds.* Particularly those shipped from other areas or countries may carry ND. Exotic bird imports, usually illegal ones, have been a source of velogenic disease in areas otherwise free of those strains.
7. *Predators*

Diagnosis

Many times a diagnosis may be made on the basis of physical observation. Nervous disorders and the presence of respiratory disease are highly suggestive of the disease. Laboratory diagnosis is the best method of determining ND is present. Tests to help with diagnosis include virus isolation, hemagglutination inhibition test, fluorescent antibody test, and the enzyme-linked immunosorbent assay (ELISA) test.

Treatment

There is no known treatment, although broad-spectrum antibiotic medication for secondary diseases will probably reduce some of the flock morbidity.

Vaccination Programs

If ND is in the area, all birds on the flockowner's premises should be vaccinated. Transmission from farm to farm is too easy to justify not vaccinating. Because of the many variations in the types of the Newcastle virus and the various form of the disease in different areas of the world, along with many types of vaccines available, any program for control should be individually designed to fit the conditions. Seek professional help before starting an ND vaccination program (see *Vaccines and Vaccination,* Chapter 25).

Velogenic Newcastle disease. There have been outbreaks of highly virulent Newcastle disease in the United States and Canada. Such outbreaks of a "hot" strain are characterized by respiratory symptoms, hemorrhagic conditions of the intestinal organs, high mortality, and a drastic drop in egg production.

Although vaccination programs commonly in use by most of the poultry industry are adequate for the usual mild variety of ND, these programs will not give adequate protection against velogenic strains of the virus. Individually applied vaccines (intranasal, ocular, or intramuscular) may provide better immunity than mass spray or drinking water vaccinations. Intramuscular-type vaccines may be given at 2 weeks of age. If there is evidence that velogenic Newcastle disease is in the area, consult the nearest diagnostic laboratory for advice and type of vaccine and vaccination program recommended. In areas where velogenic ND is not endemic, flocks infected with velogenic ND will usually have to be destroyed under governmental supervision.

27-BB. SALMONELLOSIS

There are many Salmonellae that cause disease in the chicken and some that do not cause disease in the chicken but are of public health significance. The Salmonellae of chickens are generally divided into three diseases. They are pullorum disease, caused by *Salmonella pullorum;* fowl typhoid, caused by *S. gallinarum;* and paratyphoid, caused by any of the non-chicken specific Salmonellae. It is important to note that while pullorum disease and fowl typhoid significantly limited the size and success of the chicken industry before control programs virtually eliminated them, it is now the paratyphoid organisms that threaten the chicken industry through public health concerns over Salmonellae on the meat of chickens and in eggs.

PULLORUM DISEASE

Pullorum disease is a highly contagious, egg-transmitted disease. The disease has largely been eliminated in commercial flocks through industry control programs such as the National Poultry Improvement Plan in the United States. The disease is characterized by white diarrhea in young chicks. There are asymptomatic adult carriers. It has been referred to as *bacillary white diarrhea.*

Cause

Pullorum disease is caused by a bacteria called *Salmonella pullorum.* Unless control measures are taken, the disease will become widespread and there will be high mortality.

Symptoms

Pullorum disease occurs in the chicken, turkey, pheasant, quail, pigeon, and some wild birds. In chickens, the age of the bird influences the symptoms; young chicks are affected differently than older birds. Chicks huddle together, and seem chilled. There is an acute, whitish diarrhea. Vent pasting is prevalent. Appetite drops, feathers are ruffled, and the chicks breathe with difficulty. Hock joints may become inflamed. When the infection is passed through the hatching eggs, the disease has an early onset. Death losses may begin as early as the second day. If other chicks are the source of the infection, most losses occur after 1 week of age. Death in most affected birds is rapid. Losses may run as high as 50%.

In older growing birds, and those in egg production, there are few, if any, external symptoms except for a greenish-brown diarrhea. Fertility and hatchability of the eggs laid by breeder hens may be affected. Birds that die from an acute septicemia may die rapidly without outward signs. A visual examination of the internal organs may show few changes from the normal, except for the highly mucous contents of the intestines, and on occasion, unabsorbed yolk.

In adult birds there are usually no symptoms. Localization of the organisms in the ovary sometimes causes some of the ova to atrophy. The sex organs of the male may be affected. Sometimes the heart and gallbladder show indications of a definite infection by the presence of grayish nodules, but generally a diagnosis cannot accurately be made visually from a visual observation.

Transmission

The most important route of transmission of infection is from the infected female parent, through the hatching eggs, to the newly hatched chick. Although all ova from an infected layer are not involved, enough are infected to carry the causative organisms to the next generation. A few infected chicks will soon transmit *S. pullorum* bacteria to most chicks in the incubator (hatcher) or in the pen.

Young chicks with the disease shed *S. pullorum* organisms through fecal material. This represents a major means of transmission. However, in adult birds the fecal material contains few *S. pullorum* and is not a major means of spread.

Because of the septicemia involved, cannibalism is an important means of transmission, as blood is transferred from bird to bird. Contaminated equipment may also be the source of infection. Beak-trimming machines offer a means of transferring the bacteria from bird to bird.

Diagnosis

The *S. pullorum* organism is easily isolated and identified in the laboratory. Cultures taken from such organs as the ovary, testes, heart, liver, and spleen are used to make the laboratory determinations.

Testing for Antibodies

Once a bird has been infected with the *S. pullorum* organism, antibodies specific for *S. pullorum* are present in the bloodstream. These antibodies clump with killed *S. pullorum* cells in the test antigen. Two serological tests are in general use:

Stained antigen, rapid whole blood test. This is the preferred test for chickens. It is a specially prepared and standardized mixture of killed *S. pullorum* bacterial cells, dyes, and solvents. It is commercially available. A measured drop (0.5 ml) of antigen is placed on a testing plate. Next, a vein is punctured (usually the median vein on the underside of the wing near the "elbow") and a drop of blood is picked up by a loop of wire and the blood mixed with the antigen. The plate is rotated in a circular motion to facilitate mixing and clumping. If the blood taken from the bird contains *S. pullorum* antibodies, it will clump with the bacterial cells of the antigen creating a positive reaction. Such birds are known as reactors. If no antibodies are present in the blood, the mixture remains clear, and the bird is known as a nonreactor, or is negative to the test. The simplicity of this test makes it easy to run in the field, thus getting immediate results. Individual reactors can then be immediately separated for further testing.

The tube agglutination test. This test can be used as a confirmatory test to the stained antigen, rapid whole blood test. It is run on serum and needs to be run in the laboratory.

Confirming Positive Agglutination Tests

If the agglutination test shows that certain birds are positive (reactors) these birds should be taken to a diagnostic laboratory for confirmation of the test. This is necessary because the test is not completely accurate. In such circumstances, the bird is sacrificed, and the usual cultures are taken and tests made to determine if active *S. pullorum* organisms are present. Remember, this test is not for antibodies, but rather for the presence or absence of active bacteria.

Identifying Carriers

Carriers shed *S. pullorum* bacteria through the egg to the newly hatched chick. In some areas of the world *S. pullorum* has been all but eradicated from commercial chickens. In areas where the disease is known to exist, extensive effort should be made to identify carriers. These carriers should be identified and removed from the flock. To be adequately certain that all reactors (carriers) are identified, two blood tests, no less than 6 months apart, should be conducted. If, on a test, some reactors are found, the flock should be retested after 21 days. The procedure should be continued until there are no reactors on two consecutive tests.

Why the Test Is Not Infallible

The blood test has its limitations because:

1. Cross-agglutination between organisms other than *S. pullorum*, particularly other *Salmonellae*, occurs, clouding

the validity of the test. Also, there are variant strains of *S. pullorum*, and if the antigen does not contain bacterial cells of such strains, the test will be unreliable.

2. Often the test is not sensitive enough. Marginal reactors, where the agglutination in the rapid whole-blood test is only partially complete after 2 minutes on the plate, cause erroneous readings.
3. Infected birds do not always react to the agglutination test. Some birds, particularly during the time they are in egg production, may not react to a blood test, but will react later when a pause in egg production occurs.
4. Antibodies may not have made their appearance. Once a bird becomes infected, antibodies begin to appear about a week later. These antibodies increase in number until they reach a maximum in about 3 weeks. During this 3-week period there may not be an adequate number of antibodies present in the blood of the bird to cause a reaction to either of the agglutination tests.

Choosing Which Breeder Flocks to Blood-test

As a general rule, all breeder flocks should be blood-tested for pullorum disease. However, the integrator, who uses his own chicks in his production program, may find it economically advantageous to dispense with the program. In using the primary breeder blood test, the integrator is assured that his breeder chicks at least start their life free of disease. Under modern poultry sanitation programs, there is little chance that the birds will become infected as they grow. But there is no guarantee; some integrators will "spot test" a percentage of their breeder birds as added assurance. Furthermore, they take measures to identify the breeder source of all chicks produced. If there is reason to believe some breeding flocks are transmitting active organisms, the flocks can be located easily.

Treatment

Treatment of *S. pullorum* infected chickens perpetuates the carrier state. For this reason it is not recommended.

What to do when the disease strikes. Evidence of the disease in a group of chicks is the first indication that in all probability the breeders are shedding viable bacteria through the hatching eggs. The breeders literally have infected the chicks. In such cases:

1. If the breeder flock from which the hatching eggs came can be identified, do not use eggs from this flock until the disease has been thoroughly investigated.
2. Remove any eggs from this flock in incubators.
3. Disinfect and fumigate all hatchery equipment thoroughly.
4. Blood test all the birds in the suspicious breeding flock(s). Do not use eggs from positive flocks for hatching. Recovered birds or flocks should not be saved for breeding purposes.

FOWL TYPHOID

Fowl typhoid is a septicemic disease similar to *S. pullorum,* except that mortality from typhoid may occur at any age. It is not confined to young chicks, as in the case of pullorum disease. The disease itself and its control are very similar to pullorum.

Cause

Fowl typhoid is caused by a bacterium, *Salmonella gallinarum.* In many respects, the organism acts like that of *S. pullorum.* Most species of poultry susceptible to pullorum are also susceptible to typhoid. The typhoid organism grows and develops in the chicken in a manner similar to *S. pullorum.*

External Symptoms

In comparison to pullorum, typhoid is a more slowly spreading disease. The first external symptoms are ruffled feathers, loss of appetite, and a greenish diarrhea. The comb and wattles may be pale, with an anemic-like appearance. At times, death in adult birds may be sudden, without any prior indication of illness. In untreated flocks, mortality may run as high as 50%.

Internal Symptoms

In the early stages of the disease few tissue changes occur, but in the acute stage the liver is enlarged, and may have a color from bronze to mahogany. In some cases it may be streaked. The spleen and kidneys may be enlarged. In young chicks, the egg yolk is usually unabsorbed. The liver is white and friable. The digestive tract is usually empty. As in the case of pullorum, adult carrier pullets may show atrophied ova.

Diagnosis

The diagnosis is the same as for pullorum. Agglutination tests may be used to determine if adult birds are carriers. A laboratory diagnosis is the only accurate means of determining the presence or absence of *S. gallinarum* organisms. The organism grows readily on most laboratory media and may be identified both bacteriologically and serologically

Since most antigens used for conducting the *S. pullorum* agglutination test are polyvalent (containing several pullorum variants), and since *S. gallinarum* will cross-agglutinate with *S. pullorum,* it is unnecessary to run a specific agglutination test using only *S. gallinarum* bacteria.

Control and Treatment

Control is essentially the same as for pullorum. Testing and removal for eradication is the goal. As with pullorum, treatment only perpetuates the carrier state and is not recommended.

PARATYPHOID

Beside *S. pullorum* and *S. gallinarum,* and some Arizona types, there are numerous other Salmonellae that have been isolated from domestic fowl. These have been grouped as "paratyphoids." Important ones are *S. enteritidis, S. typhimurium, S. montevideo, S. derby, S. newport,* and *S. bredeney.* They are quite stable in the environment and some can survive moderate heat for long periods. Most can live for weeks in media such as water, feed, human food, and soil.

Most paratyphoid bacteria cause little disease in chickens except for young chicks. Many can infect humans, particularly the young and old or those weakened by chronic or immunosuppressive disease. It is the potential human infection that is of the greatest significance. Some food-borne illnesses in humans have been traced to *Salmonella* infections. When these same organisms are found in or on the tissues of dressed fowl, or in eggs, suspicion is aroused that they might be the contributing source of the disease in people. This has caused the poultry industry to take extreme measures of sanitation in poultry and egg processing plants.

Only some of the paratyphoids have been shown to be egg-transmitted. *S. enteritidis* has been isolated from intact eggs and has been associated with human food-borne illness. These illnesses almost always involve mishandling of the eggs in the kitchen with time and temperature abuse. It is uncommon to find the organism in the egg. In field studies in the United States, *S. enteritidis* has been found in only about 1 in 20,000 eggs (0.005%) and then only from flocks from where the organism was isolated from the environment.

A number of paratyphoid organisms found on the surface of broilers have been associated with human food-borne disease. Reducing food-borne human disease has been the impetus behind the establishment of quality assurance programs in both the broiler and egg industries (see *Quality Assurance and Food Safety—Chicken Meat*, Chapter 44, and *Egg Quality Assurance Programs*, Chapter 62). These programs assure the production of clean breeding stock, the reduction of pests in the poultry buildings, clean feed, good biosecurity, and proper sanitation practices on production facilities and processing plants.

Cause

Paratyphoids have worldwide distribution. The cause of the specific infection is the result of infection by one of the *Salmonella* spp. in the paratyphoid group. *S. typhimurium* and *S. enteritidis* are two of the most troublesome for the poultry producer.

Symptoms

There seems to be no specific external symptoms, probably because of the wide variety of organisms involved. In some cases young chicks are listless, and there may be diarrhea, but many affected flocks show no visible evidence of the disease.

Internally, few symptoms are noticeable and none are specific. They easily may be confused with those produced by *S. pullorum*. The disease becomes pathogenic through endotoxins, a toxin produced within the organism that does not diffuse from the bacterial cell until the cell degenerates.

Transmission

In general, transmission is similar to that of *S. pullorum*, but certain modes are of more importance.

1. *Egg transmission.* Paratyphoid definitely is egg-transmitted, but eggshell penetration is more important than in the case of *S. pullorum* and *S. gallinarum*. Eggshells are abundantly covered with paratyphoid organisms as the egg passes through the cloaca. When the egg is laid, these organisms are drawn through the shell pores, an

avenue of embryo inoculation. Contaminated litter or nesting material is a major source of the organisms on eggshells.

2. *Fecal contamination.* The paratyphoid organisms are found in great numbers in the intestinal tract. This accounts for the greatest amount of transmission from bird to bird.
3. *Ovarian transmission.* Some paratyphoid organisms lodge in the ovary. This represents a possible method of transmission of the disease from parent to the newly hatched chick. In most cases, however, there is little ovarian transmission. Young chicks can be infected but will not show evidence of the disease until they are stressed.
4. *Personnel.* People are capable of transmitting paratyphoid disease. Organisms are carried on clothing and footwear from one location to another. The bacteria are capable of surviving outside the host for many weeks.
5. *Feed.* As with pullorum disease, the paratyphoid organisms may live in certain feed ingredients for long periods. This is particularly so with the animal and fish proteins. Pelleting can reduce the number of *Salmonella* organisms in the feed, but the process cannot be considered completely bactericidal.

Diagnosis

A laboratory diagnosis is necessary to ascertain the paratyphoid involved. Bacteriological methods generally are used. Although the diagnosis is more or less definite in young chicks, trying to type the organisms taken from adult birds may not always lead to conclusions, because no sickness or other evidence of the disease is present. Many times it is more practical to culture embryos after 10 days of incubation to determine any correlation with paratyphoid infection in the breeder flock.

Treatment

Most of the time, paratyphoid is not a major disease of chickens, as the disease is usually limited to the intestinal tract. Because of some cross-agglutination when the pullorum blood test is done, many adult carriers of paratyphoid are eliminated from the breeding flocks.

When carefully diagnosed cases do appear in young chicks, the use of antibiotics may be helpful (see *Medication for the Prevention and Treatment of Diseases,* Chapter 26). Treatment with antibiotics may not always be successful as many paratyphoid organisms are resistant to currently available antibiotics.

There is little evidence to show that treating the breeder flock with medications to prevent egg transmission of paratyphoid via infected ova is of any value, although some drugs may act indirectly to destroy some organisms in the intestinal tract, thus decreasing the number of bacteria deposited on the eggshell.

Prevention

Reduction or exclusion of paratyphoid organisms is very difficult because of their occurrence in many species (including man) found near production facilities. It is the goal of quality-assurance programs to try to reduce the prevalence of these organisms in flocks.

Both killed and live vaccines are available to help in the reduction of paratyphoid organisms. A gene deleted mutant is commercially available that has good cross-immunity with many paratyphoids. Other vaccines may soon be available.

Competitive exclusion products are also commercially available. These products are aimed at populating the gastrointestinal tract with adequate numbers of normal, beneficial bacteria in an effort to out-compete the paratyphoids. It is clear at this time that a multifaceted approach is necessary to control the paratyphoids.

27-CC. VIRAL ARTHRITIS (VA)

Viral arthritis (VA) (infectious tenosynovitis) is a disease primarily of meat-type chickens. It is especially significant with broilers because it affects growth, increases mortality, decreases feed efficiency, and increases condemnations in the processing plant.

Cause

Viral arthritis is caused by a *reovirus* that initially infects the synovial membranes and the tendon sheaths. The chicken is the only known host.

Symptoms

Symptoms are especially apparent in broiler chicks where there can be severe inflammation of the hock joints with lameness and stunted growth. In severe cases, the hock joint may be immobilized and the gastrocnemius tendon is enlarged and may rupture. The hock joint usually contains an exudate at the site of the swelling.

Transmission

The virus is transmitted through the feces. Horizontal transmission is the primary means of spread of the disease. The virus can persist in birds for nearly a year. Egg transmission of the virus is low, though possible.

Diagnosis

Clinical signs and laboratory identification of lesions in the tendon sheaths help make the diagnosis. Fluorescent antibody tests can identify the viral antigens in the tissues. The ELISA test can show an increase in titer further indicating VA infection. It is important to differentiate arthritis lesions from *Mycoplasma synoviae*, nutritional diseases, and bacterial infections such as *Staphylococcus* spp. Marek's disease and deformities also must be considered.

Treatment

No known treatment for an infected flock is available.

Control

Prevention of early infection is the best method of control. Thorough cleaning of brooder houses and isolation are necessary. Vaccination of breeder flocks providing maternal antibodies helps prevent disease in young chicks. Mild vaccines are also available for use in commercial chicks at 1-day of age.

28

Biosecurity on Chicken Farms

by Carol J. Cardona
and Douglas R. Kuney

Biosecurity refers to a series of practices designed to prevent disease-causing organisms from coming in contact with resident birds on the farm. It begins with isolating the farm from off-farm disease agents and continues with the isolation of individual chicken houses from agents that are on the farm. Biosecurity is the most efficient and cost-effective method of disease prevention available. Disease management and eradication are difficult and expensive alternatives to a failed disease prevention program.

28-A. COMPONENTS OF BIOSECURITY

- Isolation
- Traffic control
- Sanitation

Isolation refers to time, distance, and physical barriers that reduce or prevent entry onto the farm and or into the poultry house. Traffic control includes restricting human, equipment, and animal movement onto the farm, and movement patterns while on the farm. Sanitation refers to the cleaning and disinfection of poultry houses, people, materials, and equipment (see *Cleaning and Disinfecting of Poultry Facilities,* Chapter 29).

28-B. THE NATURE OF PATHOGENS

All biosecurity programs should take into consideration the nature of pathogens in terms of their level of infectivity for chickens and their ability to survive in the environment. Infectivity and survivability affect the level of biosecurity that should be employed and the exclusion methods that are needed. Pathogens are transmitted in the following ways:

- Fecal to oral
- Aerosol
- Mechanical vectors
- Biological vectors

Infective viruses, bacteria, and parasites transmitted by the fecal-to-oral route are shed in the feces of infected sources (birds and other animals) and then are consumed by susceptible chickens. Pathogens that are transmitted by the aerosol route travel in microscopic droplets of moisture and dust particles and are inhaled. Mechanical transmission occurs when the pathogen is carried on the surface of people and their clothing, equipment, wild animals, or insects and brought into physical contact with a susceptible chicken. Biological vectors carry the pathogen in their bodies and transmit the disease either by being consumed by the bird, by biting the bird, or through the spread of infectious particles, usually in the feces. Biological vectors are living animals that are themselves infected with a disease agent. Some examples of biological vectors that can spread disease in chicken flocks are mosquitoes (fowlpox), rats and mice (*Salmonella*), and wild-migrating water fowl (avian influenzae).

The level or extent of on farm biosecurity efforts should be based on a balance between disease risk and cost. During times of high risk, biosecurity efforts should be reinforced but can be more routine during normal risk periods. Determination of high and normal risk should be made based on disease surveillance in the surrounding area as well as the purpose and value of the flock (commercial chickens vs breeders), the age of the flock (young chickens vs older chickens near market age), and on farm management factors such as housing style (open sided vs closed environmentally controlled houses).

28-C. THE IMPORTANCE OF ISOLATION

Isolation can be considered in terms of time (the amount of time between depopulating and refilling a poultry house or farm), distance between farms or houses on a farm, and physical barriers (fences, showers, foot baths) all of which limit the spread of disease agents.

Time

There should be ample time separating succeeding flocks on a premise to prevent the transmission of pathogens. Under normal conditions, two weeks is adequate for complete cleaning and disinfection of the poultry house (see *Cleaning and Disinfecting of Poultry Facilities,* Chapter 29). However, the value of the birds in the flock and the risk of infection from previous flocks should be weighed when determining how much time needs to elapse between flocks.

As a general rule, the greater the value of the flock, the longer the downtime between flocks. As in all biosecurity program considerations, there is a balance between cost and risk. Increasing downtime directly decreases the persistence of disease-causing organisms in the poultry house and thus, reduces the risk of disease. However, increased downtime directly reduces the productive use of a poultry facility and thus, increases cost.

All-in, all-out management systems favor the elimination of disease agents that do not survive well outside of the chicken; such as mycoplasma, avian influenza virus, infectious coryza, and infectious laryngotracheitis virus. Providing at least two weeks of downtime between flocks will improve the effectiveness of disease management on the farm and reduce the time and money spent on vaccines and medications. But sometimes complete depopulation is not possible and when it is not, birds remaining on the farm should be considered potential sources of disease to incoming birds. When there is not complete depopulation of the farm between flocks or if there is not adequate downtime between flocks, then disease-causing agents may become endemic. Multiple age groups on a farm should be treated as separate units when possible. Traffic should flow from youngest to oldest birds and from healthy to sick birds.

The risk of infection from a previous flock depends, in part, on the disease agent of concern. Some disease-causing agents will persist in the environment longer than others (Table 28-1). The infection status of the preceding flock should be considered when planning the placement of a new flock. After a disease outbreak, downtime should reflect the shedding time of the agent and its survival time in the environment. Busy hatchery sched-

Table 28-1. Longevity of Disease-Causing Organisms

Disease	Lifespan Away from Birds
Infectious bursal disease	Months
Coccidiosis	Months
Fowl cholera	Weeks
Coryza	Hours to days
Marek's disease	Months
Newcastle disease	Days to weeks
Mycoplasmosis (MG, MS)	Hours to days
Salmonellosis	Weeks
Avian tuberculosis	Years
Avian influenza	Weeks to months
Infectious bronchitis	Weeks to months

Table 28-2. Biological and Mechanical Vectors That Transmit Poultry Pathogens

Disease Agent	Vector(s)
Arboviruses (encephalitis viruses)	Mosquitoes Biting insects
Avian influenza virus	Waterfowl Human traffic
Newcastle disease virus	Psittacine birds Free-flying birds Human traffic
Poxvirus	Mosquitoes Biting flies Mites
Pasteurella multocida	Free-flying birds Racoons

ules often dictate the timing of new flock placement and sometimes conflict with needed downtime. However, downtime between flocks can be increased by an early depopulation. With layers, this will be influenced by the profitability of the flock at the time.

Distance

When deciding where to place a poultry facility, distance from other birds, domestic and free flying, should be carefully considered. Wild birds, exotic pet birds, and domestic poultry are important sources of poultry pathogens because closely related species are likely to carry agents that can also infect commercial chickens. Mechanical and biological vectors that travel between groups of birds can carry pathogens between them (Table 28-2). To minimize risk, the location of other birds, including the nesting and resting grounds of wild birds, live-bird markets, and other poultry farms should be as far as possible from a new facility. Special consideration should be given to the surrounding land uses, in particular crop production that may apply poultry manure from other farms. The direction of prevailing winds, and the sources that feed any bodies of water on the premises should be considered when determining risk. The distances traveled by vertebrate pests, including free-flying birds and rodents, and invertebrate pests should also be used to limit the location of new facilities. The importance and means of limiting human traffic are discussed in the *Traffic Control* section of this chapter.

Physical Barriers

Barriers are very effective at limiting the spread of pathogens. But the successful use of barriers requires consideration of sources of disease-causing agents and their modes of transmission.

Figure 28-1. Separated Laying Flock Sites for Disease Security

Fecal-to-oral spread of pathogens occurs when the feces of an infected bird or vector is ingested by a susceptible chicken. The fecal contamination of litter and water is common in chickens kept on the floor. Cages limit this mode of spread between chickens because fecal material does not accumulate where other chickens can consume it. Fecal-to-oral transmission also occurs when there is fecal contamination of feed by vertebrate pests. Chickens in all types of housing are susceptible to this type of transmission. Feed storage structures should be secured and vegetation removed from the periphery to exclude vertebrate and invertebrate pests (see *External Parasites, Insects, and Rodents,* Chapter 12).

Aerosol transmission of disease should be considered when determining the design of poultry houses and their placement on the farm. The air intakes of one poultry house should be located away from the outflow vents of other houses.

Mechanical and biological vectors of disease should be excluded from poultry houses by the effective use of barriers. Securing buildings and removing feed and harborage for pests (including free-standing water and tall grasses) will help to create a barrier around the flock. Vertebrate vector control programs should focus on the control and exclusion of rodents and free-flying birds from the poultry house. Invertebrate vector (fly and beetle) control programs should focus on abatement from the poultry house and the immediate surrounding environment.

One of the most important vectors of disease is human traffic. Barriers should be created to prevent the transmission of disease agents into poul-

Figure 28-2. Farm Sanitation—Boot Washing

try houses by humans. Disease agents are carried on clothing, hair, exposed skin, and footwear. When entering a farm, visitors should be provided with clean protective clothing as described in section 28-D. Barriers between houses on a poultry farm can also be used to prevent the transmission of disease. A boot dipping station that includes a brush for the removal of organic material and a dip with a disinfecting agent can effectively prevent the entry of pathogens to the poultry house on boots. However, boot dipping must be done properly to be successful. First, all organic material must be removed from the surface of the boot (Figure 28-2). Failure to remove organic material will prevent even the strongest of disinfectants from inactivating pathogens. Next, the boots must be dipped into a disinfecting solution in either a bucket or saturated pad system (Figure 28-3). Disinfectants must be properly diluted to be maximally effective. In addition, solutions must be changed on a regular schedule, usually daily. In high disease risk situations, or during times when traffic is increased through the poultry house, more frequent changes of the disinfecting solution are recommended.

Equipment can be an important means of pathogen transfer between birds. Ideally, every flock should have its own dedicated equipment. When this is not possible, shared equipment must be completely cleaned and disinfected between contacts with different flocks. All racks and vehicles used to transport chickens onto the farm must be thoroughly washed, disinfected, and dried between flocks.

Figure 28-3. Disinfecting Shoes Before Entering House

28-D. TRAFFIC CONTROL

Traffic Coming onto the Farm

Personnel. People and equipment that come onto poultry facilities can mechanically introduce many potentially devastating diseases. The most common visitors to the poultry farm are the most dangerous because they are likely to have had recent contact with other poultry. Included in this group are:

1. Feed delivery trucks and their drivers
2. Field service personnel
3. Renderers
4. Vector-control personnel
5. Hatchery trucks and drivers
6. Egg trucks and drivers
7. Transport trucks and drivers
8. Veterinarians and other consultants
9. State and federal inspectors or other regulatory personnel
10. Utility company representatives
11. Equipment company repairmen

Other, less common visitors, may or may not pose a direct disease threat to poultry flocks but must be treated with equal care as they enter a poultry facility:

1. Visitors, such as school tours
2. Uninvited guests, people who are curious or lost

One of the most effective ways to control human traffic coming onto the farm is with the use of signs. Signs must be clear and posted in areas where people coming onto the premises will notice them. They can also direct visitors to stay in areas where they are welcome. Fencing with gates can help to prevent people from accidentally wandering into areas of the farm that should be off limits. Upon entry, a combination of signs and fences can be used to guide people to an office or another area where they can be checked in. Alternatively, locked gates and fences can be used to prevent all access to the farm and visitors can use an off-site intercom to contact on-farm personnel. On-farm personnel can then guide visitors onto the farm, using proper biosecurity protocols, or prevent their entry depending on the consideration of several factors (Figures 28-4 and 28-5).

People can mechanically transmit disease-causing pathogens on clothing, shoes, hair, and hands. Methods to prevent transmission of diseases by people must address these critical areas. Both cost and risk should be considered when developing a sanitation procedure for on-farm visitors. The best and most complete method for human disinfection is a complete

Figure 28-4. Farm Security—Locked Gate

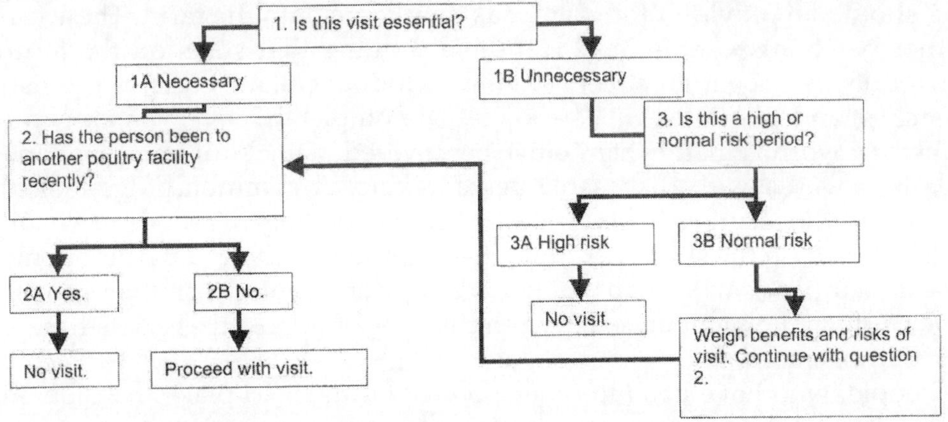

Figure 28-5. A Decision Process for Screening Farm Visitors

shower-in facility in which all clothing and jewelry is left on the outside or dirty side and new, clean clothing is issued on the farm or clean side (Figure 28-6).

To be effective, the shower must include complete wetting and shampooing of the hair. While extremely effective, the initial cost of installing shower facilities, and the time it takes personnel to shower-in—sometimes several times per day per person—may be too costly for some situations. For commercial facilities in which egg-layers or broilers are housed, cloth-

Figure 28-6. Farm Sanitation Center with Shower Facilities

ing should be provided for visitors as they come onto the farm. These can either be (1) disposable or (2) reusable clothing that stays on the farm. Coveralls that cover all street clothing including collars, cuffs, and wristwatches, and plastic or rubber boots that completely cover visitor shoes must be worn. Finally, hair must be covered with bouffant caps, and beards covered with disposable beard covers. A commonly overlooked biosecurity measure is hand washing. Because pathogens can be mechanically carried on the skin of visitors, they should be required to thoroughly wash their hands with soap and hot water prior to putting on their protective clothing. Remember, jewelry can easily be overlooked as a safe-harbor for pathogens.

People, who have been to other poultry farms or to places considered hazardous, should be turned away from the farm and asked to return another day. It is important to remember that asking the right questions of visitors will get the information necessary to make a good decision about their visit. Delivery personnel should have a logbook in which all previous visits for the day are recorded. Designated on-farm personnel should check the logbook for visits to other poultry facilities or contact with other birds. Certainly, biosecurity for visitors who come from other poultry facilities should be as rigorous as possible.

There are other situations in which people can become disease carriers. Workers, who travel to a diagnostic laboratory to submit sick birds, should not return to the farm without a shower and a complete change of clothes—preferably they should not return to work until the next day. Personnel who attend meetings with other producers must consider that they have been contaminated and should shower and change clothes before returning to the farm. Again, it is preferable that they do not return until the next day.

A large component of traffic control is the development of good habits by workers on the farm and, hence, depends greatly on their cooperation. Education about the dangers of disease agents, the objectives of the farm biosecurity plan, and about specific practices can improve compliance. Specific rules regarding high-risk activities such as visits to other poultry farms, owning their own birds at home, direct or indirect contact with live bird markets, and bird hunting should be specifically enumerated to avoid confusion and maximize compliance. Employees at all levels should be repeatedly reminded that contact with other birds should be followed by a shower and change of clothes before coming onto the farm.

Equipment. If at all possible, multiple poultry farms should not share equipment. In some cases, however, this is not practical and therefore, shared equipment must be thoroughly cleaned and disinfected between farms. For example, several poultry farms will share feed delivery trucks, live-haul and egg pick-up trucks, or manure clean-out equipment. Special precautions must be taken to avoid problems associated with equipment sharing.

Figure 28-7. Farm Sanitation—Truck Washing Facility

There are several methods for truck disinfection. The most complete and, hence, most effective, is a whole truck wash in which undercarriage, wheel wells, and the entire surface area are cleaned and disinfected (Figure 28-7). The initial cost of installing a truck wash is considerable; however, for some types of poultry facilities, the savings in disease prevention easily offsets the cost.

In some situations, disinfection will be confined to the use of a high-pressure sprayer on the truck's undercarriage, wheel wells, tires, and lower surfaces. This should be done in an entry area, away from the birds. Soap, water, and disinfectant should be applied and extra care should be taken to remove any organic material. In climates that reach below freezing temperatures, care must be taken to ensure disinfectants do not freeze.

Finally, in some high-risk situations, equipment that has previously been on a poultry facility within the preceding day or week should not be allowed access to the farm.

Traffic off the Farm

Although most biosecurity procedures are meant to protect one's own flock from disease, controlling traffic leaving the farm is an important part of being a good member of the poultry-producing community. Control of traffic off the farm is essential in high-risk disease situations. A premise with a high-risk disease should be under self-imposed quarantine.

Personnel. Personnel leaving a quarantined facility should leave disposable clothing on the farm when they depart. Additionally, visitors should take care to spray floor mats of all vehicles with disinfectant, and clean the steering wheel after leaving the premises.

Equipment. In high-risk circumstances, trucks and cars exiting a facility should be completely cleaned and disinfected. A vehicle leaving a quarantined facility should not travel to another poultry farm within 24 hours, and perhaps longer depending on the nature of the disease-causing pathogen and the risk of infection posed to other poultry.

Traffic Control on the Farm

Dirty and clean activities. The farm should be divided into dirty and clean areas. Manure handling, dead bird disposal, disposal of trapped vertebrate pests, and removal of breeding areas for vertebrate pests should all be considered dirty activities. Dirty activities will potentially put personnel in contact with disease-causing pathogens. These activities should be confined to dirty areas of the farm and must be kept separate from clean activities. Clean activities include egg handling, chick handling, movement of birds, and daily activities taking place in contact with the flock. Personnel and equipment used in dirty activities should not come into contact with personnel and equipment involved in clean activities prior to thorough cleaning and disinfection.

If possible, it is preferable that equipment and personnel not be shared between clean and dirty activities. However, even in this situation, it must be remembered that most equipment will come into contact with other equipment in the repair shop. Equipment returning to use after repair should be cleaned and disinfected before being put into clean areas of the farm.

If equipment is shared between the clean and dirty activities on the farm, then traffic should always go from the clean areas to the dirty areas on the farm and not the reverse. Complete cleaning and disinfection of equipment should occur before it travels from dirty to clean activities and personnel should shower and change clothing. It is preferable that in a day, clean activities are completed before undertaking dirty activities. After dirty activities are completed, personnel should go home.

Facilities with Multiple Flocks on the Premises

When there are multiple flocks on a single premise, there is always a danger that disease will be carried from contaminated to susceptible birds.

Strict biosecurity should be maintained between individual flocks, preferably with personnel and equipment dedicated to the flocks, and not coming into contact with any other birds. However, when this is not possible, traffic between flocks should be strictly controlled.

In order to determine the flow of traffic on the farm, one must consider the ages of the flocks, and their disease status. Traffic should always go from non-diseased chickens to those with disease and never in reverse. Additionally, traffic should go from young to older birds, because older birds are more likely to harbor disease-causing pathogens than younger birds. In addition, younger birds may be more susceptible to disease since they may not yet be immunocompetent or they may not be fully vaccinated. Personnel and equipment going from diseased to non-diseased or older to younger chickens should undergo complete cleaning and disinfection.

28-E. ROUTINE DAILY SANITATION PRACTICES

The goal of farm sanitation is to maintain a healthy environment for the chicken flock. Sanitation reduces the likelihood of poultry pathogens coming in contact with the chickens, and is, therefore, an important component of biosecurity. While sanitation includes cleaning and disinfection of poultry houses and equipment between flocks (see *Cleaning and Disinfecting of Poultry Facilities,* Chapter 29), the focus here will be on routine daily on-farm activities that contribute to maintaining a healthy environment for the chickens while they are on the farm. Daily attention should be paid to the proper management and disposal of:

- Dead birds
- Spilled feed
- Manure
- Refuse

Some mortality is normal in any population of chickens and some of this may be due to infectious causes. To reduce the spread of infectious agents, dead birds should be removed and disposed of as quickly as possible. Daily collection of mortality and disposal in a sanitary manner will prevent contact with insects, rodents, or other animals that could act as disease vectors. During outbreaks of high-risk diseases, it may be advisable to collect mortality more frequently and to use disposal methods that will inactivate the specific disease-causing agent. Depending on local laws, effective disposal can be achieved by:

- Incineration
- Deep pit burial
- Freezing for rendering pickup

- Disposal into a sealed refuse container for rendering pickup
- Composting

Chicken carcasses should never accumulate since they will attract wild or domestic animals and insects, or come in contact with water sources.

Spilled feed should not be allowed to amass in or around poultry houses. Chicken feed will attract insects, rodents, and wild birds, all of which can bring pathogens into the poultry environment. Daily attention should be given to maintaining feed delivery systems and to routine cleanup of spilled feed within the chicken house and around feed bins.

Poultry manure and litter, if not managed properly, can provide an attractive substrate for fly and other insect breeding. In addition to insects, it can serve as a habitat for rodents that can carry *Salmonella* spp. and other pathogens into the poultry house. The manure should be managed in a way as to be unattractive to flies and other insects. This primarily involves keeping manure dry (see *Fundamentals of Ventilation,* Chapter 9, and *Waste Management,* Chapter 11). Special attention should be given to preventing water system leaks and the removal of wet spots. In the case of caged layer facilities, disposal of cracked eggs into the manure should be avoided since they are an attractive food source for both insects and rodents.

The proper disposal of manure and all refuse on poultry farms is important in eliminating rodent harborage sites. Trash and manure piles are also attractive to wild shore birds like sea gulls and egrets that could carry disease agents onto the premises. When possible, the accumulation of manure or trash in open areas should be avoided. Bird crackers and other hazing devices will discourage wild birds but generally have a short period of effectiveness because the birds rapidly acclimate. In most cases, it is easier to eliminate the trash or manure than to discourage the wild birds that are attracted to it (see *External Parasites, Insects, and Rodents,* Chapter 12).

29

Cleaning and Disinfecting Poultry Facilities

by Douglas R. Kuney and Joan S. Jeffrey

Effective cleaning and disinfecting (C&D) methods can substantially decrease disease transmission by reducing pathogens in the environment below infectious levels. By reducing the number of surviving pathogens, the likelihood of disease being passed to an uninfected flock will be reduced. A thorough cleaning and removal of all organic material from the environment must always precede disinfection, so that the pathogens come in direct contact with the disinfectant.

Thorough C&D coupled with an *all-in, all-out* farm replacement is recommended to prevent disease transmission from old flocks to new ones. While *all-in, all-out* farm replacement is the most effective policy from a disease prevention standpoint, it may not always be economically feasible. Therefore, poultry producers need to customize their C&D programs to their individual situations.

Cleaning and disinfecting should also be applied to all aspects of human and mechanical traffic between farms and flocks, in addition to sanitizing poultry houses between flocks. Ideally, separate crews and equipment should be used for each flock. When this is not possible, special care should be taken to reduce the likelihood of disease transmission by effective cleaning and disinfecting of vehicles, equipment, and people (see *Biosecurity on Chicken Farms,* Chapter 28).

29-A. PREPARING THE HOUSE

Preparing the house begins with removing the previous flock and continues through removing all manure and providing a sufficient empty house down period to allow time for a die-off of residual pathogens that may have survived the C&D process. Some important steps of the process include:

- Bird removal
- Litter or manure removal
- Feed system sanitation
- Water system sanitation
- Vector control
- Housing and equipment sanitation
- Idle (down) time between flocks

Bird Removal

Removal of all birds from the depopulated house or farm is absolutely essential to breaking a disease cycle from flock to flock. *Fugitive* birds are a perfect reservoir of pathogens and will meander from house to house in their search for feed and water. Unavoidably, a few birds will escape from loading or catching crews; these birds must be caught and disposed of before the C&D process can be complete. Fugitive birds are most easily caught after sundown when they are quietly hiding from predators.

Litter and Manure Removal

Manure or litter should be completely removed from the house and transported as far away as possible. Manure and litter contain disease vectoring insects and bacteria or viruses that were shed by the previous flock. Distance from the poultry house is an important consideration because manure and litter can be very attractive to free-flying wild birds that can transmit poultry pathogens.

In contradiction to principles of disease prevention, incomplete manure or litter removal is sometimes recommended for reasons of filth fly control in caged layer facilities or economic reasons in floor bird operations. Some local environmental health departments require that cage layer facilities leave a 3 to 4 inch (7.6 to 10.2 cm) dry base of manure to aid in manure drying for the new flock (see *External Parasites, Insects, and Rodents,* Chapter 12). To reduce the cost of litter materials, some broiler producers use a built-up litter system (see *Broiler Management,* Chapter 43). In either situation, all wet manure or caked litter must be removed during the C&D process. In all cases, a complete clean-out should occur when abnormal mortality or a disease break is experienced in the previous flock.

Feed System Sanitation

All remaining feed from the previous flock should be removed from the system. Residual feed left in the house is an attractive food source for rodents and beetles, which can act as vectors for pathogens. Feed storage bins should be thoroughly cleaned to prevent mold buildup.

Feed bin boots should be disassembled to remove all feed and then a disinfectant should be applied to the bin and auger. After the bin and auger have dried thoroughly, the boot can be reassembled. Completely remove all loose and caked feed from the feed delivery system and then apply a disinfectant. Mechanical feeders can be dry cleaned with the aid of air compressors or shop vacuums and then disinfected. Feed pans should be removed from line feeders and thoroughly cleaned, paying particular attention to removing any mold buildup. After applying a disinfectant, placing the pans outside in the sun to dry will allow ultraviolet rays to aid in killing pathogens. Do not return pan feeders to the house until the total house C&D has been completed.

Water System Sanitation

Water lines should be flushed by opening the line at the end, to remove any buildup of loose slime or scale (see *Consumption and Quality of Water,* Chapter 22). All water filters should be replaced between flocks. Chlorine compounds are good disinfectants in the absence of organic debris. Chlorine is effective against bacteria and many viruses. These compounds are also more active in warm water (>65°F, 18.3°C) than in cold water. With the use of a medication proportioner, the water system can be chlorinated. Clean and remove all feed and other organic material from waterers before chlorination. The best way to distribute the chlorine solution quickly throughout the drinking system is to remove the drain plug at the end of each line or to temporarily remove the last drinker in the line allowing water to freely flow through the system. Concentrated chlorine solutions can be run through the system and left for 24 hours if the birds are not present. Caution should be used with chlorine concentrations greater than 5% because they may cause corrosion of metal equipment or parts.

Vector Control

It is important to reduce or eliminate insects, free-flying birds, and rodents as they can transmit pathogens. Their presence can contaminate the poultry environment. An insecticide application to the house following cleaning will kill many of the remaining flies or beetles. Placing rodent baits along established runways just prior to cleaning the feed system can greatly reduce rodent populations. Free-flying birds can be excluded from the houses by eliminating openings in excess of 3/4 inch (19 mm). Once a population of wild birds establishes itself in a house, getting rid of them is almost impossible. Be sure to sweep away all dead insects and remove any dead rodents before disinfecting the house (see *External Parasites, Insects, and Rodents,* Chapter 12).

Housing and Equipment Sanitation

The house should be swept from top to bottom and thoroughly dry cleaned including fan blades, louvers, lighting fixtures, curtains, and walls. Following dry cleaning it is recommended that all surfaces within the house be thoroughly washed with a detergent solution that is best applied with a high pressure sprayer. The spray should be applied with a minimum pressure of 200 pounds per square inch (psi) to dislodge any remaining debris. Cleaning should start with the upper surfaces working your way down to the floor. The detergent spray should then be followed by a high-pressure clean water rinse, to wash off residual detergent and organic material.

Finally, after the house is allowed to dry, apply a high pressure spray disinfectant—thoroughly covering all surfaces. It is important that good penetration into cracks and crevices be achieved. A second application of the disinfectant can be applied after the first has had a chance to dry. Two applications are recommended when the previous flock has had a disease history.

Caution. As with any pesticide, disinfectants must be used in accordance with label instructions. Increasing the concentration of any disinfectant is NEVER a substitute for thorough cleaning.

Idle Time Between Flocks

Idle (down) time between flocks is the final step, and one of the most important parts of C&D. After the final disinfection, close and lock the house while it is drying. Be sure to exclude all traffic including employees, and especially animal vectors. Pathogen reduction is a function of their direct contact with the disinfectant, reduction in moisture (most pathogens are killed by drying), and time. Pathogens vary in their ability to survive outside the host (see *Biosecurity on Chicken Farms,* Chapter 28), so the longer the idle period between flocks, the greater will be the reduction in pathogens. Ideally, poultry houses should be left idle for a minimum of two weeks.

29-B. CHOOSING A DISINFECTANT

Disinfectants are chemicals that kill pathogens on contact. The lethal action of disinfectants for various pathogens (viruses, bacteria, fungi, protozoa) depends on the chemical composition of the disinfectant and the

Figure 29-1. Farm Sanitation—Cleaning a Slat-floored Rearing House

type of organism. When choosing a disinfectant, consider these characteristics:

- Cost
- Efficacy (killing efficiency against viruses, bacteria, etc.)
- Activity in the presence of organic matter
- Toxicity (relative safety to man and animals)
- Residual activity
- Effect on fabrics and metals
- Activity with soap
- Solubility (acidity, alkalinity, pH)
- Contact time
- Temperature requirements

The relative importance of these characteristics will depend on the individual situation, but the efficacy of the disinfectant and its toxicity to animals are always important concerns. No disinfectant works instantaneously. All require a certain amount of contact time to be effective. Temperature and concentration of the disinfectant influences the rate of killing of microorganisms. Using disinfectants at the recommended concentration is important to their effectiveness. The activity of many disinfectants improves markedly as the temperature increases. All disinfectants are less effective in the presence of organic material. Organic matter interferes with the action of disinfectants by coating the pathogen and preventing contact with the disinfectant, forming chemical bonds with the disinfectant thereby making it inactive against organisms, or reacting chemically with and neutralizing the disinfectant. Disinfectants that are used on poultry farms can be divided into the following classes based on their chemical compositions:

- Phenols
- Hypochlorites (chlorine)
- Iodophors (iodine)
- Quaternary ammonium
- Formaldehyde
- Oxidizing agents (peroxide)
- Natural disinfecting agents

Of these, phenols, quaternary ammonium compounds, iodophors, hypochlorites, and oxidizing agents are most commonly used on poultry farms.

Phenols

Phenols are coal-tar derivatives. They have a characteristic pine-tar odor and turn milky in water. Phenols are effective antibacterial agents, and are effective against fungi and many viruses. They also retain more activity in the presence of organic material than iodine or chlorine-containing disinfectants. They are not effective against bacterial spores. Common uses in commercial production include hatchery and equipment sanitation, and footbaths.

Hypochlorites

Chlorine compounds are good disinfectants on clean surfaces, but are quickly inactivated by dirt. Chlorine is effective against bacteria and many viruses. These compounds are also much more active in warm water than in cold water. Chlorine solutions can be somewhat irritating to the skin and corrosive to metal. They are relatively inexpensive disinfectants.

Iodophors

Iodine compounds are available as iodophors, which are combinations of elemental iodine and a substance that makes the iodine soluble in water. They are good disinfectants, but do not work well in the presence of organic material. Iodophors are effective against bacteria, fungi, and many viruses. In hatcheries, iodine is used on equipment and walls, and for water disinfection. Iodine is the least toxic of the disinfectants discussed here. Many iodine products can stain clothing and porous surfaces.

Quaternary Ammonium

Quaternary ammonium compounds are generally odorless, colorless, nonirritating, and deodorizing. They also have some detergent action, and when used properly, all are good disinfectants. However, some quaternary ammonium compounds are inactivated in the presence of some soaps or soap residues, so careful product selection is important. Like most disinfectants, their antibacterial activity is reduced in the presence of organic material. Quaternary ammonium compounds are effective against bacteria and somewhat effective against fungi and viruses. These compounds are widely used in commercial hatcheries.

Formaldehyde

Formaldehyde gas is the most commonly used fumigant. Crystals of formalin are mixed with chromic acid that react to release formaldehyde into the air. In order to be effective, the house must be closed and sealed. This can be an effective disinfection method because the toxic gas can penetrate well into small cracks and crevices.

Caution. Formaldehyde gas can be extremely toxic to humans and other animals and its use may not be allowed in some areas. Always check with local governments for regulations regarding its use.

Oxidizing Agents

Hydrogen peroxide and other oxidizing agents like peracetic acid and propionic acids or acid peroxygen systems are used in commercial poultry operations. They are active against bacteria, bacterial spores, viruses, and fungi at quite low concentrations.

Natural Disinfecting Agents

The natural forces that reduce the pathogen load in the environment are important and can often be used to our advantage. These include sunlight, heat, cold, drying, and agitation. The ultraviolet rays of sunlight are tremendously potent killers of microorganisms. This can be extremely effective outside of buildings, but unfortunately the ultraviolet rays cannot pass through glass, roofs, or dust. Drying from fresh air and wind will also kill pathogens, particularly when directly exposed after the facilities have been cleaned.

A well-designed C&D program is one of the best methods we have of maintaining a farm free of disease-producing microbial agents. It is one of the most difficult tasks to accomplish, but when it is successfully done, the improvements in flock performance generally more than justify the effort.

30

Diagnostic Testing

by Carol J. Cardona and Gregg J. Cutler

Often diagnostic testing is used to determine the cause of increases in mortality, production drops, increases in condemnations, or clinical signs of disease. But diagnostic testing can also be used to validate vaccination, biosecurity, HACCP, and regulatory programs such as the National Poultry Improvement Plan. It can be used as a routine disease surveillance tool allowing the producer to detect a disease in its earliest stages, before the onset of obvious signs.

30-A. SELECTION OF SAMPLES FOR TESTING

The first step in any effective diagnostic testing program is sample selection. Selecting the wrong samples can lead to an inaccurate picture of flock health. When testing is used as a validation of management programs, birds should be randomly chosen from the flock and normal daily mortality should be included in the tested group. In flocks being sampled to diagnose a disease outbreak, healthy, sick, and recently dead birds should be included in sampling. Among the sick birds, birds that are just beginning to show clinical signs, some that are at the peak of illness, and some that are chronically ill or recovering should be selected to give as complete a picture of the disease as possible.

30-B. HOW TO COLLECT SAMPLES FOR TESTING

Tissues for Histopathology

When collecting specific tissues for histopathology, it is sometimes difficult to know how much tissue to take. This is a general guide for some of the more commonly collected organs:

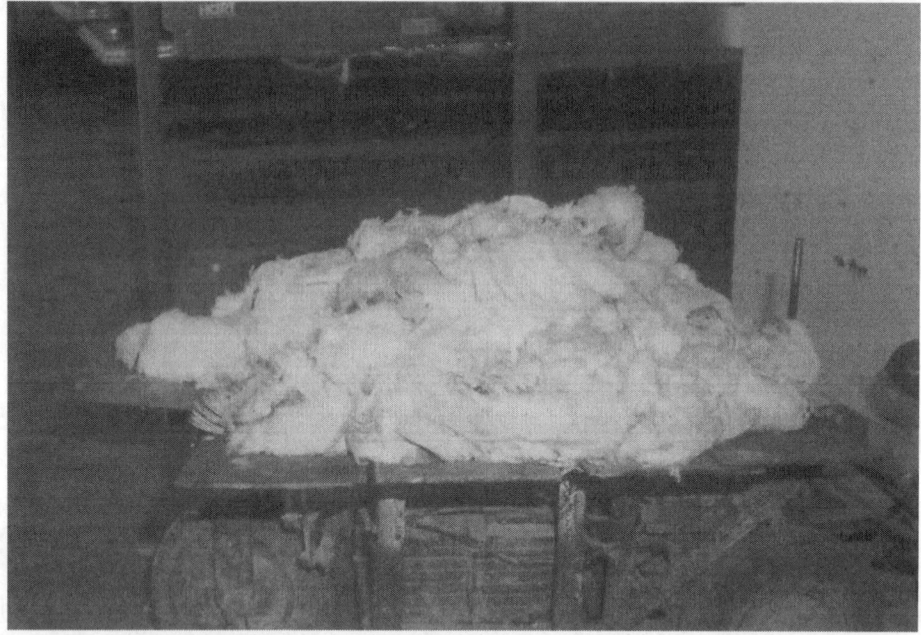

Figure 30-1. Daily Mortality of Birds for Examination

1. *Brain:* Collect the entire brain without mutilating it.
2. *Heart:* Collect the whole heart, trying to keep the heart sac intact.
3. *Liver:* Take one section (lobe) of the liver in adult birds or the whole organ in young birds.
4. *Other organs:* Take sections that are between 2 and 3 inch (6–12 mm) thick. In young birds entire organs can be collected.

Tissues collected for histopathology should be placed in a bag and refrigerated, not frozen. Freezing will damage the tissues so that they are no longer useable for histopathology. Alternatively, tissues for histopathology can be fixed in the following solution:

10% formaldehyde in water

The tissue should be fixed in a volume that is approximately ten times the size of the tissue.

Blood and Serum

Frequently, there is a need to collect blood samples from chickens in a flock. This is commonly done for the purpose of evaluating the immune status or exposure level of the flock to either a vaccine or disease agent.

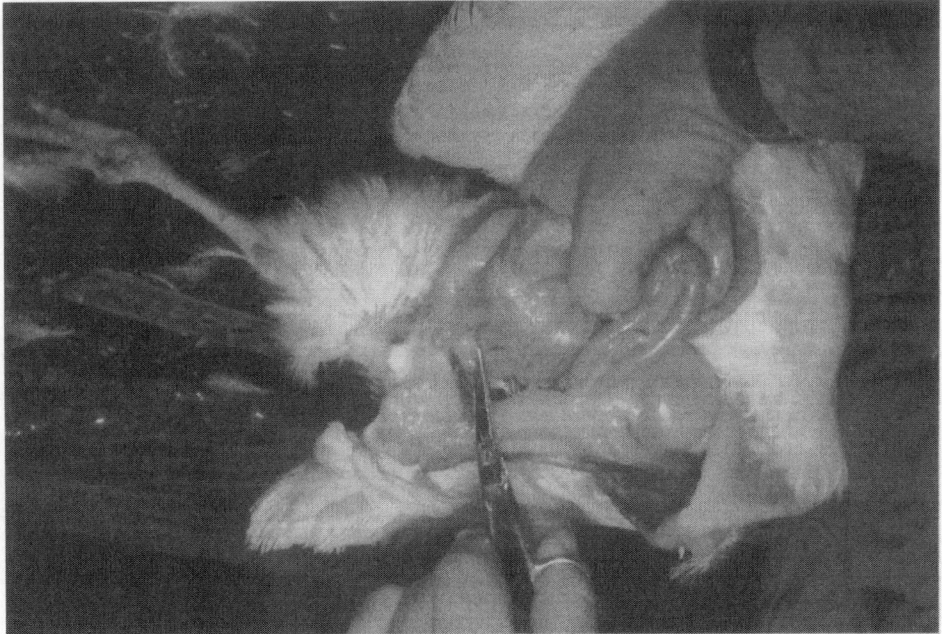

Figure 30-2. Collecting Tissues for Laboratory Examination

Drawing blood from poultry is not difficult, but does require some practice. Drawing blood from the brachial or wing vein is usually the best method for collecting blood from adult chickens. In most instances, a 20-gauge, 1- to 2-inch needle works well. A new, disposable syringe and needle should be used for each bird.

When drawing blood, there are several tips that are helpful to remember. The first is to carefully thread the needle into the vein opposite to the direction of blood flow. Veins carry blood back to the heart and, hence, in the wing vein, the needle should be pointed toward the wing tip and thus away from the body. The second tip is to be patient. Do not pull back too hard on the plunger of the syringe, as this will cause the vein to collapse and prevent blood from entering the needle. Only a slight vacuum is needed, and turning the needle or moving it up or down in the vein will sometimes help the blood flow if the vein does collapse.

Laboratory tests usually require less than 1 milliliter or cubic centimeter (cc) of serum but you should check with the laboratory first for the specific requirements of any test. Serum is the clear, yellowish fluid remaining after the whole blood has been allowed to form a clot. The serum contains the antibodies produced against vaccines and disease agents. Serum can easily be destroyed or contaminated, therefore, proper handling of the blood during and after clotting is important. After collecting the blood, it should be transferred to serum collection tubes or, alternatively, it can be left in the syringe. But, in either case, it should be placed in a nearly horizontal position until clotting has occurred. If unclotted whole blood is required, the collection tubes must contain an anticoagulant to prevent the formation of a clot.

The blood should be allowed to clot at room temperature. Do not refrigerate freshly drawn blood, expose it to sunlight, or high temperatures—temperatures in excess of 100°F (37°C). After a firm clot has formed (approximately 30 min), the containers can be placed vertically. When the serum has separated from the contracted clot, the samples should be refrigerated (not frozen). The serum should be poured off into a clean container if the samples are to be stored for more than 48 hours. Once the serum is separated from the clot, it should be refrigerated until it can be delivered to the laboratory.

Samples to Test for Disease Agents

Tissues can be used for bacterial or viral culture. When collecting for cultures, large pieces of tissue should be placed into a clean, unused plastic bag, sealed, and kept refrigerated until submitted to the laboratory. Samples should be submitted as quickly as possible. Rapid submission will improve the effectiveness of the test and it will improve turnaround times for results.

Swabs in sterile culturettes can also be used to collect samples for bacterial and viral culture. When collecting samples for bacterial cultures, it is always important to remember that the swab must be free of bacteria before attempting to culture the bird. Sterile swabs that are individually wrapped work best for this purpose. If sterile swabs are not available, it is better to submit tissues to the laboratory so that the proper cultures can be made there. Swabs for isolating viruses must be placed in a transport medium after collection. It is best to pick up the proper materials from the laboratory before attempting to collect samples. Since this requires preplanning, it is usually easiest and best to collect tissues, keep them refrigerated, and let laboratory personnel culture them.

Samples to Test for Toxins

Testing for toxins usually requires that tissues from the poisoned animal and from the potential source of the poison be submitted simultaneously. Tissues are collected from the birds as they are for culture and stored either refrigerated or frozen. Whether feed, water, or some other sample is collected as a potential source requires some detective work. The most important question is "What has changed since the birds began showing signs?" If a new load of feed was delivered, then feed is the best place to start looking for toxins.

One common problem in commercial production facilities is that by the time toxicity is detected, the birds have eaten through the feed that caused the problem. It is always a good idea to collect and set aside some feed when a load arrives, keep it refrigerated, labeled with the date, and an identifier for the load. That way, if there is poisoning, the source can be determined and the problem corrected.

30-C. PREPARING SAMPLES FOR THE LABORATORY

One of the most critical parts of submitting samples to the laboratory is making sure that the samples are properly labeled. The item being submitted, the date, the submitter's name, and a bird or flock identification should be included on every label. Alternatively, a group of items can be labeled collectively. For example, 10 bags of tissues could be labeled with a bird or flock identification, and placed in a larger bag labeled with the other information.

Every laboratory has its own forms that should be submitted with a sample. A typical form will ask for the submitter's name, source of the sample, type of sample, test requested, and some description of the problem. A relevant flock history should be supplied with the submission. A description of the clinical signs, percentage of the birds showing clinical signs, and recent flock mortality should be provided.

30-D. WHERE TO SUBMIT SAMPLES

There are several types of diagnostic laboratories that differ in their functions, facilities, and turnaround times for samples. The state diagnostic laboratory is often the first step in sample testing. These laboratories generally offer a wide variety of tests, sometimes in their own facilities and sometimes by linkages with other laboratories. For diseases of national and international importance, samples are sent for full identification to federal laboratories. The federal laboratories accept submissions only from other diagnostic laboratories, therefore, all samples should be submitted to the state laboratory, which will in turn send the sample on to the federal laboratories if necessary.

Some large poultry companies have their own private laboratories. These laboratories run more common diagnostic tests and may also be involved in testing samples from other areas of production and processing. The culture of carcasses and machinery from the slaughter plant are usually performed in these laboratories.

30-E. USING LABORATORY TESTING EFFECTIVELY

Selecting the proper samples, submitting them correctly, and sending them to a laboratory that can do the testing are all a part of using

Table 30-1. Examples of Diseases and the Tests Used to Detect Them

	Disease	Disease Agent	Diagnostic Tests
Bacterial diseases	Infectious coryza	*Haemophilus paragallinarum*	1. Necropsy & histopathology 2. Isolation & identification 3. Serology: AGP & HI
	Fowl cholera Pasteurellosis	*Pasteurella multocida*	1. Necropsy & histopathology 2. Serology: ELISA, HA, & HI 3. Isolation & identification 4. RFLP 5. PCR & sequencing
	Mycoplasmosis	*Mycoplasma* spp.	1. Necropsy & histopathology 2. Serology: ELISA, HA, & HI 3. Isolation & identification 4. RFLP 5. PCR & sequencing
	Salmonellosis Pullorum disease Fowl typhoid Paratyphoid	*Salmonella* spp.	1. Necropsy & histopathology 2. Serology: ELISA & HA 3. Isolation & identification 4. Phage typing

Viral diseases	Newcastle disease	Newcastle disease virus	1. Necropsy & histopathology 2. Serology: ELISA, HA, & HI 3. Virus isolation 4. Typing with monoclonal antibodies & chicken inoculation
	Avian influenza	Avian influenza virus	1. Necropsy & histopathology 2. Serology: AGP, ELISA, HA, & HI 3. Virus isolation 4. PCR & sequencing 5. Typing with monoclonal antibodies & chicken inoculation
	Infectious laryngotracheitis	Infectious laryngotracheitis virus	1. Necropsy & histopathology 2. Virus isolation 3. Serology: AGP, ELISA, & IFA 4. PCR & sequencing
	Infectious bursal disease	Infectious bursal disease virus	1. Necropsy & histopathology 2. Serology: ELISA 3. Virus isolation 4. PCR & sequencing
Parasitic diseases	Coccidiosis	*Eimeria* spp.	1. Necropsy & histopathology 2. Mucosal scrapings
	Mite infestation	Northern fowl mite	1. External examination 2. Microscopic examination of mites
Fungal diseases	Aspergillosis	*Aspergillus* spp.	1. Necropsy & histopathology 2. Isolation & identification
Toxic diseases	Botulism	*Clostridium botulinum*	1. Clinical examination 2. Mouse bioassay

laboratory testing effectively. Another is the appropriate use of testing. This requires a full understanding of the diagnostic tests that are available, the information that they can provide, and how to interpret the results at the farm level. Some examples of diseases and the tests that can be used to diagnose them are listed in Table 30-1.

Diagnostic results should be interpreted in light of the clinical signs observed in the flock. Necropsy, histopathology, and serology results, and the results of tests for disease agents should all be combined and used to make a coordinated diagnosis or series of diagnoses. In other words, the results obtained from any test should make sense based on the flock situation.

30-F. NECROPSY TECHNIQUES

The necropsy, or post-mortem examination is one of the most valuable diagnostic tests available. If used as a routine part of disease management programs, the necropsy of normal flock mortality can lead to early disease detection. For disease diagnosis, a necropsy done in the field can provide the preliminary diagnosis and lead to immediate remedies. A complete necropsy done in a diagnostic laboratory will take longer but can provide definitive diagnoses, information on disease severity, and how long the outbreak has been ongoing.

Tools for Necropsy

A poultry necropsy can easily be performed with a few essential tools. A pair of large, sharp poultry shears is useful in opening the body cavity (A in Figure 30-3). A smaller pair of blunt ended scissors (B in Figure 30-3) or sharp ended scissors (C in Figure 30-3) is especially helpful in opening the trachea and intestine A pair of forceps is helpful especially for collecting specimens for histopathology (D in Figure 30-3). Finally, a knife or scalpel is useful for opening joints and slicing muscles (E in Figure 30-3).

Steps in Performing a Necropsy

1. *Assess clinical signs and epidemiology of disease*

   Although the necropsy actually begins after the euthanasia of the chicken, the first step of any thorough necropsy begins with the examination of the live bird. Its mental state, its physical state, and any obvious deficiencies should be assessed. The entire flock should also be examined. If there is excessive mortality, the location of where dead birds are found in the house should be

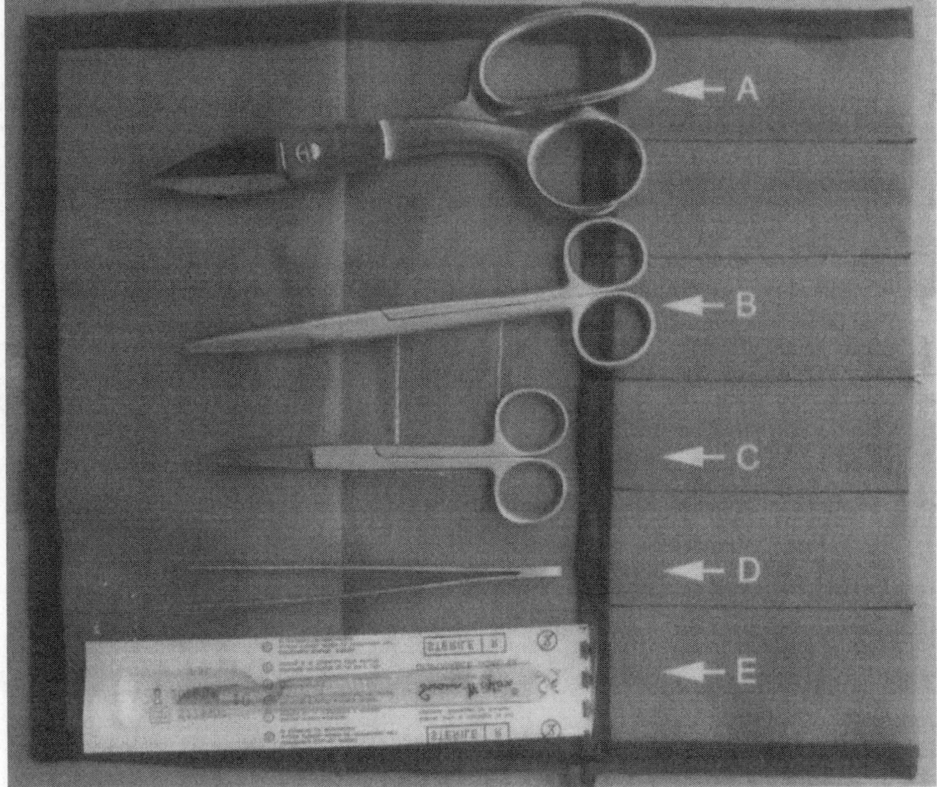

Figure 30-3. Tools for Necropsy

noted. Additionally, the age of the birds, and the season of the year should be recorded.

2. *External examination*

   The external features of the bird should be completely examined. Starting at the head and ending with the feet, examine the comb, beak, eyes, feathers, skin, vent, shanks, and feet.

3. *Opening the bird*

   Lay the chicken on its back, and disarticulate the hips so that it will lay flat. Cut the skin over the breast and pull the skin back over the breast, abdomen, and legs. This is a good opportunity to check the color, texture, and symmetry of the underlying muscles.

   The body cavity should be opened by cutting the ribs (Figure 30-4A) on each side of the breast, the collar bone on one side, and pulling the keel toward the head. At this point, the body cavity is fully visible (Figure 30-4B). Take this opportunity to evaluate the overall appearance of the organs, their color, their size, and their position

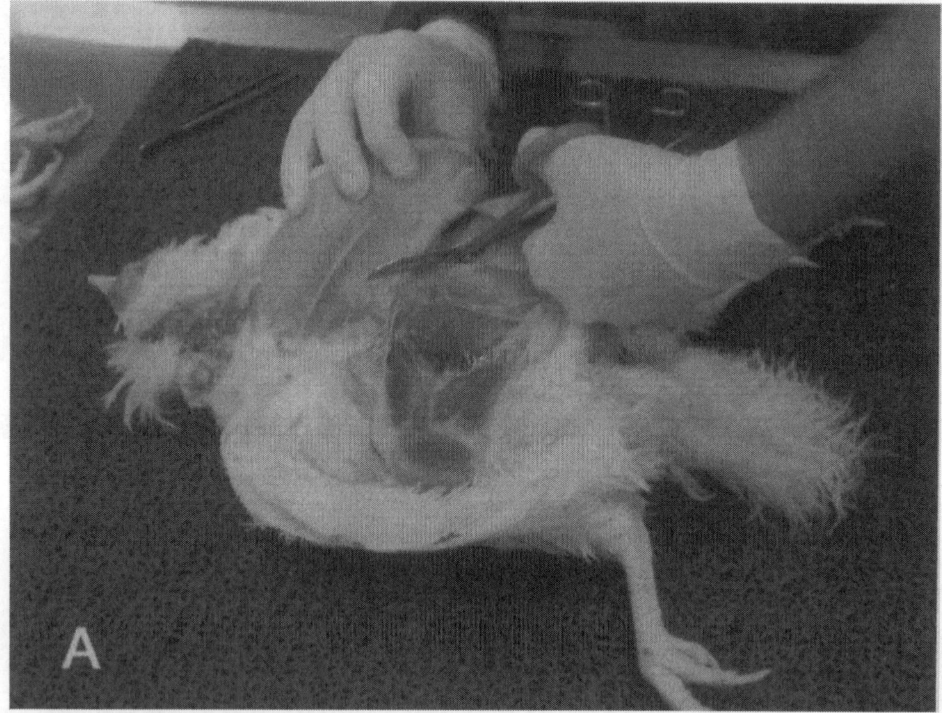

Figure 30-4(A). Preparing a Bird for Examination—Opening the Bird

in the body cavity. This is also the time to evaluate the abdominal and thoracic air sacs.

4. *Systematic examination of the abdomen*

   The abdominal organs of the digestive system include the small intestine, large intestine, pancreas, liver, and gall bladder. These organs should be carefully examined and compared to a healthy bird if they appear to be abnormal. The intestines should be opened so that the mucosal or inner surface can be observed. Examine the spleen and bursa of Fabricius of the immune system and assess their size compared to a healthy bird of similar age. The male reproductive organs, the testes and vas deferens, should be prominent in breeders but barely visible in immature chickens like broilers. The oviduct and ovary should be examined in females. The kidneys should be observed for color, symmetry, and swellings, and the ureters should be observed for urates. Finally, the kidneys should be removed to observe the sciatic plexus, the nerves that go to the legs.

5. *Systematic examination of the thorax*

   The heart should be examined for size and symmetry.

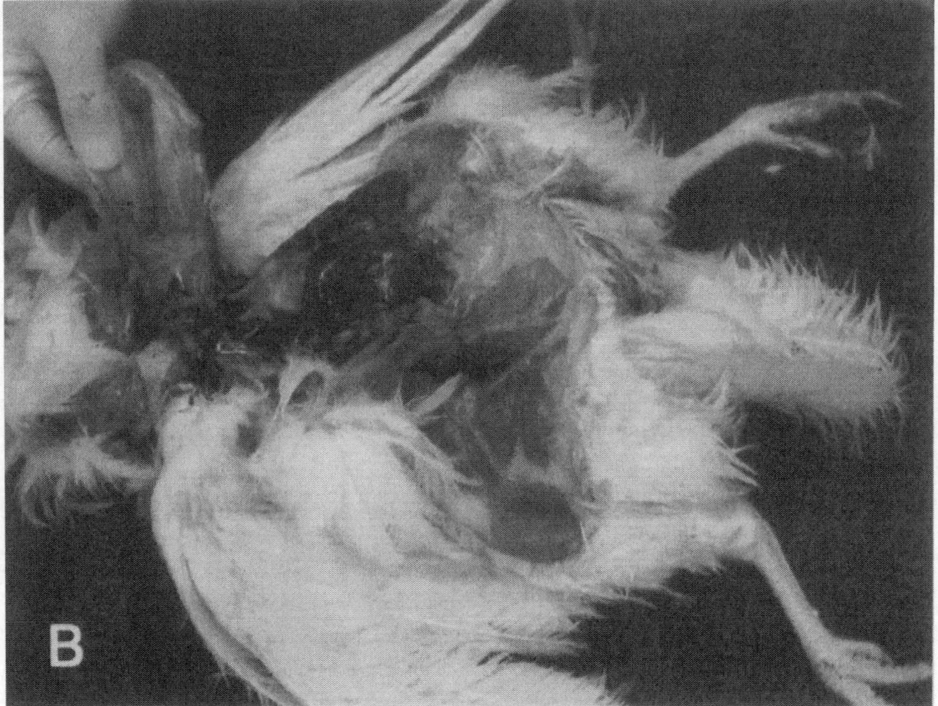

Figure 30-4(B). Preparing a Bird for Examination—Exposing the Body Cavity

The color and thickness of the heart sac should be noted. Once the heart is removed, the proventriculus and gizzard can be opened and examined. The lungs lie on either side of the backbone. The lungs should be removed and examined for color, and firmness. The ribs, keel, and thoracic spine should be observed at this time. The ribs should be somewhat flexible but smooth and free of bumps. The keel should be straight and firm.

6. *Systematic examination of the head and neck*

   The mouth should be opened and examined. The corner of the mouth should be cut with blunt-ended scissors, down the neck, thus opening the esophagus. The trachea should also be opened and the interior surface examined. The interior surfaces of the esophagus, crop, and trachea should be smooth, pale in color, and free of defects. Alongside the esophageal and tracheal tubes is the thymus, which should be assessed for size. The thymus is important in immune responses and a healthy chicken should have a prominent thymus. After the neck examination is completed, the skull should be opened. Score the skull with four even strokes making a square shape. The square cap can then be pried off with the forceps or scissors. The brain should be examined for symmetry and color.

7. *Systematic examination of the legs and wings*

   The joints should be opened with a sharp knife and checked for infection. The bone marrow can be examined by opening up at least one of the long bones. The tibias should be broken to check for caged layer fatigue or rickets. The bones of a healthy bird will snap sharply when broken.

30-G. HISTOPATHOLOGY

Histopathology refers to the study of fixed and stained tissues under a microscope. Histopathology is very useful in disease diagnosis. It can also be used to determine the recovery of a flock from an outbreak or the persistence of a disease-causing agent. Fixed tissues are embedded in paraffin, sliced very thinly, and placed on a slide. The tissue is stained and covered with a thin piece of glass, then examined by a pathologist under a microscope for indications of disease and disease agents.

Using histopathology, a trained pathologist can determine the cause of a disease by actually seeing the disease agent under magnification and the damage it has caused. For example, using standard stains, a pathologist would be able to see *Aspergillus* spp. growing in a lung nodule. However, sometimes the disease agent is hidden in the tissue. Special compounds that stain only some types of agents can be used to confirm a diagnosis. For example, silver stains are very effective for staining fungi but do not interact with bird tissues at all. A silver-stained section would help a pathologist confirm the diagnosis of fungal disease and to assess the severity of the problem. There are many types of special stains and a trained pathologist will use these routinely to differentiate disease agents and confirm diagnoses.

Immunohistochemistry can also be used to diagnose specific disease agents. Immunohistochemistry refers to the application of antibodies specific for a disease agent to a tissue section. These antibodies will react only with the agent of interest. Dyes or stains that react with the antibodies are then applied to the section. The section is then covered with a thin piece of glass and examined under a microscope. The disease agent of interest will be dyed while uninfected tissue will not. Immunohistochemistry is more specific than histopathology either with standard or special stains. For example, a specific strain of virus can be distinguished from other strains with immunohistochemistry but not with standard histopathology.

In-situ hybridization refers to the application of a labeled piece of DNA to a tissue section. This technique can be used to detect specific disease agents in tissue sections. The DNA probe will only react with DNA that has the same sequence of nucleotides, in other words, specifically with the disease agent. The covered section is examined under a microscope and the disease agent will have color while uninfected tissue will not. In-situ hybridization, like immunohistochemistry, can determine the specific strain of a disease agent. These tests may not be available in local laboratories but may be used at reference laboratories to make specific diagnoses.

30-H. TESTING FOR ANTIBODIES

A bird exposed to a foreign substance will form antibodies to that substance. The formation of antibodies is discussed in *Immunity,* Chapter 24. Antibody tests can be used to monitor vaccination programs and are a part of comprehensive health-monitoring programs.

There are several types of tests that will detect antibodies. Antibodies can cause red blood cells to agglutinate or clump. The specific name for the clumping of red blood cells is *hemagglutination* (HA) and it can be used as a test for some disease-causing agents. A variation of this test is the hemagglutination inhibition (HI) test. This test is based on the fact that specific antibodies can prevent the antibody-associated clumping of blood. Hemagglutination and hemagglutination inhibition tests are not as sensitive as the ELISA test, the standard for antibody tests. An HA or HI test will be reported as positive or negative for a series of dilutions. This means that the sample is diluted and then tested. A sample that is 1 part serum and 20 parts buffer will be a 1:20 dilution. That sample will contain more serum than a 1:320 dilution that is 1 part serum and 320 parts buffer. A serum sample that is positive at 1:320 has more antibodies or a higher antibody titer than a sample that is positive at 1:20.

The *agar gel immunoprecipitin* (AGP) test is used to detect antibodies to a specific disease-causing agent by taking advantage of the fact that antibodies interact with the specific disease agent and form a complex. These complexes can be observed in a gel. If the test serum forms a complex with the disease agent, then the sample is positive for antibodies to that agent.

The ELISA is probably the most commonly used test for antibodies. In general, it is a very sensitive and specific test. The ELISA consists of a plate coated with a disease agent. The sample serum is applied to the coated plate. If present, antibodies to the agent bind to the plate. An antibody with a visible label attached to it forms a complex with the chicken antibody bound to the disease agent when added. If the plate has color, it is a positive test. If it is colorless, the test is negative. Dilutions are used to estimate how many antibodies are present in the serum in the same way

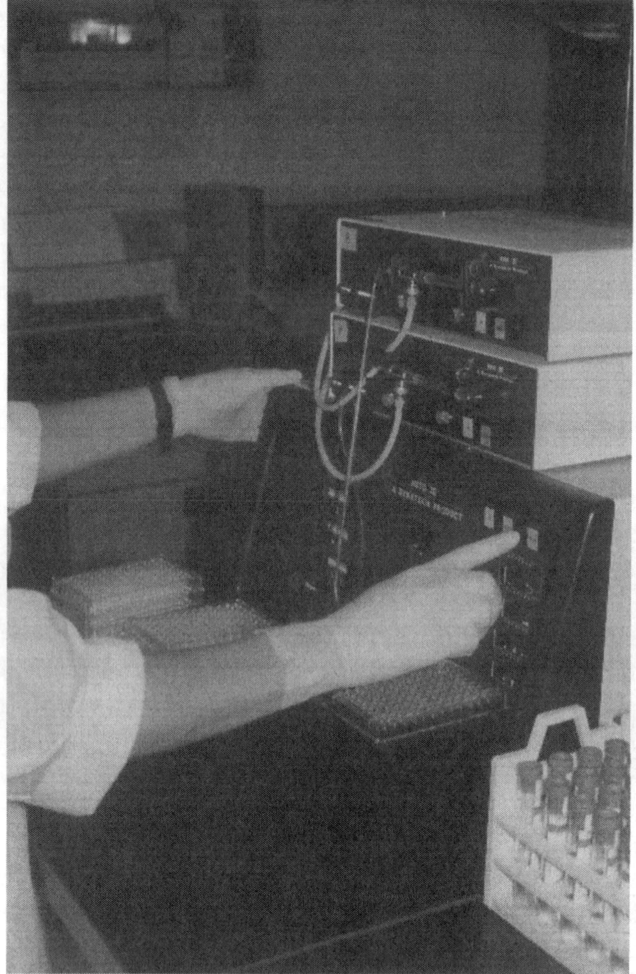

Figure 30-5. ELISA Test Machine

they are used for HA and HI tests. This estimation of how many antibodies are in a serum sample is called a titer (see *Immunity,* Chapter 24).

Interpreting ELISA Results

ELISA tests come with computer software to report their results. Some examples of typical ELISA reports are shown in Figure 30-6. The antibody titers are reported as titer groups. Titer group 1 consists of the samples that are positive at a final dilution of 1:10, titer group 2 is positive at 1:20, and so forth. The number of samples are shown on the vertical axis and the total number of samples that fall into a titer group are shown as a bar.

The results shown in Figure 30-6 demonstrate the importance of collecting paired serum samples. Figure 30-6A shows a poor immune response.

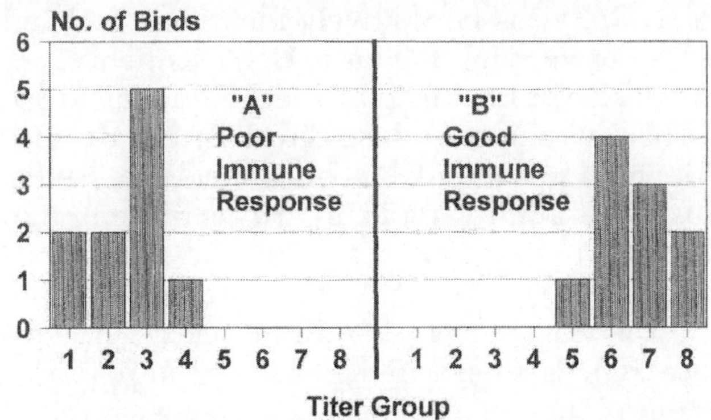

Figure 30-6 (A,B). Typical ELISA Reports—Paired Samples

The interpretation of this result depends on the appearance of the graph in Figure 30-6B. The second sample was taken 1 to 2 weeks after the first, and shows rising antibody titers. The combination of these two results is indicative of either a good vaccine response or a rapidly spreading infection. These results are typical of a variety of agents including infectious bronchitis virus and Newcastle disease virus. When beginning testing for antibodies, it is important to understand the types of results you can expect. They will vary with the test and the disease agent or vaccine.

30-I. TESTING FOR DISEASE AGENTS

Tests for disease agents have varying levels of sensitivity and provide different types of information. Interpreting these tests not only requires knowledge of the test and what it shows, but also knowledge of possible agents and their importance in disease.

Viruses

Isolation is generally the best method to identify a viral agent. After the virus is isolated, it can be identified with a variety of techniques. Some viruses cause very specific damage to embryos or cells. The sample is inoculated into fertile eggs and the embryos are examined after several days for the appearance of specific characteristics. Infectious bronchitis virus (IBV) and poxvirus can be identified this way. However, this method cannot differentiate virus strains, like the Arkansas or Delaware strains of IBV. Another method uses antibodies to identify isolated viruses. As with immunohistochemistry, specific antibodies with an attached dye can be applied to the isolated virus. The antibody will react with only a specific virus or strain of virus. The cells are examined under a microscope and if they have color, the virus is positively identified. Isolated viruses can also be detected by identifying the virus DNA sequence.

Some viruses can also be identified by a technique called *polymerase chain reaction* (PCR). This can either be done directly from the bird or after the virus is isolated. Small pieces of DNA or primers are used to specifically bind to the virus DNA in the sample. If virus is present, the two primers will bind to it, and an enzyme will amplify the virus DNA. A positive reaction will result in a piece of DNA of the proper size. The polymerase chain reaction test is very powerful and can detect very small quantities of virus. Because PCR is so sensitive, it is very important that samples for PCR not become contaminated. So, in collecting samples for PCR, it is essential that new, clean tools and containers be used. It is also very important to carefully choose a laboratory to perform this test. Skilled personnel, proper facilities, and careful handling of samples are required for reliable results.

Bacteria

The culture of bacterial agents requires the use of media that is selective for the growth of one or another disease agent. Bacteria can be partially identified by determining what type of media they require to grow and the appearance of the bacterial colonies. Further identification is made by the determination of specific reactions to chemical tests. These tests will allow the diagnostician to identify the genus and species of the bacterium.

Figure 30-7. Bacterial Culturing

Additional differentiation of bacterial strains can be done using a variety of techniques. *Restriction length polymorphism* (RFLP) analysis can be used to differentiate bacteria of the same genus and species, but which differ at the DNA level. For example, the RFLP analysis could be used to determine if the *E. coli* isolated from the slaughter plant is also in the poultry house. Additionally, PCR can be used to identify differences in phage types. *Phages* and *plasmids* are viruses that infect bacteria. These viruses can carry virulence factors that make the bacteria more damaging to the chicken. Plasmids are often the source of antibiotic resistance (see *Medication for the Prevention and Treatment of Diseases,* Chapter 26).

Once the infecting bacteria are identified, the determination of effective treatments can be made. In an *antibiotic sensitivity test,* different antibiotics are applied to a bacterial culture and their ability to stop the growth of the bacteria are determined. Bacteria may be sensitive or resistant to one or more of the antibiotics and this information will be included in a diagnostic report. A sensitivity report should guide the therapeutic use of all antibiotics.

30-J. TESTING FOR TOXINS

Toxins or poisons are chemicals, metals, or biological substances that can adversely affect health or cause injury to the chicken. Extracting and identifying chemical and metal toxins involves a series of extraction steps that physically separate the poison from the sample material. The poison can then be identified using a variety of procedures such as spectroscopy and chromatography.

The cause of botulism is botulinum toxin, which is the product of *Clostridium botulinum,* a bacterium that grows in rotting material. A *mouse bioassay* is the standard test for this toxin. The toxin is inoculated into mice, and within 3 days the mouse will show typical clinical signs if the material is positive for botulinum toxin. This test should be performed by personnel trained in detecting the specific clinical signs of poisoning in mice.

30-K. CONCLUSION

The field of disease diagnosis is rapidly expanding and changing. Technological advances have introduced many new tests in recent years and will continue to do so. However, the principle of interpreting test results in light of the observed clinical signs and necropsy findings will undoubtedly continue to be the basis of all diagnostic testing.

Section IV. Business

31

Operating a Poultry Enterprise

by Donald D. Bell

The successful operation of a poultry business is largely dependent upon the organization's structure, the policies and procedures it adopts, the communication techniques employed between departments and levels of management, its concern and respect for the needs and contributions of employees, and its ability to correctly prioritize demands from different departments. Failure in any of these areas will result in less than optimum company performance.

31-A. COMPANY STRUCTURE

Most poultry businesses today are either owner-managed or managed by a paid executive or group of executives. Large companies may have a chairman of the board of directors, a president, multiple vice presidents with responsibilities over various company divisions, a treasurer or chief financial officer, and a board consisting of department heads and in some cases, one or more outside consultants. Decisions may be made by an individual or by a vote of a larger group. Individual department heads are given a defined list of decision categories within their area of responsibility.

Typically, the general manager's principal objective is to integrate recommendations and resources from all divisions of the company into a series of policies to maximize benefits for the firm. Proposals are considered and weighed as to their impact on company costs and profits. It is important to remember that the lowest cost in a particular area does not necessarily produce the highest return for the company. The general manager must

Figure 31-1. Board of Directors Discussing Division Reports

look at the "big picture" which is the *return on invested dollar* as influenced by policies related to products and operations.

In the real world there is a trade-off between carefully defined lines of authority and the free exchange of opinions. The poultry industry tends to err on the side of military style authority and in many cases is extremely weak in allowing the free exchange of opinions. One solution is to "flatten" the organization and have fewer management levels and more "peers" who can talk more freely with each other. This system seems to work in some industries. A successful company is structured to have carefully defined lines of authority and responsibility. Opportunities for exchange of knowledge, ideas, and opinions must be encouraged among and between all levels of management.

31-B. POLICIES AND PROCEDURES

Existing policies and procedures must be developed by individual department managers and recorded in a policy and procedures manual. Such manuals must be specific with a careful description of the item so that the procedure can be duplicated by the reader. The written description should give supportive documentation for the practice with adequate references. The department manager must be able to support why this practice is used rather than another. A verification procedure must be used to assure that the application of a policy is being carried out as intended. This is known as a quality control procedure and should be performed by individuals not directly associated with the department in question.

Since economic conditions affect the choice of programs, each manager must update their program analyses whenever costs or prices change significantly. Practices and procedures are justified on their economic strengths; permanent changes in costs or prices may require significant changes in programs. This principle applies to practically all procedures.

In order to have all parties buy into the results, management must be able to communicate which goals are most important for achieving the highest possible profits for the company and which goals are of lesser importance. Goals must be profit-oriented not performance-oriented. To reinforce this profit orientation, it would be best if managers held stock, options of stock, or were paid bonuses based on profit.

31-C. NEW TECHNOLOGY

Management is constantly made aware of new products and ways of doing things. This information is obtained in magazines, books, meeting proceedings, and from visitors to the company. Some is based upon objective well-conducted research, while much is anecdotal with little if any documentation. The manager should ask for research or test results with emphasis on the thoroughness of the research, its application to the company's business and its repeatability. Even though small differences in performance may be important, it's unwise to assume that all small differences are attributable for the product or procedure being discussed.

New technology is best assessed by knowing the source of the information. Was the work done by a reputable researcher, university, or firm? Does the information apply to your type of business? Have all important criteria been evaluated or is the information limited to only one or two positive measurements with no consideration for negative results?

If a procedure or a product sounds interesting, ask for the names of other companies who've had experience with it. Talk to your consultants about the procedure or product. They may have observed the product on other farms. If you need your own experience, ask your consultant to help you design a small test which can help you evaluate the item in question. Don't run a two-house comparison if you expect small differences—two houses are always different and you won't be able to determine if differences are due to the product or to the house. If you must use a two-house test, repeat it in two or more flocks. You may even consider sponsoring research at a local university for your specific needs.

Adoption of new policies or procedures must be carefully documented by their proponents within the company and presented to the general manager when timely or to the board of directors at their regular meetings. Documentation consists of:

1. A description of the current practice
2. A description of alternative practices
3. Supporting evidence that the new practice is better for the company
4. An estimate of the economic impact if the practice is adopted
5. A suggested time-table and step-by-step procedure for applying different elements of the proposal

Independent layer companies are generally restricted in their ability to measure significant performance differences associated with various strains of chickens because of their limited number of flocks. Broiler companies, on the other hand, have hundreds of flocks which provide information relative to performance. Various products and breeds can be compared on multiple farms at the same time.

New technology should not only be borrowed from other companies, but also should be developed in-house. If there is good communication up and down the organizational chart, new ideas for research will spring up spontaneously from within the company. Independent research and cooperative research with Universities will develop cost-saving or revenue-enhancing procedures. A sure sign of a progressive company in any industry is the evidence that it is investing in developing new technology.

31-D. COMPANY PERFORMANCE—RESULTS OVER TIME

Record Management, Chapter 32, describes the various record systems required to monitor separate elements of the company. Such systems require accurate collection of data, timely analysis, comparison with company or industry standards, and communication to the proper department or individuals in the company. Departures from standards, last week results, or target values must be highlighted, questioned, and resolved. Remember, differences from standards are not necessarily bad. The appropriate business decision may be to purposely deviate from a particular standard to achieve a higher return on invested dollar.

Each department should establish annual achievable goals. Department goals should not conflict with the overall company objective of achieving the highest possible return on invested capital. Such goals should indicate the amount of improvement targeted. For example:

1. Improving layer flock egg numbers—2 eggs/hen housed to 60 weeks
2. Increasing egg weights during summer months—½ pound per case during July/August

3. Decreasing feed consumption and improving feed efficiencies—0.25 lbs / 100 hens
4. Reducing the amount of down time in the processing plant—from 10% to 8%
5. Reducing condemnations—from 1.25% to 1.15%
6. Analysis of egg breakage locations and incidence levels—from 6% to 5.5% for all eggs
7. Decreasing the rate of employee turnover—from 25% per year to 20%

Goals must be attainable and not just wish lists. Each goal should be a project in itself with in-depth studies of the issues involved. Are the data collection systems accurate so that you can truly measure change? Has every aspect of the problem been examined? Is everyone talking the same language regarding the definitions of the issue? Don't assume that everyone defines feed conversion the same. All cracked eggs are not the same—some can be seen by the unaided eye; others require a candling light. Everyone has a different definition for the production period of a flock—be specific, for example, eggs per hen housed from 20 to 60 weeks of age or body weights at 45 days.

Department managers must provide such goals to their supervisors along with periodic progress reports. If possible, progress should be graphed to facilitate discussions. At the end of the year, a summary report should be prepared with statements discussing both successes and failures with reasons for each.

If records are submitted daily to an office for immediate recording, performance changes may trigger a computer to type a warning or an alert accountant may notice the beginning of an abnormal trend. However, when collecting data on a daily basis, it is important to collect similar data at the same time each day for valid comparisons.

Problems, though, are not all of a flock nature. Accounts receivable may be higher in terms of the age of accounts, store returns of products may be increasing from a particular retailer, the hatchability of eggs in XYZ hatchery has gone down for three consecutive weeks, or the feed mill is having more and more late deliveries of feedstuffs. All of these can be monitored with modern computer software systems. Computer programmers should be told of the need for alarming devices to alert management of abnormal trends.

31-E. COMPANY PERFORMANCE—RESULTS BETWEEN COMPANIES

Progress over time is expected within a company, but a company's place within the industry is also a very important measurement for management. Such information is difficult to obtain. In years past, breeder standards were used for comparison. Today, commercial consulting companies have developed systems whereby flocks can be compared across the country and even worldwide. Systems are available which analyze the profitability of flocks within companies and by comparing them with similar flocks and similar companies. Such systems are very dependent upon a uniform understanding of terms and analytical procedures.

Large integrated broiler companies require both in-company and between-company comparisons. The first is relatively easy because of the hundreds of separate flocks and farms a broiler company has under contract. Comparison between companies is facilitated by private consulting firms who make it a business to provide the large broiler companies with performance comparisons and financial information. The key to their success has been their ability to maintain the strictest security of the sources of their data. Without these data, the industry would be floundering in a sea of statistics with little interpretation of its meaning.

31-F. THE DECISION MAKERS—A TEAM APPROACH TO SOLVING PROBLEMS

The flock caretaker or the bookkeeper may be the first person to observe a problem in its earliest stages. The flock may be listless and may be off feed, water consumption may be down, the droppings may be off color, or the flock just doesn't sound right. Most of these are visual observations and are best made by the "person in the house"—a strong argument for experienced house caretakers.

Frank communications between laborers and managers and managers and their supervisors must be encouraged. Time is of the essence and problems must be communicated as soon as there is an indication they exist. Problems which should have been reported on Friday and are allowed to aggravate over the weekend are commonly too late for correction.

Once a problem is recognized and consultation is required, each manager needs a procedure for contacting other personnel and a plan for solving the problem. A team of individuals to address feed problems may consist of the production manager, the manager of the mill, and a consulting nutritionist. A team to work on an egg quality problem may consist of the production manager, the processing manager, and the marketing manager. Depending upon the seriousness of the problem, the general manager may sit in on such discussions and contribute to the decision-making activities or simply kept advised as to the progress and correction of the problem.

31-G. PERSONNEL ISSUES

New employees need to start employment with knowledge of what their company is all about and the importance of their position in the company. Such knowledge is needed equally by the laborer and the flock manager. Without concern for the success of the company or for the causes of its possible failure, the new employee cannot have respect for his / her job or value as an individual.

Tips for managing personnel include:

1. Train them well. Few people come to a new company with sufficient knowledge to begin working the first day. Take time to acquaint them with their fellow workers, the workplace, and the type and quality of work expected. Be careful to train them in safety and regulatory matters. Have your best employees take them in hand for a day or two to point out the details of their work. Take every opportunity to expose your employees to educational sessions relevant to their work.
2. Monitor their progress. After a while, their work becomes routine and efficiencies will improve and errors will decrease. In time, a level of skills will be reached where the individual's daily output will be consistent and predicable. Monitoring, though, should still be used to avoid problems.
3. Provide for their well-being. This not only applies to their health, safety and comfort on the job, but for their families as well. Workers cannot work efficiently if they are at risk or if they are worried about their families at home.
4. Encourage suggestions. No one knows what's going on as well as those who do it every day. If employees suggest a different way of doing a particular task, listen to them and if it makes sense, give it a try. Managers should also be proficient in doing the same work. From time to time, managers should participate in various jobs to experience the problems associated with the work first hand.
5. Give opportunity for advancement. Encourage your employees to exchange jobs in different divisions of the company so that they can be multi-skilled and of greater value to the company. Make them aware of the entire scope of the company and all advancement opportunities.

6. Get to know all of your employees as individuals. Call each by name and learn the names of their family members as well. Recognize their needs when a family member becomes ill, when a close relative dies, or when they are physically exhausted.
7. Reward outstanding performance. Make an effort to discover reasons for rewarding your employees. Rewards can be in the form of incentive payments for quality work, bonuses for money-saving suggestions, or time off when special family events occur.

31-H. INFORMATION SOURCES, STORAGE, AND RETRIEVAL

Personal experience is probably the most valuable resource for solving problems. If you've experienced it and you have a good recollection of the details, you have the best background for solving the current problem. If you have a problem, you should note the details of it and the methods used to correct it. Such problem reports should be filed so that others in the company can access it when the need arises again.

Summary records of flocks, activities, and projects should be filed in appropriate files for future referral. The one page flock summary (see *Record Management,* Chapter 32) is a simple way to recollect all performance factors. A chronological listing of successive flock summaries makes an excellent data base for flock projections. Encourage your managers to maintain daily diaries of problems and solutions.

Each department should maintain an orderly information filing system which includes reference articles and reviews of specific issues relating to the department. Information files should include current breeder management guides, clippings from industry periodicals, newsletters, and proceedings from meetings.

The Internet has become an excellent source of information. Share your favorite sites with your fellow poultry producers, your consultants, and extension agents (see *Computer Applications,* Chapter 33). A list of books and periodicals on poultry and related subjects is listed in the Appendix.

31-I. GOVERNMENT, COMMUNITY, AND INDUSTRY INVOLVEMENT

Today's companies have become so large and involved they must now have departments or significant programs in areas never before considered. The number of regulations imposed upon modern agricultural companies from a myriad of government agencies oftentimes requires a full-time individual or department to keep up with the annual changes in regulations, report forms, and deadlines. Poultry companies should not be bystanders when new regulations are proposed. Poultry company staff members should be involved in suggesting or commenting on needed legislation by serving on industry committees. The industry must provide information to the legislative process (local, state, or federal officials) regarding operating procedures, economics of the industry, and other factors which may have a bearing on the issue. Of most importance, the industry needs to show the legislators what the chicken business is all about.

The enormous size of modern poultry operations requires full cooperation between the firm and the community on dozens of issues of common interest. Most important of these are those affecting the handling of wastes and how this affects the environment of the community (see *Waste Management,* Chapter 11). The introduction of the company into the community may also severely impact schools, roads, and other local services. For this reason, the company and its employees must participate in local affairs and become "good neighbors" in the community.

Industry involvement is important at all levels. Trade associations and other organizations play significant roles in marketing, legislative oversight, education and training, and funding for research and scholarships. The poultry industries could not survive without this mechanism to bring people together for the benefit of the industry.

32

Record Management

by Donald D. Bell

Today's modern poultry companies devote countless hours to the maintenance of complete records documenting hundreds of facts and figures relating to the business. Such records are necessary to remain competitive, to assist with making numerous management decisions, and to satisfy the requirements of various regulatory agencies. Records may be grouped into various categories such as financial, tax, operational, personnel, regulatory, flock performance, marketing, and accounts receivable and payable. Each division of a company has its own unique record requirements and these must be integrated into a meaningful overall system which can provide the management with all the information needed to make proper decisions affecting the company's progress and ultimate success.

32-A. TYPES OF RECORDS

Financial Records

Records which focus on finances are concerned with the investment in the current business, alternative options for new investments, interest costs, new construction and depreciation schedules, operating costs and cash flow, loan pay-offs, and long-term company objectives relative to growth.

Various ratios are used to express the net worth of a company or the financial relationships which measure the success or failure of business decisions. These include ratios of assets to liabilities, income to expenses, and profits to investments. Lending institutions, investors, and management each have distinct needs and requirements concerning the type of financial records kept by poultry companies.

Owners and investors may be mostly concerned with returns on investments (ROI). This information is necessary to evaluate different investment options. Lending institutions, on the other hand, are more concerned with various "operating ratios" which are measures of liquidity, debt vs worth, or debt coverage characteristics of the company.

Tax Records

The requirements for records associated with local, state, and federal taxing systems encompass practically all record systems, other than those on flock performance. Tax records require documentation of all purchases and expenses along with financial records of interest payments, depreciation, and payroll.

Accounting systems may use either the "cash" or "accrual" method. Cash systems allow listing of expenses during the current year which may represent usage in the following year. Accrual systems, on the other hand, require an inventory system with values for items on hand at the beginning and closing of the company's fiscal year. The accrual system is a more precise record of earnings and expenditures within a fiscal year, but it is also somewhat more complex. Poultry companies generally prefer the cash accounting system for taxation purposes as it allows more flexibility and is easier to use. Many companies use both systems, one for in-house purposes, the other for tax purposes. Advantages of one system over the other depend heavily upon the level of profits in one year versus the next and the values assigned to the beginning and closing inventories.

Operational Records

Operational records include all aspects of income, costs, efficiencies, and performance information and can apply to each element of an organization from the breeders to the processing plant. It can include the operation of the central office to the inputs and production of the feed mill. It includes preventative maintenance and flock performance. Even though many operational records are not required by law, they are extremely necessary in today's competitive business. Many of the records generated for operational analyses are associated directly with the tax accounting needs of the company and, of course, are required by law.

Further elaboration on the flock, feed mill, and processing records will be discussed in later sections of this chapter.

Personnel Records

Most of the records in this category are required by various labor laws. These include records of hours worked and corresponding payment, deductions of state and federal taxes, injuries on the job, retirement contributions, and others. In addition, companies should maintain records of performance, promotions, and training received for each of their employees.

Regulatory Records

These records comprise those required by various governmental agencies and include:

1. records of pesticide usage and handling,
2. information about the production and application of manures in nutrient management programs,
3. verification of egg and poultry processing systems relative to quality assurance and/or grading regulations,
4. feed mills usage of drugs and feed additives,
5. farm safety records,
6. records for local fly, odor, and dust control agencies.

Marketing

Accurate and timely records are a must where buyers for a company's products are concerned. Detailed contracts list the conditions under which products are bought and sold and careful adherence to these details is essential to maintain an on-going relationship that can survive the advances of competitors. Marketing requires the receipt of complex orders for different container sizes, different products, and with differing special arrangements. Such details require extreme attention to the specifics of the order, to accurate in-company communications, and to delivery arrangements. Accounts receivable and payable must be within the time frame agreed upon in the original contract.

32-B. RECORDS FOR THE FEED MILL

The feed mill manager's principal duty is to maintain a flow of quality ingredients coming into the company and a similar flow of appropriate finished feeds out to the production farms and houses. This requires a thorough knowledge of each flock's feed needs, a reliable source of timely delivered feed ingredients, and an accurate inventory system for the ingredients at the mill and for finished feeds at the individual poultry houses.

The record requirements to facilitate these functions must place considerable emphasis on the scheduling of deliveries and the quantities and varieties of products required. The forecasting of needs versus actual usage must be constantly monitored to avoid shortages or overages which could result in the deterioration of ingredient quality.

A record of feed formulas for each day and for each house should be maintained along with the weight of feed delivered to specific feed tanks and houses. These records are required by flock managers and are useful when tracking problems associated with feeding the flocks.

Current feed needs for each house and bin require constant monitoring by farm or service personnel. Orders must be called in to the mill as needed and must consider weekends, holidays, and sale of flock dates. Some companies use computerized monitoring of the feed needs for each bin based upon projections of requirements for the number of birds, age, and seasonal conditions.

Besides these records, numerous records are required by regulatory agencies. This applies to the inspection of ingredients, the use of drugs and feed additives, and the environmental regulations concerning dust emissions.

Ingredients should be routinely monitored for quality and finished feed samples should be stored for at least six months. All feed stocks should be used on a first in-first out basis.

32-C. RECORDS FOR THE PROCESSING PLANT

Processing plants, meat or eggs, are required to maintain numerous records which document:

1. the nature of products flowing into the plant (birds, eggs, and supplies),
2. the processes being performed along with information regarding malfunctions and efficiencies,
3. maintenance schedules and their status,
4. verification of required actions taken when problems occur,
5. yield, grade, and condemnation information,
6. quality control and microbiological sampling.

32-D. RECORDS FOR THE HATCHERY

To manage a hatchery efficiently, and to keep costs to a minimum, the manager must have certain records each week, after each hatch, and at the end of the month. Good management is the result of pinpointing inefficiencies and correcting them.

First, the manager must have information concerning the source of settable eggs, temperature / humidity data for the entire setting and hatching period, results of each hatch regarding the success of the hatch and the quality of the chicks produced, and follow-up information to correlate hatchery management with rearing and grow-out farm or laying house results. From time to time, complete break-outs of unhatched eggs should be made to ascertain the reasons for their failure to hatch (see *Operating the Hatchery,* Chapter 40).

The source of hatching eggs must be carefully monitored to assure at least 90% fertility, clean eggs with sound shells, and adequate egg numbers and egg size (see *Managing the Breeding Flock,* Chapter 34) for additional information concerning the breeding flock).

Secondly, a well-operated hatchery must have a strong preventive maintenance program to assure that equipment will operate as intended for the duration of each hatch with optimum conditions for the hatching eggs. Small differences in temperature or air movement can cause poor hatches and poor quality chicks. Maintenance includes not only the operation of the equipment, but also the cleanliness of the entire operation from the floor to the attic, to the insides of duct work, to behind the walls where insects and rodents may reside (see *Operating the Hatchery,* Chapter 40, for information regarding air distribution in the hatchery).

Finally, controls must be implemented to assure competitive costs. This involves a high level of utilization for the equipment and high levels of saleable quality chicks. Full scheduling of setters and hatchers with adequate time between hatches to fully sanitize the facility is essential to cost containment.

32-E. FLOCK RECORDS

1. *Rearing Records for Replacement Pullets and Broiler Breeders*

Keeping accurate records during the brooding and growing periods is an essential part of flock management. Records provide the manager with what has happened in the past, and helps him/her plan for the future. The records should include the following:

a. Strain and source of chicks
b. Vaccinations and medication information
c. Description of feeding program
d. Feed consumption by days and weeks
e. Lighting schedule
f. Body weight and uniformity by week, after 3 weeks
g. Frame measurements at 12 weeks
h. Mortality by days and weeks
i. Beak trimming description and age
j. Record of problems and observations

Many different records for raising replacement pullets can be used, but the basic record system must emphasize mortality, body weight, feed usage, and costs. Because of the relatively long rearing period of 16 to 20 weeks, weekly data are required to enable the manager to make mid-course adjustments in various programs. For example, if body weights begin to drop below standards, adjustments can be made in the feeding program or in environmental temperatures to correct these problems. Body weights should be compared to the standards for the particular strain or breed used.

Body weights for both table egg layer replacement pullets and broiler breeders should be taken weekly beginning at about 3 weeks of age (see *Managing the Breeding Flock,* Chapter 34 and *Cage Management for Raising Replacement Pullets,* Chapter 51). Record sheets which allow at least 100 individual weight entries should be used to allow for uniformity analyses. These data must be summarized immediately upon collection and results graphed to allow the manager to visualize any trends that may be taking place. Body weight monitoring is essential to know when to change feeds. Light body weights, for a particular age, usually require a longer period of time on higher nutrient diets in order to regain optimum weight standards later on.

All mortality must be recorded daily, summarized by the week and for the entire grow-out period. Data should be graphed each week and trends noted for the flock caretakers.

Cost data should be standardized to a common age. This is usually the age at housing and includes delivery costs. Each week beyond 16 weeks of age adds approximately $.10 US to the cost of replacement pullets. Per bird costs are based upon the number of "good" pullets delivered. Summaries should include the following costs: chick, feed, labor, vaccines and vaccination, medication, utilities, housing and equipment, fuel, moving costs, and other charges where necessary. Contract-grown pullets would substitute a grower payment for labor, housing, and equipment costs. Typical costs are listed in *Introduction to the US Table Egg Industry,* Table 49-3.

2. *Grow-out Records for Broiler Flocks*

The broiler industry is able to use the flock summary system because of the relatively short rearing period. Daily records are limited to a listing of mortality in each house with totals for the week. Feed consumption is usually based upon the total feed delivered minus inventory at the end of the grow-out. Body weights are often monitored near the end of the cycle in order to adjust the number of days in grow-out to the body weight needs of the processing plant. Final body weights are usually based upon truck weights for the entire flock minus the tare for the truck. Adjustments are commonly made for dead birds which were included in the final weight figures.

Determining final broiler head counts continue to be an inexact science. When attempting to determine final individual bird body weights, three separate counts can be used:

1. Grower counts—chicks delivered minus mortality.
2. Counts by the live-haul catchers.
3. Counts made by photo-cells on the processing line.

Generally, processors use the count made by the live-haul crews as they tend to be the most accurate.

Flocks may be managed by the contract grower / farm owner or the service personnel of the contracting company. Daily records should use a "check-list" system of records whereby each of the daily chores is listed and the workers check-off when they've finished each item, e.g., brooder temperatures are adjusted, waterers are checked several times a day, feeders and feed bins are inspected, dead birds are picked up, ventilation is adjusted, fans are cleaned and maintained, wet litter is replaced and so on.

Table 32-1 represents a simple one-flock spreadsheet with daily records and summaries for broiler flocks. When used properly, weekly summaries of mortality, body weight gains, and feed conversion can be compared. The summary includes total flock mortality, feed consumption and conversion, and condemnations. Costs can be summarized on another form which should include chicks, feed, fuel, contract payments, hauling costs, and miscellaneous cost expressed on a per bird and weight basis.

3. *Breeder Farm Records*

Records for breeders include age-related measurements for:

Egg production (daily and weekly)	Weekly feed consumption
Weekly samples of egg weight	Fertility
Settable eggs	Weekly body weights
Mortality (daily & weekly)	Costs

Feed consumption records may include more details such as energy and protein intake per bird and intake information for individual amino acids and minerals.

Data should be compared to farm standards and the standards from the breeder. Graphing of data will allow the manager to follow trends by trait and make corrections before they get out of control. Incubation records for hatchability and saleable chicks should be provided to the breeder manager for each flock on a current basis.

High numbers of settable eggs based upon hens housed is the goal of the breeder farm. Since the selling age is usually predetermined, the only

Table 32-1. Sample Computer Printout for a Broiler Flock Recap

Starting count: 10,000

Week	Mortality (no.) Day 1	2	3	4	5	6	7		Total Mortality Current	Total Mortality To date	Per cent Current (%)	Per cent To date (%)
1	15	20	9	8	7	10	6		75	75	0.75	0.75
2	7	6	8	9	7	5	10		52	127	0.52	1.27
3	5	6	6	4	5	7	8		41	168	0.41	1.68
4	5	7	7	6	5	3	5		38	206	0.38	2.06
5	4	4	8	3	7	6	4		36	242	0.36	2.42
6	5	4	3	6	6	7	41		72	314	0.72	3.14
7												
8												

Week	Feed purchases inventory		Feed type (lbs) A	B	C	D		Minus inventory	Net use (lb)	Feed/bird/wk (lb)	Feed Conversion* Current	Feed Conversion* To date
1			4,000					1,400	2,600	0.26	0.77	0.77
2	1,400		8,000					3,000	6,400	0.65	1.18	1.02
3	3,000			12,000				5,000	10,000	1.02	1.45	1.22
4	5,000			12,000				2,300	14,700	1.50	1.73	1.40
5	2,300				20,000			5,000	17,300	1.77	1.91	1.54
6	5,000					18,000		750	22,250	2.30	2.19	1.70
7												
8												

Body weight Age (wks)		Begin weight: (lb)			0.10 (lb)	Flock summary			
1	(7 d)	0.34	5	(35 d)	3.39	Age in days:	42	Starting count:	10,000
2	(14 d)	0.89	6	(42 d)	4.44	End Weight (lb):	4.44	Ending count:	9,686
3	(21 d)	1.59	7	(49 d)		Gain/day (lb):	0.11	Total motality (%):	3.14
4	(28 d)	2.46	8	(56 d)				Condemned (%):	1.25
								Feed conversion:	1.70
								Feed (lbs/bird):	7.56

Daily body weight gain: Age (wks)	(lb)	Age (wks)	(lb)	Summary of total inputs and production	
1	0.049	5	0.133	Total feed used (lb):	73,250
2	0.079	6	0.150	Total weight of poultry delivered (lb):	43,006
3	0.100	7		Calculated feed conversion: (lb feed: lb poultry)	1.70
4	0.124	8			

Feed summary	Feed type	lbs purchased	lbs/survivor
	A	12,000	1.24
	B	24,000	2.48
	C	20,000	2.06
	D	17,250	1.78
			Total: 7.56

* Feed conversion: current = feed consumed for the week divided by body weight gain for the week.
to date = feed consumed to date divided by the body weight at end of the week.

way these goals can be met is to get the flock into production with adequately sized eggs as soon as possible, maintain egg production rates comparable to the breeder standards, keep the females well fertilized, and follow proper egg handling procedures for sound and clean egg shells. The model broiler complex discussed in *A Model Integrated Broiler Firm,* Chapter 42, suggests 165 settable eggs as a standard of egg production to 66 weeks of age.

4. Layer Farm Records

Layer records are probably the most detailed of all flock records. Data are collected daily and are usually summarized on a weekly and flock basis. Records commonly include:

Egg production	Daily intake of specific nutrients
Mortality	Body weight
Egg weight	Water consumption
Egg quality	House temperatures
Feed consumption	Costs
Feed conversion	Profitability indices

Table 32-2 illustrates a weekly record which compares one week with the previous week and results to date.

Some measurements have more economic value than others. The profitability index (see *Egg Production and Egg Weight Standards for Table-Egg Layers,* Chapter 55) probably has the highest relationship to flock profits since it considers egg numbers, egg weight, feed consumption, mortality, and typical costs and prices for feed and eggs. Other measures associated with flock profits (in order of ranking to profits) include:

Total weight of eggs to 60 weeks of age per hen housed
Eggs per hen housed to 60 weeks of age
Hen-day egg production rate to 60 weeks of age
Average daily egg mass
Egg to feed ratio

Commercial egg producers must evaluate the size and quality of eggs produced. Egg weights should be taken at weekly intervals and compared with breeder standards. Egg grade-outs for egg size and quality should be monitored both by hand candling and through the normal grading procedure in the processing plant. Modern egg complexes commonly find it difficult to sample egg quality in their in-line processing system since eggs are commonly blended from several houses of differing ages. To overcome this problem, it is recommended that samples of eggs from each flock be

Table 32-2. Sample Printout for a Table-Egg Laying Flock—Weekly Record

Weekly Individual Flock Summary (From 20 weeks of age)

Flock: A Week: 30 Week Ending Date: 3/10/01

Strain: XYZ Hens Housed at 20 Weeks: 10,000

Flock Data	This Week	Last Week	To Date
Starting count	9,820	9,840	10,000
Number died	20	20	200
Number culled	0	0	0
Ending count	9,800	9,820	9,800
Mortality (% hen-day)	0.20	0.20	2.02
Mortality (% hen-housed)	0.20	0.20	2.00
Hen-days	68,670	68,810	693,000
Body weight (lb)	3.45	3.41	3.45
Egg Production			
Total eggs	63,516	62,784	452,160
Total dozens	5,293	5,232	37,680
Hen-day production (%)	92.5	91.2	65.2
Eggs per hen-housed	6.35	6.28	45.2
Egg Size			
Ounces per dozen	22.8	22.5	21.5
Pounds per case	42.8	42.2	40.4
Daily egg mass (g)[1]	49.9	48.5	33.2
Egg mass/hen-housed (kg)[1]	0.349	0.340	2.300
Feed			
Feed consumed (lb)	14,568	14,352	135,838
Per 100 hens/day (lb)	21.2	20.9	19.6
Per hen-housed (lb)	1.46	1.44	13.58
Per dozen (lb)	2.80	2.70	3.61
lb feed/lb eggs	2.07	2.10	2.68
Daily kilocalories/hen (ME)	270	266	250
Daily protein/hen (g)[1]	17.3	17.1	16.0
Summary			
Hen-day egg production (%)	92.5	91.2	65.2
Av egg weight (oz/doz)	22.8	22.5	21.5
Mortality (% hen-day)	0.20	0.20	2.00
Daily feed intake (lb/100)	21.2	20.9	19.6
Feed/doz (lb)	2.80	2.70	3.61
Feed:egg ratio	2.07	2.10	2.68
Daily egg mass (g)[1]	49.9	48.5	33.2

[1] Metric system is suggested.

measured periodically either by hand or with the processing equipment at the end of the day.

Different strains of birds have significantly different egg size patterns, internal egg quality, and shell strength. Measurement of these traits, as well as those mentioned earlier, is essential when comparing strains or breeds.

32-F. OTHER RECORDS

1. *Grow-out Summary*

This is a record of flock performance from 1 day of age to the age of sale (started pullet or broiler) or for replacements when moved to breeder or layer facilities. It is a complete week-by-week summary of a flock's performance—usually all on one page. In this record, standards of performance may be included. This makes it possible for the poultry producer to make regular comparisons between actual results and the standard, and for flock-by-flock comparisons.

2. *Summary Graphs*

As weekly data are inserted into various summary reports, visual interpretation of results becomes more difficult. There are too many figures and each person has his/her own favorite measurements. To get a better picture of flock performance, the figures on the summary record should be transferred to a graph. Furthermore, the graph should also have the standards for the factors measured. Variations in actual production from the standards will be quickly evident. Graph forms are usually available from the prime breeder.

3. *Layer and Breeder Recaps*

A summary of actual flock production should be made at the end of the laying period. A simple computer spread sheet can be constructed to summarize six or more key items of performance by week along with an overall summary at the bottom (see Table 32-3).

32-G. COMPUTERIZED RECORDS

The concentration of ownership and multiple flock firms require more timely and comprehensive record systems. Managers and owners need

Table 32-3. Layer Flock Recap (sample spreadsheet)

Flock: A — Strain: XYZ

Number housed: 10,000 — Date housed: 12/19/00

Age (wks)	Hen-day egg production (%)	Mortality (%/wk)	Pounds feed/100 hens/day	Feed conversion (lb/doz)	Egg weight (lb/case)	Eggs per hen housed to date
20	14.2	0.08	17.5		34.8	1.0
21	33.5	0.08	17.9	6.4	36.5	3.3
22	56.3	0.08	19.2	4.1	37.7	7.3
23	73.8	0.10	19.9	3.2	39.2	12.4
24	84.2	0.11	21.0	3.0	40.6	18.3
25	88.9	0.12	21.4	2.9	41.6	24.5
26	90.7	0.13	21.9	2.9	42.6	30.8

Summary
(20 to 60 weeks of age)

Hen-day egg production (%)	81.3	Av case weight (lbs)	46.2
Eggs per hen housed	226.5	Av egg weight (g)	58.3
Feed/day per 100 hens (lb)	22.6	Av weekly mortality (%)	0.164
Feed/dozen (lb)	3.34	Total egg mass (kg/hen housed)	13.2
Feed:egg (ratio)	2.16	Av daily egg mass (g)	47.4

Additional columns or rows can be added to include other measurements or weeks of production.

to make *daily* decisions based on flock performance and environmental conditions. They cannot wait for weekly summaries.

Egg production, mortality, and feed consumption records are now analyzed daily with the use of in-house computer systems. Records are available within minutes of their collection and managers can make decisions regarding tomorrow's feed formulas. In many cases, flock performance, environmental measurements, egg and poultry processing data, and feed formulations are all integrated into one network for analysis. Some sophisticated systems can relate two or more measurements and even suggest possible causes for various problems.

Computers allow immediate comparisons with previous flocks and standards and of previous periods for the same flock. Flock histories are easily prepared and records may be as detailed as required. Table 32-3 illustrates a typical weekly record for a table egg flock. Computers are excellent tools to speed up the analysis of records, but they do not replace the need for daily observations for flock comfort and conditions by well trained personnel.

Some data may be automatically entered into a record system with proper equipment/computer interfaces. These may include brood-grow house temperatures for various locations, egg counts per row of cages, daily scale weights for feed and calculated feed consumption, ammonia

Figure 32-1. Computers Are Universally Used on Modern Poultry Farms

Figure 32-2. The Data Generated on Commercial Farms Is Often More than Management Can Absorb

levels in the brooder house, dead bird data with the use of in-house keypads, cooler temperatures, and others (see *Computer Applications*, Chapter 33).

32-H. CAPITALIZING PULLET COSTS ON THE ACCRUAL SYSTEM

When the accrual system of bookkeeping is used, it is necessary to capitalize the cost involved with growing the pullets to be used for commercial egg production or pullet and cockerel costs for breeder replacement flocks. As capitalized growing costs are offset by inventory values, there is no profit or loss involved with the growing process. Once capitalized, however, the amount must be reduced during the period of egg production, usually by a procedure of charging an equal amount to each dozen of commercial eggs or hatching eggs produced.

Charging an equal amount to each dozen is the preferred method if projected and actual production are similar. Any problems with mortality or production, though, will result in significant error. Some producers prefer to capitalize pullet expenditures on a time basis. This procedure allocates a share of the pullet cost to the eggs produced during a given week. This system, though, results in a fluctuating cost per dozen as production varies over the course of one or two cycles of production.

The amortization cost per dozen eggs is the "pullet cost" at capitalization less the salvage value at the end of the laying year, divided by the number of dozen eggs the bird will lay during her production cycle after her amortization date. Today, the age used for determining pullet costs will vary from 16 to 24 weeks depending upon the age at housing, the accounting method used, and the individual company's needs for assigning costs.

33

Computer Applications

by Gene M. Pesti

Computers have had a great impact on all aspects of modern society, including the poultry industry. Three uses of computers have greatly impacted poultry husbandry and management: data collection and management, enterprise optimization, and telecommunications.

33-A. DATA COLLECTION AND MANAGEMENT

Like any modern business, the various firms producing poultry meat and eggs rely heavily on computers for data storage and manipulation. Computer software systems are necessary for managing the business side of poultry production. Accounting and financial officers of poultry businesses may use general accounting software. Also, in many cases specialized software written for the various aspects of poultry meat and egg production is available. For instance, many firms use off-the-shelf software available for general inventory control and monitoring. Many also use specialized software written specifically for inventory control in feed mills and on-farm storage.

Farm Level Data Collection and Monitoring

Computer controlled systems are available to monitor and maintain the environment in both layer and broiler houses. Special sensors are used to monitor such things as inside and outside temperature, relative humidity, and in a few instances ammonia levels in buildings. Microcomputers are used to process these data and to control ventilation systems to provide the desired environment. These microcomputer-based systems can monitor water flow to detect leaks and can even measure the electrical usage by a series of lights to tell when a bulb is burned out. The performance of the birds can be monitored in the poultry house by electronic devices that continuously weigh growing birds, and by counters that can count eggs as they are collected. Hourly and daily feed intake can be monitored by the row or for an individual house. House conditions can be viewed locally or remotely, often from the growers' residence or the service person's office (see Figure 33-1).

Computer Use in the Processing Plant

Broiler processing. Computerized monitoring and inventory control software is especially important in broiler processing plants that produce many different products each day. The weights of individual carcasses can be measured shortly after the birds exit the chill tanks for routing based on size. In this way the proper number of carcasses can be prepared to meet the specific marketing needs for any particular day. Records are kept of the type and age and location of each product.

Linear programming models, like those developed by Rahman and Bender (1971) can be used to choose the mix of products that maximize profits based on the value of the various parts vs whole birds. Computerized systems in the processing plant can also print store weight and pricing labels for individual packages and labels for trays to see that products get on the appropriate truck to be delivered to retail outlets in the shortest time possible

Egg packaging. Egg packaging and processing plants today are also highly dependent upon computers to coordinate supplies with orders, to categorize the eggs available into weight categories which match the needs of buyers, and to get the most value from the eggs that are processed. Analysis of cartoned egg and undergrade data are used to help the flock manager determine optimum selling and molting ages. Sophisticated equipment which can identify and remove eggs with broken shells, dirty eggs, and eggs with blood spots rely heavily on the use of complex computer software programs.

Computer Use in Feed Mills and Feed Delivery

The milling and mixing operations of modern feed mills are controlled by computers. Batching systems control the transfer of ingredients from storage bins into and out of mixers through pellet mills and into other storage bins. Feed inventory control systems are now capable of tracking feed consumption by flock to predict when deliveries of each feed will be necessary. Thus feed mill and farm inventories can be integrated into the same systems. Feed formulation programs communicate directly with batching systems that control feed mills.

Computerized Performance Comparison Summaries

There are several companies in the United States that collect and summarize technical and economic performance data for egg and broiler meat production. Data on every aspect of production are collected and submitted by subscribing companies. The data from the various companies are combined into a large computer database and summarized. The subscribing firms are given summaries of the entire database with only their firm identified. This allows producers to compare their efficiency with their competitors and identify their strengths and areas that need improvement.

Poultry software. Software for poultry applications are in many forms and represent many different needs. A directory of applications for poultry (Bell, 1996) includes some 90 programs from 16 vendors. Worldwide applications would number in the hundreds. Several dozen companies specialize in generic and specific software production to meet the needs of the worldwide poultry industry.

Software programs are written as "stand-alone" programs which can be utilized by anyone with the correct computer system (DOS, Windows, Mac) and those that require commercially available support software programs such as spreadsheets. In the latter, the software is usually simple templates (layouts) which "automate" various analyses.

Applications include programs for feed formulation, flock performance records, cash flow and projections, feed inventory control, transportation and efficient use of trucking, business records, processing plant operations, flock scheduling, egg size projections and value comparisons, and an assortment of specific programs designed to analyze or compare different management options. Listed below are specific examples.

- *Flock performance.* Flock managers are required to keep numerous records in order to successfully manage their flocks. These records include body weight, feed consumption, egg numbers, egg weight, mortality. and others. Most of the data are collected daily, weekly, and as batch summaries. To be of maximum use, reports must be available to management within 24 hours of their collection, and with departures from normal standards highlighted. With such masses of data and the need for timely reporting, computer use is essential.

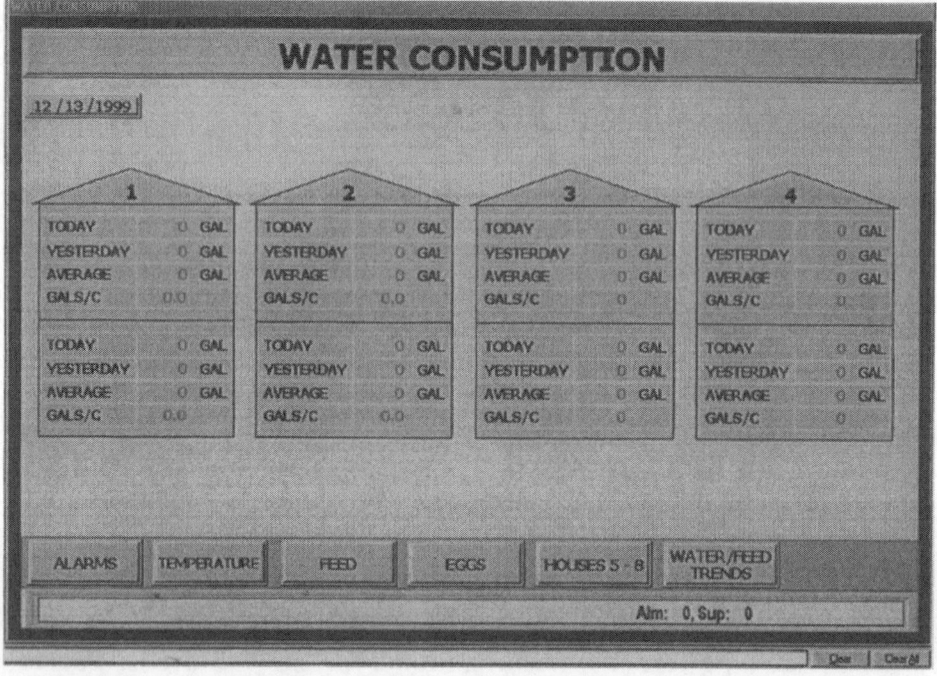

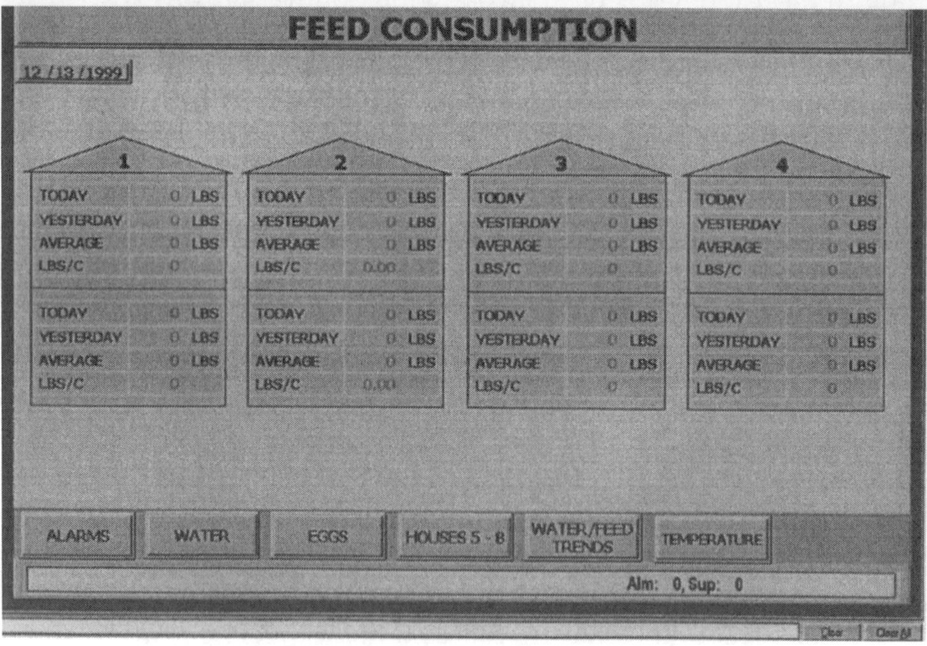

Figure 33-1. Sample Computer Screens for Various Measures of Flock Performance (courtesy of EYEdeal Manager)

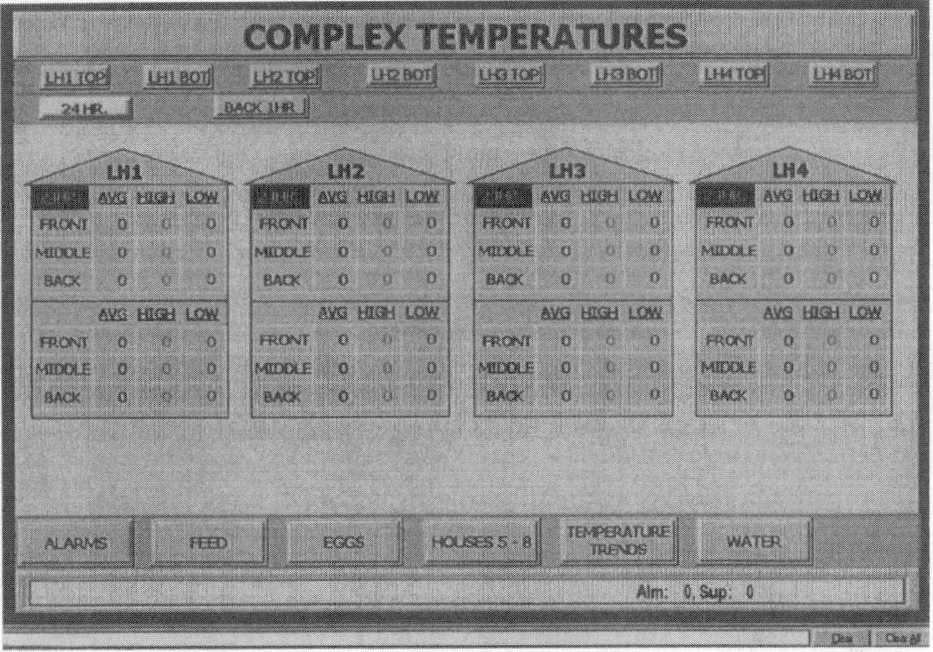

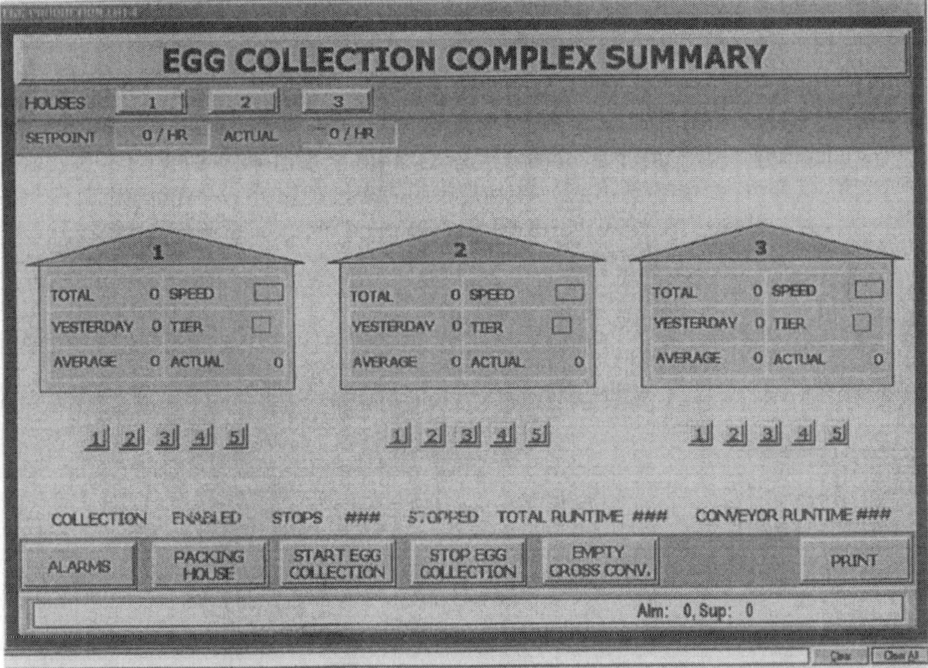

Figure 33-1. (continued)

In *Record Management,* Chapter 32, Table 32-2 illustrates a weekly summarization record for a layer flock with a comparison to the previous week and a summary to date. Graphic depiction of performance with color printing is also used to illustrate abnormal performance.

- *Single and multiple flock projections.* Projections and cash flow reports are useful in budgeting financial needs throughout the course of a flock's life. Many banking institutions require such reports before they will agree to finance an enterprise. Projections of performance and prices are also useful to determine molting and sale policies in the egg industry. With fluctuating prices and the need to molt or sell flocks, projection techniques can be extremely useful in coordinating high production with high prices thereby improving overall flock profits. This is a form of "enterprise optimization." Multiple flock projections can also give marketing firms an advance summary of product availability by size and quantity for dates in the future (see Table 33-1A, B).
- *Economic analyses of performance.* Traditional performance analyses concentrate on measurable results with little attention given to their overall economic significance. New techniques of analysis enable a manager to balance various performance results with representative prices and costs to determine the net economic effects of various management systems and evaluation of flocks. Actual data for egg production, mortality, egg weight, and feed consumption are entered into a spreadsheet with results expressed as egg income minus feed costs per hen-housed. A sample is shown in Table 33-2A, B.

Table 33-1A. Sample Layer Flock Projection (abbreviated version)*

Date	Wk	Begin Count	% Hen-day Production	¢ per Dozen	$ Income	Feed Consumption (lbs/hen-day)	Feed Cost (¢/dozen)
1–7	21	10,000	43.0	32.6	820	0.160	33.5
1–14	22	9,990	63.4	36.7	1,350	0.167	23.7
1–21	23	9,980	79.7	40.4	1,870	0.174	19.7
1–28	24	9,970	86.9	43.2	2,180	0.182	18.8
2–4	25	9,960	89.5	45.6	2,370	0.189	19.0
2–11	26	9,950	92.0	47.3	2,520	0.196	19.2
2–18	27	9,940	93.7	48.8	2,650	0.203	19.5
2–25	28	9,930	95.0	49.9	2,740	0.211	19.9
3–4	29	9,920	94.7	50.7	2,770	0.218	20.7
3–11	30	9,910	94.3	51.4	2,800	0.225	21.5

* Actual page has 19 columns. The full report has 29 pages.

Table 33-1B. Sample Egg Size Distribution (abbreviated version)*

Date	Wk	Total Cases	% Jumbo	% X-large	% Large	% Medium	% Small	% Undergrades
1–7	21	80	6	0	1	36	51	2
1–14	22	120	5	0	3	51	37	2
1–21	23	150	4	0	8	61	24	2
1–28	24	170	4	0	16	64	14	2
2–4	25	170	3	1	26	60	8	2
2–11	26	180	3	1	35	54	5	2
2–18	27	180	2	3	44	46	3	2
2–25	28	180	2	5	51	38	2	2
3–4	29	180	1	8	56	31	1	2
3–11	30	180	1	11	59	25	1	2

* Actual page has 19 columns.

Table 33-2A. Sample Economic Analysis of Performance (abbreviated version)*

Week	% Died	% Egg Production	Egg Weight (lb/case)	Feed Consumption (lb/hen-day)	Eggs per Hen to Date	$ EI-FC** per HH/wk	$ EI-FC to Date
19	0.05	18.9	37.0	0.180	1.1		
20	0.04	33.5	38.0	0.185	2.3	−0.02	−0.02
21	0.03	63.9	39.0	0.191	6.8	0.05	0.03
22	0.05	78.8	40.0	0.192	12.3	0.10	0.12
23	0.05	88.7	41.5	0.196	18.5	0.14	0.26
24	0.07	93.2	41.5	0.204	25.0	0.15	0.41
25	0.07	94.2	43.0	0.201	31.6	0.17	0.57
—	—	—	—	—	—	—	—
55	0.14	85.9	50.0	0.226	217.5	0.16	5.64
56	0.25	85.4	50.5	0.224	223.2	0.16	5.79
57	0.17	84.6	50.0	0.216	228.8	0.16	5.94
58	0.15	83.8	50.5	0.232	234.3	0.15	6.08
59	0.20	84.6	50.8	0.216	239.9	0.16	6.22
60	0.22	83.2	51.3	0.238	245.4	0.14	6.36

* Actual print-out has 2 pages with 21 columns.
** Egg income minus feed cost per hen housed.

Table 33-2B. Summary of Results (60 weeks of age)

Hen-day (%)	87.7	Wkly mortality (%)	0.144	Egg value (¢/doz)	52.7
Eggs per hen housed	245.4	Egg wt/egg (g)	59.1	Feed cost (¢/doz)	21.6
Feed/day (lb)	0.210	Egg wt/case (lb)	46.9	$ EI-FC per hen housed	6.36
Feed/dozen (lb)	2.88	Egg mass/HH (kg)	14.51		

- *Replacement pullet costs.* Software is available to either budget costs of raising a flock of replacement pullets in advance or to record the costs actually incurred. A simple modification would allow side-by-side comparison of the budget with the actual.
- *Egg size distribution and egg value.* Egg size is affected by age, breed / strain, nutrition, temperature, and the time of the year. Projection programs (see Table 33-1A, B) are available to project the effects of age and season on weight. This, in turn, is used to generate egg size distribution for any weight-class definition with egg values associated with either bulk egg weight or egg size category. Such information is extremely useful in marketing programs.

33-B. ENTERPRISE OPTIMIZATION

Computers are capable of much more than collecting and summarizing data. Computer programs capable of optimizing growth and performance are available and their use is growing in poultry production.

- *Least-cost feed formulation* as applied to most linear programming models assumes that the most profitable levels of nutrients to feed are already known. More sophisticated computer models of broiler performance have been developed that determine the most profitable levels of nutrients to feed. These models include the influences of environment, genetics, and nutrition on bird growth and performance. Since genetics and the bird's environment affect responses to different nutrient concentrations, these models can become quite complex (see *Feed Formulation and the Computer,* Chapter 21).
- *Performance optimization models* require a very extensive knowledge of how each aspect of production affects overall profitability. There are many important interactions that have to be explicitly stated and quantified. The best example of such an interaction is evident in choosing genetic stocks. Stocks selected for high meat yield may not lay as many eggs as others. Thus performance optimization models should be able to predict not only increased profits from birds with target carcass weights reaching the processing plant earlier, they must also predict how decreased egg numbers impact breeder hen costs and placement scheduling.

- *Layer replacement.* The determination of a replacement policy for commercial egg farms requires the analysis of dozens of options and factors which affect performance. Whether or not a farm should sell their flocks after a year of lay or prolong its life by molting to a second or even third laying cycle will depend upon the relative performance between cycles and cost factors for the time periods under consideration. Computer modeling is used to determine the optimum policy. Effective use of these principles require careful input of representative flock performance data with realistic price and cost standards. Existing software analyzes numerous options and expresses its solution in terms of monetary returns per year per hen-housed.

33-C. TELECOMMUNICATIONS

The telecommunications revolution that started in the 1980's with the Internet exploded into the 1990's with the development of the World Wide Web. All aspects of poultry production have been affected by improvements in telecommunications, some more than others dependent on the willingness of individual managers. The most obvious aspect of improvements in telecommunications may be the pagers seen on service persons' belts and the cellular telephones so commonly used, but the most important may be the information provided directly into consumers' homes. Microcomputers serve as the window to the Internet.

The Internet

Thanks to e-mail, it is now as easy and inexpensive to leave a note for a colleague across the country (or world) as it is to leave a note for someone in the next office. That might be revolutionary enough, but the same system is being used for the discussion of production problems and sharing of experiences on a worldwide basis. Several discussion, or "chat" rooms, have been set up specifically for poultry interests, including poultry health, broiler breeders, and fanciers. The poultry industry has a long history of cooperation and information sharing; the Internet promises to facilitate more cooperation and quicker responses to problems in the future.

The World Wide Web (www)

The so-called *Information Super Highway* is providing several ways to facilitate poultry and egg marketing: electronic, or cyber, marketplaces and marketing information. The electronic marketplaces provide a place for buyers and sellers to gather in cyberspace and do business. These marketplaces serve to help companies sell surpluses and buy to meet contract obligations when shortages occur.

The World Wide Web is also providing access for companies to deliver information directly into consumers' businesses and homes *at the instant the consumers want the information.* When consumers need information on food safety or an interesting new recipe to try, they can access it immediately through their home computer and modem. There are excellent web sites available from the American Egg Board, National Chicken Council, and several poultry-producing companies providing such information for consumers. Many other sites have a variety of recipes for consumers to try. As larger and larger proportions of consumers gain www access, the potential for consumer education will continue to increase.

Information Acquisition

Besides consumer information, there is a great deal of information of interest to poultry producers on the www. There are web sites with information on the people, organization, and policies and reports of the US Department of Agriculture, for instance. Other sites deal with educational programs at various universities and meetings dealing with agriculture. Others provide information on such varied subjects as Hazard Analysis and Critical Control Point (HACCP) programs and getting the most value from poultry manure.

Another important www site is *Medline,* furnished by the National Library of Medicine. *Medline* contains a searchable database with much of the world's scientific literature in the areas of biology and medicine. Both the *Poultry Science* and *British Poultry Science* journals listings are included. *Medline* makes it possible for anyone with a microcomputer and modem to read thousands of abstracts of scientific papers from their home or office.

It is beyond the scope of this book to list pertinent web addresses, especially since web addresses frequently change. At the time of this writing there are good lists of links (web site names and address) at several major US Universities teaching poultry science. They should remain a good place to get started looking for information on the web. To find them, search for the University's main home page. It should have links to the individual departments.